Lecture Notes in Computer Science 10944

Commenced Publication in 1973
Founding and Former Series Editors:
Gerhard Goos, Juris Hartmanis, and Jan van Leeuwen

FoLLI Publications on Logic, Language and Information
Subline of Lectures Notes in Computer Science

More information about this series at http://www.springer.com/series/7407

Lawrence S. Moss · Ruy de Queiroz
Maricarmen Martinez (Eds.)

Logic, Language, Information, and Computation

25th International Workshop, WoLLIC 2018
Bogota, Colombia, July 24–27, 2018
Proceedings

Springer

Editors
Lawrence S. Moss
Department of Mathematics
Indiana University
Bloomington, IN
USA

Maricarmen Martinez
Universidad de los Andes
Bogotá
Colombia

Ruy de Queiroz
Centro de Informática
Univ Federal de Pernambuco (UFPE)
Recife, Pernambuco
Brazil

ISSN 0302-9743   ISSN 1611-3349 (electronic)
Lecture Notes in Computer Science
ISBN 978-3-662-57668-7   ISBN 978-3-662-57669-4 (eBook)
https://doi.org/10.1007/978-3-662-57669-4

Library of Congress Control Number: 2018947331

LNCS Sublibrary: SL1 – Theoretical Computer Science and General Issues

Printed on acid-free paper

This Springer imprint is published by the registered company Springer-Verlag GmbH, DE
part of Springer Nature
The registered company address is: Heidelberger Platz 3, 14197 Berlin, Germany

Preface

The 25th Workshop on Logic, Language, Information, and Computation, WoLLIC 2018, was held July 24–27, 2018, in Bogotá, Colombia. WoLLIC is a series of workshops that started in 1994 with the aim of fostering interdisciplinary research in pure and applied logic. The vision for the conference is to provide an annual forum large enough to provide meaningful interactions between logic and the sciences related to information and computation, and yet small enough to allow for concrete and useful interaction among participants.

There were 30 submissions. Each submission was reviewed by three or more Program Committee members. The committee decided to accept 16 papers. This volume contains the papers that were presented at the meeting. Selected contributions will be published (after a new round of reviewing) as a special post-conference issue of the *Journal of Logic, Language, and Information.*

The program also included six invited talks, by Katalin Bimbó, Xavier Caicedo, José Meseguer, Elaine Pimentel, Guillermo Simari, and Renata Wassermann. The invited speakers are represented in this volume either by a full paper or by a title and abstract of their talk. In addition, Katalin Bimbó, José Meseguer, and Elaine Pimentel provided additional tutorial lectures as part of the meeting.

We are grateful to the institutions who provided scientific sponsorship of our meeting: the Association for Symbolic Logic (ASL), the Interest Group in Pure and Applied Logics (IGPL), the Association for Logic, Language and Information (FoLLI), the European Association for Theoretical Computer Science (EATCS), the European Association for Computer Science Logic (EACSL), the Sociedade Brasileira de Computação (SBC), and the Sociedade Brasileira de Lógica (SBL).

We thank everyone who submitted a paper to our meeting. We are grateful to all the members of the Program Committee for their hard work, and also the many subreviewers who also helped evaluate papers. As a result of their efforts, several papers were substantially improved, and every paper was improved in some way. We are especially grateful to the Universidad de los Andes for its support, and to its Departamento de Matemáticas for hosting. We thank the local organizers for all of their work to make our conference a success.

May 2018

Lawrence S. Moss
Ruy de Queiroz
Maricarmen Martinez

Organization

Program Committee

Siddharth Bhaskar	Haverford College, USA
Torben Braüner	Roskilde University, Denmark
Hazel Brickhill	University of Kobe, Japan
Valeria De Paiva	Nuance Communications, USA
	and University of Birmingham, UK
Ruy De Queiroz	Universidade Federal de Pernambuco, Brazil
Michael Detlefsen	University of Notre Dame, USA
Juliette Kennedy	University of Helsinki, Finland
Sophia Knight	Uppsala University, Sweden
Alex Kruckman	Indiana University Bloomington, USA
Maricarmen Martinez	Universidad de los Andes, Colombia
Friederike Moltmann	CNRS-IHPST, France
Lawrence S. Moss	Indiana University, Bloomington, USA
Cláudia Nalon	University of Brasília, Brazil
Sophie Pinchinat	IRISA Rennes, France
David Pym	University College London, UK
Revantha Ramanayake	Vienna University of Technology, Austria
Giselle Reis	Carnegie Mellon University, Qatar
Jeremy Seligman	The University of Auckland, Australia
Yanjing Wang	Peking University, China
Fan Yang	University of Helsinki, Finland

Additional Reviewers

Zeinab Bakhtiari	Sonia Marin
Philippe Balbiani	Michael Norrish
John Baldwin	Aybüke Özgün
Olivier Bournez	Victor Pambuccian
Richard Crouch	Gerard Renardel de Lavalette
Jeremy Dawson	Philippe Schnoebelen
Hans van Ditmarsch	François Schwarzentruber
Harley Eades III	Gert Smolka
Fredrik Engström	Nicolas Tabareau
Jie Fan	Alasdair Urquhart
John Gallagher	Andrew Withy
Guido Governatori	Elia Zardini
Loig Jezequel	Shengyang Zhong
Tadeusz Litak	Martin Ziegler
Feifei Ma	

Abstracts of Invited Talks

Inhabitants of Intuitionistic Implicational Theorems

Katalin Bimbó ⓘ

Department of Philosophy, University of Alberta, Edmonton,
AB T6G 2E7, Canada
bimbo@ualberta.ca
http://www.ualberta.ca/~bimbo

Abstract. The aim of this paper is to define an algorithm that produces a *combinatory inhabitant* for an implicational theorem of intuitionistic logic *from a proof in a sequent calculus*. The algorithm is applicable to standard proofs, that exist for every theorem, moreover, non-standard proofs can be straightforwardly transformed into standard ones. We prove that the resulting combinator inhabits the simple type for which it is generated.

Generic Models on Sheaves

Xavier Caicedo

Universidad de los Andes, Bogot
xcaicedo@uniandes.edu.co

Abstract. Sheaves over topological spaces, introduced by Weyl and developed by Leray, Cartan and Lazard, play a fundamental role in Analysis, Geometry, and Algebra. Prototypical examples are the sheaf of germs of continuous real valued functions on a topological space, or Grothendieck's sheaf of local rings over the spectrum of the ring. Grothendieck moved from spaces to "sites" in order to generalize the cohomology of modules, following Weil's tip to prove his conjectures on finite fields. Lawvere observed in the 60's and it was established in firm ground by Makkai, Reyes and Joyal that the category of sheaves over a site (today a Grothendieck topos) is a mathematical universe of varying objects, governed by an inner logic and a heterodox mathematics. These inner logics lie between classical logic and Brouwer-Heyting intuitionistic logic proposed in the first third of last century as an alternative to Fregean logic and the Cantorian foundations of mathematics. New logical operations arise naturally in this context, particularly in the case of connectives, which go beyond Grotendieck topologies. We will sketch with minimum categorical apparatus the main features of mathematics and logic on sheaves, and concentrate on generic models which establishes a bridge between intuitionistic and classical logic, encompassing the uses of forcing in set theory and the basic model theoretic constructions of classic finitary or infinitary logic.

Symbolic Reasoning Methods in Rewriting Logic and Maude

José Meseguer

Department of Computer Science, University of Illinois at Urbana-Champaign,
Urbana, USA
meseguer@illinois.edu

Abstract. Rewriting logic is both a logical framework where many logics can be naturally represented, and a semantic framework where many computational systems and programming languages, including concurrent ones, can be both specified and executed. Maude is a declarative specification and programming language based on rewriting logic. For reasoning about the logics and systems represented in the rewriting logic framework *symbolic methods* are of great importance. This paper discusses various symbolic methods that address crucial reasoning needs in rewriting logic, how they are supported by Maude and other symbolic engines, and various applications that these methods and engines make possible. Because of the generality of rewriting logic, these methods are widely applicable: they can be used in many areas and can provide useful reasoning components for other reasoning engines.

A Semantical View of Proof Systems

Elaine Pimentel

Departamento de Matemática, UFRN, Natal, Brazil
elaine.pimentel@gmail.com

Abstract. In this work, we explore proof theoretical connections between sequent, nested and labelled calculi. In particular, we show a semantical characterisation of intuitionistic, normal and non-normal modal logics for all these systems, via a case-by-case translation between labelled nested to labelled sequent systems.

Dynamics of Knowledge: Belief Revision and Argumentation

Guillermo R. Simari

Artificial Intelligence Research and Development Laboratory (LIDIA),
Department of Computer Science and Engineering,
Institute for Computer Science and Engineering (UNS-CONICET),
Universidad Nacional del Sur (UNS), Bahía Blanca, Argentina
grs@cs.uns.edu.ar

Abstract. The exploration of the relationships between Belief Revision and Computational Argumentation has led to significant contributions for both areas; several techniques employed in belief revision are being studied to formalize the dynamics of argumentation frameworks, and the capabilities of the argumentation-based defeasible reasoning are being used to define belief change operators. By briey considering the fundamental ideas of both areas it is possible to examine some of the mutually beneficial cross-application in different proposals that model reasoning mechanisms that combine contributions from the two domains.

In the field of Knowledge Representation and Reasoning, the two areas of Belief Revision and Computational Argumentation consider the problem of inconsistency in knowledge bases. Techniques designed by researchers investigating in Belief Revision have put forward formal tools that help in obtaining a consistent knowledge base from an inconsistent one. Argumentation frameworks tolerate inconsistency obtaining a consistent set of conclusions through the use of specialized argumentation-based inference mechanisms.

During the last decades, research in Belief Revision has examined diverse forms of belief change producing formalisms on the dynamics of knowledge that reflect different intuitions. In 1985, the paper by Alchourrón, Gärdenfords, and Makinson, commonly known as AGM, provided a firm foundation for the area coalescing different investigations that were active at that time. An interesting point regarding this research community is how diverse interests met to make contributions with engaging problems and offered possible solutions; these researchers came, and continue coming, from Philosophy, Formal Logic, Economics, Computer Science, and Artificial Intelligence, among others.

Despite the fact that human beings have been concerned with argumentation processes since antiquity acknowledging its importance, the advent of computational approaches to this activity contributed to give further momentum to the research. Several contributions fueled this impulse since the publication of Toulmin's book *The Uses of Argument* in 1958, but the computational approaches began to appear twenty years later. Of particular importance to this line of research was the work of Jon Doyle on *Truth Maintenance Systems* introducing the issue of how something is believed. By describing a dynamic framework to maintain a knowledge base, Doyle's work could also be considered as a one of the first were Belief Revision and Computational Argumentation met.

Computational Argumentation offers frameworks for defeasible reasoning, i.e., reasoning that could be challenged, by creating dialectical inference mechanisms that contrast arguments in favor and against a particular claim. This kind of reasoning is tolerant to inconsistency becoming useful in many areas concerning decision making, offering the characteristic of being able of explaining conclusions in a way that is understandable.

Belief revision and Computational Argumentation can be regarded as two complementary forms of reasoning; on the one hand, belief revision formalizes methods to explore and comprehend how a set of beliefs should be rationally changed, on the other hand, argumentative reasoning looks for justifications to accept beliefs. From this complementarity, it is possible to build better architectures for reasoning agents.

Revising System Specifications
in Temporal Logic

RenataWassermann (iD)

Computer Science Department, Institute of Mathematics
and Statistics, University of São Paulo, Brazil
renata@ime.usp.br

Abstract. Although formal system verification has been around for many years, little attention was given to the case where the specification of the system has to be changed. This may occur due to a failure in capturing the client's requirements or due to some change in the domain (think for example of banking systems that have to adapt to different taxes being imposed). We are interested in having methods not only to verify properties, but also to suggest how the system model should be changed so that a property would be satisfied. For this purpose, we will use techniques from the area of Belief Revision.

Belief revision deals with the problem of changing a knowledge base in view of new information. In the last thirty years, several authors have contributed with change operations and ways of characterising them. However, most of the work concentrates on knowledge bases represented using classical propositional logic. In the last decade, there have been efforts to apply belief revision theory to description and modal logics.

In this talk, I will analyse what is needed for a theory of belief revision which can be applied to the temporal logic CTL. In particular, I will illustrate different alternatives for formalising the concept of revision of CTL specifications. Our interest in this particular logic comes both from practical issues, since it is used for software specification, as from theoretical issues, as it is a non-compact logic and most existing results rely on compactness.

This talk is based mostly on joint work with Aline Andrade and Paulo de Tarso Guerra Oliveira.

Keywords: Belief change · Temporal logic · Model revision

Contents

Inhabitants of Intuitionistic Implicational Theorems

Katalin Bimbó$^{(\boxtimes)}$ (iD)

Department of Philosophy, University of Alberta, Edmonton,
AB T6G 2E7, Canada
bimbo@ualberta.ca
http://www.ualberta.ca/~bimbo

Abstract. The aim of this paper is to define an algorithm that produces a *combinatory inhabitant* for an implicational theorem of intuitionistic logic *from a proof in a sequent calculus*. The algorithm is applicable to standard proofs, that exist for every theorem, moreover, non-standard proofs can be straightforwardly transformed into standard ones. We prove that the resulting combinator inhabits the simple type for which it is generated.

Keywords: Combinator · Curry–Howard correspondence
Intuitionistic logic · Sequent calculus

1 Introduction

A connection between implicational formulas and combinatory and λ-terms has been well known for more than half a century. A link is usually established between a natural deduction system and λ-terms, or axioms and combinators (or λ-terms). Moreover, the natural deduction system is usually defined as a type-assignment calculus.

Our interest is in *sequent calculi*, on one hand, and in *combinatory terms*, on the other. A natural deduction type-assignment system with combinators very closely resembles an axiomatic calculus. Neither of the latter two kinds of proof systems is highly suitable for decidability proofs, nor they provide much control over the shape of proofs. There have been certain attempts to use sequent calculi in type assignment systems. However, we believe that no algorithm has been defined so far for the extraction of combinatory inhabitants for intuitionistic theorems from sequent calculus proofs thereof.[1]

The next section briefly overviews some other approaches that link λ-terms and sequent calculi. Section 3 introduces a sequent calculus for $J^T_{\rightarrow}$. This sequent calculus falls into the line of sequent calculi that were initiated by the non-associative Lambek calculus. The bulk of the paper, Sect. 4 is devoted to the description of the extraction algorithm, which has several stages, and then, to the proof of its correctness. We conclude the paper with a few remarks.

---

[1] The paper [9] deals with generating HRM^n_n terms and combinatory inhabitants from proofs of $T_{\rightarrow}$ theorems.

© Springer-Verlag GmbH Germany, part of Springer Nature 2018
L. S. Moss et al. (Eds.): WoLLIC 2018, LNCS 10944, pp. 1–24, 2018.
https://doi.org/10.1007/978-3-662-57669-4_1

2   Some Other Approaches

We should note that there have been earlier attempts to define a connection between sequent calculus proofs and λ-terms. Girard's idea—when he introduced his sequent calculus *LJT*—was to add just a minimal amount of structure to antecedents which are otherwise sets of formulas. A special location (the so-called "stoup") is reserved for a single formula, but it may be left empty. All the rules of *LJT*, except the rule called "mid-cut," affect a formula in the special location of at least one premise. The effect of this focusing on a formula in most of the proof steps is the *exclusion of many proofs* that could be constructed in a more usual formulation of a sequent calculus for implicational intuitionistic logic. *Adding structure to sets* is a step in the right direction in our view, but *LJT* does not go far enough. (See [17] for a detailed presentation of *LJT*.)

Another approach is to define a type-assignment system in a sequent calculus form. Barendregt and Ghilezan in [2] introduced λL (and λL^{cf}) to find a correspondence between λ-terms and implicational intuitionistic theorems. It is natural to start with λN in a horizontal presentation (which resembles sequent calculus notation). Then the antecedent is a context (or a set of type-assignments to variables) in which a certain type-assignment to a (possibly) complex λ-term can be derived. The core question is how the left introduction rule for $\rightarrow$ should look like. [2] combines *substitution* and *residuation*: given that $M: \mathcal{A}$ and $x: \mathcal{B}$, if there is a $y: \mathcal{A} \rightarrow \mathcal{B}$, then yM can replace x. Combinators were originally introduced by Schönfinkel in [20] to eliminate the need for substitution, which is a complex operation in a system with variable binding operations (such as λ or $\forall$). Thus reintroducing substitution seems to be a step backward. A more serious complaint about λL is that the $\rightarrow$ introduction rule essentially requires the λ operator. The λ could be replaced by the λ^*, but the latter would have to be deciphered on a case-by-case basis (i.e., by running a λ^* algorithm on every concrete term). In sum, this approach differs from ours in that it uses type-assignments and λ's.

Arguably, Girard's as well as Barendregt and Ghilezan's calculi are directly motivated by a natural deduction calculus (such as *NJ*) for intuitionistic logic. On the other hand, a connection between *sequent calculus rules* and *combinators* was discovered by Curry; see, for instance, his [11]. It is easy to see that $\mathcal{A} \rightarrow \mathcal{B} \rightarrow \mathcal{A}$, the principal type schema of the combinator K cannot be proved without some kind of thinning rule on the left. However, such informal observations about rules and matching combinators cannot be made precise as long as sequents are based on sequences or even sets of formulas, which is inherent both in [17] and in [2].

3   A New Sequent Calculus for $J_{\rightarrow}$

Function application is not an associative operation—unlike conjunction and disjunction are in intuitionistic logic. The analogy between combinators and structural rules in a sequent calculus cannot be made precise without replacing the usual associative structural connective with a non-associative one. Of course, the use of a structural connective that is not stipulated to be associative is by no means new; Lambek's [18] already includes such a connective. Combinators have been explicitly incorporated into a sequent calculus by Meyer in [19].[2]

Taking the reduction patterns of combinators seriously helped to formulate sequent calculi for the relevance logic of positive ticket entailment, $\mathbf{T}_+$. Giambrone's calculi in [15,16] utilized insights concerning B, B' and W in the form of structural rules. Bimbó [3] defined sequent calculi for $\mathbf{T}^t_\to$ and $\mathbf{T}^{\circ t}_+$ (among other logics), in which special rules involving t are added. It is easy to show that t is a left identity for fusion (i.e., intensional conjunction) in $\mathbf{T}$, hence, a special instance of the thinning rule may be added in a sequent calculus. However, the latter rule is not the only one needed that involves t, if we algebraize $\mathbf{T}^{\circ t}_+$ with an equivalence relation defined from implicational theorems. The introduction of additional structural rules specific to t, on the other hand, requires us to refine the proof of the cut theorem, which could otherwise proceed by double induction for a multicut rule or by triple induction for the single cut rule.[3]

A connection between K, C and W, on one hand, and (left) structural rules, on the other hand, was noted by Curry in [10,11], who is probably the first to use combinators in the labels of the thinning, the permutation and the contraction rules. The operational rules of a sequent calculus leave their trace in a sequent in the form of a formula with a matching connective or quantifier; however, structural rules leave more subtle vestige behind. (This is not quite true for the thinning rules, which add a formula into a sequent. However, the formula is arbitrary, which means that there is no single kind of a formula that indicates that an application of a thinning rule has taken place.) *Structurally free logics* that were invented by Dunn and Meyer in [13] introduce a combinator. The "formulas-as-types" slogan is often used to connect typable combinators or λ-terms and (implicational) formulas. "Combinators-as-formulas" could be a similar catchphrase for structurally free logics. To illustrate the idea, we use *contraction* and the combinator M. The axiom for M is $\mathrm{M}x \rhd xx$.[4]

$$\frac{\mathfrak{A}[\mathfrak{B};\mathfrak{B}] \vdash C}{\mathfrak{A}[\mathfrak{B}] \vdash C} \ (\mathrm{M}\vdash) \qquad\qquad \frac{\mathfrak{A}[\mathfrak{B};\mathfrak{B}] \vdash C}{\mathfrak{A}[\mathrm{M};\mathfrak{B}] \vdash C} \ (\mathrm{M}\vdash)$$

The $(\mathrm{M}\vdash)$ rule on the left is contraction (on structures), whereas the other $(\mathrm{M}\vdash)$ rule is the *combinatory rule*. For the latter, we could omit the label $(\mathrm{M}\vdash)$,

---

[2] This paper was known for a long time only to a select group of relevance logicians, but nowadays, it is freely available online.

[3] See [1] for information on relevance logics, in general. Dunn [12] introduced a sequent calculus for $\mathbf{R}_+$, whereas [7,8] introduced and used a sequent calculus for $\mathbf{R}^t_\to$. See also [6] for a comprehensive overview of a variety of sequent calculi.

[4] In combinatory terms, parentheses are omitted by assuming association to the left; for example, $xz(yz)$ is a shorthand for $((xz)(yz))$. For more on combinatory logic, see for example [4]. Brackets in a sequent indicate a hole in a structure in the antecedent, with the result of a replacement put into brackets in the lower sequent.

because M appears in the sequent itself. Combinatory rules supplant structural rules, hence, the label "structurally free." The implicational types of typable combinators are provable in structurally free logic—with an appropriately chosen set of combinatory rules. The following is a pointer toward this (with some of the more obvious steps omitted or compressed).[5]

$$
\begin{array}{c}
\vdots \\
\hline
\mathcal{D} \to \mathcal{A} \to \mathcal{B} \to \mathcal{C}; \mathcal{D}; \mathcal{A}; \mathcal{B} \vdash \mathcal{C} \\
\hline
\mathsf{C}; (\mathcal{D} \to \mathcal{A} \to \mathcal{B} \to \mathcal{C}; \mathcal{D}); \mathcal{B}; \mathcal{A} \vdash \mathcal{C} \\
\hline
\mathsf{B}; \mathsf{C}; \mathcal{D} \to \mathcal{A} \to \mathcal{B} \to \mathcal{C}; \mathcal{D}; \mathcal{B}; \mathcal{A} \vdash \mathcal{C} \\
\hline
\mathsf{B}; \mathsf{C} \vdash (\mathcal{D} \to \mathcal{A} \to \mathcal{B} \to \mathcal{C}) \to (\mathcal{D} \to \mathcal{B} \to \mathcal{A} \to \mathcal{C})
\end{array} \quad \vdash \to\text{'s}
$$

The combinators B and C are the standard ones with axioms $\mathsf{B}xyz \triangleright x(yz)$ and $\mathsf{C}xyz \triangleright xzy$. The principal (simple) type schema of BC is $(\mathcal{A} \to \mathcal{B} \to \mathcal{C} \to \mathcal{D}) \to \mathcal{A} \to \mathcal{C} \to \mathcal{B} \to \mathcal{D}$. In structurally free logics, other (than simple) types can be considered; for instance, $\mathsf{M} \vdash \mathcal{A} \to (\mathcal{A} \circ \mathcal{A})$ is easily seen to be provable, when $\circ$ (fusion) is included with the usual rules. However, our aim is to find a way to obtain a combinator that inhabits a simple type from a sequent calculus proof—without making combinators into formulas and without introducing a type assignment system.

There are several sequent calculi for intuitionistic logic including its various fragments, but most often they use an associative comma in the antecedent (and possibly in the succedent too).[6] We define a *new sequent calculus* $LJ^T_\to$ for the implicational fragment of intuitionistic logic with the truth constant $\boldsymbol{T}$.

Definition 1. The *language* of $LJ^T_\to$ contains a denumerable sequence of propositional variables $\langle p_i \rangle_{i \in \omega}$, the arrow connective $\to$ and $\boldsymbol{T}$.

The set of *well-formed formulas* is inductively defined from the set of propositional variables with $\boldsymbol{T}$ added, by the clause: "If $\mathcal{A}$ and $\mathcal{B}$ are well-formed formulas, then so is $(\mathcal{A} \to \mathcal{B})$."

The set of *structures* is inductively defined from the set of well-formed formulas by the clause: "If $\mathfrak{A}$ and $\mathfrak{B}$ are structures, then so is $(\mathfrak{A}; \mathfrak{B})$."

A *sequent* is a structure followed by $\vdash$, and then, by a well-formed formula.

REMARK 1. We will occasionally use some (notational) conventions. For example, we might omit some parentheses as mentioned previously, or we might call well-formed formulas simply formulas. Again, we will use [ ] for a context (or hole) in a structure, as usual. The use of a semi-colon as the structural connective is motivated by our desire to consider the structural connective as an analog of function application.

---

[5] Parentheses in the antecedent of a sequent are restored as in combinatory terms, but parentheses in simple types are restored by association to the right.

[6] See [14] as well as [10,11] for such calculi.

REMARK 2. We already mentioned M, B and C together with their axioms. Further combinators that we will use in this paper are I, B$'$, T, W, K and S. Their axioms, respectively, are I$x \triangleright x$, B$'xyz \triangleright y(xz)$, T$xy \triangleright yx$, W$xy \triangleright xyy$, K$xy \triangleright x$ and S$xyz \triangleright xz(yz)$.

Definition 2. The *sequent calculus* $LJ^T_\to$ comprises the axiom and the rules listed below.

$$\mathcal{A} \vdash \mathcal{A} \ \text{(I)}$$

$$\frac{\mathfrak{A} \vdash \mathcal{A} \quad \mathfrak{B}[\mathcal{B}] \vdash \mathcal{C}}{\mathfrak{B}[\mathcal{A} \to \mathcal{B}; \mathfrak{A}] \vdash \mathcal{C}} \ (\to \vdash) \qquad \frac{\mathfrak{A}; \mathcal{A} \vdash \mathcal{B}}{\mathfrak{A} \vdash \mathcal{A} \to \mathcal{B}} \ (\vdash \to)$$

$$\frac{\mathfrak{A}[\mathfrak{B}; (\mathfrak{C}; \mathfrak{D})] \vdash \mathcal{A}}{\mathfrak{A}[(\mathfrak{B}; \mathfrak{C}); \mathfrak{D}] \vdash \mathcal{A}} \ (\mathsf{B} \vdash) \qquad \frac{\mathfrak{A}[\mathfrak{C}; (\mathfrak{B}; \mathfrak{D})] \vdash \mathcal{A}}{\mathfrak{A}[(\mathfrak{B}; \mathfrak{C}); \mathfrak{D}] \vdash \mathcal{A}} \ (\mathsf{B}' \vdash)$$

$$\frac{\mathfrak{A}[(\mathfrak{B}; \mathfrak{D}); \mathfrak{C}] \vdash \mathcal{A}}{\mathfrak{A}[(\mathfrak{B}; \mathfrak{C}); \mathfrak{D}] \vdash \mathcal{A}} \ (\mathsf{C} \vdash) \qquad \frac{\mathfrak{A}[\mathfrak{C}; \mathfrak{B}] \vdash \mathcal{A}}{\mathfrak{A}[\mathfrak{B}; \mathfrak{C}] \vdash \mathcal{A}} \ (\mathsf{T} \vdash)$$

$$\frac{\mathfrak{A}[(\mathfrak{B}; \mathfrak{C}); \mathfrak{C}] \vdash \mathcal{A}}{\mathfrak{A}[\mathfrak{B}; \mathfrak{C}] \vdash \mathcal{A}} \ (\mathsf{W} \vdash) \qquad \frac{\mathfrak{A}[\mathfrak{B}; \mathfrak{B}] \vdash \mathcal{A}}{\mathfrak{A}[\mathfrak{B}] \vdash \mathcal{A}} \ (\mathsf{M} \vdash)$$

$$\frac{\mathfrak{A}[\mathfrak{B}] \vdash \mathcal{A}}{\mathfrak{A}[\mathfrak{B}; \mathfrak{C}] \vdash \mathcal{A}} \ (\mathsf{K} \vdash) \qquad \frac{\mathfrak{A}[\mathfrak{B}] \vdash \mathcal{A}}{\mathfrak{A}[T; \mathfrak{B}] \vdash \mathcal{A}} \ (\boldsymbol{T} \vdash)$$

REMARK 3. The two rules $(\mathsf{T} \vdash)$ and $(\boldsymbol{T} \vdash)$ have a somewhat similar label, but they are clearly different. Both combinators and sentential constants have had their practically standard notation for many years, and so both T's are entrenched. From the point of view of structurally free logics, however, $(\boldsymbol{T} \vdash)$ is like $(\mathsf{I} \vdash)$, a rule that simply inserts the identity combinator on the left of a structure.

The rules may be grouped into operational and structural rules. The former group consists of $(\to \vdash)$, $(\vdash \to)$ and $(\boldsymbol{T} \vdash)$. The structural rules are all labeled with combinators, that is, they are $(\mathsf{B} \vdash)$, $(\mathsf{B}' \vdash)$, $(\mathsf{C} \vdash)$, $(\mathsf{T} \vdash)$, $(\mathsf{W} \vdash)$, $(\mathsf{M} \vdash)$ and $(\mathsf{K} \vdash)$.

REMARK 4. The structural rules clearly suffice for intuitionistic logic; indeed, they all together constitute a redundant set of structural rules. The rules $(\mathsf{T} \vdash)$, $(\mathsf{M} \vdash)$ and $(\mathsf{K} \vdash)$ are permutation, contraction and thinning, moreover, they *operate on structures*, not formulas; thus, in a sense, they are more powerful than the usual versions of these rules. (Of course, in a usual sequent calculus for intuitionistic logic, the notion of a structure cannot be replicated. However, contraction, permutation and thinning could affect a sequence of formulas at once, because those versions of the rules are admissible.) The semi-colon is not associative, but $(\mathsf{B} \vdash)$ is the left-to-right direction of associativity. The other direction of associativity can be obtained by several applications of $(\mathsf{T} \vdash)$ and an application of $(\mathsf{B} \vdash)$. Finally, we note that each of the combinators that appear in the labels to the rules are *proper combinators* (as it is understood in combinatory logic), and they are definable from the combinatory base $\{ \mathsf{S}, \mathsf{K} \}$.

The notion of a *proof* is the usual one for sequent calculi. $\mathcal{A}$ is a *theorem* iff $T \vdash \mathcal{A}$ has a proof. The T cannot be omitted, because $\vdash \mathcal{A}$ is not a sequent in $LJ_{\rightarrow}^{T}$, let alone it has a proof.

The set of rules is not minimal in the sense that some rules could be omitted, because they are derivable from others. As an example, $(T \vdash)$ is easily seen to be derivable by first applying $(K \vdash)$ with T in place of $\mathfrak{C}$, and then, applying $(T \vdash)$ to permute T to the left of $\mathfrak{B}$. The abundance of rules facilitates shorter proofs and finding shorter inhabitants.

$J_{\rightarrow}$ can be thought of as an "S–K calculus," because the principal type schemas of those combinators suffice as axioms (with detachment as a rule) for an axiom system. Similarly, in a combinatory type-assignment system, starting with type schemas assigned to S and K allows us to prove all the theorems of $J_{\rightarrow}$ with a combinator attached to each. A widely known theorem is that S and K are sufficient to define any function—in the sense of combinatorial definability.

However, proofs in a sequent calculus are built differently than in an axiomatic calculus, which is one of the advantages sequent calculi supply. Thus, while we included a rule labeled $(K \vdash)$, there is no $(S \vdash)$ rule. The three left structural rules in LJ emulate (to some extent) the effect of the combinators K, C and W. In LJ, both $(C \vdash)$ (the left permutation rule) and $(W \vdash)$ (the left contraction rule) would have to be used to prove the type of S. It is not possible to define S from merely C and W, because neither has an associative effect.

EXAMPLE 5. The principal type schema of S is $(\mathcal{A} \rightarrow \mathcal{B} \rightarrow \mathcal{C}) \rightarrow (\mathcal{A} \rightarrow \mathcal{B}) \rightarrow \mathcal{A} \rightarrow \mathcal{C}$. The following is a proof of this formula in $LJ_{\rightarrow}^{T}$, which shows the usefulness of some of the redundant structural rules—permitting several steps to be collapsed into single steps.

$$
\cfrac{
  \cfrac{
    A \vdash A \quad \cfrac{
      \cfrac{B \vdash B \quad C \vdash C}{B \rightarrow C; B \vdash C} \rightarrow\vdash
    }{(\mathcal{A} \rightarrow \mathcal{B} \rightarrow \mathcal{C}; \mathcal{A}); \mathcal{B} \vdash \mathcal{C}} \rightarrow\vdash
  }{
  \cfrac{
    \cfrac{
      \cfrac{
        \cfrac{(\mathcal{A} \rightarrow \mathcal{B} \rightarrow \mathcal{C}; \mathcal{B}); \mathcal{A} \vdash \mathcal{C}}{(\mathcal{A} \rightarrow \mathcal{B} \rightarrow \mathcal{C}; (\mathcal{A} \rightarrow \mathcal{B}; \mathcal{A})); \mathcal{A} \vdash \mathcal{C}} \rightarrow\vdash
      }{((\mathcal{A} \rightarrow \mathcal{B} \rightarrow \mathcal{C}; \mathcal{A} \rightarrow \mathcal{B}); \mathcal{A}); \mathcal{A} \vdash \mathcal{C}} B\vdash
    }{(\mathcal{A} \rightarrow \mathcal{B} \rightarrow \mathcal{C}; \mathcal{A} \rightarrow \mathcal{B}); \mathcal{A} \vdash \mathcal{C}} W\vdash
  }{((T; \mathcal{A} \rightarrow \mathcal{B} \rightarrow \mathcal{C}); \mathcal{A} \rightarrow \mathcal{B}); \mathcal{A} \vdash \mathcal{C}} T\vdash
  } \, C\vdash
}{T \vdash (\mathcal{A} \rightarrow \mathcal{B} \rightarrow \mathcal{C}) \rightarrow (\mathcal{A} \rightarrow \mathcal{B}) \rightarrow \mathcal{A} \rightarrow \mathcal{C}} \vdash\rightarrow\text{'s}
$$

REMARK 6. We can show easily that the $(S \vdash)$ rule is a derived rule of $LJ_{\rightarrow}^{T}$.

$$
\cfrac{
  \cfrac{
    \cfrac{
      \cfrac{\mathfrak{A}[\mathfrak{B}; \mathfrak{D}; (\mathfrak{C}; \mathfrak{D})] \vdash \mathcal{A}}{\mathfrak{A}[\mathfrak{B}; \mathfrak{D}; (\mathfrak{D}; \mathfrak{C})] \vdash \mathcal{A}} (T\vdash)
    }{\mathfrak{A}[\mathfrak{B}; \mathfrak{D}; \mathfrak{D}; \mathfrak{C}] \vdash \mathcal{A}} (B\vdash)
  }{\mathfrak{A}[\mathfrak{B}; \mathfrak{D}; \mathfrak{C}] \vdash \mathcal{A}} (W\vdash)
}{\mathfrak{A}[\mathfrak{B}; \mathfrak{C}; \mathfrak{D}] \vdash \mathcal{A}} (C\vdash)
$$

Our goal is to find combinatory inhabitants. We hasten to note that despite the combinatory labels in the proof in Example 5 and in the derived $(S\vdash)$ rule, we do not have a combinatory inhabitant on the horizon. WBC is a fancy (and complex) ternary identity combinator, whereas CBW is not even a proper combinator. CWBT is a quaternary combinator that is often denoted by D, but it is not a duplicator and has no permutative effect. TBWC $\triangleright_1$ WBC, that is, TBWC is an even longer ternary identity combinator. Of course, there are further combinatory terms that can be formed from one occurrence of each of W, B and C, and S is definable from these three combinators. (For example, it is easy to verify that B(BW)(BC(BB)) defines S.) However, we do not need to consider all the combinators with exactly one W, B and C to see the point we are emphasizing here, namely, that we cannot simply convert the labels for the rules into a combinatory inhabitant.

The lack of a direct match between a definition of S and the rules that are used in the proof of the principal type schema of S is perhaps not very surprising, because the $(\to\vdash)$ rule bears a slight resemblance to the affixing rule, which is a composite of the suffixing and prefixing rules. The latter are connected to B$'$ and B, respectively.

The cut rule is of paramount interest in any sequent calculus. $LJ_{\to}^T$ is a single right-handed calculus, but it has structured antecedents; therefore, we formulate the cut rule as follows.

$$\frac{\mathfrak{A}\vdash\mathcal{C}\qquad\mathfrak{B}[\mathcal{C}]\vdash\mathcal{B}}{\mathfrak{B}[\mathfrak{A}]\vdash\mathcal{B}}\ \text{cut}$$

Theorem 3. *The cut rule is* admissible *in* $LJ_{\to}^T$.

Proof. The proof is a standard multi-inductive proof. Here we take as the parameters of the induction the rank of the cut and the degree of the cut formula. We include only some sample steps.

1. First, let us define a multi-cut rule. We will indicate one or more occurrences of a structure $\mathcal{C}$ in the antecedent of the right premise by $[\mathcal{C}]\cdots[\mathcal{C}]$, with the result of replacement indicated as before.

$$\frac{\mathfrak{A}\vdash\mathcal{C}\qquad\mathfrak{B}[\mathcal{C}]\cdots[\mathcal{C}]\vdash\mathcal{B}}{\mathfrak{B}[\mathfrak{A}]\cdots[\mathfrak{A}]\vdash\mathcal{B}}\ \text{multi-cut}$$

The multi-cut rule encompasses the cut rule by permitting a single occurrence of $\mathcal{C}$. On the other hand, multi-cut is a derived rule, because as many applications of cut (with the same left premise) as the number of the selected occurrences of $\mathcal{C}$ yields the lower sequent of multi-cut.[7]

2. The only connective is $\to$, and if the rank of the cut formula is 2, then an interesting case is when the premises are by the $(\to\vdash)$ and $(\vdash\to)$ rules. The cut is replaced by two cuts, each of lower degree; the latter justifies the change in the

---

[7] The multi-cut rule is not the same rule as the mix rule, and its use in the proof of the cut theorem is not necessary. It is possible to use triple induction with a parameter called contraction measure as in [5], for example.

proof. The given and the modified proofs are as follows. Due to the stipulation about the rank, this case has a single occurrence of the cut formula in the right premise.[8]

$$\vdash\!\rightarrow \dfrac{\mathfrak{A}; A \vdash B}{\mathfrak{A} \vdash A \rightarrow B} \quad \dfrac{\mathfrak{B} \vdash A \quad \mathfrak{C}[B] \vdash C}{\mathfrak{C}[A \rightarrow B; \mathfrak{B}] \vdash C} \ \rightarrow\!\vdash$$
$$\dfrac{}{\mathfrak{C}[\mathfrak{A}; \mathfrak{B}] \vdash C} \ \text{(cut)}$$

$$\text{(cut)} \ \dfrac{\mathfrak{B} \vdash A \quad \mathfrak{A}; A \vdash B}{\dfrac{\mathfrak{A}; \mathfrak{B} \vdash B \qquad \mathfrak{C}[B] \vdash C}{\mathfrak{C}[\mathfrak{A}; \mathfrak{B}] \vdash C}} \ \text{(cut)}$$

3. To illustrate the interaction of the cut and a structural rule, we detail the situation when the right rank of the cut is >1, and the right premise is by $(\mathsf{B}'\vdash)$. It is important to note that the structural rules are applicable to structures. For a quick example to show this, let $\mathfrak{B}, \mathfrak{C}$ and $\mathfrak{D}$ be formulas, and let $\mathfrak{C}$ be an occurrence of the cut formula C, in the proof below. If $\mathfrak{E}$ were not an atomic structure (which is possible), then the transformation of the proof would not result in a proof, if the structural rule could not be applied to complex structures.

$$\dfrac{\mathfrak{E} \vdash C \quad \dfrac{\mathfrak{A}[\mathfrak{C}; (\mathfrak{B}; \mathfrak{D})][C] \cdots [C] \vdash A}{\mathfrak{A}[\mathfrak{B}; \mathfrak{C}; \mathfrak{D}][C] \cdots [C] \vdash A} \ (\mathsf{B}'\vdash)}{\mathfrak{A}[\mathfrak{B}; \mathfrak{C}; \mathfrak{D}][\mathfrak{E}] \cdots [\mathfrak{E}] \vdash A} \ \text{(multi-cut)}$$

$$\dfrac{\dfrac{\mathfrak{E} \vdash C \quad \mathfrak{A}[\mathfrak{C}; (\mathfrak{B}; \mathfrak{D})][C] \cdots [C] \vdash A}{\mathfrak{A}[\mathfrak{C}; (\mathfrak{B}; \mathfrak{D})][\mathfrak{E}] \cdots [\mathfrak{E}] \vdash A} \ \text{(multi-cut)}}{\mathfrak{A}[\mathfrak{B}; \mathfrak{C}; \mathfrak{D}][\mathfrak{E}] \cdots [\mathfrak{E}] \vdash A} \ (\mathsf{B}'\vdash)$$

We do not include the other cases here in order to keep our paper focused. ⊰

REMARK 7. We mentioned that the $(T\vdash)$ rule is like the $(\mathsf{I}\vdash)$ rule, which in turn is like the $(t\vdash)$ rule. In some sequent calculi, especially, in sequent calculi for various *relevance logics*, it should be considered what happens when the cut formula is t. The clearest way to prove the cut theorem is to use a separate induction to show that cuts on t are eliminable. In $LJ^T_{\rightarrow}$, this problem does not arise at all, because the $(T\vdash)$ rule is a special instance of the $(\mathsf{K}\vdash)$ rule.

Theorem 4. *If A is a theorem of $J_{\rightarrow}$, then A is a theorem of $LJ^T_{\rightarrow}$.*

Proof. The outline of the proof is as follows. $A \rightarrow B \rightarrow A$ is easily shown to be a theorem of $LJ^T_{\rightarrow}$, and we have the proof in Example 5. If $T \vdash A$ and $T \vdash A \rightarrow B$ are provable sequents, then $T \vdash B$ is a theorem, because from $T \vdash A \rightarrow B$ we can obtain $T; A \vdash B$ by cut, and then, by another cut, we have $T; T \vdash B$, and by $(\mathsf{M}\vdash)$, we get $T \vdash B$. ⊰

---

[8] We omit ∴'s from above the top sequents, which does not mean that the leaves are instances of the axiom; some of them are obviously not.

The converse of this claim is obviously false, simply because T is not in the language of $J_\rightarrow$. However, in intuitionistic logic all theorems are provably equivalent, because $\mathcal{A} \rightarrow \mathcal{B} \rightarrow \mathcal{A}$ is a theorem. Hence, we can define $J_\rightarrow^T$ by adding a single axiom to $J_\rightarrow$ (after extending its language). (A3) says that T is the top element of the word algebra of $LJ_\rightarrow^T$.

(A3) $\mathcal{A} \rightarrow T$

Theorem 5. *If $\mathcal{A}$ is a theorem of $LJ_\rightarrow^T$, then $\mathcal{A}$ is a theorem of $J_\rightarrow^T$.*

Proof. The proof is more elaborate than that of the previous theorem, but follows the usual lines for such proofs. Our sequent calculus can be extended by $\wedge$ with the usual rules (which originated in Gentzen's work). The resulting sequent calculus, $LJ_\wedge^T$ enjoys the admissibility of the cut rule. The proof of the cut theorem allows us to conclude that $LJ_\wedge^T$ is a conservative extension of $LJ_\rightarrow^T$. The axiomatic calculus can be extended with $\wedge$ too, and that can be shown to be conservative, for example, by relying on semantics for $J_\wedge^T$ and $J_\rightarrow^T$.

Once the framework is set up on both sides, we define a function that is applicable to sequents in $LJ_\wedge^T$, and yields a formula in the common language of these calculi. Lastly, we use an induction on the height of a proof in $LJ_\wedge^T$ to show that the axiom and the rules can be emulated in $J_\wedge^T$. (The detailed proof is quite lengthy, but straightforward; hence, we omit the details.) ∴

REMARK 8. The constant T has a double role in $LJ_\rightarrow^T$. First of all, it is an atomic formula, which can be a subformula of a more complex formula. However, we introduced this constant primarily to keep track of the structure of antecedents. There is no difficulty in determining whether T does or does not occur in a formula.

Corollary 6. *A formula $\mathcal{A}$, which contains no occurrences of T, is a theorem of $LJ_\rightarrow^T$ iff it is a theorem of $J_\rightarrow$.*

This claim follows from the three previous theorems and their proofs. Most importantly, the cut theorem guarantees that the *subformula property* holds. That is, if $\mathcal{A}$ is a formula, which occurs in a sequent in the proof of $\mathfrak{A} \vdash \mathcal{B}$, then $\mathcal{A}$ is a subformula of a formula in the sequent $\mathfrak{A} \vdash \mathcal{B}$. If T occurs in a proof of the theorem $\mathcal{A}$ (and it does by the definition of a theorem), then it is a subformula of a formula in $T \vdash \mathcal{A}$. By stipulation, $\mathcal{A}$ is T-free, therefore, the only occurrence of T in $T \vdash \mathcal{A}$ is T itself.

As a result, we can consider T-free theorems of $J_\rightarrow^T$ in $LJ_\rightarrow^T$, without any problems. All the occurrences of T will turn out to be ancestors of the T in the root sequent.

4 Standard Proofs, Cafs and an Inhabitant

Sequent calculi maintain the premises at each step in a derivation. Derivations in axiomatic calculi limit the manipulation of the premises to the addition of new premises. In natural deduction calculi, one can not only add new premises to a derivation, but it is possible to decompose premises and it is also possible to incorporate premises into more complex formulas (thereby eliminating them from among the premises).

Sequent calculi offer the most flexibility and the strictest control over the premises (as well as over the conclusions). Thinning on the left allows the addition of new premises, whereas the most typical way to eliminate a premise is via the implication introduction rule on the right-hand side. The possibility to tinker with the premises and carry them along from step to step while they are not affected by an application of a rule is advantageous.

However, exactly the mentioned features of sequent calculi lead to certain complications in the process of extracting combinatory inhabitants from proofs. For example, in a sufficiently large sequent, we might be able to apply rules to different formulas without interfering with the applicability of another rule. The combinatory counterpart of this *indeterminacy* is when the inhabitant cannot differentiate between proofs in which such pairs of applications of rules are permuted. Of course, the permutability of certain rules is a phenomenon that appears not only in $LJ_{\rightarrow}^T$, but in other sequent calculi too. Still, with ; structuring antecedents, permutability may be less expected than in sequent calculi with flat sequents.

Another potential problem is that some theorems may have more than one proof. If the two proofs become subproofs in a larger proof, for example, by applications of the $(\rightarrow \vdash)$ rule, then the $(W\vdash)$ or the $(M\vdash)$ rule may be applied to contract two complex antecedents, which contain theorems that are (potentially) inhabited by distinct combinators. Contraction may be thought of as *identification*, but it is not possible to identify distinct combinators. There are ways to deal with this problem; we mention two of them.

First, we can apply an abstraction algorithm to obtain terms of the form $Z_1x_1 \ldots x_n$ and $Z_2x_1 \ldots x_n$. It is straightforward to further obtain $Z_3x_1 \ldots x_n$, which reduces to $Z_1x_1 \ldots x_n(Z_2x_1 \ldots x_n)$. While the x's are variables in the term we started with, it well might be that that term is not of the form $x_1 \ldots x_n$. Intuitionistic logic has full permutation, that is, $(\mathcal{A} \rightarrow \mathcal{B} \rightarrow \mathcal{C}) \rightarrow \mathcal{B} \rightarrow \mathcal{A} \rightarrow \mathcal{C}$ is a theorem. In combinatory logic, this means that there is a combinatory term that is isomorphic to the starting term possibly except combinators appearing in front of the variables. In other words, instead of an x_i, there may be an occurrence of Z_ix_i. A nice aspect of this approach is that a larger class of proofs can be subjected to the inhabitant extraction procedure. On the other hand, a drawback is that it is less clear how to systematize the various abstraction steps. It is clear from combinatory logic that Z_1, Z_2, Z_3 as well as all the necessary Z_i's exist, but it is less clear how to describe an algorithm that is applicable independently of the shape of the starting term, and preferably, efficient too. Of course, as a last resort, we could always use some variant of λ^*.

Second, we can delineate the kind of proofs that we consider. In particular, we can completely avoid the complication caused by contraction on structures that correspond to distinct combinators if we exclude proofs in which $(M\vdash)$

or (W⊢) is applied to a complex structure. A careful reader may immediately notice that even if contraction rules are applied to single formulas only, it may happen that those formulas are inhabited by distinct combinators; hence, the restriction seems not to be sufficient. However, as it will become clear later in this section, we do not run into any problem with combinatory terms that are of the form $Z_1 x_1 (Z_2 x_1)$. Of course, the limitation of proofs is acceptable, if it does not exclude theorems of $J_\rightarrow$ from our consideration. We can apply the same insight from combinatory logic as in the previous approach, however, now we talk about the proofs themselves—before the inhabitant extraction is applied.

Given a proof in which (M⊢) or (W⊢) has been applied with a complex $\mathfrak{B}$ or $\mathfrak{C}$, respectively, we can apply (B⊢), (B'⊢), (T⊢) and (C⊢) together with (M⊢) and (W⊢) restricted to single formulas and produce the same lower sequent that resulted from the application of (M⊢) or (W⊢).

Lemma 7. *If $\mathcal{A}$ is a theorem of $J_\rightarrow$, then there is a proof of $\mathcal{A}$ in $LJ_\rightarrow^T$ in which applications of (M⊢) and (W⊢) (if there are any) involve formulas in place of $\mathfrak{B}$ and $\mathfrak{C}$, respectively.*

Proof. The proof is by induction on the structure of the antecedents, $\mathfrak{B}$ in case of (M⊢) and $\mathfrak{C}$ in case of (W⊢).
1. If $\mathfrak{B}$ in (M⊢) or $\mathfrak{C}$ in (W⊢) is $\mathcal{D}$, respectively, then the claim is true.
2. Let us assume that $\mathfrak{C}$ is $\mathfrak{C}_1; \mathfrak{C}_2$.

$$
\cfrac{
\cfrac{
\cfrac{
\mathfrak{A}[\mathfrak{B};(\mathfrak{C}_1;\mathfrak{C}_2);(\mathfrak{C}_1;\mathfrak{C}_2)] \vdash \mathcal{A}
}{
\mathfrak{A}[\mathfrak{C}_1;(\mathfrak{B};(\mathfrak{C}_1;\mathfrak{C}_2));\mathfrak{C}_2] \vdash \mathcal{A}
}(B'\vdash)
}{
\mathfrak{A}[\mathfrak{C}_1;(\mathfrak{B};\mathfrak{C}_1;\mathfrak{C}_2);\mathfrak{C}_2] \vdash \mathcal{A}
}(B\vdash)
}{
\mathfrak{A}[\mathfrak{C}_1;(\mathfrak{B};\mathfrak{C}_1);\mathfrak{C}_2;\mathfrak{C}_2] \vdash \mathcal{A}
}(B\vdash)
$$

$$\vdots$$

$$
\cfrac{
\mathfrak{A}[\mathfrak{C}_1;(\mathfrak{B};\mathfrak{C}_1);\mathfrak{C}_2] \vdash \mathcal{A}
}{
\mathfrak{A}[\mathfrak{B};\mathfrak{C}_1;\mathfrak{C}_1;\mathfrak{C}_2] \vdash \mathcal{A}
}(B'\vdash)
$$

$$\vdots$$

$$
\cfrac{
\cfrac{
\cfrac{
\cfrac{
\mathfrak{A}[\mathfrak{B};\mathfrak{C}_1;\mathfrak{C}_2] \vdash \mathcal{A}
}{
\mathfrak{A}[\mathfrak{C}_2;(\mathfrak{B};\mathfrak{C}_1)] \vdash \mathcal{A}
}(T\vdash)
}{
\mathfrak{A}[\mathfrak{C}_2;(\mathfrak{C}_1;\mathfrak{B})] \vdash \mathcal{A}
}(T\vdash)
}{
\mathfrak{A}[\mathfrak{C}_1;\mathfrak{C}_2;\mathfrak{B}] \vdash \mathcal{A}
}(B'\vdash)
}{
\mathfrak{A}[\mathfrak{B};(\mathfrak{C}_1;\mathfrak{C}_2)] \vdash \mathcal{A}
}(T\vdash)
$$

The thicker lines and :'s indicate where the hypothesis of the induction is applied. $\mathfrak{C}_1$ and $\mathfrak{C}_2$ are proper substructures of $(\mathfrak{C}_1;\mathfrak{C}_2)$; hence, (W⊢) can be applied. We note that the structure of the bottom sequent is exactly the desired one, because the last four steps restore the association of $\mathfrak{C}_1$ and $\mathfrak{C}_2$.

3. Let us assume that $\mathfrak{B}$ is $\mathfrak{B}_1; \mathfrak{B}_2$.

$$\frac{\dfrac{\mathfrak{A}[\mathfrak{B}_1; \mathfrak{B}_2; (\mathfrak{B}_1; \mathfrak{B}_2)] \vdash \mathcal{A}}{\mathfrak{A}[\mathfrak{B}_1; (\mathfrak{B}_1; \mathfrak{B}_2); \mathfrak{B}_2] \vdash \mathcal{A}} \, (\mathsf{B'\vdash})}{\mathfrak{A}[\mathfrak{B}_1; \mathfrak{B}_1; \mathfrak{B}_2; \mathfrak{B}_2] \vdash \mathcal{A}} \, (\mathsf{B\vdash})$$

$$\vdots$$

$$\mathfrak{A}[\mathfrak{B}_1; \mathfrak{B}_2; \mathfrak{B}_2] \vdash \mathcal{A}$$

$$\vdots$$

$$\mathfrak{A}[\mathfrak{B}_1; \mathfrak{B}_2] \vdash \mathcal{A}$$

Once again, the places where the inductive hypothesis is applied are indicated by thick lines and ⋮'s. $\mathfrak{B}_1$ and $\mathfrak{B}_2$ are proper substructures of $\mathfrak{B}_1; \mathfrak{B}_2$; hence, the piecewise contractions can be performed by ($\mathsf{M\vdash}$) and ($\mathsf{W\vdash}$).   ∴

It is easy to see that an application of the ($\mathsf{K\vdash}$) rule, in which $\mathfrak{C}$ is a compound antecedent can be dissolved into successive applications of ($\mathsf{K\vdash}$) in each of which $\mathfrak{C}$ is a formula.

NOTE 9. If $\mathcal{A}$ is a theorem of $J_\to$, then $\mathcal{A}$ is of the form $\mathcal{B}_1 \to \cdots \mathcal{B}_n \to p$, for some $\mathcal{B}_1, \ldots, \mathcal{B}_n$ and p.

We know that $J_\to$ is consistent, hence, no propositional variable is a theorem. Then, a theorem $\mathcal{A}$ contains at least one implication, but the consequent of $\ldots$ the consequent of $\ldots$ the consequent of the formula must be a propositional variable, because there are finitely many $\to$'s in $\mathcal{A}$. We will refer to $\mathcal{B}_1, \ldots, \mathcal{B}_n$ as the *antecedents* of $\mathcal{A}$. Of course, a formula and a sequent are related but different kinds of objects, and so there is no danger that the two usages of the term "antecedent" become confusing.

In order to gain further control over the shape of proofs, we will use the observation we have just stated together with the next lemma to narrow the set of proofs that we will deal with.

Lemma 8. *If* $\mathcal{B}_1 \to \cdots \mathcal{B}_n \to p$ *is a theorem of* $J_\to$, *then there is a proof of the sequent* $\mathbf{T} \vdash \mathcal{B}_1 \to \cdots \mathcal{B}_n \to p$ *in which the last n steps are consecutive applications of* ($\vdash \to$).

Proof. We prove that if there is an application of a rule after ($\to \vdash$), which introduces one of the antecedents, then that can be permuted with the application of the ($\to \vdash$) rule. We do not (try to) prove, and it obviously does not hold in general, that different applications of the ($\vdash \to$) rule can be permuted with each other.[9] (We only consider a couple of cases, and omit the rest of the details.)
1. Let us assume that ($\mathsf{C\vdash}$) follows ($\vdash \to$). The given and the modified proof segments are as follows. (Again, we omit the ⋮'s to save some space.)

---

[9] $\mathcal{B}_i$ is one of $\mathcal{B}_1, \ldots, \mathcal{B}_n$, whereas $\mathcal{C}$ is p (if $\mathcal{B}_i$ is $\mathcal{B}_n$) or $\mathcal{B}_{i+1} \to \cdots \mathcal{B}_n \to p$ (otherwise).

$$\dfrac{\dfrac{\mathfrak{A}[\mathfrak{B};\mathfrak{D};\mathfrak{C}]; \mathcal{B}_i \vdash C}{\mathfrak{A}[\mathfrak{B};\mathfrak{D};\mathfrak{C}] \vdash \mathcal{B}_i \to C} \; (\vdash \to)}{\mathfrak{A}[\mathfrak{B};\mathfrak{C};\mathfrak{D}] \vdash \mathcal{B}_i \to C} \; (C\vdash) \qquad \dfrac{\dfrac{\mathfrak{A}[\mathfrak{B};\mathfrak{D};\mathfrak{C}]; \mathcal{B}_i \vdash C}{\mathfrak{A}[\mathfrak{B};\mathfrak{C};\mathfrak{D}]; \mathcal{B}_i \vdash C} \; (C\vdash)}{\mathfrak{A}[\mathfrak{B};\mathfrak{C};\mathfrak{D}] \vdash \mathcal{B}_i \to C} \; (\vdash \to)$$

2. Let us suppose that the rule that is to be permuted upward is $(T\vdash)$.

$$\dfrac{\dfrac{\mathfrak{A}[\mathfrak{C};\mathfrak{B}]; \mathcal{B}_i \vdash C}{\mathfrak{A}[\mathfrak{C};\mathfrak{B}] \vdash \mathcal{B}_i \to C} \; (\vdash \to)}{\mathfrak{A}[\mathfrak{B};\mathfrak{C}] \vdash \mathcal{B}_i \to C} \; (T\vdash) \qquad \dfrac{\dfrac{\mathfrak{A}[\mathfrak{C};\mathfrak{B}]; \mathcal{B}_i \vdash C}{\mathfrak{A}[\mathfrak{B};\mathfrak{C}]; \mathcal{B}_i \vdash C} \; (T\vdash)}{\mathfrak{A}[\mathfrak{B};\mathfrak{C}] \vdash \mathcal{B}_i \to C} \; (\vdash \to)$$

3. Now, let the rule be $(M\vdash)$.

$$\dfrac{\dfrac{\mathfrak{A}[\mathfrak{B};\mathfrak{B}]; \mathcal{B}_i \vdash C}{\mathfrak{A}[\mathfrak{B};\mathfrak{B}] \vdash \mathcal{B}_i \to C} \; (\vdash \to)}{\mathfrak{A}[\mathfrak{B}] \vdash \mathcal{B}_i \to C} \; (M\vdash) \qquad \dfrac{\dfrac{\mathfrak{A}[\mathfrak{B};\mathfrak{B}]; \mathcal{B}_i \vdash C}{\mathfrak{A}[\mathfrak{B}]; \mathcal{B}_i \vdash C} \; (M\vdash)}{\mathfrak{A}[\mathfrak{B}] \vdash \mathcal{B}_i \to C} \; (\vdash \to)$$

4. Let us assume that the rule after $(\vdash \to)$ is $(K\vdash)$.

$$\dfrac{\dfrac{\mathfrak{A}[\mathfrak{B}]; \mathcal{B}_i \vdash C}{\mathfrak{A}[\mathfrak{B}] \vdash \mathcal{B}_i \to C} \; (\vdash \to)}{\mathfrak{A}[\mathfrak{B};\mathfrak{C}] \vdash \mathcal{B}_i \to C} \; (K\vdash) \qquad \dfrac{\dfrac{\mathfrak{A}[\mathfrak{B}]; \mathcal{B}_i \vdash C}{\mathfrak{A}[\mathfrak{B};\mathfrak{C}]; \mathcal{B}_i \vdash C} \; (K\vdash)}{\mathfrak{A}[\mathfrak{B};\mathfrak{C}] \vdash \mathcal{B}_i \to C} \; (\vdash \to) \qquad \therefore$$

REMARK 10. The proof of Lemma 7 relied on applications of some of the structural rules, and it did not involve the $(\vdash \to)$ rule at all. Therefore, the two restrictions on the set of proofs—based on Lemmas 7 and 8—are orthogonal. That is, we know that every theorem of $J_\to$ has a proof that is both free of contractions in which the contracted structures are complex and that ends with the successive introduction of the antecedents of the theorem.

Definition 9. Let $\mathcal{A}$ be a theorem of $J_\to$. A proof $\mathscr{P}$ of $\mathcal{A}$ is *standard* iff $\mathscr{P}$ has no applications of $(M\vdash)$ or $(W\vdash)$ or $(K\vdash)$ in which the subalterns in the rules are complex, and $\mathscr{P}$ ends with n applications of $(\to \vdash)$, when $\mathcal{A}$ is $\mathcal{B}_1 \to \cdots \mathcal{B}_n \to p$.

REMARK 11. Every theorem of $J_\to$ has a standard proof in $LJ_\to^T$, however, not every proof is standard. We should emphasize that standard proofs are *normal* objects. Clearly, every theorem of $J_\to$ has a normal proof, furthermore, a proof that is not normal can be transformed into a normal one. However, the existence of standard proofs for theorems cannot be strengthened to a statement of *unique* existence. For instance, the $(T\vdash)$ and $(M\vdash)$ rules can always embellish a proof, and so can the $(T\vdash)$ rule.

The general form of standard proofs is as follows.

$$\dfrac{\dfrac{\vdots}{\dfrac{T; \mathcal{B}_1; \mathcal{B}_2; \ldots; \mathcal{B}_n \vdash p}{T; \mathcal{B}_1 \vdash \mathcal{B}_2 \to \cdots \mathcal{B}_n \to p} \; (\vdash \to)\text{'s}}}{T \vdash \mathcal{B}_1 \to \mathcal{B}_2 \to \cdots \mathcal{B}_n \to p} \; (\vdash \to)$$

Definition 10. Given a standard proof of a theorem $\mathcal{B}_1 \to \cdots \mathcal{B}_n \to p$, we call the sequent $T; \mathcal{B}_1; \ldots; \mathcal{B}_n \vdash p$ followed by $(\vdash \to)$'s the *source sequent.*

Given a standard proof, the part of the proof below the source sequent is uniquely determined. Moreover, for our purposes, the applications of the $(\vdash \to)$ rule are not very important. Hence, we will mainly ignore the rest of a standard proof below the source sequent; indeed, we will concentrate on the proof of the source sequent itself.

The extraction of a combinatory inhabitant starts with a given standard proof, and as a result of processing the standard proof, we will obtain *one combinatory inhabitant.*

REMARK 12. It is always guaranteed that our procedure yields a combinatory inhabitant that has the theorem as its type, however, the procedure does not guarantee that the combinatory inhabitant has the theorem as its *principal type.* In axiomatic calculi, there is a way to ensure that principal types are generated. Starting with axioms (rather than axiom schemas), applications of *condensed detachment* produce theorems in which no propositional variables are identified unless that move is inevitable for the application of the detachment rule. There is no similar mechanism built into sequent calculi, though it might be an interesting question to ask what conditions would be sufficient to guarantee that the theorem is the principal type schema of the combinatory inhabitant we generate.

For the sake of transparency, we divide the whole procedure into *three parts.* First, we want to *trace* how formulas move around in a proof. This is especially pivotal in $LJ_{\to}^T$, because of the plenitude of structural rules. Second, we will introduce a new sort of objects, that we call *caf*'s. By using caf's, we recuperate combinators in accordance with the applications of the rules that bear combinatory labels. Third, we will apply a simple BB′I-*abstraction algorithm*, which will produce the combinatory inhabitant from the caf's that replace the formulas in the source sequent.

4.1 Tracing Occurrences

Curry's observation about a certain connection between structural rules and combinators remained dormant for decades—until Dunn and Meyer in [13] revived this idea. As we already mentioned, they not only established a precise link between structural rules and combinators, but they replaced all the structural rules with *combinatory rules.* The latter kind of rules differ from rules like the structural rules in $LJ_{\to}^T$ by introducing a combinator as a formula into the lower sequent of the rule. The formulation of $LJ_{\to}^T$ drew some inspiration from structurally free logics, but we retained the combinators in the labels of the structural rules (instead of including combinators into the set of formulas).

Following Curry, it is customary to provide an *analysis* for a sequent calculus, which is a classification of the formulas in the rules according to the role that they play in a proof.[10] We give only a partial analysis here, which focuses on

---

[10] We would have included the analysis earlier, if we would have included all the details of the proof of the cut theorem.

occurrences of formulas that are passed down from the upper sequent to the lower sequent in an application of a rule.

Definition 11. The notion of an *ancestor* of a formula that occurs in a lower sequent of a rule is defined rule by rule as follows.
(1) $(\to \vdash)$: $\mathcal{C}$ (in the upper sequent) is the ancestor of $\mathcal{C}$ (in the lower sequent), and any formula occurring in $\mathfrak{A}$ or $\mathfrak{B}$ in the upper sequent is the ancestor of its copy in the lower sequent.
(2) $(\vdash \to)$: A formula occurring in $\mathfrak{A}$ in the upper sequent is an ancestor of its copy within $\mathfrak{A}$ in the lower sequent.
(3) $(\mathsf{B}\vdash)$, $(\mathsf{B}'\vdash)$, $(\mathsf{C}\vdash)$ and $(\mathsf{T}\vdash)$: $\mathcal{A}$ in the upper sequent is the ancestor of $\mathcal{A}$ in the lower sequent. A formula occurring in $\mathfrak{A}$, $\mathfrak{B}$, $\mathfrak{C}$ or $\mathfrak{D}$ in the upper sequent is an ancestor of its copy in the identically labeled structure in the lower sequent.
(4) $(\mathsf{W}\vdash)$: $\mathcal{A}$ is the ancestor of $\mathcal{A}$, and a formula in $\mathfrak{A}$ or $\mathfrak{B}$ in the upper sequent is an ancestor of the same formula occurring in the same structure in the lower sequent. A formula occurring in either copy of $\mathfrak{C}$ in the upper sequent is an ancestor of the same formula occurring in $\mathfrak{C}$ in the lower sequent.
(5) $(\mathsf{M}\vdash)$: $\mathcal{A}$ is the ancestor of $\mathcal{A}$, and a formula occurring in $\mathfrak{A}$ in the upper sequent is an ancestor of the same formula occurring in $\mathfrak{A}$ in the lower sequent. A formula occurring in either of the two $\mathfrak{B}$'s in the upper sequent is an ancestor of the same formula occurring in the same position in the structure $\mathfrak{B}$ in the lower sequent.
(6) $(\mathsf{K}\vdash)$: $\mathcal{A}$ is the ancestor of $\mathcal{A}$, and any formula in $\mathfrak{A}$ or $\mathfrak{B}$ is the ancestor of that formula in the identically labeled structure in the lower sequent.
(7) $(\boldsymbol{T}\vdash)$: $\mathcal{A}$ is the ancestor of $\mathcal{A}$, and a formula that occurs in the structure $\mathfrak{A}$ or $\mathfrak{B}$ in the upper sequent is the ancestor of the formula that is of the same shape and has the same location within $\mathfrak{A}$ and $\mathfrak{B}$, respectively, in the lower sequent.

REMARK 13. Certain formulas do not have ancestors at all, while some other formulas have exactly one ancestor (even if the same formula has several occurrences in a sequent), and yet another formulas have exactly two ancestors. The part of the analysis that we included here is completely oblivious to the emergence of complex formulas from their immediate subformulas by applications of the connective rules.

A sequent calculus proof—by definition—does not include a labeling of the proof tree with codes for the rules that were applied. However, there is no difficulty in decorating the proof tree, and in most cases, the decoration is *unique*. The possibility of multiple labels arises in the case of some structural rules when certain structures that we denoted by distinct letters turn out to be identical. Nonetheless, the following is straightforward; hence, we do not provide a proof for it.

Claim 12. *Given a proof in $LJ^T_{\to}$, it is* decidable *which rules could have been applied at each proof step in the proof tree.*

To put a very wide bound on how many decorations can be added at a step, we can multiply the length of a sequent by the number of rules in $LJ^T_{\to}$—still a finite number. This estimate is very much off target, because several pairs of rules cannot be unified at all.

Definition 13. We define a *preferred decoration* of a proof tree by (1)–(6).

(1) If the rule applied is ($\rightarrow$ $\vdash$) or ($\vdash$ $\rightarrow$), then the appropriate label is the preferred decoration, which is attached to the step.[11]
(2) If the rule applied is either (B$\vdash$) or (B'$\vdash$), that is, $\mathfrak{B}$ is $\mathfrak{C}$, then the preferred decoration is (B$\vdash$).
(3) If the rule applied is either (W$\vdash$) or (M$\vdash$), that is, $\mathfrak{B}$ is $\mathfrak{C}$, then the preferred decoration is (M$\vdash$).
(4) If the rule applied is either (K$\vdash$) or ($\boldsymbol{T}\vdash$), that is, $\mathfrak{B}$ and $\mathfrak{C}$ are $\boldsymbol{T}$, then the preferred decoration is (K$\vdash$).
(5) If the rule applied is either (C$\vdash$) or (T$\vdash$), that is, $\mathfrak{C}$ and $\mathfrak{D}$ are $\mathfrak{B}$, then the preferred decoration is (T$\vdash$).
(6) If there is a unique structural rule that is applied, then its label is the preferred decoration.

If a sequent contains several formulas that are of the same shape, then an ambiguity may arise as to where the rule was applied. For instance, if the upper sequent is $\mathcal{A};\mathcal{A};(\mathcal{A};\mathcal{A}) \vdash \mathcal{B}$, and it is identical to the lower sequent, then this could be an application of (T$\vdash$) in three different ways. In such a case, we assume that the rule has been applied with the least possible scope, and at the leftmost place.[12]

Based on the notion of an ancestor together with the notion of preferred decoration we define an algorithm, which starts off the process of turning sequents into pairs of combinatory terms.

Definition 14. Let a standard proof be given, with the root sequent being the source sequent, which is of the shape $\boldsymbol{T};\mathcal{B}_1;\ldots;\mathcal{B}_n \vdash p$. The formulas in the source sequent, and iteratively, in the sequents above it, are *represented* and *replaced* by variables according to (1)–(5).

(1) The source sequent becomes $x_0;x_1;\ldots;x_n \vdash x_{n+1}$.
(2) The formulas (i.e., formula occurrences) in the (B$\vdash$), (B'$\vdash$), (C$\vdash$) and (T$\vdash$) rules are in one-one correspondence between the lower and upper sequents. Thus, the formulas in the upper sequent are replaced by variables according to this correspondence. In the (K$\vdash$) and ($\boldsymbol{T}\vdash$) rules, the same is true, if $\mathfrak{C}$ and $\boldsymbol{T}$ are omitted. That is, the formulas in the upper sequent are ancestors of formulas in the lower sequent, and they are replaced by the variables that stand for them in the lower sequent.

---

[11] No application of ($\rightarrow$ $\vdash$) and ($\vdash$ $\rightarrow$) can be unified with an application of any other rule, as it is easy to see.
[12] Of course, this is a futile step to start with, and for instance, in a proof-search tree we would prune proofs to forbid such happenings. However, our present definition for a standard proof does not exclude proofs that contain identical sequents.

(3) In the $(W \vdash)$ and $(M \vdash)$ rules, some formulas have two ancestors. The variables replace the formulas in the upper sequent according to the ancestor relation.

(4) In the $(\vdash \rightarrow)$ rule, the principal formula of the rule has no ancestor, but it is also absent from the upper sequent. On the other hand, the subalterns are new formulas, which means that they have to be replaced by new variables. If m is the greatest index on a variable used up to this point, then the subaltern, which is the consequent of the arrow formula, is replaced by x_{m+1}. The subaltern, which is the antecedent of the arrow formula, is replaced by x_{m+2}. All the other formulas are replaced by the variable that stands for the formula of which they are an ancestor.

(5) In the $(\rightarrow \vdash)$ rule, the structures $\mathfrak{A}$ and $\mathfrak{B}$ are the same in the upper and lower sequents, so are the $\mathcal{C}$'s. Here the replacement is carried out according to the one-one correspondence. The immediate subformulas of the arrow formula are handled as in the previous case. The consequent formula is replaced by a new variable with the index $m + 1$, if m is the highest index on any variable so far. The antecedent formula is replaced by x_{m+2}.

It is obvious that the definition of the ancestor relation together with the above algorithm guarantees that all formulas are represented by an indexed variable. The proof tree has now been transformed into a tree, in which only x's occur, however, there is an isomorphism between the two trees, and so we continue to modify the new tree by assuming that we can use the information contained in the proof tree itself.

4.2   Formulas Turned into Caf's

As we already mentioned, structurally free logics include combinators into the set of atomic formulas. We do not take combinators to be formulas, but we want to augment the tree of variables with combinators. The variables stand for *formula occurrences*, and we think of them as proxies for identifiable formulas.

Definition 15. The set of *combinatorially augmented formulas* (*caf*'s, for short) is inductively defined by (1)–(2).

(1) If X is a formula, then X is a caf.
(2) If X is a caf and Z is a combinator, then ZX is a caf.

The ;'s remained in the variable tree, and it is straightforward to define the inductive set generated by ; from the set of x's or from the set of caf's. We expand the range of objects that can instantiate the meta-variables $\mathfrak{A}, \mathfrak{B}, \mathfrak{C}, \dots$ to include the variables and the caf's. Context always disambiguates what the structures are composed of.

Using the indexed x's in place of the formula occurrences, we can insert combinators into the tree. However, we need some facts from combinatory logic to make the procedure smooth and easily comprehensible. The following is well known.

Claim 16. $Z(MN)$ *is* weakly equivalent *to* $BZMN$. *In general,* $Z(M_1 \dots M_n)$ *is weakly equivalent to* $\underbrace{B(\dots(B\,Z)\dots)}_{n}M_1 \dots M_n$.

The impact of the first part of the claim is that compound caf's can be equivalently viewed as being of the form $Z\mathcal{A}$, for some Z and the formula $\mathcal{A}$ which is the atomic caf component of the compound caf. In other words, if we want to add Z_1 to a caf $Z_2 x$, then we can add Z_2 and immediately move on to BZ_1Z_2x.

REMARK 14. The impact of the second part of this claim is that we can use sufficiently many regular compositors to disassociate a complex combinatory term and to position a combinator *locally* on a caf (rather than on a compound structure), when the structures affected by a structural rule are complex. We will summarily refer to the utilization of this observation by the label (Bs) (or we will use it tacitly).

Now, we define an algorithm that inserts combinators into our tree. The algorithm starts at the leaves of the tree and proceeds in a top-down fashion, level by level. The highest level of the tree contains one or more leaves, each of which is an instance of the axiom.

REMARK 15. The ancestor relation is *inherited* by the caf's in an obvious way. Indeed, after the replacement of formula occurrences with indexed variables, the ancestor relation is simpler to spot. If X is a caf in the upper sequent, which is of the form Zx_i, and x_i is a variable in the lower sequent, then the formula occurrence (represented by x_i) in the upper sequent is an *ancestor* of the formula occurrence x_i. As a *default step*, we assume that the lower x_i is changed to Zx_i, and then possibly, similar manipulations are performed on the other caf's too.

We assume that we have the original proof tree with its preferred decoration at hand.

Definition 17. Given a tree of variables, *combinators* are inserted according to (1)–(10), after the default copying of caf's from the upper to the lower sequents.

(1) If the rule applied is $(T \vdash)$, then no combinator is inserted.[13]
(2) If the rule applied is $(K \vdash)$, then K is added to $\mathfrak{B}$. If $\mathfrak{B}$ is a complex antecedent, then (Bs) is applied to attach K to the leftmost caf in $\mathfrak{B}$.
(3) If the rule applied is $(W \vdash)$, then we distinguish four subcases. (i) From $\mathfrak{B}; x; x$ we get $W\mathfrak{B}; x$. (ii) From $\mathfrak{B}; Zx; x$ we get $BW(B'Z)\mathfrak{B}; x$. (iii) From $\mathfrak{B}; x; Zx$ we get $BW(B(B'Z))\mathfrak{B}; x$. (iv) From $\mathfrak{B}; Z_1x; Z_2x$ we get $BW(B(B'Z_2)(BB'Z_1))\mathfrak{B}; x$. If $\mathfrak{B}$ is complex in any of the subcases, then (Bs) is applied too in order to position the combinator on the leftmost caf within $\mathfrak{B}$.

---

[13] The variable that is introduced at this step cannot be anything else than x_0, that we will later on turn into I.

(4) If the rule applied is $(\mathsf{M} \vdash)$, then we distinguish four subcases. (i) From $x; x$ we get $\mathsf{M}x$. (ii) From $\mathsf{Z}x; x$ we get $\mathsf{WZ}x$. (iii) From $x; \mathsf{Z}x$ we get $\mathsf{W}(\mathsf{B'Z})x$. Lastly, (iv) from $\mathsf{Z}_1 x; \mathsf{Z}_2 x$ we get $\mathsf{W}(\mathsf{B}(\mathsf{B'Z}_2)\mathsf{Z}_1)x$.

(5) If the rule applied is $(\mathsf{C} \vdash)$, then C is added to $\mathfrak{B}$. Should it be necessary—due to $\mathfrak{B}$'s being complex—(Bs) is applied to move C to the leftmost caf.

(6) If the rule applied is $(\mathsf{T} \vdash)$, then T is added to $\mathfrak{B}$, possibly, with (Bs).

(7) If the rule applied is $(\mathsf{B} \vdash)$, then B is added to $\mathfrak{B}$, with (Bs), if needed.

(8) If the rule applied is $(\mathsf{B'} \vdash)$, then we proceed as in (7), but insert $\mathsf{B'}$.

(9) If the rule applied is $(\vdash \rightarrow)$, then we distinguish two subcases. (i) If the caf standing in for the antecedent of the principal formula is x, then no combinator is inserted. (ii) If the sequent is of the form $\mathfrak{A}; \mathsf{Z}x_i \vdash x_j$, then $\mathsf{B'Z}$ is added to $\mathfrak{A}$ (if atomic), or to its leftmost caf using (Bs) (if $\mathfrak{A}$ is complex).

(10) If the rule applied is $(\rightarrow \vdash)$, then there are two subcases to consider. (i) The antecedent of the principal arrow formula is x, then no combinator is introduced. (ii) If the caf standing in for the subaltern in the right premise is $\mathsf{Z}x$, then the caf Y representing the principal formula, is modified to $\mathsf{B}ZY$.

The algorithm is well-defined, because at each step there is exactly one (sub)case that is applicable. Once the source sequent is reached, we have accumulated the information about the combinatory inhabitant in a somewhat dispersed form.

4.3 BB′I-Abstraction

The source sequent has the general form $Z_0 x_0; Z_1 x_1; \ldots; Z_n x_n \vdash x_{n+1}$, where any of the Z's may be absent. We want to consider the antecedent as a combinatory term, namely, as $Z_0 \mathsf{I}(Z_1 x_1) \ldots (Z_n x_n)$. The base $\{\mathsf{S}, \mathsf{K}\}$ is combinatorially complete, and so is the base $\{\mathsf{B}, \mathsf{B'}, \mathsf{C}, \mathsf{T}, \mathsf{W}, \mathsf{M}, \mathsf{K}\}$. For instance, I is definable as $\mathsf{CKB'}$. That is, there is no problem with obtaining a Z_{n+1} such that $Z_{n+1} x_1 \ldots x_n$ is weakly equivalent to the previous combinatory term.

However, we want to utilize the insight that the term resulting from the source sequent has a very *special form*. First, we note that if Z_0 is empty, then it is sufficient to consider $Z_1 x_1 \ldots (Z_n x_n)$. Our general idea is to move the Z's systematically to the front, and then to disassociate them from the rest of the term. We achieve this by applying $\mathsf{B'}$- and B-expansion steps.

Definition 18. Given the term $Z_0 \mathsf{I}(Z_1 x_1) \ldots (Z_n x_n)$, the BB′I-*abstraction* is defined by replacing the term with that in (1), and then with the term in (2).

(1) $\mathsf{B'}Z_n(\mathsf{B'}Z_{n-1} \ldots (\mathsf{B'}Z_1(Z_0 \mathsf{I})x_1) \ldots x_{n-1})x_n$

(2) $\underbrace{\mathsf{B}(\mathsf{B} \ldots (\mathsf{B'}Z_n) \ldots)}_{n-1}\underbrace{(\mathsf{B}(\mathsf{B} \ldots (\mathsf{B'}Z_{n-1}) \ldots) \ldots}_{n-2}$
$$(\mathsf{B}(\mathsf{B'}Z_2)(\mathsf{B'}Z_1(Z_0 \mathsf{I}))) \ldots)x_1 x_2 \ldots x_{n-1} x_n$$

REMARK 16. Of course, the term in (2) by itself is what we want, however, moving through the term in (1), it becomes obvious that the BB′I-abstraction yields a term that is weakly equivalent to the term obtained from the source sequent.

Before providing a justification for the whole procedure in the form of a correctness theorem, we return to Example 5 from Sect. 3. It so happens that that proof is a standard proof.

EXAMPLE 17. We give here the tree of variables, which is isomorphic to the proof tree—save that we completely omit the part of the tree below the source sequent.

$$
\cfrac{
  x_3 \vdash x_8 \qquad
  \cfrac{
    \cfrac{x_5 \vdash x_{10} \qquad x_9 \vdash x_4}{x_7; x_5 \vdash x_4}\ (\rightarrow\vdash)
  }{x_1; x_3; x_5 \vdash x_4}\ (\rightarrow\vdash)
}{
  \cfrac{
    x_3 \vdash x_6 \qquad
    \cfrac{x_1; x_3; x_5 \vdash x_4}{x_1; x_5; x_3 \vdash x_4}\ (\mathsf{C}\vdash)
  }{
    \cfrac{x_1; (x_2; x_3); x_3 \vdash x_4}{
      \cfrac{x_1; x_2; x_3; x_3 \vdash x_4}{
        \cfrac{x_1; x_2; x_3 \vdash x_4}{x_0; x_1; x_2; x_3 \vdash x_4}\ (\boldsymbol{T}\vdash)
      }\ (\mathsf{W}\vdash)
    }\ (\mathsf{B}\vdash)
  }\ (\rightarrow\vdash)
}
$$

Next is the tree that results by the algorithm that inserts combinators into the tree of variables.

$$
\cfrac{
  x_3 \vdash x_8 \qquad
  \cfrac{\cfrac{x_5 \vdash x_{10} \qquad x_9 \vdash x_4}{x_7; x_5 \vdash x_4}\ (\rightarrow\vdash)}{x_1; x_3; x_5 \vdash x_4}\ (\rightarrow\vdash)
}{
  \cfrac{x_3 \vdash x_6 \qquad \cfrac{x_1; x_3; x_5 \vdash x_4}{\mathsf{C}x_1; x_5; x_3 \vdash x_4}\ (\mathsf{C}\vdash)}{
    \cfrac{\mathsf{C}x_1; (x_2; x_3); x_3 \vdash x_4}{\cfrac{\mathsf{BBC}x_1; x_2; x_3; x_3 \vdash x_4}{\cfrac{\mathsf{B(BW)(BBC)}x_1; x_2; x_3 \vdash x_4}{x_0; \mathsf{B(BW)(BBC)}x_1; x_2; x_3 \vdash x_4}\ (\boldsymbol{T}\vdash)}\ (\mathsf{W}\vdash)}\ (\mathsf{B}\vdash)
  }\ (\rightarrow\vdash)
}
$$

Despite the fact that there is no separate $(\mathsf{S}\vdash)$ rule, we have a simple proof of the principal type of S. Z_0 is absent, hence, we get $\mathsf{I}(\mathsf{B(BW)(BBC)}x_1)x_2x_3$ as the combinatory term from the source sequent. This term immediately reduces to $\mathsf{B(BW)(BBC)}x_1x_2x_3$, which is already in normal form with the combinator disassociated from the variables.[14] We omit the simple verification that $\mathsf{B(BW)(BBC)}$ is weakly equivalent to S. Alternatively, one can easily check that $(\mathcal{A} \rightarrow \mathcal{B} \rightarrow \mathcal{C}) \rightarrow (\mathcal{A} \rightarrow \mathcal{B}) \rightarrow \mathcal{A} \rightarrow \mathcal{C}$ is a type of $\mathsf{B(BW)(BBC)}$.

REMARK 18. It may be noted that the term WBC was right in spirit. The three atomic combinators in that term appear in the term that resulted (as we indicate by underlining): $\mathsf{B(B\underline{W})(B\underline{B}\,\underline{C})}$.

---

[14] Incidentally, the combinator that we gave after Example 5 is different. It may be an interesting question how to find sequent calculus proofs given a combinatory inhabitant for a simple type.

4.4   Correctness

The *formulas-as-types* slogan proved to be fruitful, but instead of the separation and recombination of terms and types, we want to view *formulas as terms*. Informally, it is probably clear at this point that the combinatory inhabitant that we extracted from a standard proof has the theorem as its type. However, we will prove the correctness of the whole procedure rigorously. First, we officially turn structures comprising caf's into combinatory terms. That is, we assume that the x's are proxies for certain formula occurrences, and then we reuse them as real variables in combinatory terms.

Definition 19. The *sharpening* operation, denoted by $^{\#}$, is defined by (1)–(4)— relative to a standard proof with caf's in place of formulas.

(1) $x_0^{\#}$, that is, $T^{\#}$ is I.
(2) If $i \neq 0$, then $x_i^{\#}$, that is, $\mathcal{A}^{\#}$ is x_i.
(3) If X is a compound caf Zx_i, then $X^{\#}$ is $Z(x_i)^{\#}$, that is, $Z(\mathcal{A})^{\#}$.
(4) $(\mathfrak{A};\mathfrak{B})^{\#}$ is $(\mathfrak{A}^{\#}\mathfrak{B}^{\#})$.

REMARK 19. We have three trees, the proof tree we started with, the tree of variables and the tree of caf's. The variables stand for certain formula occurrences. The sharpening operation applied to the antecedent and the succedent in the third tree yields a pair of combinatory terms.

It is easy to see that the following claim, which is important for the consistency in the use of the variables, is true.

Claim 20. *If x_i occurs in the tree of variables that is obtained from the proof tree of a standard proof, then x_i represents occurrences of only one formula throughout the tree.*

In other words, it can happen that different occurrences of one formula turn into different variables, but no variable stands for more than one formula.

Given a sharpened sequent $\mathfrak{A}^{\#} \vdash \mathcal{A}^{\#}$, we can determine which variables occur in it; we focus on the left-hand side of the turnstile. The variables ensure that we do not loose track of the formula occurrences across sequents, but otherwise, they simply get as their type the formula for which they stand.

Definition 21. Given the sharpened sequent $\mathfrak{A}^{\#} \vdash \mathcal{A}^{\#}$, let the *context* Δ be defined as $\Delta = \{\, x_i \colon \mathcal{B} \mid x_i \in \mathrm{fv}(\mathfrak{A}^{\#}) \wedge \mathcal{B}^{\#} = x_i \,\}$. The *interpretation* of the sequent is $\Delta \Vdash \mathfrak{A}^{\#} \colon \mathcal{A}$.

EXAMPLE 20. The source sequent in our example is $x_0; \mathsf{B}(\mathsf{BW})(\mathsf{BBC})x_1; x_2; x_3 \vdash x_4$. If we display the formulas in the caf's, then the source sequent looks like the following: $T; \mathsf{B}(\mathsf{BW})(\mathsf{BBC})\mathcal{A} \to \mathcal{B} \to \mathcal{C}; \mathcal{A} \to \mathcal{B}; \mathcal{A} \vdash \mathcal{C}$. The sharpening operation turns the antecedent of the sequent into $\mathsf{I}(\mathsf{B}(\mathsf{BW})(\mathsf{BBC})x_1)x_2x_3$. Then, $\mathrm{fv}(\mathfrak{A}^{\#}) = \{x_1, x_2, x_3\}$ and $\Delta = \{x_1 \colon \mathcal{A} \to \mathcal{B} \to \mathcal{C}, x_2 \colon \mathcal{A} \to \mathcal{B}, x_3 \colon \mathcal{A}\}$. Finally, the interpretation of the source sequent is

$$\{\, x_1 \colon \mathcal{A} \to \mathcal{B} \to \mathcal{C}, x_2 \colon \mathcal{A} \to \mathcal{B}, x_3 \colon \mathcal{A} \,\} \Vdash \mathsf{I}(\mathsf{B}(\mathsf{BW})(\mathsf{BBC})x_1)x_2x_3 \colon \mathcal{C}.$$

REMARK 21. We hasten to point out that interpreting the sequents of caf's through statements that closely resemble type-assignment statements is merely a convenience, and it could be completely avoided. However, we presume that it helps to understand the correctness proof for our construction.

Lemma 22. *Given a standard proof that has been transformed into a tree of caf's, the interpretation of each sequent $\mathfrak{A} \vdash \mathcal{A}$ in the tree is correct in the sense that if the combinators are replaced by suitable instances of their principal type schemas, and the variables stand for the formulas as indicated in the context as well as function application is detachment, then $\mathfrak{A}^{\#}$ is $\mathcal{A}$.*

Proof. The proof of this lemma is rather lengthy, hence, we include only two subcases here.

1. If the rule is $(\mathsf{K}\vdash)$, then we have $\Delta \Vdash (\mathfrak{A}[\mathfrak{B}])^{\#} : \mathcal{A}$. $\mathfrak{B}^{\#}$ is a subterm of $(\mathfrak{A}[\mathfrak{B}])^{\#}$, hence, in Δ, it computes to a formula, let us say $\mathcal{B}$. $\mathfrak{C}$ may, in general, introduce a new variable, that is, Δ is expanded to Δ'. (We have already pointed out that all the Δ's match one formula to an x.) In the context Δ', $\mathfrak{C}^{\#}$ is assigned a formula, namely, $\mathfrak{C}^{\#} : \mathcal{C} \in \Delta'$—whether $\mathfrak{C}^{\#}$ is new in Δ' or not. If we choose for K's type $\mathcal{B} \to \mathcal{C} \to \mathcal{B}$, then $\mathsf{K}\mathfrak{B}^{\#}\mathfrak{C}^{\#}$ gets the type $\mathcal{B}$. Then, we have that $\Delta' \Vdash (\mathfrak{A}[\mathfrak{B}; \mathfrak{C}])^{\#} : \mathcal{A}$, as we had to show.

2. Let us consider the rule $(\mathsf{C}\vdash)$. We suppose that $\Delta \Vdash (\mathfrak{A}[\mathfrak{B}; \mathfrak{D}; \mathfrak{C}])^{\#} : \mathcal{A}$. There is no change in the context in an application of the $(\mathsf{C}\vdash)$ rule, because $\mathrm{fv}((\mathfrak{A}[\mathfrak{B}; \mathfrak{D}; \mathfrak{C}])^{\#}) = \mathrm{fv}((\mathfrak{A}[\mathfrak{B}; \mathfrak{C}; \mathfrak{D}])^{\#})$. The term $(\mathfrak{B}; \mathfrak{D}; \mathfrak{C})^{\#}$ is a subterm of $(\mathfrak{A}[\mathfrak{B}; \mathfrak{D}; \mathfrak{C}])^{\#}$, and so it computes to a formula, let us say $\mathcal{E}$. Similarly, $\mathfrak{B}^{\#}$, $\mathfrak{D}^{\#}$ and $\mathfrak{C}^{\#}$ yield some formulas, respectively, $\mathcal{D} \to \mathcal{C} \to \mathcal{E}$, $\mathcal{D}$ and $\mathcal{C}$. The principal type schema of C is $(\mathcal{A} \to \mathcal{B} \to \mathcal{C}) \to \mathcal{B} \to \mathcal{A} \to \mathcal{C}$. Taking the instance where $\mathcal{A}$ is $\mathcal{D}$, $\mathcal{B}$ is $\mathcal{C}$ and $\mathcal{C}$ is $\mathcal{E}$, we get that $\mathsf{C}\mathfrak{B}^{\#}$ is $\mathcal{C} \to \mathcal{D} \to \mathcal{E}$. Then further, $\mathsf{C}\mathfrak{B}^{\#}\mathfrak{C}^{\#}$ is $\mathcal{D} \to \mathcal{E}$, and finally $\mathsf{C}\mathfrak{B}^{\#}\mathfrak{C}^{\#}\mathfrak{D}^{\#}$ is $\mathcal{E}$. Placing back the term into the hole in $(\mathfrak{A}[])^{\#}$, we obtain that $\Delta \Vdash (\mathfrak{A}[\mathfrak{B}; \mathfrak{C}; \mathfrak{D}])^{\#} : \mathcal{A}$.   ⋰

We have established that up to the BB'I-abstraction the combinators inserted into the caf's in the source sequent are correct. The next lemma provides the last step in the proof of correctness.

Lemma 23. *Let* $\{ x_1 : \mathcal{A}_1, \ldots, x_n : \mathcal{A}_n \} \Vdash \mathsf{Z}_0\mathsf{I}(\mathsf{Z}_1 x_1) \ldots (\mathsf{Z}_n x_n) : \mathcal{A}_{n+1}$. *Then,*

$$\{ x_1 : \mathcal{A}_1, \ldots, x_n : \mathcal{A}_n \} \Vdash \underbrace{\mathsf{B}(\mathsf{B}\ldots(\mathsf{B}'\mathsf{Z}_n)\ldots)}_{n-1}\underbrace{(\mathsf{B}(\mathsf{B}\ldots(\mathsf{B}'\mathsf{Z}_{n-1})\ldots)}_{n-2}\ldots$$

$$\mathsf{B}(\mathsf{B}'\mathsf{Z}_2)(\mathsf{B}'\mathsf{Z}_1(\mathsf{Z}_0\mathsf{I})))\ldots)x_1 x_2 \ldots x_{n-1}x_n : \mathcal{A}_{n+1}.$$

Proof. The lemma is a special case of B- and B'-expansions, and their well-known properties; hence, we omit the details.   ⋰

5 Conclusions

We have shown that there is a way to extend the Curry–Howard correspondence to connect *sequent calculus proofs* of intuitionistic implicational theorems and *combinatory inhabitants* over the base $\{ \mathsf{B}, \mathsf{B}', \mathsf{C}, \mathsf{T}, \mathsf{W}, \mathsf{M}, \mathsf{K}, \mathsf{I} \}$. A similar approach has been shown in [9] to be applicable to $LT^t_{\rightarrow}$, the implicational fragment of ticket entailment, and we conjecture that the approach can be adapted to other implicational logics that extend the relevance logic $\mathbf{TW}_{\rightarrow}$.

Acknowledgments. I would like to thank the program committee and the organizers of WoLLIC 2018 for inviting me to speak at the conference in Bogotá, Colombia.

References

1. Anderson, A.R., Belnap, N.D., Dunn, J.M.: Entailment: The Logic of Relevance and Necessity, vol. II. Princeton University Press, Princeton (1992)
2. Barendregt, H., Ghilezan, S.: Lambda terms for natural deduction, sequent calculus and cut elimination. J. Funct. Program. **10**, 121–134 (2000)
3. Bimbó, K.: Relevance logics. In: Jacquette, D. (ed.) Philosophy of Logic. Handbook of the Philosophy of Science (D. Gabbay, P. Thagard, J. Woods, eds.), vol. 5, pp. 723–789. Elsevier (North-Holland), Amsterdam (2007). https://doi.org/10.1016/B978-044451541-4/50022-1
4. Bimbó, K.: Combinatory Logic: Pure, Applied and Typed. Discrete Mathematics and its Applications. CRC Press, Boca Raton (2012). https://doi.org/10.1201/b11046
5. Bimbó, K.: The decidability of the intensional fragment of classical linear logic. Theor. Comput. Sci. **597**, 1–17 (2015). https://doi.org/10.1016/j.tcs.2015.06.019
6. Bimbó, K.: Proof Theory: Sequent Calculi and Related Formalisms. Discrete Mathematics and its Applications. CRC Press, Boca Raton (2015). https://doi.org/10.1201/b17294
7. Bimbó, K., Dunn, J.M.: New consecution calculi for $R^t_{\rightarrow}$. Notre Dame J. Form. Logic **53**(4), 491–509 (2012). https://doi.org/10.1215/00294527-1722719
8. Bimbó, K., Dunn, J.M.: On the decidability of implicational ticket entailment. J. Symb. Log. **78**(1), 214–236 (2013). https://doi.org/10.2178/jsl.7801150
9. Bimbó, K., Dunn, J.M.: Extracting $\mathsf{BB'IW}$ inhabitants of simple types from proofs in the sequent calculus $LT^t_{\rightarrow}$ for implicational ticket entailment. Log. Universalis **8**(2), 141–164 (2014). https://doi.org/10.1007/s11787-014-0099-z
10. Curry, H.B.: A Theory of Formal Deducibility. Notre Dame Mathematical Lectures, no. 6. University of Notre Dame Press, Notre Dame (1950)
11. Curry, H.B.: Foundations of Mathematical Logic. McGraw-Hill Book Company, New York (1963). (Dover, New York, NY, 1977)
12. Dunn, J.M.: A 'Gentzen system' for positive relevant implication (abstract). J. Symb. Log. **38**(2), 356–357 (1973). https://doi.org/10.2307/2272113
13. Dunn, J.M., Meyer, R.K.: Combinators and structurally free logic. Log. J. IGPL **5**(4), 505–537 (1997). https://doi.org/10.1093/jigpal/5.4.505
14. Gentzen, G.: Investigations into logical deduction. Am. Philos. Q. **1**(4), 288–306 (1964)
15. Giambrone, S.: Gentzen systems and decision procedures for relevant logics. Ph.D. thesis, Australian National University, Canberra, ACT, Australia (1983)
16. Giambrone, S.: TW_+ and RW_+ are decidable. J. Philos. Log. **14**, 235–254 (1985)
17. Herbelin, H.: A λ-calculus structure isomorphic to Gentzen-style sequent calculus structure. In: Pacholski, L., Tiuryn, J. (eds.) CSL 1994. LNCS, vol. 933, pp. 61–75. Springer, Heidelberg (1995). https://doi.org/10.1007/BFb0022247

18. Lambek, J.: On the calculus of syntactic types. In: Jacobson, R. (ed.) Structure of Language and its Mathematical Aspects, pp. 166–178. American Mathematical Society, Providence (1961)
19. Meyer, R.K.: A general Gentzen system for implicational calculi. Relevance Log. Newsl. **1**, 189–201 (1976)
20. Schönfinkel, M.: On the building blocks of mathematical logic. In: van Heijenoort, J. (ed.) From Frege to Gödel. A Source Book in Mathematical Logic, pp. 355–366. Harvard University Press, Cambridge (1967)

Symbolic Reasoning Methods in Rewriting Logic and Maude

José Meseguer$^{(\boxtimes)}$

Department of Computer Science, University of Illinois at Urbana-Champaign,
Urbana, USA
meseguer@illinois.edu

Abstract. Rewriting logic is both a logical framework where many logics can be naturally represented, and a semantic framework where many computational systems and programming languages, including concurrent ones, can be both specified and executed. Maude is a declarative specification and programming language based on rewriting logic. For reasoning about the logics and systems represented in the rewriting logic framework *symbolic methods* are of great importance. This paper discusses various symbolic methods that address crucial reasoning needs in rewriting logic, how they are supported by Maude and other symbolic engines, and various applications that these methods and engines make possible. Because of the generality of rewriting logic, these methods are widely applicable: they can be used in many areas and can provide useful reasoning components for other reasoning engines.

Keywords: Symbolic computation · Rewriting logic · Maude

1 Introduction

Rewriting Logic is a simple logic extending equational logic. A rewrite theory is a triple

$$\mathcal{R} = (\Sigma, E, R)$$

where Σ is a typed signature of function symbols, E is a set of (possibly conditional) Σ-equations (so that (Σ, E) is an equational theory), and R is a collection of (possibly conditional) rewrite rules of the form:

$$l \rightarrow r \ if \ \phi$$

where l and r are Σ-terms, and ϕ is a condition or "guard" that must be satisfied for an instance of the rule to fire. This means that rewriting happens *modulo* the equations E. The inference rules for the logic as well as its model theory can be found in [25,93]. A comprehensive survey of rewriting logic and its applications can be found in [95], and a comprehensive bibliography as of 2012 can be found in [89].

© Springer-Verlag GmbH Germany, part of Springer Nature 2018
L. S. Moss et al. (Eds.): WoLLIC 2018, LNCS 10944, pp. 25–60, 2018.
https://doi.org/10.1007/978-3-662-57669-4_2

1.1   Rewriting Logic as a Logical and Semantic Framework

Any logic worth its salt should support precise reasoning in some domain. What is the reasoning domain for rewriting logic? There are two but—because of the many ways in which logic and computation are intertwined—the boundary between them is often in the eyes of the beholder:

- **Logically**, rewriting logic is a *logical framework* [88,133], and, more specifically, a *meta-logical framework* [19]. In this reading, a rewrite theory $\mathcal{R} = (\Sigma, E, R)$ specifies a *logic* whose *formulas*—perhaps also its sequents—are specified as the elements of the algebraic data type (initial algebra) $T_{\Sigma/E}$ defined by the equational theory (Σ, E), and where the logic's inference rules are specified by the rewrite rules R, where a rewrite rule $l \to r$ *if* ϕ is *displayed* as an inference rule:

$$\frac{l}{r}$$

  having ϕ as its *side condition*.
- **Computationally**, what a rewrite theory $\mathcal{R} = (\Sigma, E, R)$ specifies is a *concurrent system*, whose *states* are the elements of the algebraic data type $T_{\Sigma/E}$, and whose *concurrent transitions* are specified by the rewrite rules R.

As already mentioned, whether a rewrite theory $\mathcal{R}$ does have a logical or a computational meaning can easily be in the eyes of the beholder. For example, the *same* rewrite theory $\mathcal{R}$ may simultaneously specify a theory in the multiplicative ($\otimes$) fragment of propositional linear logic, *and* a place/transition Petri net.

1.2   Maude

Maude [31] is a high-performance implementation of rewriting logic. Since a program in Maude is just a rewrite theory (or just an equational theory in Maude's functional sublanguage), Maude is a very simple language that can be explained on a paper napkin to a Freshman having no previous knowledge of it (I have done it over coffee, using three paper napkins to be exact). However, Maude is both highly expressive and versatile and very high level, affording in most cases *isomorphic* representations of the logics or concurrent systems being modeled (no Turing machine encodings, please!). For example, Maude representations for entities as different as: (i) a Petri net [134], (ii) hereditarily finite (HF) sets [42], (iii) Barendregt's lambda cube [133], (iv) Girard's Linear Logic [88], (v) Milner's π-Calculus [132], (vi) the Java programming language semantics [51], and (vii) asynchronous digital circuits [76], all have Maude representations that are *isomorphic* to their original, textbook ones, both syntactically and behaviorally.

Maude is also a high-performance language. For example, in a recent detailed benchmarking of 15 well-known algebraic, functional and object-oriented languages by Hubert Garabel and his collaborators at INRIA Rhône-Alpes, Haskell and Maude were the two languages showing the highest overall performance [59]. In fact, from this perspective, Maude is an attractive *declarative programming language* in which to efficiently program many applications.

For the purposes of this paper, the key thing to note is that, as further explained in Sect. 3, Maude does support not just the standard logical inference for equational or rewrite theories by *rewriting modulo equations* (as well as breadth-first search and LTL model chekcking for concurrent systems specified as rewrite theories [31]), but also a wide range of *symbolic execution* methods that make it possible not just to execute logics and systems, but to do various forms of *symbolic reasoning* about the logics or systems specified in Maude.

1.3 Symbolic Reasoning Methods

Rewriting modulo equations *is* of course a *symbolic* reasoning method. But I would like to use "symbolic reasoning" in a stronger sense. The key difference to hang on to is that between *finite/concrete* and *infinite/symbolic*. A term like $\{a, \{b, 7\}, \varnothing\}$ is a *concrete* hereditarily *finite* set involving urelements a, b and 7 as part of its tree structure. Instead, a term like $\{a, \{b, 7\}, \varnothing\} \cup X$, where X is a variable ranging over all hereditarily finite sets, is a finitary *symbolic* description of the *infinite* set (no longer hereditarily finite) of all hereditarily finite sets that contain the set $\{a, \{b, 7\}, \varnothing\}$ as a subset.

Of course, symbolic descriptions do not always specify an infinite set of objects: BDDs do not. But most often they do. For example: (i) a formula φ (interpreted in a chosen infinite model), (ii) a regular expression, and (iii) a pattern with logical variables such as the above pattern $\{a, \{b, 7\}, \varnothing\} \cup X$, all describe infinite sets of objects, namely: (i) the set defined by φ in the model, (ii) the regular language defined by the given regular expression, and (iii) the infinite family of HF sets described by the above pattern. In all cases, the power of a symbolic description in *this* sense is that it *describes an infinite (or huge) set of objects in a compact, finitary way.*

Now that I have made the meaning of my terms clear, let me distinguish four levels involved in symbolic reasoning:

1. **Symbolic Representations.** The above examples (i)–(iii) illustrate different ways of symbolically representing an infinite set of objects.
2. **Symbolic Methods/Algorithms.** Once some form of symbolic representation is chosen, suitable methods are used to manipulate/transform/combine such representations to obtain new ones or gain further information. For example: (i) unification of terms with variables as symbolic representations of their infinite sets of instances provides a symbolic way of *intersecting* such sets; (ii) automata or regular expression operations likewise provide ways of performing Boolean and other operations on regular languages; and (iii) Boolean operations, and/or quantification, on formulas interpreted in an infinite model provide a finitary way or combining the infinite sets they define and of specifying related sets.
3. **Symbolic Reasoning Engines.** At the level of *inference systems* there is no sharp distinction between this level and Level (2): a unification algorithm can, if insisted upon, be regarded as a reasoning engine. But there are two practical distinctions. First of all, by an "engine" we usually mean not just

an *algorithm*—described, say by an inference system—but an *implementation*. Second, even if we keep to the level of inference systems, there is also a pragmatic distinction between a more specialized subtask and an overall reasoning task. For example, a superposition-based first-order theorem prover is a symbolic reasoning *engine* that uses unification as one of its reasoning methods/algorithms.

4. **Applications of Symbolic Reasoning.** This is what reasoning engines are for. They are often described by areas, and can be more or less generic or specialized. For example, we may have areas such as: (i) general purpose (resp. inductive) theorem proving; (ii) program analysis by symbolic execution; (iii) general purpose symbolic LTL model checking of infinite-state system; (iv) symbolic reachability analysis of security properties for cryptographic protocols; (v) partial evaluation of functional programs; (vi) SMT-based verification by model checking or theorem proving, (vii) various specialized formal verification areas such as termination or confluence checking, invariant verification, and so on.

1.4 Symbolic Methods for Rewriting Logic in a Nutshell

Now that I have made clear what I mean by "symbolic methods," and how this level is different from that of engines or applications, let me focus on six such kinds of methods that are very relevant for symbolic reasoning in rewriting logic, are supported by Maude or other Maude-based engines and prototypes, and are used to solve symbolic reasoning problems in various application areas. The six kinds of methods in question are the following (no references are given here for these methods; references are given in Sect. 2):

1. **Tree Automata and Term Pattern Methods.** Given any rewrite theory $\mathcal{R} = (\Sigma, E, R)$ or equational theory (Σ, E), we often need to reason about *infinite sets of ground terms*. Tree automata, and also term *patterns* (i.e., terms with logical variables), provide finite descriptions of such sets. Methods in this area include algorithms for performing various Boolean operations on such sets of terms and performing various decidable tests (e.g., emptiness).

2. **Equational Unification and Generalization.** Unification allows us to solve systems of equations in a term algebra. But since where we often want to reason is not in a term algebra, but in a term algebra modulo the *equations* E of our given rewrite theory $\mathcal{R} = (\Sigma, E, R)$ or equational theory (Σ, E), we often need to solve such equations *modulo E*. For specific equational theories E, decidable algorithms are known; but for general E only semi-decidable semi-algorithms are known. Can we do better?

   Generalization is the dual of unification. When E-unifying terms t and t' we look for a term u and substitution σ such that $t\sigma =_E u =_E t'\sigma$. Instead, when we want to E-*generalize* two term patterns t and t' we look for a term g and substitutions σ, τ of which they are instances up to E-equality, i.e., such that $g\sigma =_E t$ and $g\tau =_E t'$. Sometimes, generalization is described as "anti-unification." But why using such a confusing and ugly name when a clear and natural one is at hand?

3. **Methods for Satisfiability Modulo a Theory.** The use of decision procedures to decide satisfiability of formulas in specific decidable theories is one of the most powerful symbolic methods at the heart of many model checkers and theorem provers. It is also a very successful technology boosting the range of applications and the scalability of formal verification. Since rewrite theories $\mathcal{R} = (\Sigma, E, R)$ and equational theories (Σ, E) *are* theories, they often contain decidable subtheories, particularly when they are interpreted with an *initial semantics*, since the algebraic data type of states $T_{\Sigma/E}$ *is* the initial algebra of the category of all (Σ, E)-algebras. Therefore, having satisfiability decision procedures for $T_{\Sigma/E}$, or for subclasses of formulas in a decidable (initially interpreted) subtheory, can greatly extend the symbolic reasoning capabilities of rewriting logic.

4. **Symbolic Reachability Analysis.** A rewrite theory $\mathcal{R} = (\Sigma, E, R)$ specifies a concurrent system. Specifically, its concurrent transitions. But this means that an existential $\mathcal{R}$-sentence of the form

$$\exists x_1, \ldots, x_n \ t \to^* t'$$

where t and t' are terms with logical variables describing infinite sets of concrete states, is a *reachability formula* asking a very simple question: is there a concrete instance of t from which we can reach a concrete instance of t' after performing a finite (possibly zero) number of transitions using $\mathcal{R}$? When $\mathcal{R}$ represents a logic's inference system and t' is the true formula $\top$, we are just searching for a proof of an existential closure $\exists \ t$; and when t' is the false formula $\bot$, a proof of the above existential formula is usually interpreted as a *proof by refutation* of a universal closure $\forall \neg t$. This is how it is interpreted in logic programming and in resolution theorem proving.

Narrowing is both a unification-based and a symbolic-evaluation-based method to solve reachability existential formulas of the above form.

Rewriting modulo SMT is an alternative symbolic reachability analysis method based on representing infinite sets of states as constrained patterns $u \mid \varphi$, where the satisfiability of the constraint φ can be decided by an SMT solver.

5. **Symbolic LTL Model Checking of Infinite-State Systems.** When the pattern t' in a reachability existential formula $\exists x_1, \ldots, x_n \ t \to^* t'$ represents a set of *bad* states, such as a set of unsafe states or an attack to a cryptographic protocol, then the set of ground states described by t' is a *coinvariant*, that is, the Boolean complement of an invariant, or *good* set of states. But the narrowing-based symbolic methods that can perform symbolic reachability analysis to reason about violations of invariants can be seamlessly extended to *symbolic LTL and LTLR model checking* methods that, under suitable conditions, can verify, not just invariants, but general LTL formulas (and even more general LTLR formulas, see Sect. 2.5) for infinite-state concurrent systems.

6. **Constructor-Based Reachability Logic.** Constructor-based reachability logic generalizes Hoare logic in two ways: (i) it can be used not just to reason about the partial correctness of programs in a conventional programming

language, but also about the partial correctness of concurrent systems specified by rewriting logic theories; and (ii) although in both Hoare logic and
in reachability logic formulas have a *precondition* and a *postcondition*, in a
reachability logic formula its "postcondition" does not necessarily specify a
set of *terminating* states, but can instead specify a set of *intermediate* states.
The extra generality of constructor-based reachability logic to handle rewrite
theories, and not just conventional code, means that it supports not just code
verification but also verification of concurrent system *designs*.

1.5   Ecumenical Scope

So what? Why should one care about all this if one is not interested in rewriting
logic? I would answer as follows: (1) one may always choose to care or not
care about *anything*, but (2) we cannot avoid the consequences of our own free
decisions. This paper is not written primarily for rewriting logic specialists, but
for researchers interested in logic, programming languages, formal methods, and
automated deduction. Therefore it has what might be called, somewhat tongue-
in-cheek, and *ecumenical* aim and scope. The reason why I think this paper
can be useful within this formal reasoning *oecumene*, and why I have taken
the trouble of writing it, is that the methods I present are in my view of very
wide applicability and should be of interest to many other researchers. I believe
that they can provide useful components for many different symbolic engines.
To me this is not surprising at all: the contrary situation would be surprising.
Since rewriting logic is a very general logical and semantic framework, methods
develop for it enjoy a similar high level of generality that tends to make them
widely applicable.

As I show in Sect. 2, the methods themselves often arose in various automated
deduction areas outside rewriting logic. But here they have become generalized
because of the simultaneous need for greater generality in at least three orthogonal dimensions:

1. **From Unsorted to Order-Sorted.** Untyped treatments are of theoretical
   and academic interest, but quite useless for computing purposes. Types (and
   subtypes) are needed both for programming and to correctly model almost
   anything, including stacks and vector spaces (more on this in Sect. 2.1).
2. **From Free to Modulo Equations.** Terms, that is, the element of a free
   algebra $T_\Sigma(X)$ for function symbols Σ, are of course tree data structures.
   But many other data structures, in particular the data structures needed to
   model many concurrent states, are *not* trees at all. The great power of order-
   sorted algebraic data types, that is, of initial algebra semantics, is that the
   initial algebras $T_{\Sigma/E \uplus B}$ of convergent equational theories $(\Sigma, E \uplus B)$ with
   structural axioms B and Church-Rosser equations E modulo B (more on
   this in Sect. 2.1) can model *any* data type of interest. This is of course an
   indemonstrable thesis, just as Church's thesis is; but the evidence for it is
   very strong. It is even stronger for data types specified as initial algebras in
   membership equational logic [94], also supported by rewriting logic [25] and
   Maude [31].

3. **From Equational to Non-equational.** Many symbolic techniques, including, for example, term rewriting and narrowing, have originated in equational logic and have for a long time only been used in their original equational setting. But in rewriting logic the meaning of both rewriting and narrowing need not be equational at all (more on this in Sect. 2.4). This is intrinsic in the very notion of a rewrite theory, which is typically of the form $\mathcal{R} = (\Sigma, E \uplus B, R)$, with $(\Sigma, E \uplus B)$ convergent. This means that in $\mathcal{R}$ rewriting and narrowing happens at *two* levels: (i) *equational rewriting and narrowing* with the left-to-right oriented equations E, and (i) *non-equational rewriting and narrowing* with the rules R; in both cases *modulo* equations.

Paper Organization. The rest of this paper is organized as follows. Section 2 summarizes how the above-mentioned symbolic reasoning methods (1)–(6) substantially extend the reasoning capabilities of requiting logic. Section 3 then gives a brief summary of how Maude and other Maude-based symbolic engines use these methods as key components; and Sect. 4 summarizes various application areas that benefit from them. Conclusions are given in Sect. 5.

2 Symbolic Methods for Rewriting Logic

2.1 Tree Automata and Term Pattern Methods

Tree automata and term pattern methods are both all about symbolically describing infinite sets of ground terms. But which ones? In rewriting logic, to begin with, we need to describe sets of *states* for the concurrent system specified by a rewrite theory $\mathcal{R} = (\Sigma, G, R)$. For both a rewrite theory $\mathcal{R} = (\Sigma, G, R)$ and an equational theory (Σ, G), the equations G may include some commonly-occurring axioms such as: associativity (A), commutativity (C), associativity-commutativity (AC), identity (U), and so on. In fact, in Maude [31] such, so-called *equational attributes*, can be automatically declared by the user with the `assoc`, `comm`, and `id:` keywords, so that rewriting happens *modulo* such axioms. This has a strong bearing on the kinds of sets of terms we need to reason about. To explain this, a short digression on convergence and constructors is needed.

Convergence and Constructors in a Nutshell. The above equational axioms are important because, for many applications, the equations G in practice decompose as a disjoint union $G = E \uplus B$, where B is some combination or associativity and/or commutativity and/or identity axioms for some symbols in Σ, and E are equations that, when oriented from left to right as rewrite rules, are *convergent*[1] modulo B, i.e., confluent (also called Church-Rosser), terminating, and coherent modulo B. For a simple example of a convergent equational theory $(\Sigma, E \uplus C)$, let $\Sigma = \{0, s, +\}$ be the Peano signature for natural numbers with 0, successor s and plus $+$, E the equations $0 + y = y$ and $s(x) + y = s(x + y)$ defining

---

[1] For a very general formulation of convergence, including conditional equations E, see [84].

addition, and C the commutativity axiom $x + y = y + x$. Under the convergence assumption the Church-Rosser Theorem modulo B (see, e.g., [75,96]) makes G-equality decidable by rewriting with the oriented equations E as rewrite rules modulo B, so that given any Σ-terms t, t', we have the equivalence:

$$E \uplus B \vdash t = t' \iff t!_{E,B} =_B t'!_{E,B}$$

where $=_B$ denotes provable B-equality, and $t!_{E,B}$ denotes the *normal form* of t by simplifying it with the oriented equations E modulo B until no more simplification steps are possible. In fact, Maude's *functional modules* are always assumed to be convergent theories in the above sense, or to be at least confluent and coherent modulo axioms B.

In practice, more, very useful information is often available about a convergent equational theory $(\Sigma, E \uplus B)$: there is a subsignature $\Omega \subseteq \Sigma$ of *constructor* operators so that the "data elements" of the algebraic data type $T_{\Sigma/E \uplus B}$ defined by $(\Sigma, E \uplus B)$ are precisely the Ω-terms. Instead, the other function symbols in $\Sigma - \Omega$ have the meaning of functions that are *defined* by the equations E modulo B, whose inputs are constructor-built data elements and whose *result* is also "data," i.e., an Ω-term. More precisely, this means that: (i) Ω has the same sorts as Σ, (ii) we have an isomorphism $T_{\Sigma/E \uplus B}|_\Omega \cong T_{\Omega/B_\Omega}$ (where $B_\Omega \subseteq B$ are axioms[2] satisfied by some constructor operators in Ω), and (iii) for every *ground* term $t \in T_\Sigma$, its normal form $t!_{E,B}$ is always an Ω-term. For example, for $(\Sigma, E \uplus C)$ the above convergent specification of natural number addition, the constructors are of course $\Omega = \{0, s\}$, and $B_\Omega = \varnothing$.

Now that I have sketched the meaning of convergent equational theories and of constructors, it becomes meaningful to ask the key relevant question: In the context of a rewrite theory $\mathcal{R} = (\Sigma, G, R)$ or an equational theory (Σ, G), where $G = E \uplus B$ and $(\Sigma, E \uplus B)$ is convergent and has constructors Ω satisfying above properties (i)–(iii), what is the best way of symbolically describing *subsets* of the initial algebra $T_{\Sigma/E \uplus B}$ (that is, subsets of the set of *states*, in case we are considering $\mathcal{R}$)? And the answer is obvious: given the isomorphism $T_{\Sigma/E \uplus B}|_\Omega \cong T_{\Omega/B_\Omega}$, their simplest description is as subsets of the initial algebra of constructors T_{Ω/B_Ω}. Certain classes of such subsets are then describable by either *tree automata* methods, or by *term pattern* methods.

Equational Tree Automata. For most rewrite theories $\mathcal{R} = (\Sigma, E, R)$ or equational theories (Σ, E), the *signature* Σ of function symbols is *order-sorted* [62]. That is, such signatures support sorts (types), subsorts (subtypes), and subsort polymorphism, so that a symbol such as $+$ may have several subsort-polymorphic typings for, say, sorts *Nat*, *Int*, *Rat*, and so on. But, as pointed out

---

[2] In some cases, the algebra of constructors may have the more general form $T_{\Omega/E_\Omega \uplus B_\Omega}$. For example, the data elements may be *sets*, which, together with axioms B_Ω of associativity and commutativity (AC) of set union, may also have an *idempotency* equation in E_Ω which is applied modulo AC to put all set expressions in normal form. All I say applies also to this more general case, but tree automata and/or pattern methods may be more complex in the presence of both axioms and equations for constructors.

by Comon long ago [35], an order sorted signature and a tree automaton are one and the same thing (*la même chose!* An example is given below). Under slightly different guises (for tree automata the guise is that of a rewrite theory having ground term left- and right-hand sides in its rules, see below) what these equivalent representations describe are sets of terms for what are called *states* in automata parlance and *sorts* in signature parlance. In summary, therefore, tree automata (see [34] for a general reference) allow us to describe infinite sets of terms (the so-called regular tree languages) in a symbolic, finitary fashion, and to perform various operations (including Boolean ones) as well as various tests on such sets in a decidable way.

All this means that a regular tree language is a subset $L \subseteq T_\Sigma$. But the above discussion has made it clear that we need not just sets like this but, more generally, sets of the form $L \subseteq T_{\Sigma/E \uplus B}$. These are of course much more complicated sets; but under the convergence and constructor assumptions explained above, they have a much simpler descriptions as sets $L \subseteq T_{\Omega/B_\Omega}$. So, if we stick to *regular* sets, all we need to handle such sets are so-called *equational tree automata*, that is, rewrite theories of the form $\mathcal{A} = (\Omega \uplus S, B_\Omega, R)$, where B_Ω are equational axioms such as combinations of associativity and/or commutativity and/or identity axioms, and R are rewrite rules whose left- and right-hand sides are ground terms having a particularly simple format: either $f(q_1, \ldots, q_n) \to q$ for some $f \in \Omega$, or $q \to q'$, where $q_1, \ldots, q_n, q, q'$, etc. are the so-called *states* of the automaton $\mathcal{A}$ belonging to a set S of constants. Note that the q_i are indeed states of the automaton's rewrite theory $\mathcal{A}$. But, of course, what we use each state $q \in S$ of $\mathcal{A} = (\Omega \uplus S, B, R)$ for, is for describing a *set* $L_q \subseteq T_{\Omega/B_\Omega}$ *of states* in the rewrite theory $\mathcal{R} = (\Sigma, E \uplus B, R')$, under the convergence and constructor assumptions for $(\Sigma, E \uplus B)$. Here is a simple example that can alternatively be viewed as an order-sorted signature or as a tree automaton. The constructors for the Peano natural numbers can be specified in a more expressive *order-sorted* signature Ω with two *sorts*, *Nat* and *NzNat* (for non-zero naturals), a *subsort inclusion NzNat < Nat*, a constant 0 of sort *Nat*, and a successor function $s : Nat \to NzNat$. The (essentially isomorphic) tree automaton is a rewrite theory $\mathcal{A} = (\Omega \uplus S, \varnothing, R)$ with *states*, $S = \{Nat, NzNat\}$, and whose rules R are: $0 \to Nat$, $s(Nat) \to NzNat$, and $NzNat \to Nat$, the last one a so-called "epsilon rule" corresponding to the subsort inclusion $NzNat < Nat$. Then, the set of ground terms $\{s(0), s(s(0)), \ldots, s^n(0), \ldots\}$ is simultaneously the set of terms $T_{\Omega,NzNat}$ of sort *NzNat* for the order-sorted signature Ω, and the regular language associated to state *NzNat* in the tree automaton $\mathcal{A} = (\Omega \uplus S, \varnothing, R)$ (where in $\mathcal{A}$ the signature Ω is viewed as an *unsorted* signature).

The long and short of it is that tree automata, when generalized to the notion of *equational tree automata with constraints* [69,72] can perform many of the operations of standard tree automata, including various Boolean operations and the crucial *emptiness check*. The case where some operator $f \in \Omega$ is associative but not commutative is well-known to be undecidable. However, machine learning techniques can often learn a regular language that make the problem decidable [69]. In all Maude tools and applications using these equational tree automata (see Sects. 3–4), these learning techniques have been so effective that, *de facto*, the undecidability of the associative-only case becomes a non-issue.

Term Pattern Operations. Tree automata are *a* way of describing sets of terms, but definitely not the simplest. The simplest is the logic programming way of describing a set of terms by one or several *patterns*, that is, by terms with logical variables. The description "logical" adds nothing whatsoever to the variables in question; it is just a hint that they can become instantiated by substitution. To put it briefly, what a term pattern t *denotes* is the set of ground terms:

$$[\![t]\!] =_{def} \{t\sigma \in T_\Sigma \mid \sigma \in [X \to T_\Sigma]\}$$

where X is the set of variables appearing in t, and the function set $[X \to T_\Sigma]$ is of course the set of ground substitutions for those variables. For example, for $\Sigma = \{0, s, +\}$, the term pattern $x + s(y)$ denotes the set $[\![x + s(y)]\!]$ of all ground terms instantiating the variables x and y. A seminal paper in this area is the one by Lassez and Marriott [81]. The key point is that terms patterns can be symbolically operated on to perform *Boolean* operations on the (typically infinite) sets of ground terms they denote. The only operation where some restrictions apply is *set difference*. Exactly as for tree automata, where linear terms (i.e., terms without repeated variables) describe regular languages and are therefore tree-automata-definable, the required *sufficient condition* for set difference to be symbolically describable is linearity of the patterns involved (this can be relaxed somewhat, as explained in [81]).

But the results by Lassez and Marriott and the subsequent literature always assume an *unsorted* (untyped) signature Σ. This is an antediluvian notion, both for programming languages, where typing is *de rigueur* in modern programming languages, and for formal reasoning purposes, where order-sorted reasoning has been shown to be much more expressive and efficient than unsorted reasoning. This is because unsorted reasoning for the *same* problems trivially handled in an order-sorted setting forces the introduction of *predicates* to describe the missing sorts, and therefore pushes the entire reasoning into the realm of first-order theorem proving. It is of course always possible to stick to an unsorted setting; but the literature on order-sorted reasoning amply shows that there are no defensible rational grounds for favoring unsorted reasoning over order-sorted reasoning. Besides making reasoning about sorts trivial—as opposed to pushing such reasoning into predicate logic, order-sorting makes specifications and programs very natural and expressive, and solves many partiality issues impossible to solve in an untyped or even simply-typed setting. Indeed, order-sorted typing is widely used in many declarative programming and specification languages, e.g., [21,31,56,60,61,63,103,127], in resolution theorem proving, e.g., [27,32,121,137], in artificial intelligence, e.g., [33,55,127], and in unification theory, e.g., [50,70,77,99,128]. Furthermore, it has a smooth, conservative extension to a higher-order typing discipline supporting subtypes [68,87].

Given the above considerations, and given the fact that the sets of terms we need to reason about are of the form $L \subseteq T_{\Omega/B_\Omega}$ corresponding to *sets of states* in a concurrent system specified by a rewrite theory $\mathcal{R} = (\Sigma, E \uplus B, R)$ (under convergence and constructor assumptions for $(\Sigma, E \uplus B)$) whose signatures Σ and Ω are *order-sorted*, what we obviously need are pattern operations where the patterns themselves are terms with logical variables in an *order-sorted* signature. The greater expressiveness of order-sorted patterns has as a consequence the somewhat surprising fact that terms are no longer closed under term difference, even assuming they are linear [101]. However, this can be easily remedied: any order-sorted signature Σ can be automatically and conservatively extended into a more expressive one $\Sigma \cup \Sigma^\#$ so that *all* term pattern operations defined in [81] for the unsorted case smoothly extend to the order-sorted case. Furthermore, using the order-sorted information in the patterns makes these operations more efficient than if only many-sorted information is used [81]. Since, as already pointed out, an order-sorted signature and a tree automaton are *la même chose*, what linear order-sorted term patterns define are sublanguages of languages that are themselves tree-atomata-definable. As explained in Sects. 5.3 and 6 of [101], this brings order-sorted patterns and tree automata very close to each other in expressive power. But patterns have the added advantage of allowing a much more direct way of defining languages. This becomes crucially important when sets of terms have to be manipulated by other inference system; for example, as state predicates in reachability logic (see the constrained constructor patterns explained below, as well as the summary of reachability logic in Sect. 2.6).

Making Patterns More Expressive. Recall again that the sets of terms we need to symbolically describe in general have the form $L \subseteq T_{\Omega/B_\Omega}$. This suggests increasing the expressive power of patterns, possibly at the expense of losing some operations or the decidability of some questions, and is analogous to how equational tree automata generalize standard tree automata to be able to describe certain sets of the form $L \subseteq T_{\Omega/B_\Omega}$. A quite general notion of *constrained constructor pattern* proposed in [125] assumes a convergent equational theory $(\Sigma, E \uplus B)$ with a subsignature of constructors[3] (Ω, B_Ω) and allows constrained patterns of the form $u \mid \varphi$, where φ is a quantifier-free Σ-formula. The set of terms defined by a constrained constructor pattern $u \mid \varphi$ is:

$$[\![u \mid \varphi]\!] =_{def} \{[u\rho]_{B_\Omega} \in T_{\Omega/B_\Omega} \mid \rho \in [X \to T_\Omega] \wedge E \cup B \models \varphi\rho\}.$$

Finite sets of such patterns are then closed under finite unions and intersections, which can be performed symbolically, assuming a B_Ω-unification algorithm [125].

2.2 Equational Unification and Generalization

Some material in this Section is adapted from [2,66,98].

Equational Unification. Syntactic unification (no equational axioms) goes back to Alan Robinson [112]. For equational unification valuable surveys

---

[3] The same remarks as in Footnote 2 apply here: the signature of constructors could more generally be a convergent theory of the form $(\Omega, E_\Omega \uplus B_\Omega)$. Unification modulo $E_\Omega \uplus B_\Omega$ can still be performed in practice to intersect constrained patterns because in many cases $(\Omega, E_\Omega \uplus B_\Omega)$ has the finite variant property (see Sect. 2.2).

include [10,11,123]. As already explained, given an equational theory (Σ, E), *E-unification* is the process of solving systems of equations (i.e., conjunctions of equations) in the term algebra modulo E, denoted $T_{\Sigma/E}(X)$, where the set X of variables is assumed countably infinite. Given a conjunction of Σ-equations ϕ, an *E-unifier* is a solution of ϕ in $T_{\Sigma/E}(X)$, that is, a substitution σ such that for each $t = t'$ in ϕ we have $t\sigma =_E t'\sigma$. The first crucial distinction is between E-unification *algorithms* and E-unification *semi-algorithms*. In turn, an E-unification *algorithm* can be *finitary* or *infinitary*. A finitary E-unification *algorithm* either returns a complete set of *most general* E-unifiers for any system of Σ-equations ϕ (where "most general" means that all other E-unifiers are substitution instances of the returned ones up to E-equality), or indicates that no solution exists. For example, for B associative-commutative (AC) axioms, there is a finitary unification algorithm. An infinitary E-unification *algorithm* generates a possibly infinite complete set of most general E-unifiers for any system of Σ-equations ϕ, or stops in finite time indicating that no solution exists. For example, for B associative-only (A) axioms, there is an infinitary unification algorithm. This means that having an E-unification algorithm makes *unifiability*, that is, the question of whether a system of equations ϕ has any solution at all *decidable*. For example, for C the commutativity axiom $x + y = y + x$ the equation $x + s(y) = x' + y'$ has two C-unifiers, namely, $\{x = x', y' = s(y)\}$, and $\{x' = s(y), x = y'\}$.

By contrast, an E-unification *semi-algorithm*, when given a system of Σ-equations ϕ will: (i) generate a complete set of most general E-unifiers for any system of Σ-equations ϕ *if* some solution exists, but may loop forever when no solution exists. Therefore, if only an E-unification semi-algorithm exists, E-unifiability becomes *undecidable* in general. Unification theory has had for a long time *generic* E-unification *semi-algorithms*, namely, *narrowing-based* [52,73,74,126] and *transformation-based* [58,129] ones. But one important drawback of these semi-algorithms is that, as just mentioned, E-unifiability becomes undecidable. These makes such semi-algorithms useless for *deciding* unifiability, i.e., satisfiability of formulas $\varphi \equiv \bigvee_i \bigwedge G_i$ in both the free (Σ, E)-algebra $T_{\Sigma/E}(X)$ and the initial (Σ, E)-algebra $T_{\Sigma/E}$, *unless* they can be proved *terminating*. For convergent equational theories (Σ, E) (i.e., convergent but with no equational axioms $(B = \varnothing)$ some termination results for narrowing-based unification, mostly centered on special input theories for the *basic narrowing* strategy [73], do exist for some quite restrictive classes of rules R (see [3,4], and references there, for a comprehensive and up-to-date treatment). Instead, the more general case of termination for narrowing-based unification for convergent equational theories $(\Sigma, E \uplus B)$ has been a real *terra incognita* until very recently, because *negative* results, like the impossibility of using basic narrowing when B is a set of associative-commutative (AC) axioms [36], seemed to dash any hopes not just of termination, but even of efficient implementation. Many of these limitations have now disappeared thanks to the *folding variant narrowing* algorithm [47,50]. Let me first give a brief explanation of what variants and the finite variant property are, and then summarize the current state of the art about variant-based equational unification after [50]. Details about narrowing are explained later, in Sect. 2.4.

Variants in a Nutshell. Given a convergent equational theory $(\Sigma, E \cup B)$, a *variant* [30,36,50] of a Σ-term t is a pair (u, θ) where θ is a substitution, and u is the normal form of the term instance $t\theta$ by the rewrite rules E modulo B. Intuitively, the variants of t are the fully simplified *patterns* to which the instances of t can be simplified with the oriented equations E modulo B. Some simplified instances are of course more general (as patterns) than others. $(\Sigma, E \cup B)$ has the *finite variant property* (FVP) in the Comon-Delaune sense [36] iff any Σ-term t has a *finite* set of most general variants. For example, the addition equations $E = \{x+0 = x, x+s(y) = s(x+y)\}$ are *not* FVP, since $(x+y, id), (s(x+y_1), \{y \mapsto s(y_1)\}), (s(s(x+y_2)), \{y \mapsto s(s(y_2))\}), \ldots, (s^n(x+y_n), \{y \mapsto s^n(y_n)\}), \ldots$, are all *incomparable* variants of $x+y$. Instead, the Boolean equations $G = \{x \vee \top = \top, x \vee \bot = x, x \wedge \top = x, x \wedge \bot = \bot\}$ *are* FVP. For example, the most general variants of $x \vee y$ are: $(x \vee y, id), (x, \{y \mapsto \bot\})$, and $(\top, \{y \mapsto \top\})$.

Now that the notion of variant has been sketched, let me summarize the current state of the art about variant-based equational unification after [50]:

1. When B has a finitary unification algorithm, folding variant narrowing with convergent oriented equations E modulo B will terminate on any input term (including unification problems expressed in an extended signature) iff $E \uplus B$ has the *finite variant property* in the Comon-Delaune sense [36].
2. No other complete narrowing strategy can terminate more often than folding variant narrowing; in particular, basic narrowing (when applicable, e.g., $B = \emptyset$) terminates strictly less often.
3. FVP is a semi-decidable property, and when it actually holds it can be easily checked by folding variant narrowing, assuming convergence of the oriented equations E modulo B and a finitary B-unification algorithm [30].
4. As further explained in Sect. 3, folding variant narrowing and variant-based unification for theories $E \uplus B$, where B can be any combination of associativity, commutativity and identity axioms are well supported by Maude [41].

There are by now papers, e.g., [36,45,46], many cryptographic protocol specifications, e.g., [28,46,64,118,138], and several verification tools, e.g., [28,46,118], demonstrating that FVP equational theories are omni-present in cryptographic protocol verification and that variant-based unification and narrowing are very general and effective formal reasoning methods to verify such protocols. As further explained in Sect. 2.3 the applications of varian unification are not at all restricted to cryptographic protocols: reasoning about many other algebraic data types can greatly benefit from variant unification methods.

Domain-Specific vs. Theory-Generic Equational Unification. Standard E-unification algorithms are *domain-specific*, that is, the set E of equations is *fixed* once and forall, and then an E-unification algorithm is developed. For example AC-unification and A-unification algorithms are obviously domain-specific. Instead, an equational unification algorithm is *theory-generic* if it provides an E-unification algorithm not for a fixed E, but for an infinite class of theories (Σ, E). Since defining and implementing equational unification algorithms is a difficult and labor-intensive task, theory-generic unification methods are very useful, because: (i) they greatly expand the class of theories for which equational unification algorithms exist, and (ii) they make unification algorithms *user-definable*: a user only needs to check that his/her equational theory is in the class supported by a theory-generic algorithm, and can then use such a theory-generic algorithm to obtain a unification algorithm for his/her chosen theory *for free*. For example, this is the great advantage of folding variant narrowing as a theory-generic unification algorithm for the infinite class of FVP theories.

Equational Generalization. Generalization is a symbolic method used for many tasks, including theorem proving, and automatic techniques for program analysis, synthesis, verification, compilation, refactoring, test case generation, learning, specialisation, and transformation; see, e.g., [24, 26, 57, 79, 80, 82, 105, 108]. Generalization is the dual of unification. Given terms t and t', a *generalizer* of t and t' is a term t'' of which t and t' are substitution instances. The dual of a *most general unifier* (mgu) is that of a *least general generalizer* (lgg); that is, a generalizer that is more specific than any other generalizer. For example, the lgg of the terms $0 + 0$ and $0 + s(0)$ is $0 + x$. For untyped signatures and for syntactic unification and generalization, if they exist, the mgu and its dual, the lgg, are *unique* (up to variable renaming). However, as soon as order-sortedness and/or reasoning modulo equations E are considered, as already discussed, there is no more *a* unique mgu. Instead there is typically a *complete set* of most general order-sorted unifiers, perhaps modulo equations E. In a completely analogous way, there is not just a unique lgg but a complete set of least general generalizers.

In the literature on machine learning and partial evaluation, least general generalization is also known as *most specific generalization* (msg) and *least common anti-instance* (lcai) [102]. Least general generalization was originally introduced by Plotkin in [109], see also [111]. Indeed, Plotkin's work originated from the consideration in [110] that, since unification is useful in automatic deduction by the resolution method, its dual might prove helpful for induction.

Somewhat surprisingly, unlike order-sorted unification, equational unification, and order-sorted equational unification, which all three had been thoroughly investigated in the literature—see, e.g., [10, 99, 120, 123, 128]—until the appearance of [2] there seems to have been no previous, systematic treatment of order-sorted generalization, equational generalization, and order-sorted equational generalization, although some specific syntactic order-sorted, or unsorted equational cases had been studied.

The relevance of [2] as a symbolic technique for rewriting logic is that it supports reasoning about generalization in a setting that is both order-sorted and modulo equations E, and does so in modular way. Specifically, the work in [2] provides:

1. An order-sorted generalization algorithm that computes a finite, complete and minimal set of *least general generalizers*.
2. A modular equational generalization algorithm in the sense that it combines different generalization algorithms—one for each kind of equational axiom

such as: (i) no axioms, (ii) associativity, (iii) commutativity, and (iv) identity, each specified by its inference rules, into an overall algorithm that is *modular* in the precise sense that the *combination* of different equational axioms for different function symbols is automatic and seamless: the inference rules can be applied to generalization problems involving each symbol, with no need whatsoever for any changes or adaptations. This is similar to, but much simpler and easier than, modular methods for combining E-unification algorithms, e.g., [10]. An interesting result is that associative generalization is finitary, whereas associative unification is infinitary.
3. An order-sorted modular equational generalization algorithm which combines and refines the inference rules given in (1) and (2).

2.3   Methods for Satisfiability Modulo a Theory

Some material in this Section is adapted from [66,98].

The use of decision procedures for theories axiomatizing data structures and functions commonly occurring in software and hardware systems is currently one of the most effective methods at the heart of state-of-the-art theorem provers and model checkers. In the late 70's and early 80's *combination methods* were developed by Nelson and Oppen [106] and Shostak [122] to achieve satisfiability in combinations of decidable theories. In this century, SAT-solving technology has been synergistically combined with satisfiability procedures for decidable theories, an approach pioneered independently by a number of different groups [6,9,18,53,54,104] and distilled in the influential DPLL(T) architecture [107]. This approach has been key to the success of SMT solving, as witnessed by a vast literature on the subject.

The key distinction between *domain-specific* versus *theory-generic* algorithms, already discussed in the context of equational unification in Sect. 2.2, reappears here for the exact same reasons. The points is that, in practice, the above technologies are usually applied to concrete, domain-specific decision procedures. The upshot of this is that one needs to have algorithms and implementations for each of the theories supported by the given SMT solver, which requires a non-trivial effort and in any case limits at any given time each SMT solver to support a *finite* (and in practice not very large) library of theories that it can handle (plus of course their Nelson-Oppen combinations when possible). Although very powerful, the domain-specific approach lacks *extensibility*: any theory outside those supported by an SMT solver (and their combinations) is out of reach for a tool user. Obviously, *theory-generic*—i.e., not for a single theory, but for an infinite class of theories—satisfiability decision procedures, can make input theories *user-definable* and make an SMT solver's repertoire of individual decidable theories potentially *infinite* and easily specifiable by the tool's *users*, as opposed to its implementers. Of course, the theory-generic approach does not exclude the domain-specific one: it *complements* it, thus achieving extensibility.

Variant-Based Satisfiability in a Nutshell. In software verification practice, the decidable theories that are needed are not esoteric ones, but eminently concrete: typically theories deciding satisfiability of fomulas in various *data types*, such as integers, rationals, arrays, lists, and so on. For verifying programs in a conventional programming language having a fix repertoire of data types, the domain-specific SMT solving approach can be quite useful, although it may not address all the verification needs (for example, heap-intensive programs are more challenging). But for declarative languages with arbitrary user-definable algebraic data types, the domain specific-approach is much more limited. In Maude and other languages such as OBJ [63], Cafe OBJ [56], ELAN [23], and ASF+SDF [37], the user can define *any* data type as the initial algebra $T_{\Sigma/E \uplus B}$ of a user-defined convergent equational theory $(\Sigma, E \uplus B)$. There is therefore *no hope* for a domain-specific approach to satisfiability of data types for these declarative languages.

The key idea of *variant-based satisfiability* [67,98,124] is to lift the theory-generic approach to equational unification by folding variant narrowing [50] already discussed in Sect. 2.2 to a likewise theory-generic decision procedure for satisfiability of quantifier-free (QF) formulas in the infinite class of user-definable algebraic data types $T_{\Sigma/E \uplus B}$ specified by convergent theories $(\Sigma, E \uplus B)$ that are FVP. The key method to achieve satisfiability is by reduction to the initial algebra of constructors. Recall from Sect. 2.1 that for most algebraic data types we typically have a subtheory of constructors $(\Omega, E_\Omega \uplus B_\Omega) \subseteq (\Sigma, E \uplus B)$ such that $T_{\Sigma/E \uplus B}|_\Omega \cong T_{\Omega/E_\Omega \uplus B_\Omega}$. Variant-based satisfiability uses variant-based unification to reduce satisfiability of QF formulas in $T_{\Sigma/E \uplus B}$ to satisfiability of QF formulas in the much simpler data type of constructors $T_{\Omega/E_\Omega \uplus B_\Omega}$. Furthermore, the theory-generic requirement of *OS-compactness* (see below for an explanation) for the constructor specification $(\Omega, E_\Omega \uplus B_\Omega)$ automatically ensures decidability of QF formulas in $T_{\Omega/E_\Omega \uplus B_\Omega}$. In particular, in the very common case when only axioms B_Ω are present, decidability of QF formulas in any order-sorted algebra of the form T_{Ω/B_Ω} for B_Ω any combination of associativity and/or commutativity and/or identity axioms (except associativity without commutativity) the OS-compactness requirement is met [98]. Variant satisfiability is not the only possible approach to theory-generic satisfiability: two alternative approaches based on: (i) superposition theorem proving [5,7,22,44,78,85,86,135], and (ii) procedures axiomatized by formulas with triggers [40] are discussed in [98]. However, the variant-based approach is naturally aligned with user-definable algebraic data types, whereas the other two approaches are not so directly aligned with algebraic data types.

OS-Compactness in a Nutshell. OS-compactness applies particularly to equational theories for constructor terms. Assuming for simplicity that all sorts in a convergent theory $(\Omega, E_\Omega \cup B_\Omega)$ have an infinite number of ground terms of that sort which are all different modulo the equations $E_\Omega \cup B_\Omega$, then *OS-compactness* of $(\Omega, E_\Omega \cup B_\Omega)$ means that any conjunction of disequalities $\bigwedge_{1 \leq i \leq n} u_i \neq v_i$ such that $E_\Omega \cup B_\Omega \not\vdash u_i = v_i$, $1 \leq i \leq n$, is *satisfiable* in the initial algebra $T_{\Omega/E_\Omega \cup B_\Omega}$. For example, $(\{0, s\}, \varnothing)$ is OS-compact, where $\{0, s\}$ are the usual natural number constructors. Thus, $s(x) \neq s(y) \wedge 0 \neq y$ *is* satisfiable in $T_{\{0,s\}}$.

A Variant Satisfiability Example. (Naturals with $+$, max, min and $\dot{-}$) [98]. Consider the following unsorted specification $\mathcal{N}_+ = (\{0,1,+\}, ACU)$ of constructors, specifying the data type of natural numbers with addition, where ACU are the axioms $(x + y) + z = x + (y + z), x + y = y + x$, and $x + 0 = x$. As explained above, this theory is OS-compact and therefore has decidable QF satisfiability. Consider now the convergent theory $\mathcal{N}_{+,max,min,\dot{-}}$ obtained by adding to $\mathcal{N}_+$ the max and min operators max, min : $Nat\ Nat \to Nat$ and the "monus" operator $\_\dot{-}\_$: $Nat\ Nat \to Nat$ as defined functions, and the commutativity axioms for max and min, which are respectively defined by the equations $max(n, n + m) = n + m, min(n, n + m) = n$. Likewise, $\dot{-}$ is defined by the equations, $n \dot{-} (n + m) = 0, (n + m) \dot{-} n = m$. $\mathcal{N}_{+,max,min,\dot{-}}$ is FVP (this can be easily checked in Maude). Furthermore, $\mathcal{N}_{+,max,min,\dot{-}}$ protects the OS-compact constructor subtheory $\mathcal{N}_+$ (in the sense that the constructor reduct of the initial algebra for $\mathcal{N}_{+,max,min,\dot{-}}$ is the initial algebra for $\mathcal{N}_+$, as explained in Sect. 2.1). This makes QF satisfiability in the initial algebra of $\mathcal{N}_{+,max,min,\dot{-}}$ decidable by variant satisfiability. For example, $n \dot{-} m = 0 \vee m \dot{-} n = 0$ is a theorem in the initial algebra of $\mathcal{N}_{+,max,min,\dot{-}}$, because its negation $n \dot{-} m \neq 0 \wedge m \dot{-} n \neq 0$ is equisatisfiable with the disjunction of constructor conjunctions $(0 \neq 0 \wedge m \neq 0) \vee (0 \neq 0 \wedge m' \neq 0)$, which is unsatisfiable in the initial algebra of $\mathcal{N}_+$.

2.4 Symbolic Reachability Analysis

As already mentioned in Sect. 1.4, if $\mathcal{R} = (\Sigma, E, R)$ specifies a concurrent system, a *reachability formula* has the general form

$$\exists x_1, \ldots, x_n\ t \to^* t'$$

where t and t' are term patterns with logical variables describing infinite sets of concrete states and we want to know if a ground instance of t can reach a ground instance of t'. Let me discuss two methods to perform this kind of reachability analysis symbolically: narrowing, and rewriting modulo SMT.

Narrowing in a Nutshell. Narrowing originated in the efforts to add equational reasoning to resolution theorem proving (see [73, 126]) and was subsequently generalized to narrowing modulo equational axioms [74]. It sounds a bit *recherché*, but it is very easy to explain. It is, essentially, the *symbolic evaluation* method par excellence. Consider a trivial but illuminating example, namely, the equational definition of addition in Peano notation by the following equations E: $0 + y = y$ and $s(x) + y = s(x + y)$. Since these equations are convergent (see Sect. 2.1 for an explanation of convergence), we can orient them as rewrite rules $R(E)$: $0 + y \to y$ and $s(x) + y \to s(x + y)$. The need for symbolic evaluation arises when there is not enough information to perform standard evaluation: we can evaluate $s(s(0)) + s(s(0))$ to $s(s(s(s(0))))$ by just three applications of the above two rewrite rules. But we cannot evaluate a symbolic expression like $n + m$ in this standard sense. However, by performing steps of *unification* instead of just doing *term matching*, we can *symbolically evaluate* $n + m$ in various ways: (i) by solving the equation $n + m = 0 + y$ we can instantiate $n + m$ to $0 + m$, which can then be rewritten by the first rule to m. This is called a *narrowing step*, denoted $n + m \leadsto_{R(E)} m$. Likewise, by solving the equation $n + m = s(x) + y$ we get in the same manner a narrowing step $n + m \leadsto_{R(E)} s(n' + m)$ using the second rule. When a variable is reached, like in the step $n + m \leadsto_{R(E)} m$, no further narrowing is performed; but in the case of the step $n + m \leadsto_{R(E)} s(n' + m)$, we can now narrow the subterm $n' + m$ exactly as we did so for $n + m$, and we can keep going! In this way we get a *narrowing tree* that records all the *most general ways* in which our original expression $n + m$ can be symbolically evaluated. But narrowing can be performed in a more powerful way *modulo axioms B*. For example, if we had declared $+$ to be *commutative* (C) and had performed unification modulo C, we would get additional narrowing steps such as $n + m \leadsto_{R(E)/C} n$ and $n + m \leadsto_{R(E)/C} s(m' + n)$, where the subindex $/C$ indicates that we are narrowing *modulo* commutativity.

Equational Narrowing vs. Non-equational Narrowing. Narrowing was studied for symbolic evaluation of convergent equational theories such as in the above example. In particular, as already mentioned in Sect. 2.2, narrowing with the oriented equations E of a convergent equational theory $(\Sigma, E \uplus B)$ modulo B (where B itself has a unification algorithm) provides a semi-algorithm for $E \uplus B$-unification, and if $(\Sigma, E \uplus B)$ is FVP an actual $E \uplus B$-unification algorithm by folding variant narrowing [50]. I call this, traditional form of narrowing *equational narrowing*, to distinguish it from *non-equational narrowing*, which is the analogous form of symbolic execution for a rewrite theory $\mathcal{R} = (\Sigma, E, R)$, where the rules R do *not* have an equational meaning: what they specify are *transitions* in a concurrent system. Of course, the rules R specifying such transitions need not be confluent or terminating *at all*. To the best of my knowledge, the first systematic study of non-equational narrowing was carried out in [92]. The intuitive meaning of narrowing is similar in both cases: symbolic execution. But for a rewrite theory $\mathcal{R} = (\Sigma, E, R)$ what are symbolically executed are the *concurrent transitions* of the system specified by $\mathcal{R}$, of course *modulo* the equations E. Therefore, we get narrowing steps of the form $t \leadsto_{R/E} t'$ by performing E-unification with the lefthand sides of the rules in R. And what the term t being narrowed describes is a typically infinite set of *states* in such a system. Therefore, non-equational narrowing is all about symbolic reachability analysis. A simple and not very restrictive condition on $\mathcal{R}$ of being *topmost* (essentially that all narrowing steps $t \leadsto_{R/E} t'$ take place at the top of term t) ensures that narrowing is a *complete* symbolic method for reachability analysis [92]. That is, the formula $\exists x_1, \ldots, x_n \, t \rightarrow^* t'$ holds in $\mathcal{R}$ if and only if narrowing can find an instance of t that can reach an instance of t' by a sequence of transitions with R modulo E. See Sect. 6 in [41] for several simple, yet non-trivial examples of how narrowing-based symbolic reachability analysis can be used to get answers to various symbolic queries in the context of a rewrite theory specifying a meta-interpreter for grammars.

Conditional Equational Narrowing with Constrained Patterns. As mentioned in Sect. 1.4, we can view symbolic reachability analysis from two complementary viewpoints: (i) *computationally*, the existential formula

$$\exists x_1, \ldots, x_n \; t \rightarrow^* t'$$

asks for instance states of t that can reach an instance state of t'; but (ii) *logically*, when $t' = \top$ it asks for a proof of the existential closure $\exists \, t$. Consider in particular the case where t is a conjunction ϕ of equalities for a convergent equational theory $(\Sigma, E \uplus B)$ whose equations E can be conditional, where conditions in equations satisfy a natural determinism requirement (see [29] for details). Then we can associate to $(\Sigma, E \uplus B)$ a conditional rewrite theory $(\Sigma^\wedge, B, R(E))$, where: (i) $\Sigma^\wedge$ extends Σ by adding a fresh new sort *Conj* with a constant $\top$, a binary conjunction operator[4] $\_ \wedge \_$, and equality operators $\_ \overset{?}{=} \_ : s \; s \rightarrow \mathit{Conj}$ for each top sort s in Σ; (ii) for each conditional equation $u = v$ *if* ψ in E there is a conditional rule $u \rightarrow v$ *if* ψ in $R(E)$; and (iii) we further add to $R(E)$ rewrite rules $x \overset{?}{=} x \rightarrow \top$ and $\top \wedge C \rightarrow C$ specifying the meaning of conjunction. Then we can express the $E \uplus B$-unification problem for ϕ as the symbolic reachability problem

$$\exists x_1, \ldots, x_n \; \phi \rightarrow^* \top$$

in the associated rewrite theory $(\Sigma^\wedge, B, R(E))$, which can be solved by *conditional narrowing*. The crucial issue, however, is what to do with the condition ψ of a rule $u \rightarrow v$ *if* ψ when using it to solve the system of equations ϕ by narrowing. In principle, *before* performing a narrowing step with rule $u \rightarrow v$ *if* ψ with substitution θ to solve ϕ, we would first have to use the rules in $R(E)$ to narrow the condition's instance $\psi\theta$ searching for a solution of it. But this can be hopelessly inefficient. The alternative proposed in [29] is to use *constrained patterns* modulo B, in the sense of Sect. 2.1, so that the rule's condition ψ is *not* solved before a narrowing step but is instead added (instantiated by the given narrowing substitution θ) to the new constrained pattern obtained by a narrowing step as an additional constraint. That is, we start, not with the goal ϕ, but with the trivially constrained pattern $\phi \mid \top$, and, to narrow ϕ to ϕ' with a rule $u \rightarrow v$ *if* ψ and substitution θ, we perform the constrained narrowing step $\phi \mid \top \rightsquigarrow_{R(E)/B} \phi' \mid \top \wedge (\psi\theta)$. We then continue in the same way further narrowing the goal $\phi' \mid \top \wedge (\psi\theta)$, which, using the rule $\top \wedge C \rightarrow C$, simplifies to the goal $\phi' \mid \psi\theta$. By computing the substitution μ obtained by composing the substitutions for each of the steps along a narrowing path from $\phi \mid \top$ that reaches a term of the form $\top \mid \eta$ for some constraint η, we obtain a so-called *constrained unifier* [29] of the form $\mu \mid \eta$, where μ is a $E \uplus B$-unifier of ϕ, but subject to the condition that the constraint η is itself solvable. Constrained unifiers provide a complete set of $E \uplus B$-solutions for ϕ in the expected way [29]. Furthermore, the same methodology can be applied in a recursive manner to solve the unifier's constraint η as a new $E \uplus B$-unification problem; but this is always postponed until after an actual constrained unifier $\mu \mid \eta$ has already been found.

Rewriting Modulo SMT. As explained in Sect. 2.1, a constrained pattern $t \mid \varphi$ denotes the set of ground instances of t such that the constrain φ is satisfied.

---

[4] To avoid wasteful computations we can further make the binary operator $\wedge$ *frozen* in its second argument, which forbids rewriting under that second argument. See [95] for a detailed explanation of frozen arguments and [29] for its use in narrowing.

In general, it may be undecidable whether φ is satisfiable. But we may be in much more favorable circumstances if φ is SMT solvable. Furthermore, if the conditions ψ in the rules $u \to v$ *if* ψ of the given rewrite theory $\mathcal{R} = (\Sigma, E, R)$ are themselves *SMT* solvable, we may be in an even more favorable situation. The key idea about *rewriting modulo SMT* [116] and related approaches, e.g., [8], is to solve *constrained reachability problems* of the form:

$$\exists x_1, \ldots, x_n \; t \mid \varphi \to^* t' \mid \varphi'$$

for a topmost rewrite theory $\mathcal{R} = (\Sigma, B \uplus E_0, R)$ such that: (i) there is a sub-signature $\Sigma_0 \subseteq \Sigma$, typically with fewer sorts, such that the equation E_0 are all Σ_0-equations; (ii) B consists only of equational axioms involving operators in the signature $\Sigma - \Sigma_0$; (iii) the Σ_0-reduct $T_{\Sigma/E}|_{\Sigma_0}$ of the initial algebra $T_{\Sigma/E}$ is isomorphic to the initial algebra T_{Σ_0/E_0}; (iv) satisfiability of QF Σ_0-formulas in T_{Σ_0/E_0} is decidable; (v) the rules R are of the form $u \to v$ *if* ψ with ψ a QF Σ_0-formula and where any new variables in v must belong to sorts in Σ_0; and (vi) t and t' do not have any Σ_0-subterms, except variables, and the only variables appearing in the pattern $t \mid \varphi$ belong to sorts in Σ_0.

Under these circumstances, it is possible to *symbolically rewrite* the initial pattern state $t \mid \varphi$ to obtain a *complete reachability analysis method*, called *rewriting modulo SMT*, to semi-decide if an instance of the target pattern $t' \mid \varphi'$ can be reached. Essentially, given a rule $u \to v$ *if* ψ in R, where without loss of generality we may assume that u and v do not have any Σ_0-subterms, except variables and u is linear, i.e., has no repeated variables (otherwise, such Σ_0-subterms or non-linear variables in u can be abstracted away and moved to the condition ψ), a constrained term $w \mid \eta$ can be rewritten by this rule if: (a) it can be rewritten in the standard[5] sense with matching substitution θ and (b) the Σ_0-constraint $\eta \wedge \psi\theta$ is satisfiable in T_{Σ_0/E_0}. See [116] for several examples illustrating the usefulness of rewriting modulo SMT.

An important problem is that of *state space reduction*. That is, how to drastically reduce the number of constrained patterns obtained from an initial pattern state $t \mid \varphi$ by this method in an attempt to reach the target pattern state $t' \mid \varphi'$. The recent work by Bae and Rocha provides an elegant and effective solution to this problem by means of the technique of so-called *guarded terms* [17].

2.5 Symbolic LTL Model Checking of Infinite-State Systems

As mentioned in Sect. 1.4, the need for symbolic LTL model checking of rewrite theories arises naturally as an outgrowth of symbolic reachability analysis, were we can only analyze (violations of) invariants. In general, we want to verify richer LTL properties including, but not limited to, invariants. In both cases the state space explored is *symbolic*, in the sense that the states are term patterns denoting not a single state, but a typically infinite set of concrete states. And in

---

[5] Here "standard" should be taken with a grain of salt, since extra variables with sorts in Σ_0 are allowed in the rule's righthand side v. These extra variables will be mapped by θ to new, *fresh* variables.

both cases symbolic execution by *narrowing* with the rules R of the (topmost) rewrite theory $\mathcal{R} = (\Sigma, E \uplus B, R)$ modulo $E \uplus B$ is the way of performing transition steps.

The work on narrowing-based LTL symbolic model checking of infinite-state concurrent systems specified as rewrite theories in [13–15, 49] assumes that the topmost rewrite theory $\mathcal{R} = (\Sigma, E \uplus B, R)$ specifying the Kripke structure to be verified (therefore also the state predicates) is such that $(\Sigma, E \uplus B)$ is a convergent and FVP equational theory, where B has a finitary unification algorithm. For simple examples the symbolic state space may already be finite; but in general it is not so. The work in [13–15, 49] develops several techniques that can be combined to try to make the symbolic state space finite, including:

1. **Folding Abstraction.** This is both related to equational generalization (see Sect. 2.2), and to what is called *subsumption* in resolution theorem proving. The idea is the following: when computing a new symbolic state t' by a narrowing step $t \rightsquigarrow_{R/E \uplus B} t'$ with a rule in $\mathcal{R} = (\Sigma, E \uplus B, R)$, we check whether there is already a symbolic state u in the current state space more general than t' modulo B. That is, such that there is a B-matching substitution α with $t' =_B u\alpha$. In that case, the new symbolic state t' is not added. Instead, it is "folded into" or "merged with" the more general symbolic state u. This produces a state space that is more abstract than the original one and in some cases finite. However, as shown in [13], this abstraction is always *faithful* for LTL safety properties, in the sense that it will never generate spurious counterexamples for such properties.

2. **Equational Abstraction.** The idea here is to identify symbolic states by considering a more abstract rewrite theory $\mathcal{R}/A = (\Sigma, E \uplus A \uplus B, R)$, where new abstraction equations A have been added to $\mathcal{R}$. Although in general this can produce spurious counterexamples, easily checkable conditions on A can ensure that $\mathcal{R}$ and $\mathcal{R}/A$ are *bisimilar*, and therefore that no such spurious counterexamples are added.

3. **Predicate Abstraction.** This method, developed in [15], is more widely applicable, in the sense that the convergent equational theory $(\Sigma, E \uplus B)$ in $\mathcal{R} = (\Sigma, E \uplus B, R)$ need not be FVP. It is in fact a good example of a symbolic method in its own right, since symbolic techniques are crucially used to generate the predicate abstraction. This method is automatic, and can be used to generate increasingly more accurate abstractions in an abstraction/refinement manner.

4. **Bounded LTL Model Checking.** Even if by some of the abstraction methods (1)–(2) a finite symbolic state space has already been obtained, it is not always possible for a user to *know* that this is so. Therefore, a practical approach is to ensure that the state space if finite by performing a symbolic LTL *iterated bounded model checking*, where the depth k of paths from the initial state to other symbolic states is iteratively incremented until a certain bound, or until reaching a fixpoint if one exists.

In [14], the above symbolic model checking approach has been extended to verify temporal properties richer than LTL ones. Specifically, properties expressible in the logic LTLR [91] obtained by adding to LTL action properties specified by *spatial action patterns* that describe *where* in a symbolic state a certain kind of action described by one of the rewrite rules in the rewrite theory $\mathcal{R}$ has taken place. The papers [13–15, 49] contain interesting examples illustrating the usefulness of narrowing-based symbolic LTL and LTLR model checking.

2.6   Constructor-Based Reachability Logic

The material in this section is adapted from [125]. As already explained in Sect. 1.4, reachability logic generalizes Hoare logic. The main applications of reachability logic to date have been as a *language-generic* logic of programs [117, 130, 131]. In these applications, a $\mathbb{K}$ specification of a language's operational semantics by means of rewrite rules is assumed as the language's "golden semantic standard," and then a correct-by-construction reachability logic for a language so defined is automatically obtained [131]. This method has been shown effective in proving a wide range of programs in real programming languages specified within the $\mathbb{K}$ Framework.

Although the foundations of reachability logic are very general [130, 131], such general foundations do not provide a straightforward answer to the following non-trivial questions: (1) Could a reachability logic be developed to verify not just conventional programs, but also *distributed system designs and algorithms* formalized as *rewrite theories* in rewriting logic [93, 95]? And (2) if so, what would be the most natural way to conceive such a *rewrite-theory-generic* logic? Since $\mathbb{K}$ specifications are a special type of rewrite theories [100], a satisfactory answer to questions (1)–(2) would move the verification game from the level of verifying *code* to that of verifying *both code and distributed system designs*. Since the cost of design errors can be several orders of magnitude higher than that of coding errors, questions (1) and (2) are of practical software engineering interest.

Building on prior ideas in [83], the answer given by *constructor-based reachability logic* [125] to questions (1)–(2) tries to achieve four main goals: (i) to handle a broad class of rewrite theories and therefore of concurrent infinite-state systems that can be verified using reachability logic, including theories specifying *open systems* that interact with an external environment; (ii) to support in particular the verification of *invariants*; (iii) to maximize the applicability of *symbolic methods*—including equational unification, narrowing, and SMT solving—so as to obtain a mechanization of the logic as effective as possible; and (iv) to achieve all this as simply as possible.

Constructor-Based Reachability Logic in a Nutshell. We want to deductively verify reachability properties of a rewrite theory $\mathcal{R} = (\Sigma, E \uplus B, R)$ having a convergent equational theory $(\Sigma, E \uplus B)$ with a subsignature of constructors (Ω, B_Ω). Such properties are expressed as *reachability formulas*. A reachability formula has the form $A \to^{\circledast} B$, where A and B are state predicates. Consider the easier to explain case where the formula has no parameters, i.e., $vars(A) \cap vars(B) = \varnothing$. We interpret such a formula in the initial model $\mathcal{T}_\mathcal{R}$ of the rewrite theory $\mathcal{R} = (\Sigma, E \uplus B, R)$ whose states, thanks to convergence and constructors, can be represented as B_Ω-equivalence classes $[u]$ of ground Ω-terms, and where a state transition $[u] \to_{\mathcal{R}} [v]$ holds iff $\mathcal{R} \vdash u \to v$ according to the rewriting logic inference system [93,95]. As a first approximation, $A \to^\circledast B$ is a Hoare logic *partial correctness* assertion of the form[6] $\{A\}\mathcal{R}\{B\}$, but with the slight twist that B need not hold of a terminating state, but just *somewhere along the way*. To be fully precise, $A \to^\circledast B$ holds in $\mathcal{T}_{\mathcal{R}}$ iff for each state $[u_0]$ satisfying A and each terminating sequence $[u_0] \to_{\mathcal{R}} [u_1] \ldots \to_{\mathcal{R}} [u_{n-1}] \to_{\mathcal{R}} [u_n]$ there is a j, $0 \le j \le n$ such that $[u_j]$ satisfies B. A key question is how to choose a good language of state predicates like A and B. Here is where the potential for increasing the logic's automation resides. This version of reachability logic is called *constructor-based*, because the state predicates A and B in a reachability formula $A \to^\circledast B$ are positive (only $\vee$ and $\wedge$) combinations of *constrained constructor patterns* of the form $u \mid \varphi$, where u is a *constructor* term. That is, the atomic formulas in state predicates are exactly the constrained constructor patterns already discussed in Sect. 2.1. This is crucially important, because the initial algebra $\mathcal{T}_{\Omega/B_\Omega}$ of constructor terms is typically *much simpler* than $\mathcal{T}_{\Sigma/E}$, and this can be systematically exploited for matching, unification, narrowing, and satisfiability purposes to automate large portions of reachability logic. See [125] for various examples illustrating the usefulness of constructor-based reachability logic.

3 Maude-Based Symbolic Reasoning Engines

The symbolic methods for rewriting logic that I have described in Sect. 2 are supported by various symbolic reasoning engines and/or prototype systems. I first describe several such engines and then summarize some the, still experimental, prototypes that have also been developed for more recent techniques.

3.1 Symbolic Reasoning Engines

The list below may be incomplete. I may have inadvertently failed to mention some engines. I apologize in advance for any omissions.

Maude. Maude itself, besides its support for equational and non-equational rewriting, and for explicit state breadth first search reachability analysis, LTL, and LTLR [16] model checking, supports the following *symbolic methods*:

- *Order-sorted unification modulo any combination of associativity and/or commutativity and/or identity axioms* [41].
- *Computation of variants and variant unification*, not only for FVP convergent theories, where both computations are finite, but also for general convergent theories where variants and unifiers can be incrementally computed by folding variant narrowing [41].
- *Narrowing-based symbolic reachability analysis* [41].

---

[6] The notation $\{A\}\mathcal{R}\{B\}$, and the relation to Hoare logic are explained in [125].

Maude-NPA. The Maude-NRL Protocol Analyzer [46,64,65,138,139], performs symbolic reachability analysis of cryptographic protocols in the Dolev-Yao model, in which crypto algorithms are modeled by function symbols that may satisfy equational theories. The network is assumed to be controlled by an intruder who is trying to break the protocol, and who may be in league with some of the principals. It uses *backwards narrowing*, that is, it uses the protocol's transition rules in reverse, starting from a constrained term pattern describing a class of attack states and searching for an initial state. It analyzes security and authentication properties of such protocols specified as rewrite theories whose equational part is assumed to be convergent and FVP. All the symbolic methods for unification, variant unification, and narrowing-based symbolic reachability analysis (with folding) described in Sect. 2 play a crucial role in Maude-NPA. Cryptographic protocol verification for an unbounded number of sessions is known to be undecidable. However, Maude-NPA, besides supporting folding abstraction, uses other powerful symbolic state space reduction techniques [48] that can make the state space finite for many protocols without sacrificing the completeness of the analysis. This means that in such cases if an attack is not found no such attack exists in the Dolev-Yao model. To the best of my knowledge Maude-NPA is the most general cryptographic protocol analysis tool to date, in the sense that it can analyze protocols *modulo* any FVP equational properties of the cryptographic functions. It is well-known that attacks that would never be found when the formal analysis does not take such algebraic properties into consideration do indeed exist in real protocols and can be mounted by an attacker exploiting such algebraic properties.

Tamarin [90] is a theorem prover for security protocol verification developed at ETH's Information Security Group. It supports falsification as well as unbounded verification of security protocols in the Dolev-Yao model with respect to temporal first-order properties. Such protocols are specified as multiset rewrite theories (i.e., as rewrite theories modulo AC). The equational theory of messages models Diffie-Hellman exponentiation and can be combined with a user-defined subterm-convergent equational theory for other cryptographic functions. Subterm convergent theories are a special class of convergent FVP theories of the form (Σ, E), that is, where $B = \emptyset$. Recently, Tamarin has been extended in several ways to support several new features, including: (i) bilinear pairing in a built-in way [119], (ii) verification of observational equivalence, to check properties such as anonymity and unlinkability [20], (iii) convergent FVP theories of the form (Σ, E), that is, where $B = \emptyset$ [38], and (iv) built-in support for exclusive or [39]. Tamarin is a state-of the art security prover that has verified, or has found attacks in, a wide range of protocols. Two of the symbolic methods mentioned in Sect. 2 are quite important in Tamarin, namely: (i) equational unification, and (ii) variant computation. Tamarin uses Maude as a component for some of its symbolic reasoning computations.

CETA and Maude's Sufficient Completeness Checker (SCC). Maude can be compiled with the CETA library for equational tree automata computations developed by Joseph Hendrix in C++ as part of his Ph.D. thesis at the

University of Illinois at Urbana-Champaign [72]. Maude's Sufficient Completeness Checker [71,72] uses CETA to verify the sufficient completeness of Maude's modules. That is, sufficient completeness, i.e., the property that constructors, in the sense defined in Sect. 2.1, are really constructors is reduced to an emptiness check for a certain equational tree automaton associated to the module's signature and equations.

ACUOS. The ACUOS tool at the Technical University of Valencia is written in Maude and supports order-sorted equational generalization modulo any combination of associativity and/or commutativity and/or identity axioms in the sense explained in Sect. 2.2 and more fully documented in [2].

The Maude Church-Rosser and Coherence Checker. This tool [43] can be used to automatically perform two checks for Maude modules: (i) the local confluence of equations modulo given axioms B under a termination assumption, proving that the module's equational theory is convergent in the sense explained in Sect. 2.1, and (ii) the coherence of the module's rules R with respect to the equations E modulo the axioms B, which is essential for the executability of a rewrite theory [43,136]. Order-sorted unification modulo axioms B consisting of any combination of associativity and/or commutativity and/or identity axioms, equational narrowing modulo such axioms, and certain forms of variant computation are some of the symbolic methods used in this tool.

The Maude LTL Logical Model Checker. The narrowing-based symbolic LTL model checking methods described in Sect. 2.5 are supported by this Maude-based tool developed by Kyungmin Bae as part of his Ph.D. thesis at the University of Illinois at Urbana-Champaign [12]. The tool is partly implemented in Maude and partly in C++ for greater efficiency.

The $\mathbb{K}$ Reachability Logic Tool. This tool [117,130,131], developed by Andrei Stefanescu together with Grigore Rosu and other members of the $\mathbb{K}$ Group at the University of Illinois at Urbana-Champaign, uses reachability logic as a language-generic logic of programs to verify programs in a programming language whose semantics is specified by rewrite rules in $\mathbb{K}$. In fact, under very general assumptions, the reachability logic and prover of a language are automatically obtained from its $\mathbb{K}$ specification [131]. This method has been shown effective in proving a wide range of programs in real programming languages specified within the $\mathbb{K}$ Framework. It has also been an important inspiration for developing the constructor-based reachability logic for verifying properties of rewrite theories presented in Sect. 2.6.

3.2 Prototypes

The following prototypes correspond to experimental work advancing and/or using several of the symbolic methods I have described in Sect. 2. The list below may be incomplete. I may have inadvertently failed to mention some prototype tools. I apologize in advance for any omissions.

1. **Maude's Invariant Analyzer.** This prototype tool [113,115] developed by Camilo Rocha as part of his Ph.D. thesis at the University of Illinois at Urbana-Champaign [114] is written in Maude and provides a *deduction-based*, as opposed to a model-checking-based, approach to reachability analysis of rewrite theories and their safety properties. It crucially uses symbolic techniques such as order-sorted equational unification and non-equational narrowing.

2. **Term Pattern Tool.** This prototype tool [101] has been developed by Stephen Skeirik and José Meseguer at the University of Illinois at Urbana-Champaign. It is written in Maude and supports for the first time order-sorted term-pattern operations in the exact sense explained in Sect. 2.1.

3. **Partial Evaluation Tool.** This prototype tool [1], developed at the Technical University of Valencia by María Alpuente, Angel Cuenca-Ortega, Santiago Escobar and José Meseguer and written in Maude, is the first partial evaluator for order-sorted equational programs modulo axioms I am aware of. Several of the symbolic techniques I have described in Sect. 2 play a crucial role, including: (i) order-sorted equational unification, (ii) order-sorted equational generalization, and (iii) folding variant narrowing.

4. **Rewriting Modulo SMT Prototypes.** Two prototypes supporting the rewriting modulo SMT symbolic method for reachability analysis explained in Sect. 2.4 have been developed in Maude. The first was developed by Camilo Rocha as part of his Ph.D. thesis at the University of Illinois at Urbana-Champaign [114] and was used in the experiments reported in [116]. The second [17], developed by Camilo Rocha and Kyungmin Bae, supports the notion of *guarded terms* to achieve important state space reductions.

5. **Variant Satisfiability Tool.** This prototype tool [101] has been developed by Stephen Skeirik and José Meseguer at the University of Illinois at Urbana-Champaign. It is written in Maude and supports variant satisfiability computations for initial algebras of FVP theories having an OS-compact constructor subtheory in the sense explained in Sect. 2.3.

6. **Variant Satisfiability with Predicates Tool.** This prototype tool [66], developed at the Technical University of Valencia by Raúl Gutiérrez and José Meseguer and written in Maude, supports variant satisfiability for initial algebras of FVP equational theories that, due to the presence of predicates among its constructors, may fail to satisfy the OS-compactness requirement explained in Sect. 2.3.

7. **Reachability Logic Tool for Rewrite Theories.** This prototype tool [83], developed in Maude by Dorel Lucanu, Vlad Rusu, Andrei Arusoaie and David Nowak, was the first tool that considered a reachability logic for proving properties of rewrite theories. Besides using some of the SMT solving symbolic techniques discussed in Sect. 2.3, it also used a technique akin to rewriting modulo SMT to reason deductively about state transitions. As mentioned in Sect. 2.6, this work stimulated the subsequent development of the constructor-based version of reachability logic.

8. **Constructor-Based Reachability Logic Tool.** This prototype tool [125] has been developed in Maude at the University of Illinois at Urbana-

Champaign by Stephen Skeirik, Andrei Stefanescu and José Meseguer. It is still a prototype undergoing further development but, as reported in [125], it has been already used to verify properties for a suite of rewrite theories. It has also been used by students in a Program Verification course at the University of Illinois at Urbana-Champaign. Many of the symbolic techniques discussed in Sect. 2 play a crucial role in this prototype, including: (i) constrained term patterns, (ii) order-sorted equational unification, (iii) variant computations, (iv) variant unification, and (v) variant satisfiability.

4   Some Application Areas

The list below is probably incomplete. To some extent it has been partially anticipated by the previous discussion on tools and prototypes. But here are some of the application areas that, among others, are either already benefitting from the symbolic techniques that I have described in this paper, or are ripe for benefitting from them:

- *Theorem Proving Verification.*
- *Symbolic Model Checking Verification.*
- *Verification of Programs in Conventional Languages based on their Rewriting Logic Semantics.*
- *Cryptographic Protocol Analysis.*
- *Symbolic Verification of Cyber-Physical Systems.*
- *Partial Evaluation.*

5   Conclusions

I have explained how symbolic methods in six different but related areas play a crucial role in rewriting logic, are for the most part well supported by Maude-based tools and prototypes, and are applied in several important areas. My aim has been clearly ecumenical: I believe that these methods are widely applicable and can be used as components of many different formal reasoning systems.

As usual much work remains ahead. For example, several of the prototypes I mention in Sect. 3.2 should be further advanced and should become mature tools. More applications and case studies should likewise be developed to both test and improve these symbolic methods and to develop new ones. One theme I have had no chance to touch upon, but that I would like to briefly mention, is that of *generalized rewrite theories*. The point is that, to take full advantage of the symbolic reasoning possibilities I have discussed, the notion of rewrite theory should itself be generalized as explained in [97]. A related point, also explained in [97], is that new *symbolic executability conditions* are needed for rewrite theories. However, the good news is that *transformation techniques* can substantially expand the range of theories that can be symbolically executed and analyzed in practice.

Acknowledgments. I thank the organizers of WoLLIC for kindly giving me the opportunity of presenting these ideas in Bogotá. As the references make clear, these ideas have been developed in joint work with a large number of collaborators and former or present students, and in dialogue with many other colleagues. I cannot mention them all and apologize in advance for this; but I would like to mention and cordially thank in particular: María Alpuente, Kyungmin Bae, Andrew Cholewa, Angel Cuenca-Ortega, Francisco Durán, Steven Eker, Santiago Escobar, Raúl Gutiérrez, Joseph Hendrix, Dorel Lucanu, Salvador Lucas, Narciso Martí-Oliet, Catherine Meadows, César A. Muñoz, Hitoshi Ohsaki, Camilo Rocha, Grigore Rosu, Vlad Rusu, Sonia Santiago, Ralf Sasse, Andrei Stefanescu, Carolyn Talcott, Prasanna Thati, and Fan Yang. This work has been partially supported by NRL under contract number N00173-17-1-G002.

References

1. Alpuente, M., Cuenca-Ortega, A., Escobar, S., Meseguer, J.: Partial evaluation of order-sorted equational programs modulo axioms. In: Hermenegildo, M.V., Lopez-Garcia, P. (eds.) LOPSTR 2016. LNCS, vol. 10184, pp. 3–20. Springer, Cham (2017). https://doi.org/10.1007/978-3-319-63139-4_1
2. Alpuente, M., Escobar, S., Espert, J., Meseguer, J.: A modular order-sorted equational generalization algorithm. Inf. Comput. **235**, 98–136 (2014)
3. Alpuente, M., Escobar, S., Iborra, J.: Termination of narrowing revisited. Theor. Comput. Sci. **410**(46), 4608–4625 (2009)
4. Alpuente, M., Escobar, S., Iborra, J.: Modular termination of basic narrowing and equational unification. Log. J. IGPL **19**(6), 731–762 (2011)
5. Armando, A., Bonacina, M.P., Ranise, S., Schulz, S.: New results on rewrite-based satisfiability procedures. ACM Trans. Comput. Log. **10**(1), 4:1–4:51 (2009)
6. Armando, A., Castellini, C., Giunchiglia, E.: SAT-based procedures for temporal reasoning. In: Biundo, S., Fox, M. (eds.) ECP 1999. LNCS (LNAI), vol. 1809, pp. 97–108. Springer, Heidelberg (2000). https://doi.org/10.1007/10720246_8
7. Armando, A., Ranise, S., Rusinowitch, M.: A rewriting approach to satisfiability procedures. Inf. Comput. **183**(2), 140–164 (2003)
8. Arusoaie, A., Lucanu, D., Rusu, V.: Symbolic execution based on language transformation. Comput. Lang. Syst. Struct. **44**, 48–71 (2015)
9. Audemard, G., Bertoli, P., Cimatti, A., Korniłowicz, A., Sebastiani, R.: A SAT based approach for solving formulas over Boolean and linear mathematical propositions. In: Voronkov, A. (ed.) CADE 2002. LNCS (LNAI), vol. 2392, pp. 195–210. Springer, Heidelberg (2002). https://doi.org/10.1007/3-540-45620-1_17
10. Baader, F., Snyder, W.: Unification theory. In: Handbook of Automated Reasoning. Elsevier (1999)
11. Baader, F., Siekmann, J.H.: Unification theory. In: Handbook of Logic in Artificial Intelligence and Logic Programming, vol. 2, pp. 41–126. Oxford University Press (1994)
12. Bae, K.: Rewriting-based model checking methods. Ph.D. thesis, University of Illinois at Urbana-Champaign (2014)
13. Bae, K., Escobar, S., Meseguer, J.: Abstract logical model checking of infinite-state systems using narrowing. In: Rewriting Techniques and Applications (RTA 2013). LIPIcs, vol. 21, pp. 81–96. Schloss Dagstuhl-Leibniz-Zentrum fuer Informatik (2013)

14. Bae, K., Meseguer, J.: Infinite-state model checking of LTLR formulas using narrowing. In: Escobar, S. (ed.) WRLA 2014. LNCS, vol. 8663, pp. 113–129. Springer, Cham (2014). https://doi.org/10.1007/978-3-319-12904-4_6

15. Bae, K., Meseguer, J.: Predicate abstraction of rewrite theories. In: Dowek, G. (ed.) RTA 2014. LNCS, vol. 8560, pp. 61–76. Springer, Cham (2014). https://doi.org/10.1007/978-3-319-08918-8_5

16. Bae, K., Meseguer, J.: Model checking linear temporal logic of rewriting formulas under localized fairness. Sci. Comput. Program. **99**, 193–234 (2015)

17. Bae, K., Rocha, C.: Guarded terms for rewriting modulo SMT. In: Proença, J., Lumpe, M. (eds.) FACS 2017. LNCS, vol. 10487, pp. 78–97. Springer, Cham (2017). https://doi.org/10.1007/978-3-319-68034-7_5

18. Barrett, C.W., Dill, D.L., Stump, A.: Checking satisfiability of first-order formulas by incremental translation to SAT. In: Brinksma, E., Larsen, K.G. (eds.) CAV 2002. LNCS, vol. 2404, pp. 236–249. Springer, Heidelberg (2002). https://doi.org/10.1007/3-540-45657-0_18

19. Basin, D., Clavel, M., Meseguer, J.: Rewriting logic as a metalogical framework. ACM Trans. Comput. Log. **5**, 528–576 (2004)

20. Basin, D., Dreier, J., Sasse, R.: Automated symbolic proofs of observational equivalence. In: Ray, I., Li, N., Kruegel, C. (eds.) Proceedings of the 2015 ACM SIGSAC Conference on Computer and Communications Security, pp. 1144–1155. ACM (2015)

21. Bidoit, M., Mosses, P.D. (eds.): CASL User Manual - Introduction to Using the Common Algebraic Specification Language. LNCS, vol. 2900. Springer, Heidelberg (2004). https://doi.org/10.1007/b11968

22. Bonacina, M.P., Echenim, M.: On variable-inactivity and polynomial $\mathcal{T}$-satisfiability procedures. J. Log. Comput. **18**(1), 77–96 (2008)

23. Borovanský, P., Kirchner, C., Kirchner, H., Moreau, P.E.: ELAN from a rewriting logic point of view. Theor. Comput. Sci. **285**, 155–185 (2002)

24. Boyer, R., Moore, J.: A Computational Logic. Academic Press, Cambridge (1980)

25. Bruni, R., Meseguer, J.: Semantic foundations for generalized rewrite theories. Theor. Comput. Sci. **360**(1–3), 386–414 (2006)

26. Bulychev, P.E., Kostylev, E.V., Zakharov, V.A.: Anti-unification algorithms and their applications in program analysis. In: Pnueli, A., Virbitskaite, I., Voronkov, A. (eds.) PSI 2009. LNCS, vol. 5947, pp. 413–423. Springer, Heidelberg (2010). https://doi.org/10.1007/978-3-642-11486-1_35

27. Bürckert, H.-J. (ed.): A Resolution Principle for a Logic with Restricted Quantifiers. LNCS, vol. 568. Springer, Heidelberg (1991). https://doi.org/10.1007/3-540-55034-8

28. Chadha, R., Ciobâcǎ, Ş., Kremer, S.: Automated verification of equivalence properties of cryptographic protocols. In: Seidl, H. (ed.) ESOP 2012. LNCS, vol. 7211, pp. 108–127. Springer, Heidelberg (2012). https://doi.org/10.1007/978-3-642-28869-2_6

29. Cholewa, A., Escobar, S., Meseguer, J.: Constrained narrowing for conditional equational theories modulo axioms. Sci. Comput. Program. **112**, 24–57 (2015)

30. Cholewa, A., Meseguer, J., Escobar, S.: Variants of variants and the finite variant property. Technical report, CS Dept. University of Illinois at Urbana-Champaign, February 2014. http://hdl.handle.net/2142/47117

31. Clavel, M., Durán, F., Eker, S., Lincoln, P., Martí-Oliet, N., Meseguer, J., Talcott, C.: All About Maude - A High-Performance Logical Framework. LNCS, vol. 4350. Springer, Heidelberg (2007). https://doi.org/10.1007/978-3-540-71999-1

32. Cohn, A.G.: A more expressive formulation of many sorted logic. J. Autom. Reason. **3**(2), 113–200 (1987)
33. Cohn, A.G.: Taxonomic reasoning with many-sorted logics. Artif. Intell. Rev. **3**(2–3), 89–128 (1989)
34. Comon, H., Dauchet, M., Gilleron, R., Löding, C., Jacquemard, F., Lugiez, D., Tison, S., Tommasi, M.: Tree automata techniques and applications (2007). http://www.grappa.univ-lille3.fr/tata, Accessed 12 Oct 2007
35. Comon, H.: Equational formulas in order-sorted algebras. In: Paterson, M.S. (ed.) ICALP 1990. LNCS, vol. 443, pp. 674–688. Springer, Heidelberg (1990). https://doi.org/10.1007/BFb0032066
36. Comon-Lundh, H., Delaune, S.: The finite variant property: how to get rid of some algebraic properties. In: Giesl, J. (ed.) RTA 2005. LNCS, vol. 3467, pp. 294–307. Springer, Heidelberg (2005). https://doi.org/10.1007/978-3-540-32033-3_22
37. van Deursen, A., Heering, J., Klint, P.: Language Prototyping: An Algebraic Specification Approach. World Scientific, Singapore (1996)
38. Dreier, J., Duménil, C., Kremer, S., Sasse, R.: Beyond subterm-convergent equational theories in automated verification of stateful protocols. In: Maffei, M., Ryan, M. (eds.) POST 2017. LNCS, vol. 10204, pp. 117–140. Springer, Heidelberg (2017). https://doi.org/10.1007/978-3-662-54455-6_6
39. Dreier, J., Hirschi, L., Radomirovic, S., Sasse, R.: Automated unbounded verification of stateful cryptographic protocols with exclusive OR. In: Accepted at Computer Security Foundations (CSF) (2018)
40. Dross, C., Conchon, S., Kanig, J., Paskevich, A.: Adding decision procedures to SMT solvers using axioms with triggers. J. Autom. Reason. **56**(4), 387–457 (2016)
41. Durán, F., Eker, S., Escobar, S., Martí-Oliet, N., Meseguer, J., Talcott, C.: Associative unification and symbolic reasoning modulo associativity in Maude. In: Preproceedings of WRLA 2018, Thessaloniki, Greece, April 2018. (Distributed in Electronic Form by the ETAPS 2018 Organizers). Proceedings version to appear in LNCS
42. Durán, F., Meseguer, J., Rocha, C.: Proving ground confluence of equational specifications modulo axioms. Technical report, CS Dept., University of Illinois at Urbana-Champaign, March 2018. http://hdl.handle.net/2142/99548. Shorter version to appear in Proceedings of the WRLA 2018. Springer LNCS
43. Durán, F., Meseguer, J.: On the Church-Rosser and coherence properties of conditional order-sorted rewrite theories. J. Algebraic Log. Program. **81**, 816–850 (2012)
44. Echenim, M., Peltier, N.: An instantiation scheme for satisfiability modulo theories. J. Autom. Reason. **48**(3), 293–362 (2012)
45. Erbatur, S., Escobar, S., Kapur, D., Liu, Z., Lynch, C.A., Meadows, C., Meseguer, J., Narendran, P., Santiago, S., Sasse, R.: Asymmetric unification: a new unification paradigm for cryptographic protocol analysis. In: Bonacina, M.P. (ed.) CADE 2013. LNCS (LNAI), vol. 7898, pp. 231–248. Springer, Heidelberg (2013). https://doi.org/10.1007/978-3-642-38574-2_16
46. Escobar, S., Meadows, C., Meseguer, J.: Maude-NPA: cryptographic protocol analysis modulo equational properties. In: Aldini, A., Barthe, G., Gorrieri, R. (eds.) FOSAD 2007-2009. LNCS, vol. 5705, pp. 1–50. Springer, Heidelberg (2009). https://doi.org/10.1007/978-3-642-03829-7_1
47. Escobar, S., Sasse, R., Meseguer, J.: Folding variant narrowing and optimal variant termination. In: Ölveczky, P.C. (ed.) WRLA 2010. LNCS, vol. 6381, pp. 52–68. Springer, Heidelberg (2010). https://doi.org/10.1007/978-3-642-16310-4_5

48. Escobar, S., Meadows, C., Meseguer, J., Santiago, S.: State space reduction in the Maude-NRL protocol analyzer. Inf. Comput. **238**, 157–186 (2014)
49. Escobar, S., Meseguer, J.: Symbolic model checking of infinite-state systems using narrowing. In: Baader, F. (ed.) RTA 2007. LNCS, vol. 4533, pp. 153–168. Springer, Heidelberg (2007). https://doi.org/10.1007/978-3-540-73449-9_13
50. Escobar, S., Sasse, R., Meseguer, J.: Folding variant narrowing and optimal variant termination. J. Algebraic Log. Program. **81**, 898–928 (2012)
51. Farzan, A., Chen, F., Meseguer, J., Roşu, G.: Formal analysis of Java programs in JavaFAN. In: Alur, R., Peled, D.A. (eds.) CAV 2004. LNCS, vol. 3114, pp. 501–505. Springer, Heidelberg (2004). https://doi.org/10.1007/978-3-540-27813-9_46
52. Fay, M.: First-order unification in an equational theory. In: Proceedings of the 4th Workshop on Automated Deduction, pp. 161–167 (1979)
53. Filliâtre, J.-C., Owre, S., Rue*B, H., Shankar, N.: ICS: integrated Canonizer and solver? In: Berry, G., Comon, H., Finkel, A. (eds.) CAV 2001. LNCS, vol. 2102, pp. 246–249. Springer, Heidelberg (2001). https://doi.org/10.1007/3-540-44585-4_22
54. Flanagan, C., Joshi, R., Ou, X., Saxe, J.B.: Theorem proving using lazy proof explication. In: Hunt Jr., W.A., Somenzi, F. (eds.) CAV 2003. LNCS, vol. 2725, pp. 355–367. Springer, Heidelberg (2003). https://doi.org/10.1007/978-3-540-45069-6_34
55. Frisch, A.M.: The substitutional framework for sorted deduction: fundamental results on hybrid reasoning. Artif. Intell. **49**(1–3), 161–198 (1991)
56. Futatsugi, K., Diaconescu, R.: CafeOBJ Report. World Scientific, Singapore (1998)
57. Gallagher, J.P.: Tutorial on specialisation of logic programs. In: Proceedings of the 1993 ACM SIGPLAN Symposium on Partial Evaluation and Semantics-Based Program Manipulation, PEPM 1993, pp. 88–98. ACM, New York (1993)
58. Gallier, J.H., Snyder, W.: Complete sets of transformations for general E-unification. Theor. Comput. Sci. **67**(2&3), 203–260 (1989)
59. Garavel, H., Tabikh, M.A., Arrada, I.S.: Benchmarking implementations of term rewriting and pattern matching in algebraic, functional, and object-oriented languages. In: Preproceedings of WRLA 2018, Thessaloniki, Greece, April 2018. (Distributed in electronic form by the ETAPS 2018 Organizers). Proceedings version to appear in LNCS
60. Goguen, J., Meseguer, J.: Equality, types, modules and (why not?) generics for logic programming. J. Log. Program. **1**(2), 179–210 (1984)
61. Goguen, J., Meseguer, J.: Unifying functional, object-oriented and relational programming with logical semantics. In: Shriver, B., Wegner, P. (eds.) Research Directions in Object-Oriented Programming, pp. 417–477. MIT Press, Cambridge (1987)
62. Goguen, J., Meseguer, J.: Order-sorted algebra I: equational deduction for multiple inheritance, overloading, exceptions and partial operations. Theor. Comput. Sci. **105**, 217–273 (1992)
63. Goguen, J., Winkler, T., Meseguer, J., Futatsugi, K., Jouannaud, J.P.: Introducing OBJ. In: Goguen, J., Malcolm, G. (eds.) Software Engineering with OBJ: Algebraic Specification in Action, pp. 3–167. Kluwer, Dordrecht (2000)
64. González-Burgueño, A., Santiago, S., Escobar, S., Meadows, C., Meseguer, J.: Analysis of the IBM CCA security API protocols in Maude-NPA. In: Chen, L., Mitchell, C. (eds.) SSR 2014. LNCS, vol. 8893, pp. 111–130. Springer, Cham (2014). https://doi.org/10.1007/978-3-319-14054-4_8

65. González-Burgueño, A., Santiago, S., Escobar, S., Meadows, C., Meseguer, J.: Analysis of the PKCS#11 API using the Maude-NPA tool. In: Chen, L., Matsuo, S. (eds.) SSR 2015. LNCS, vol. 9497, pp. 86–106. Springer, Cham (2015). https://doi.org/10.1007/978-3-319-27152-1_5

66. Gutiérrez, R., Meseguer, J.: Variant-based decidable satisfiability in initial algebras with predicates. To appear in Proceedings of LOPSTR 2017. Springer LNCS 2018

67. Gutiérrez, R., Meseguer, J.: Variant-based decidable satisfiability in initial algebras with predicates. Technical report, University of Illinois at Urbana-Champaign, June 2017. http://hdl.handle.net/2142/96264

68. Haxthausen, A.E.: Order-sorted algebraic specifications with higher-order functions. Theor. Comput. Sci. **183**(2), 157–185 (1997)

69. Hendrix, J., Ohsaki, H., Viswanathan, M.: Propositional tree automata. In: Pfenning, F. (ed.) RTA 2006. LNCS, vol. 4098, pp. 50–65. Springer, Heidelberg (2006). https://doi.org/10.1007/11805618_5

70. Hendrix, J., Meseguer, J.: Order-sorted equational unification revisited. Electr. Notes Theor. Comput. Sci. **290**, 37–50 (2012)

71. Hendrix, J., Meseguer, J., Ohsaki, H.: A sufficient completeness checker for linear order-sorted specifications modulo axioms. In: Furbach, U., Shankar, N. (eds.) IJCAR 2006. LNCS (LNAI), vol. 4130, pp. 151–155. Springer, Heidelberg (2006). https://doi.org/10.1007/11814771_14

72. Hendrix, J.D.: Decision procedures for equationally based reasoning. Ph.D. thesis, University of Illinois at Urbana-Champaign (2008). http://hdl.handle.net/2142/10967

73. Hullot, J.-M.: Canonical forms and unification. In: Bibel, W., Kowalski, R. (eds.) CADE 1980. LNCS, vol. 87, pp. 318–334. Springer, Heidelberg (1980). https://doi.org/10.1007/3-540-10009-1_25

74. Jouannaud, J.-P., Kirchner, C., Kirchner, H.: Incremental construction of unification algorithms in equational theories. In: Diaz, J. (ed.) ICALP 1983. LNCS, vol. 154, pp. 361–373. Springer, Heidelberg (1983). https://doi.org/10.1007/BFb0036921

75. Jouannaud, J.P., Kirchner, H.: Completion of a set of rules modulo a set of equations. SIAM J. Comput. **15**, 1155–1194 (1986)

76. Katelman, M., Keller, S., Meseguer, J.: Rewriting semantics of production rule sets. J. Log. Algebraic Program. **81**(7–8), 929–956 (2012)

77. Kirchner, C.: Order-sorted equational unification. Technical report 954, INRIA Lorraine & LORIA, Nancy, France (1988)

78. Kirchner, H., Ranise, S., Ringeissen, C., Tran, D.K.: On superposition-based satisfiability procedures and their combination. In: Van Hung, D., Wirsing, M. (eds.) ICTAC 2005. LNCS, vol. 3722, pp. 594–608. Springer, Heidelberg (2005). https://doi.org/10.1007/11560647_39

79. Kitzelmann, E., Schmid, U.: Inductive synthesis of functional programs: an explanation based generalization approach. J. Mach. Learn. Res. **7**, 429–454 (2006)

80. Kutsia, T., Levy, J., Villaret, M.: Anti-unification for unranked terms and hedges. In: Schmidt-Schauß, M. (ed.) Proceedings of the 22nd International Conference on Rewriting Techniques and Applications, RTA 2011. LIPIcs, Novi Sad, Serbia, 30 May–1 June 2011, vol. 10, pp. 219–234. Schloss Dagstuhl - Leibniz-Zentrum fuer Informatik (2011)

81. Lassez, J.L., Marriott, K.: Explicit representation of terms defined by counter examples. J. Autom. Reason. **3**(3), 301–317 (1987)

82. Lu, J., Mylopoulos, J., Harao, M., Hagiya, M.: Higher order generalization and its application in program verification. Ann. Math. Artif. Intell. **28**(1–4), 107–126 (2000)
83. Lucanu, D., Rusu, V., Arusoaie, A., Nowak, D.: Verifying reachability-logic properties on rewriting-logic specifications. In: Martí-Oliet, N., Ölveczky, P.C., Talcott, C. (eds.) Logic, Rewriting, and Concurrency. LNCS, vol. 9200, pp. 451–474. Springer, Cham (2015). https://doi.org/10.1007/978-3-319-23165-5_21
84. Lucas, S., Meseguer, J.: Normal forms and normal theories in conditional rewriting. J. Log. Algebric Methods Program. **85**(1), 67–97 (2016)
85. Lynch, C., Morawska, B.: Automatic decidability. In: Proceedings of the LICS 2002, p. 7. IEEE Computer Society (2002)
86. Lynch, C., Tran, D.-K.: Automatic decidability and combinability revisited. In: Pfenning, F. (ed.) CADE 2007. LNCS (LNAI), vol. 4603, pp. 328–344. Springer, Heidelberg (2007). https://doi.org/10.1007/978-3-540-73595-3_22
87. Martí-Oliet, N., Meseguer, J.: Inclusions and subtypes II: higher-order case. J. Log. Comput. **6**, 541–572 (1996)
88. Martí-Oliet, N., Meseguer, J.: Rewriting logic as a logical and semantic framework. In: Gabbay, D., Guenthner, F. (eds.) Handbook of Philosophical Logic, 2nd edn, pp. 1–87. Kluwer Academic Publishers, Dordrecht (2002). First published as SRI Technical report SRI-CSL-93-05, August 1993
89. Martí-Oliet, N., Palomino, M., Verdejo, A.: Rewriting logic bibliography by topic: 1990–2011. J. Log. Algebric Program. **81**(7–8), 782–815 (2012). https://doi.org/10.1016/j.jlap.2012.06.001
90. Meier, S., Schmidt, B., Cremers, C., Basin, D.: The TAMARIN prover for the symbolic analysis of security protocols. In: Sharygina, N., Veith, H. (eds.) CAV 2013. LNCS, vol. 8044, pp. 696–701. Springer, Heidelberg (2013). https://doi.org/10.1007/978-3-642-39799-8_48
91. Meseguer, J.: The temporal logic of rewriting: a gentle introduction. In: Degano, P., De Nicola, R., Meseguer, J. (eds.) Concurrency, Graphs and Models. LNCS, vol. 5065, pp. 354–382. Springer, Heidelberg (2008). https://doi.org/10.1007/978-3-540-68679-8_22
92. Meseguer, J., Thati, P.: Symbolic reachability analysis using narrowing and its application to the verification of cryptographic protocols. J. High.-Order Symb. Comput. **20**(1–2), 123–160 (2007)
93. Meseguer, J.: Conditional rewriting logic as a unified model of concurrency. Theor. Comput. Sci. **96**(1), 73–155 (1992)
94. Meseguer, J.: Membership algebra as a logical framework for equational specification. In: Presicce, F.P. (ed.) WADT 1997. LNCS, vol. 1376, pp. 18–61. Springer, Heidelberg (1998). https://doi.org/10.1007/3-540-64299-4_26
95. Meseguer, J.: Twenty years of rewriting logic. J. Algebraic Log. Program. **81**, 721–781 (2012)
96. Meseguer, J.: Strict coherence of conditional rewriting modulo axioms. Theor. Comput. Sci. **672**, 1–35 (2017)
97. Meseguer, J.: Generalized rewrite theories and coherence completion. Technical report, University of Illinois Computer Science Department, March 2018. http://hdl.handle.net/2142/99546. Shorter version to appear in Proceedings of WRLA 2018, Springer LNCS
98. Meseguer, J.: Variant-based satisfiability in initial algebras. Sci. Comput. Program. **154**, 3–41 (2018)
99. Meseguer, J., Goguen, J., Smolka, G.: Order-sorted unification. J. Symb. Comput. **8**, 383–413 (1989)

100. Meseguer, J., Rosu, G.: The rewriting logic semantics project: a progress report. Inf. Comput. **231**, 38–69 (2013)
101. Meseguer, J., Skeirik, S.: Equational formulas and pattern operations in initial order-sorted algebras. Formal Asp. Comput. **29**(3), 423–452 (2017)
102. Mogensen, T.Æ.: Glossary for partial evaluation and related topics. High.-Order Symbol. Comput. **13**(4), 355–368 (2000)
103. Mosses, P.D. (ed.): CASL Reference Manual. The Complete Documentation of the Common Algebraic Specification Language. LNCS, vol. 2960. Springer, Heidelberg (2004). https://doi.org/10.1007/b96103
104. de Moura, L., Rueß, H.: Lemmas on demand for satisfiability solvers. In: Proceedings of the Fifth International Symposium on the Theory and Applications of Satisfiability Testing (SAT 2002), May 2002
105. Muggleton, S.: Inductive logic programming: issues, results and the challenge of learning language in logic. Artif. Intell. **114**(1–2), 283–296 (1999)
106. Nelson, G., Oppen, D.C.: Simplification by cooperating decision procedures. ACM Trans. Program. Lang. Syst. **1**(2), 245–257 (1979)
107. Nieuwenhuis, R., Oliveras, A., Tinelli, C.: Solving SAT and SAT modulo theories: from an abstract Davis-Putnam-Logemann-Loveland procedure to DPLL(T). J. ACM **53**(6), 937–977 (2006)
108. Pfenning, F.: Unification and anti-unification in the calculus of constructions. In: Proceedings, Sixth Annual IEEE Symposium on Logic in Computer Science, Amsterdam, The Netherlands, 15–18 July 1991, pp. 74–85. IEEE Computer Society (1991)
109. Plotkin, G.: A note on inductive generalization. In: Machine Intelligence, vol. 5, pp. 153–163. Edinburgh University Press (1970)
110. Popplestone, R.: An experiment in automatic induction. In: Machine Intelligence, vol. 5, pp. 203–215. Edinburgh University Press (1969)
111. Reynolds, J.: Transformational systems and the algebraic structure of atomic formulas. Mach. Intell. **5**, 135–151 (1970)
112. Robinson, J.A.: A machine-oriented logic based on the resolution principle. J. Assoc. Comput. Mach. **12**, 23–41 (1965)
113. Rocha, C., Meseguer, J.: Proving safety properties of rewrite theories. In: Corradini, A., Klin, B., Cîrstea, C. (eds.) CALCO 2011. LNCS, vol. 6859, pp. 314–328. Springer, Heidelberg (2011). https://doi.org/10.1007/978-3-642-22944-2_22
114. Rocha, C.: Symbolic reachability analysis for rewrite theories. Ph.D. thesis, University of Illinois at Urbana-Champaign (2012)
115. Rocha, C., Meseguer, J.: Mechanical analysis of reliable communication in the alternating bit protocol using the Maude invariant analyzer tool. In: Iida, S., Meseguer, J., Ogata, K. (eds.) Specification, Algebra, and Software. LNCS, vol. 8373, pp. 603–629. Springer, Heidelberg (2014). https://doi.org/10.1007/978-3-642-54624-2_30
116. Rocha, C., Meseguer, J., Muñoz, C.A.: Rewriting modulo SMT and open system analysis. J. Log. Algebraic Methods Program. **86**, 269–297 (2017)
117. Rosu, G., Serbanuta, T.: An overview of the K semantic framework. J. Log. Algebraic Program. **79**(6), 397–434 (2010)
118. Schmidt, B., Meier, S., Cremers, C.J.F., Basin, D.A.: Automated analysis of Diffie-Hellman protocols and advanced security properties. In: Proceedings of the CSF 2012, pp. 78–94. IEEE (2012)

119. Schmidt, B., Sasse, R., Cremers, C., Basin, D.: Automated verification of group key agreement protocols. In: Proceedings of the 2014 IEEE Symposium on Security and Privacy, SP 2014, pp. 179–194. IEEE Computer Society, Washington, D.C. (2014)

120. Schmidt-Schauss, M.: Unification in many-sorted equational theories. In: Siekmann, J.H. (ed.) CADE 1986. LNCS, vol. 230, pp. 538–552. Springer, Heidelberg (1986). https://doi.org/10.1007/3-540-16780-3_118

121. Schmidt-Schauß, M. (ed.): Computational Aspects of an Order-Sorted Logic with Term Declarations. LNCS, vol. 395. Springer, Heidelberg (1989). https://doi.org/10.1007/BFb0024065

122. Shostak, R.E.: Deciding combinations of theories. J. ACM **31**(1), 1–12 (1984)

123. Siekmann, J.H.: Unification theory. J. Symb. Comput. **7**(3/4), 207–274 (1989)

124. Skeirik, S., Meseguer, J.: Metalevel algorithms for variant satisfiability. J. Log. Algebraic Methods Program. **96**, 81–110 (2018)

125. Skeirik, S., Stefanescu, A., Meseguer, J.: A constructor-based reachability logic for rewrite theories. Technical report, University of Illinois Computer Science Department, March 2017. http://hdl.handle.net/2142/95770. Shorter version to appear in Proceedings of LOPSTR 2107, Springer LNCS 2018

126. Slagle, J.R.: Automated theorem-proving for theories with simplifiers commutativity, and associativity. J. ACM **21**(4), 622–642 (1974)

127. Smolka, G., Aït-Kaci, H.: Inheritance hierarchies: semantics and unification. J. Symb. Comput. **7**(3/4), 343–370 (1989)

128. Smolka, G., Nutt, W., Goguen, J., Meseguer, J.: Order-sorted equational computation. In: Nivat, M., Aït-Kaci, H. (eds.) Resolution of Equations in Algebraic Structures, vol. 2, pp. 297–367. Academic Press, Cambridge (1989)

129. Snyder, W.: A Proof Theory for General Unification. Birkhäuser, Boston (1991)

130. Ştefănescu, A., Ciobâcă, Ş., Mereuta, R., Moore, B.M., Şerbănută, T.F., Roşu, G.: All-path reachability logic. In: Dowek, G. (ed.) RTA 2014. LNCS, vol. 8560, pp. 425–440. Springer, Cham (2014). https://doi.org/10.1007/978-3-319-08918-8_29

131. Stefanescu, A., Park, D., Yuwen, S., Li, Y., Rosu, G.: Semantics-based program verifiers for all languages. In: Proceedings of the OOPSLA 2016, pp. 74–91. ACM (2016)

132. Stehr, M.O.: CINNI - a generic calculus of explicit substitutions and its application to λ-, σ- and π-calculi. ENTCS **36**, 70–92 (2000). Proceedings of the 3rd International Workshop on Rewriting Logic and Its Applications

133. Stehr, M.-O., Meseguer, J.: Pure type systems in rewriting logic: specifying typed higher-order languages in a first-order logical framework. In: Owe, O., Krogdahl, S., Lyche, T. (eds.) From Object-Orientation to Formal Methods. LNCS, vol. 2635, pp. 334–375. Springer, Heidelberg (2004). https://doi.org/10.1007/978-3-540-39993-3_16

134. Stehr, M.-O., Meseguer, J., Ölveczky, P.C.: Rewriting logic as a unifying framework for Petri nets. In: Ehrig, H., Padberg, J., Juhás, G., Rozenberg, G. (eds.) Unifying Petri Nets. LNCS, vol. 2128, pp. 250–303. Springer, Heidelberg (2001). https://doi.org/10.1007/3-540-45541-8_9

135. Tushkanova, E., Giorgetti, A., Ringeissen, C., Kouchnarenko, O.: A rule-based system for automatic decidability and combinability. Sci. Comput. Program. **99**, 3–23 (2015)

136. Viry, P.: Equational rules for rewriting logic. Theor. Comput. Sci. **285**, 487–517 (2002)

137. Walther, C.: A mechanical solution of Schubert's steamroller by many-sorted resolution. Artif. Intell. **26**(2), 217–224 (1985)
138. Yang, F., Escobar, S., Meadows, C., Meseguer, J., Narendran, P.: Theories of homomorphic encryption, unification, and the finite variant property. In: Proceedings of the PPDP 2014, pp. 123–133. ACM (2014)
139. Yang, F., Escobar, S., Meadows, C.A., Meseguer, J., Santiago, S.: Strand spaces with choice via a process algebra semantics. In: Proceedings of the 18th International Symposium on Principles and Practice of Declarative Programming (PPDP), Edinburgh, United Kingdom, 5–7 September 2016, pp. 76–89. ACM (2016)

A Semantical View of Proof Systems

Elaine Pimentel$^{(\boxtimes)}$

Departamento de Matemática, UFRN, Natal, Brazil
elaine.pimentel@gmail.com

Abstract. In this work, we explore proof theoretical connections between sequent, nested and labelled calculi. In particular, we show a semantical characterisation of intuitionistic, normal and non-normal modal logics for all these systems, via a case-by-case translation between labelled nested to labelled sequent systems.

1   Introduction

The quest of finding *analytic* proof systems for different logics has been the main research topic for proof theorists since Gentzen's seminal work [5]. One of the best known formalisms for proposing analytic proof systems is Gentzen's *sequent calculus.* While its simplicity makes it an ideal tool for proving meta-logical properties, sequent calculus is not expressive enough for constructing analytic calculi for many logics of interest. The case of modal logic is particularly problematic, since sequent systems for such logics are usually not modular, and they mostly lack relevant properties such as separate left and right introduction rules for the modalities. These problems are often connected to the fact that the modal rules in such calculi usually introduce more than one connective at a time, e.g. as in the rule k for modal logic K:

$$\frac{B_1,\ldots,B_n \vdash A}{\Box B_1,\ldots,\Box B_n \vdash \Box A}\ \text{k}$$

One way of solving this problem is by considering extensions of the sequent framework that are expressive enough for capturing these modalities using separate left and right introduction rules. This is possible e.g. in *labelled sequents* [18] or in *nested sequents* [1]. In the labelled sequent framework, usually the semantical characterisation is explicitly added to sequents. In the nested framework in contrast, a single sequent is replaced with a tree of sequents, where successors of a sequent (nestings) are interpreted under a given modality. The nesting rules of these calculi govern the transfer of formulae between the different sequents, and they are *local*, in the sense that it is sufficient to transfer only one formula at a time. As an example, the labelled and nested versions for the *necessity right rule* ($\Box_R$) are

E. Pimentel—Funded by CNPq, CAPES and the project FWF START Y544-N23.

L. S. Moss et al. (Eds.): WoLLIC 2018, LNCS 10944, pp. 61–76, 2018.
https://doi.org/10.1007/978-3-662-57669-4_3

$$\frac{\mathcal{R}, xRy, X \vdash Y, y:A}{\mathcal{R}, X \vdash Y, x:\Box A} \ \Box_R^l \qquad \frac{\Gamma \vdash \Delta, [\cdot \vdash A]}{\Gamma \vdash \Delta, \Box A} \ \Box_R^n$$

where y is a fresh variable in the $\Box_R^l$ rule. Reading bottom up, while the labelled system creates a new variable y related to x via a relation R and changes the label of A to y, in $\Box_R^n$ a new nesting is created, and A is moved there. It seems clear that nestings and semantical structures are somehow related. Indeed, a direct translation between proofs in labelled and nested systems for the modal logic of provability (a.k.a. the Gödel-Löb provability logic) is presented in [6], while in [4] it is shown how to relate nestings with Kripke structures for intuitionistic logic (via indexed tableaux systems). In this work, we show this relationship for intuitionistic logic and some normal modal logics, using only sequent based systems.

Since nested systems have been also proposed for other modalities, such as the non-normal ones [2], an interesting question is whether this semantical interpretation can be generalised to other systems as well. In [15] a labelled approach was used for setting the grounds for proof theory of some non-normal modal systems based on *neighbourhood semantics*. In parallel, we have proposed [10] modular systems based on nestings for several non-normal modal logics. We will relate these two approaches for the logics M and E, hence clarifying the nesting-semantics relationship for such logics.

Finally, in [11], we showed that a class of nested systems can be transformed into sequent systems via a linearisation procedure, where sequent rules can be seen as nested *macro-rules*. By relating nested and sequent systems, we are able to extend the semantical interpretation also to the sequent case, hence closing the relationship between systems and shedding light on the semantical interpretation of several sequent based systems.

Organisation and Contributions. Section 2 presents the basic notation for sequent systems; Sect. 3 presents nested systems and summarizes the results for their sequentialisation; Sect. 4 presents the basic notation for labelled systems; Sects. 5, 6 and 7 show the results under the particular views of intuitionistic, normal and non-normal logics; Sect. 8 concludes the paper.

2   Sequent Systems

Contemporary proof theory started with Gentzen's work [5], and it has had a continuous development with the proposal of several proof systems for many logics.

Definition 1. *A* sequent *is an expression of the form $\Gamma \vdash \Delta$ where Γ (the* antecedent*) and Δ (the* succedent*) are finite sets of formulae. A sequent calculus* (SC) *consists of a set of rule schemas, of the form*

$$\frac{S_1 \quad \cdots \quad S_k}{S} \ r$$

$$\dfrac{\Gamma \vdash A, \varDelta \quad \Gamma, B \vdash \varDelta}{\Gamma, A \to B \vdash \varDelta} \to_L \qquad \dfrac{\Gamma, A \vdash B}{\Gamma \vdash A \to B} \to_R \qquad \dfrac{\Gamma, A, B \vdash \varDelta}{\Gamma, A \wedge B \vdash \varDelta} \wedge_L$$

$$\dfrac{\Gamma \vdash A, \varDelta \quad \Gamma \vdash B, \varDelta}{\Gamma \vdash A \wedge B, \varDelta} \wedge_R \qquad \dfrac{\Gamma, A, \vdash \varDelta \quad \Gamma, B \vdash \varDelta}{\Gamma, A \vee B \vdash \varDelta} \vee_L \qquad \dfrac{\Gamma \vdash A, B, \varDelta}{\Gamma \vdash A \vee B, \varDelta} \vee_R \qquad \dfrac{}{\Gamma, \bot \vdash \varDelta} \bot_L$$

Fig. 1. Multi-conclusion intuitionistic calculus $\mathsf{SC}_{\mathsf{mLJ}}$.

where the sequent S is the conclusion inferred from the premise sequents $S_1, \ldots, S_k$ in the rule r. If the set of premises is empty, then r is an axiom. An instance of a rule is a rule application.

A derivation is a finite directed tree with nodes labelled by sequents and a single root, axioms at the top nodes, and where each node is connected with the (immediate) successor nodes (if any) according to the application of rules. The height of a derivation is the greatest number of successive applications of rules in it, where an axiom has height 0.

In this work we will consider only *fully structural* sequent systems, *i.e.* allowing freely applications of the schemas init and W bellow

$$\dfrac{}{\Gamma, P \vdash P, \varDelta} \text{ init} \qquad \qquad \dfrac{\Gamma \vdash \varDelta}{\Gamma, \Gamma' \vdash \varDelta, \varDelta'} \text{ W}$$

where P is atomic.

As an example, Fig. 1 presents $\mathsf{SC}_{\mathsf{mLJ}}$ [12], a multiple conclusion sequent system for propositional intuitionistic logic. The rules are exactly the same as in classical logic, except for the implication right rule, that forces all formulae in the succedent of the conclusion sequent to be previously weakened. This guarantees that, on applying the $(\to_R)$ rule on $A \to B$, the formula B should be proved assuming *only* the pre-existent antecedent context extended with the formula A, creating an interdependency between A and B.

3 Nested Systems

Nested systems [1,16] are extensions of the sequent framework where a single sequent is replaced with a tree of sequents.

Definition 2. *A nested sequent is defined inductively as follows:*

(i) if $\Gamma \vdash \varDelta$ is a sequent, then it is a nested sequent;
(ii) if $\Gamma \vdash \varDelta$ is a sequent and $G_1, \ldots, G_k$ are nested sequents, then $\Gamma \vdash \varDelta, [G_1], \ldots, [G_k]$ is a nested sequent.

A nested system (NS) consists of a set of inference rules acting on nested sequents.

For readability, we will denote by $\Gamma, \varDelta$ sequent contexts and by Λ sets of nestings. In this way, every nested sequent has the shape $\Gamma \vdash \varDelta, \Lambda$ where elements of Λ

$$\frac{S\{\Gamma \vdash \Delta, A, \Lambda\} \quad S\{\Gamma, B \vdash \Delta, \Lambda\}}{S\{\Gamma, A \to B \vdash \Delta, \Lambda\}} \to^n_L \quad \frac{S\{\Gamma \vdash \Delta, \Lambda, [A \vdash B]\}}{S\{\Gamma \vdash A \to B, \Delta, \Lambda\}} \to^n_R$$

$$\frac{S\{\Gamma, A, B \vdash \Delta, \Lambda\}}{S\{\Gamma, A \wedge B \vdash \Delta, \Lambda\}} \wedge^n_L \quad \frac{S\{\Gamma \vdash A, \Delta, \Lambda\} \quad S\{\Gamma \vdash B, \Delta, \Lambda\}}{S\{\Gamma \vdash A \wedge B, \Delta, \Lambda\}} \wedge^n_R$$

$$\frac{S\{\Gamma, A \vdash \Delta, \Lambda\} \quad S\{\Gamma, B \vdash \Delta, \Lambda\}}{S\{\Gamma, A \vee B \vdash \Delta, \Lambda\}} \vee^n_L \quad \frac{S\{\Gamma \vdash A, B, \Delta, \Lambda\}}{S\{\Gamma \vdash A \vee B, \Delta, \Lambda\}} \vee^n_R$$

$$\frac{S\{\Gamma \vdash \Delta, \Lambda, [\Gamma', A \vdash \Delta', \Lambda']\}}{S\{\Gamma, A \vdash \Delta, \Lambda, [\Gamma' \vdash \Delta', \Lambda']\}} \text{lift}^n \quad \frac{}{S\{\Gamma, \bot \vdash \Delta, \Lambda\}} \bot^n_L$$

Fig. 2. Nested system $\mathsf{NS}_{\mathsf{mLJ}}$.

have the shape $[\Gamma' \vdash \Delta', \Lambda']$ and so on. We will denote by Υ an arbitrary nested sequent.

Application of rules and schemas in nested systems will be represented using *holed contexts*.

Definition 3. *A nested-holed context is a nested sequent that contains a hole of the form $\{\ \}$ in place of nestings. We represent such a context as $S\{\ \}$. Given a holed context and a nested sequent Υ, we write $S\{\Upsilon\}$ to stand for the nested sequent where the hole $\{\ \}$ has been replaced by $[\Upsilon]$, assuming that the hole is removed if Υ is empty and if S is empty then $S\{\Upsilon\} = \Upsilon$. The depth of $S\{\ \}$, denoted by $\mathrm{dp}\,(S\{\ \})$, is the number of nodes on a branch of the nesting tree of $S\{\ \}$ of maximal length.*

For example, $(\Gamma \vdash \Delta, \{\ \})\{\Gamma' \vdash \Delta'\} = \Gamma \vdash \Delta, [\Gamma' \vdash \Delta']$ while $\{\ \}\{\Gamma' \vdash \Delta'\} = \Gamma' \vdash \Delta'$.

The definition of application of nested rules and derivations in a NS are natural extensions of the one for SC, only replacing sequents by nested sequents. In this work we will assume that nested systems are *fully structural*, *i.e.*, including the following nested versions for the initial axiom and weakening[1]

$$\frac{}{S\{\Gamma, P \vdash \Delta, P, \Lambda\}} \text{init}^n \quad \frac{S\{\Gamma \vdash \Delta, \Lambda\}}{S\{\Gamma, \Gamma' \vdash \Delta, \Delta', \Lambda, \Lambda'\}} \mathsf{W}^n$$

Figure 2 presents the $\mathsf{NS}_{\mathsf{mLJ}}$ [4], a nested system for mLJ.

3.1 Sequentialising Nested Systems

In [11] we identified general conditions under which a nested calculus can be transformed into a sequent calculus by restructuring the nested sequent derivation (proof) and shedding extraneous information to obtain a derivation of the same formula in the sequent calculus. These results were formulated generally

---

[1] All over this text, we will use n as a superscript, etc for indicating "nested". Hence *e.g.*, $\to^n_R$ will be the designation of the implication right rule in the nesting framework.

so that they apply to calculi for intuitionistic, normal and non-normal modal logics. Here we will briefly explain the main ideas in that work.

First of all, we restrict our attention to *shallow directed* nested systems, in with rules are restricted so to falling in one of the following mutually exclusive schemas:

i. *sequent-like* rules

$$\frac{S\{\Gamma_1 \vdash \Delta_1\} \quad \cdots \quad S\{\Gamma_k \vdash \Delta_k\}}{S\{\Gamma \vdash \Delta\}}$$

ii *nested-like* rules

    ii.a *creation rules*

$$\frac{S\{\Gamma \vdash \Delta, [\Gamma_1 \vdash \Delta_1]\}}{S\{\Gamma \vdash \Delta\}}$$

    ii.b *upgrade rules*

$$\frac{S\{\Gamma' \vdash \Delta', [\Gamma_1' \vdash \Delta_1']\}}{S\{\Gamma \vdash \Delta, [\Gamma_1 \vdash \Delta_1]\}}$$

The nesting in the premise of a creation rule is called the *auxiliary nesting*.

The following extends the definition of permutability to the nested setting.

Definition 4. *Let* NS *be shallow directed,* r_1, r_2 *be applications rules and* Υ *be a nested sequent. We say that* r_2 *permutes down* r_1 *($r_2 \downarrow r_1$) if, for every derivation in which* r_1 *operates on* Υ *and* r_2 *operates on one or more of* r_1*'s premises (but not on auxiliary formulae/nesting of* r_1*), there exists another derivation of* Υ *in which* r_2 *operates on* Υ *and* r_1 *operates on zero or more of* r_2*'s premises (but not on auxiliary formulae/nesting of* r_2*). If* $r_2 \downarrow r_1$ *and* $r_1 \downarrow r_2$ *we will say that* r_1, r_2 *are permutable, denoted by* $r_1 \updownarrow r_2$*. Finally,* NS *is said fully permutable if* $r_1 \updownarrow r_2$ *for any pair of rules.*

Finally, the next result shows that fully permutable, shallow directed systems can be sequentialised.

Theorem 5. *Let* NS *be fully permutable, shallow and directed. There is a normalisation procedure of proofs in* NS *transforming maximal blocks of applications of nested-like rules into sequent rules.*

Next an example of such procedure.

Example 6. In $\mathsf{NS_{mLJ}}$, a nested block containing the creation rule $\to_R^n$ and the upgrade rule $\mathtt{lift}^n$ has the shape

$$\frac{\dfrac{S\{\Gamma' \vdash \Delta', [\Gamma, A \vdash B]\}}{S\{\Gamma, \Gamma' \vdash \Delta', [A \vdash B]\}} \;\mathtt{lift}^n}{S\{\Gamma, \Gamma' \vdash \Delta', A \to B\}} \;\to_R^n$$

Observe that $\mathtt{lift}^n$ maps a left formula into itself and there are no context relations on right formulae. Hence the corresponding sequent rule is

$$\frac{\Gamma, A \vdash B}{\Gamma \vdash A \to B} \ \to R$$

which is the implication right rule for mLJ. That is, sequentialising the nested system $\mathsf{NS}_{\mathsf{mLJ}}$ (Fig. 2) results in the sequent system mLJ (Fig. 1).

4   Labelled Proof Systems

While it is widely accepted that nested systems carry the Kripke structure on nestings for intuitionistic and normal modal logics, it is not clear what is the relationship between nestings and semantics for other systems. For example, in [9] we presented a linear nested system [8] for linear logic, but the interpretation of nestings for this case is still an open problem.

In this work we will relate labelled nested systems [6] with labelled systems [18]. While the results for intuitionistic and some normal modal logics are not new [4,6], we give a complete different approach for these results, and present the first semantical interpretation for nestings in non-normal modal logics. In this section we shall recall some of the terminology for labelled systems.

Labelled Nested Systems. Let SV a countable infinite set of *state variables* (denoted by $x, y, z, \ldots$), disjoint from the set of propositional variables. A *labelled formula* has the form $x : A$ where $x \in \mathsf{SV}$ and A is a formula. If $\Gamma = \{A_1, \ldots, A_k\}$ is a set of formulae, then $x : \Gamma$ denotes the set $\{x : A_1, \ldots, x : A_k\}$ of labelled formulae. A (possibly empty) set of relation terms (*i.e.* terms of the form xRy, where $x, y \in \mathsf{SV}$) is called a *relation set*. For a relation set $\mathcal{R}$, the *frame* $Fr(\mathcal{R})$ defined by $\mathcal{R}$ is given by $(|\mathcal{R}|, \mathcal{R})$ where $|\mathcal{R}| = \{x \mid xRy \in \mathcal{R} \text{ or } yRx \in \mathcal{R} \text{ for some } y \in \mathsf{SV}\}$. We say that a relation set $\mathcal{R}$ is *treelike* if the frame defined by $\mathcal{R}$ is a tree or $\mathcal{R}$ is empty.

Definition 7. *A labelled nested sequent* LbNS *is a labelled sequent* $\mathcal{R}, X \vdash Y$ *where*

1. $\mathcal{R}$ *is treelike;*
2. *if* $\mathcal{R} = \emptyset$ *then* X *has the form* $x : A_1, \ldots, x : A_k$ *and* Y *has the form* $x : B_1, \ldots, x : B_m$ *for some* $x \in SV$;
3. *if* $\mathcal{R} \neq \emptyset$ *then every state variable* y *that occurs in either* X *or* Y *also occurs in* $\mathcal{R}$.

A labelled nested sequent calculus *is a labelled calculus whose initial sequents and inference rules are constructed from* LbNS.

As in [6], labelled nested systems can be automatically generated from nested systems.

Definition 8. *Given* $\Gamma \vdash \Delta$ *and* $\Gamma' \vdash \Delta'$ *sequents, we define* $(\Gamma \vdash \Delta) \otimes (\Gamma' \vdash \Delta')$ *to be* $\Gamma, \Gamma' \vdash \Delta, \Delta'$. *For a state variable* x, *define the mapping* $\mathbb{TL}_x$ *from* NS *to* LbLNS *as follows*

$$\mathbb{TL}_x(\Gamma \vdash \Delta, [\Upsilon_1], \dots, [\Upsilon_n]) = xRx_1, \dots, xRx_n, (x:\Gamma \vdash x:\Delta) \otimes$$
$$\mathbb{TL}_{x_1}(\Upsilon_1) \otimes \dots \otimes \mathbb{TL}_{x_n}(\Upsilon_n)$$
$$\mathbb{TL}_x([\Gamma \vdash \Delta]) \qquad = x:\Gamma \vdash x:\Delta$$

with all state variables pairwise distinct.

For the sake of readability, when the state variable is not important, we will suppress the subscript, writing $\mathbb{TL}$ instead of $\mathbb{TL}_x$. We will shortly illustrate the procedure of constructing labelled nestings using the mapping $\mathbb{TL}$. Consider the following application of the rule $\rightarrow_R$ of Fig. 2:

$$\frac{S\{\Gamma \vdash \Delta, \Lambda, [A \vdash B]\}}{S\{\Gamma \vdash \Delta, A \rightarrow B, \Lambda\}} \rightarrow_R^n$$

Applying $\mathbb{TL}$ to the conclusion we obtain $\mathcal{R}, X \vdash Y, x:A \rightarrow B$ where the variable x label formulae in two components of the NS, and X, Y are sets of labelled formulae. Applying $\mathbb{TL}$ to the premise we obtain $\mathcal{R}, xRy, X, y:A \vdash Y, y:B$ where y is a fresh variable (*i.e.* different from x and not occurring in X, Y). We thus obtain an application of the LbLNS rule

$$\frac{\mathcal{R}, xRy, X, y:A \vdash Y, y:B}{\mathcal{R}, X \vdash Y, x:A \rightarrow B} \ \mathbb{TL}(\rightarrow_R^n)$$

Some rules of the labelled nested system LbNS$_{\text{mLJ}}$ are depicted in Fig. 3.

The following result follows readily by transforming derivations bottom-up [6].

Theorem 9. *The mapping* $\mathbb{TL}_x$ *preserves open derivations, that is, there is a 1-1 correspondence between derivations in a nested sequent system* NS *and in its labelled translation* LbNS.

Labelled Sequent Systems. In the labelled sequent framework, a semantical characterisation of a logic is explicitly added to sequents via the labelling of formulae [3,13–15,18]. In the case of world based semantics, the forcing relation $x \Vdash A$ is represented as the labelled formula $x:A$ and sequents have the form $\mathcal{R}, X \vdash Y$, where $\mathcal{R}$ is a relation set and X, Y are multisets of labelled formulae.

The rules of the labelled calculus G3I are obtained from the inductive definition of validity in a Kripke frame (Fig. 4a), together with the rules describing a partial order, presented in Fig. 4b. Note that the anti-symmetry rule does not need to be stated directly since, for any x, the formula $x = x$ is equivalent to true and hence can be erased from the left side of a sequent.

$$\dfrac{\mathcal{R}, xRy, X, y{:}A \vdash Y, y{:}B}{\mathcal{R}, X \vdash Y, x{:}A \to B} \; \mathrm{TL}(\to_R^n) \qquad \dfrac{\mathcal{R}, X, \vdash x{:}A, Y \quad X, x{:}B \vdash Y}{\mathcal{R}, X, x{:}A \to B \vdash Y} \; \mathrm{TL}(\to_L^n)$$

$$\dfrac{\mathcal{R}, X, x{:}A, x{:}B \vdash Y}{\mathcal{R}, X, x{:}A \wedge B \vdash Y} \; \mathrm{TL}(\wedge_L^n) \qquad \dfrac{\mathcal{R}, X \vdash x{:}A, Y \quad X \vdash x{:}B, Y}{\mathcal{R}, X \vdash x{:}A \wedge B, Y} \; \mathrm{TL}(\wedge_R^n)$$

$$\dfrac{\mathcal{R}, X, x{:}A, \vdash Y \quad X, x{:}B \vdash Y}{\mathcal{R}, X, x{:}A \vee B \vdash Y} \; \mathrm{TL}(\vee_L^n) \qquad \dfrac{\mathcal{R}, X \vdash x{:}A, x{:}B, Y}{\mathcal{R}, X \vdash x{:}A \vee x{:}B, Y} \; \mathrm{TL}(\vee_R^n)$$

$$\dfrac{}{\mathcal{R}, X, x{:}A \vdash x{:}A, Y} \; \mathrm{TL}(\mathrm{init}^n) \qquad \dfrac{}{\mathcal{R}, X, x{:}\bot \vdash Y} \; \mathrm{TL}(\bot_L^n) \qquad \dfrac{\mathcal{R}, xRy, X, y{:}A \vdash Y}{\mathcal{R}, xRy, X, x{:}A \vdash Y} \; \mathrm{TL}(\mathrm{lift}^n)$$

Fig. 3. Labelled nested system $\mathsf{LbNS_{mLJ}}$.

$$\dfrac{\mathcal{R}, x \leq y, X, x{:}A \to B \vdash y{:}A, Y \quad \mathcal{R}, x \leq y, X, y{:}B \vdash Y}{\mathcal{R}, x \leq y, X, x{:}A \to B \vdash Y} \to_L^t \qquad \dfrac{xRx, \mathcal{R}, X \vdash Y}{\mathcal{R}, X \vdash Y} \; \mathrm{Ref}$$

$$\dfrac{\mathcal{R}, x \leq y, X, y{:}A \vdash Y, y{:}B}{\mathcal{R}, X \vdash Y, x{:}A \to B} \to_R^t \qquad \dfrac{}{\mathcal{R}, X, x \leq y, x{:}P \vdash Y, y{:}P} \; \mathrm{init}^t \qquad \dfrac{xRz, xRy, yRz, \mathcal{R}, X \vdash Y}{xRy, yRz, \mathcal{R}, X \vdash Y} \; \mathrm{Trans}$$

(a) y is fresh in $\to_R$ and P is atomic in init.   (b) Relation rules.

Fig. 4. Some rules of the labelled system G3I

5   Intuitionistic Logic

In this section we will relate various proof systems for intuitionistic logic by applying the results presented in the last sections.

Theorem 10. *All rules in* $\mathsf{NS_{mLJ}}$ *are height-preserving invertible and* $\mathsf{NS_{mLJ}}$ *is fully permutable.*

Proof. The proofs of invertibility are by induction on the depth of the derivation, distinguishing cases according to the last applied rule. Permutability of rules is proven by a case-by-case analysis, using the invertibility results.   □

The results in the previous sections entail the following.

Theorem 11. *Systems* $\mathsf{NS_{mLJ}}$, mLJ *and* $\mathsf{LbNS_{mLJ}}$ *are equivalent.*

Observe that the proof uses *syntactical arguments only*, differently from *e.g.* [4,8].
    For establishing a comparison between labels in G3I and $\mathsf{LbNS_{mLJ}}$, first observe that applications of rule Trans in G3I can be restricted to the leaves (*i.e.* just before an instance of the initial axiom). Also, since weakening is admissible in G3I and the monotonicity property: $x \Vdash A$ and $x \leq y$ implies $y \Vdash A$ is derivable in G3I (Lemma 4.1 in [3]), the next result follows.

Lemma 12. *The following rules are* derivable *in* G3I *up to weakening.*

$$\dfrac{\mathcal{R}, X, x{:}A \to B \vdash x{:}A, Y \quad \mathcal{R}, X, x{:}B \vdash Y}{\mathcal{R}, X, x{:}A \to B \vdash Y} \to_{L'} \qquad \dfrac{}{\mathcal{R}, X, x{:}P \vdash Y, x{:}P} \; \mathrm{init}'$$

Moreover, the rule

$$\frac{\mathcal{R}, x \leq y, X, y\!:\!A \vdash Y}{\mathcal{R}, x \leq y, X, x\!:\!A \vdash Y} \ \text{lift}'$$

is admissible in G3I.

Proof. For the derivable rules, just note that

$$\frac{\mathcal{R}, X, x \leq x, x\!:\!A \rightarrow B \vdash Y, x\!:\!A \quad \mathcal{R}, X, x \leq x, x\!:\!B \vdash Y}{\dfrac{\mathcal{R}, X, x \leq x, x\!:\!A \rightarrow B \vdash Y}{\mathcal{R}, X, x\!:\!A \rightarrow B \vdash Y} \ \text{Ref}} \rightarrow_L^t$$

and

$$\frac{\overline{\mathcal{R}, x \leq x, X, x\!:\!P \vdash Y, x\!:\!P}}{\mathcal{R}, X, x\!:\!P, \vdash Y, x\!:\!P} \ \text{init}^t}{\text{Ref}}$$

$\square$

Using an argument similar to the one in [6], it is easy to see that, in the presence of the primed rules shown above, the relational rules are admissible. Moreover, labels are preserved.

Theorem 13. G3I *is label-preserving equivalent to* LbNS$_{\text{mLJ}}$.

That is, nestings in NS$_{\text{mLJ}}$ and LNS$_{\text{mLJ}}$ correspond to worlds in the Kripke structure where the sequent is valid and this is the semantical interpretation of the nested system for intuitionistic logic [4].

Observe that, since mLJ derivations are equivalent to normal NS$_{\text{mLJ}}$ derivations, the semantical analysis for LNS$_{\text{mLJ}}$ also hold for mLJ, that is, an application of the $\rightarrow_R$ rule over $\Gamma \vdash A \rightarrow B$ in mLJ corresponds to creating a new world w in the Kripke structure and setting the forcing relation to A, B and all the formulae in Γ.

In what follows, we will show how this approach on different proof systems can be smoothly extended to normal as well as non-normal modalities, using propositional classical logic as the base logic.

6 Normal Modal Logics

The next natural step on investigating the relationship between frame semantics and nested sequent systems is to consider modal systems.

The normal modal logic K is obtained from classical propositional logic by adding the unary modal connective $\square$ to the set of classical connectives, together with the necessitation rule and the K axiom (see Fig. 5 for the Hilbert-style axiom schemata) to the set of axioms for propositional classical logic.

$$\mathsf{K}\ \square(A \to B) \to (\square A \to \square B) \qquad \frac{A}{\square A}\ \mathsf{nec} \qquad \mathsf{D}\ \neg\square\bot \qquad \mathsf{T}\ \square A \to A \qquad \mathsf{4}\ \square A \to \square\square A$$

Fig. 5. Modal axiom K, necessitation rule nec and extensions D, T, 4.

$$\frac{S\{\Gamma, A \vdash B, \varDelta, \Lambda\}}{S\{\Gamma \vdash A \to B, \varDelta, \Lambda\}} \to_R^n \qquad \frac{S\{\Gamma \vdash \varDelta, [\Gamma', A \vdash \varDelta'], \Lambda\}}{S\{\Gamma, \square A \vdash \varDelta, [\Gamma' \vdash \varDelta'], \Lambda\}} \square_L^n \qquad \frac{S\{\Gamma \vdash \varDelta, [\vdash A]\}}{S\{\Gamma \vdash \varDelta, \square A, \Lambda\}} \square_R^n$$

Fig. 6. Nested system $\mathsf{NS_K}$. The rules $\to_L^n, \wedge_R^n, \wedge_L^n, \vee_R^n, \vee_L^n$ and $\bot_L^n$ are the same as in Fig. 2.

The nested framework provides an elegant way of formulating modal systems, since no context restriction is imposed on rules. Figure 6 presents the modal rules for the nested sequent calculus $\mathsf{NS_K}$ for the modal logic K [1,16].

Observe that there are two rules for handling the box operator ($\square_L$ and $\square_R$), which allows the treatment of one formula at a time. Being able to separate the left/right behaviour of the modal connectives is the key to modularity for nested calculi [8,17]. Indeed, K can be modularly extended by adding to $\mathsf{NS_K}$ the nested corresponding to other modal axioms. In this paper, we will consider the axioms D, T and 4 (Fig. 5). Figure 7 shows the modal nested rules for such extensions: for a logic $\mathsf{K\mathcal{A}}$ with $\mathcal{A} \subseteq \{\mathsf{D}, \mathsf{T}, \mathsf{4}\}$ the calculus $\mathsf{NS_{K\mathcal{A}}}$ extends $\mathsf{NS_K}$ with the corresponding nested modal rules.

Note that rule t^n is actually a sequent-like rule. On the other hand, $\square_R^n$ and d^n are creation rules while $\square_L^n$ and $\mathsf{4}^n$ are upgrade rules. It is straightforward to verify that $\mathsf{NS_{K\mathcal{A}}}$ is shallow directed and fully permutable. Moreover, a nested block containing the application of one of the creation rules and possible several applications of the upgrade rules has one of the following shapes

$$\frac{\dfrac{S\{\Gamma' \vdash \varDelta', [\square\Gamma_4, \Gamma_\mathsf{K} \vdash A]\}}{S\{\square\Gamma_4, \square\Gamma_\mathsf{K}, \Gamma' \vdash \varDelta', [\vdash A]\}}\ \square_L^n, \mathsf{4}^n}{S\{\square\Gamma_4, \square\Gamma_\mathsf{K}, \Gamma' \vdash \varDelta', \square A\}}\ \square_R^n \qquad \frac{\dfrac{S\{\Gamma' \vdash \varDelta', [\square\Gamma_4, \Gamma_\mathsf{K}, A \vdash]\}}{S\{\square\Gamma_4, \square\Gamma_\mathsf{K}, \Gamma' \vdash \varDelta', [A \vdash]\}}\ \square_L^n, \mathsf{4}^n}{S\{\square\Gamma_4, \square\Gamma_\mathsf{K}, \square A, \Gamma' \vdash \varDelta'\}}\ \mathsf{d}^n$$

where $\square_L^n$ acted in the context Γ_K and $\mathsf{4}^n$ in the context Γ_4. Observe that $\mathsf{4}^n$ maps a boxed left formula into itself, $\square_L^n$ maps left formulae into the boxed versions and there are no context relations on right formulae. Hence sequentialising the nested system $\mathsf{NS_{K\mathcal{A}}}$ (Fig. 7) results in the sequent system $\mathsf{SC_{K\mathcal{A}}}$ (shown as rule schemas in Fig. 8).

Finally, Definition 8 of Sect. 4 can be extended to the normal modal case in a trivial way, resulting in the labelled nested system $\mathsf{LbNS_{K\mathcal{A}}}$ (Fig. 9).

Theorem 14. *Systems* $\mathsf{NS_{K\mathcal{A}}}$, $\mathsf{SC_{K\mathcal{A}}}$ *and* $\mathsf{LbNS_{K\mathcal{A}}}$ *are equivalent.*

Figures 10a and b present the modal and relational rules of $\mathsf{G3K\mathcal{A}}$ [14], a sound and complete labelled sequent system for $\mathsf{K\mathcal{A}}$.

The next results follow the same lines as the ones in Sect. 5.

Lemma 15. *The rules* $\mathsf{TL(d}^n), \mathsf{TL(t}^n), \mathsf{TL(4}^n)$ *are derivable in* $\mathsf{G3K\mathcal{A}}$.

$$\frac{S\{\Gamma \vdash \Delta, [A \vdash], \Lambda\}}{S\{\Gamma, \Box A \vdash \Delta, \Lambda\}} \; \mathsf{d}^n \qquad \frac{S\{\Gamma, A \vdash \Delta, \Lambda\}}{S\{\Gamma, \Box A \vdash \Delta, \Lambda\}} \; \mathsf{t}^n \qquad \frac{S\{\Gamma \vdash \Delta, [\Gamma', \Box A \vdash \Delta'], \Lambda\}}{S\{\Gamma, \Box A \vdash \Delta, [\Gamma' \vdash \Delta'], \Lambda\}} \; \mathsf{4}^n$$

$$\mathsf{NS_{K\mathcal{A}}}: \quad \{\Box_R^n, \Box_L^n\} \cup \mathcal{A} \quad \text{for } \mathcal{A} \subseteq \{\mathsf{D}, \mathsf{T}, \mathsf{4}\}$$

Fig. 7. Nested sequent rules for extensions of K.

$$\frac{\Gamma \vdash A}{\Box \Gamma \vdash \Box A} \; \mathsf{k} \qquad \frac{\Gamma, A \vdash \Delta}{\Gamma, \Box A \vdash \Delta} \; \mathsf{t} \qquad \frac{\Gamma, A \vdash}{\Box \Gamma, \Box A \vdash} \; \mathsf{d} \qquad \frac{\Box \Gamma_4, \Gamma_K \vdash A}{\Box \Gamma_4, \Box \Gamma_K \vdash \Box A} \; \mathsf{k4} \qquad \frac{\Box \Gamma_4, \Gamma_K, A \vdash}{\Box \Gamma_4, \Box \Gamma_K, \Box A \vdash} \; \mathsf{d4}$$

$$\mathsf{SC_K} \; \{\mathsf{k}\} \qquad \mathsf{SC_{KT}} \; \{\mathsf{k,t}\} \qquad \mathsf{SC_{KD}} \; \{\mathsf{k,d}\} \qquad \mathsf{SC_{K4}} \; \{\mathsf{k4}\} \quad \mathsf{SC_{KD4}} \; \{\mathsf{d4}\}$$

Fig. 8. Modal sequent rules for normal modal logics $\mathsf{SC_{K\mathcal{A}}}$, for $\mathcal{A} \subseteq \{\mathsf{T}, \mathsf{D}, \mathsf{4}\}$.

Proof. The proof is straightforward. For example, for KD

$$\frac{\cfrac{\mathcal{R}, X, xRy, y : A \vdash Y}{\mathcal{R}, X, xRy, x : \Box A \vdash Y} \; \Box_L^t}{\mathcal{R}, X, x : \Box A \vdash Y} \; \text{Ser}$$

$\square$

Theorem 16. *$\mathsf{G3K\mathcal{A}}$ is label-preserving equivalent to $\mathsf{LbNS_{K\mathcal{A}}}$.*

Proof. That every provable sequent in $\mathsf{LbNS_{K\mathcal{A}}}$ is provable in $\mathsf{G3K\mathcal{A}}$ is a direct consequence of Lemma 15. For the other direction, observe that rule relational rules can be restricted so to be applied just before a $\Box_L^t$ rule. $\square$

This means that labels in $\mathsf{NS_{K\mathcal{A}}}$ represent worlds in a Kripke-frame, and this extends the results in [6] for modal logic of provability to normal modal logics $\mathsf{K\mathcal{A}}$.

7 Non-normal Modal Systems

We now move our attention to non-normal modal logics, i.e., modal logics that are not extensions of K. In this work, we will consider the *classical modal logic* E and the *monotone modal logic* M. Although our approach is general enough for considering nested, linear nested and sequent systems for several extensions of such logics (such as the *classical cube* or the *modal tesseract* – see [10]), there are no satisfactory labelled sequent calculi in the literature for such extensions.

For constructing nested calculi for these logics, the sequent rules should be decomposed into their different components. However, there are two complications compared to the case of normal modal logics: the need for (1) a mechanism for capturing the fact that exactly one boxed formula is introduced on the left hand side; and (2) a way of handling multiple premises of rules. The first problem is solved by introducing the indexed nesting $[\cdot]^e$ to capture a state where a sequent rule has been partly processed; the second problem is solved by making

$$\frac{\mathcal{R}, xRy, X, y{:}A \vdash Y}{\mathcal{R}, xRy, X, x{:}\Box A \vdash Y}\ \mathsf{TL}(\Box_L^n) \qquad \frac{\mathcal{R}, xRy, X \vdash Y, y{:}A}{\mathcal{R}, X \vdash Y, x{:}\Box A}\ \mathsf{TL}(\Box_R^n) \qquad \frac{\mathcal{R}, X, x{:}A \vdash Y}{\mathcal{R}, X, x{:}\Box A \vdash Y}\ \mathsf{TL}(\mathsf{t}^n)$$

$$\frac{\mathcal{R}, xRy, X, y{:}A \vdash Y}{\mathcal{R}, X, x{:}\Box A \vdash Y}\ \mathsf{TL}(\mathsf{d}^n) \qquad \frac{\mathcal{R}, xRy, X, y{:}\Box A \vdash Y}{\mathcal{R}, xRy, X, x{:}\Box A \vdash Y}\ \mathsf{TL}(4^n)$$

Fig. 9. Modal rules for labelled indexed nested system $\mathsf{LbNS}_{\mathsf{K}\mathcal{A}}$.

$$\frac{\mathcal{R}, xRy, y : A, \Gamma \vdash \Delta}{\mathcal{R}, xRy, x : \Box A, \Gamma \vdash \Delta}\ \Box_L^t \qquad\qquad \frac{\mathcal{R}, xRx, \Gamma \vdash \Delta}{\mathcal{R}, \Gamma \vdash \Delta}\ \mathsf{Ref} \qquad \frac{\mathcal{R}, xRz, \Gamma \vdash \Delta}{\mathcal{R}, xRy, yRz, \Gamma \vdash \Delta}\ \mathsf{Trans}$$

$$\frac{\mathcal{R}, xRy, \Gamma \vdash \Delta, y : A}{\mathcal{R}, \Gamma \vdash \Delta, x : \Box A}\ \Box_R^t \qquad\qquad\qquad \frac{\mathcal{R}, xRy, \Gamma \vdash \Delta}{\mathcal{R}, \Gamma \vdash \Delta}\ \mathsf{Ser}$$

(a) Modal rules. (b) Modal relational rules. y is fresh in Ser.

Fig. 10. Some rules of the labelled system $\mathsf{G3K}\mathcal{A}$.

the nesting operator $[\cdot]^e$ *binary*, which permits the storage of more information about the premises. Figure 11 presents a unified nested system for logics $\mathsf{NS_E}$ and $\mathsf{NS_M}$.

$\mathsf{NS_E}$ and $\mathsf{NS_M}$ are fully permutable but, since the nested-like rule $\Box_L^{en}$ has two premises, it does not fall into the definitions of shallowness/directedness. However, since propositional rules cannot be applied inside the indexed nestings, the modal rules naturally occur in blocks. Hence the nested rules correspond to macro-rules equivalent to the sequent rules in Fig. 12 for $\mathsf{SC_E}$ and $\mathsf{SC_M}$.

Finally, using the labelling method in Sect. 4, the rules in Fig. 11 correspond to the rules in Fig. 13, where xNy and $xN_e(y_1, y_2)$ are relation terms capturing the behaviour of the nestings $[\cdot]$ and $[\cdot]^e$ respectively.

The semantical interpretation of non-normal modalities E, M can be given via *neighbourhood semantics*, that smoothly extends the concept of Kripke frames in the sense that accessibility relations are substituted by a family of neighbourhoods.

Definition 17. *A neighbourhood frame is a pair* $\mathcal{F} = (W, N)$ *consisting of a set* W *of worlds and a neighbourhood function* $N : W \to \wp(\wp W)$. *A neighbourhood model is a pair* $\mathcal{M} = (\mathcal{F}, \mathcal{V})$, *where* $\mathcal{V}$ *is a valuation. We will drop the model symbol when it is clear from the context.*

The truth description for the box modality in the neighbourhood framework is

$$w \Vdash \Box A \text{ iff } \exists X \in N(w).[(X \Vdash^\forall A) \wedge (A \lhd X)] \qquad (1)$$

where $X \Vdash^\forall A$ is $\forall x \in X.x \Vdash A$ and $A \lhd X$ is $\forall y.[(y \Vdash A) \to y \in X]$. The rules for $\Vdash^\forall$ and $\lhd$ are obtained using the geometric rule approach [15] and are depicted in Fig. 14.

$$\frac{\Gamma \vdash \Delta, \Lambda, [\vdash B; B \vdash \cdot]^\Theta}{\Gamma \vdash \Delta, \Lambda, \Box B} \; \Box_R^{en} \qquad \frac{\Gamma, \Box A, \vdash \Delta, \Lambda, [\Sigma, A \vdash \Pi] \quad \Gamma, \Box A, \vdash \Delta, \Lambda, [\Omega \vdash \Theta, A]}{\Gamma, \Box A \vdash \Delta, \Lambda, [\Sigma \vdash \Pi; \Omega \vdash \Theta]^\Theta} \; \Box_L^{en}$$

$$\frac{\Gamma \vdash \Delta, \Lambda, [\Sigma \vdash \Pi; \Omega, \bot \vdash \Theta]^\Theta}{\Gamma \vdash \Delta, \Lambda, [\Sigma \vdash \Pi; \Omega \vdash \Theta]^\Theta} \; M^n$$

Fig. 11. Modal rules for systems $\mathsf{NS_E}$ and $\mathsf{NS_M}$.

$$\frac{A \vdash B \quad B \vdash A}{\Box A \vdash \Box B} \; \mathsf{E} \qquad \frac{A \vdash B}{\Box A \vdash \Box B} \; \mathsf{M}$$

Fig. 12. Modal sequent rules for non-normal modal logics $\mathsf{SC_E}$ and $\mathsf{SC_M}$.

If the neighbourhood frame is monotonic (i.e. $\forall X \subseteq W$, if $X \in N(w)$ and $X \subseteq Y \subseteq W$ then $Y \in N(w)$), it is easy to see [15] that (1) is equivalent to

$$w \Vdash \Box A \text{ iff } \exists X \in N(w).X \Vdash^\forall A. \tag{2}$$

This yields the set of labelled rules presented in Fig. 15, where the rules are adapted from [15] by collapsing invertible proof steps. Intuitively, while the box left rules create a fresh neighbourhood to x, the box right rules create a fresh world in this new neighbourhood and move the formula to it.

Theorem 18. G3E *(resp.* G3M*) is label-preserving equivalent to* $\mathsf{LbNS_E}$ *(resp.* $\mathsf{LbNS_M}$*).*

Proof. Let π be a normal proof of $\mathcal{N}, X \vdash Y$ in $\mathsf{LbNS_E}$. An instance of the blocked derivation

$$\frac{\overset{\pi_1}{\mathcal{N}, xNy_1, X, y_1 : A, y_2 : B \vdash Y, y_1 : B \quad \mathcal{N}, xNy_2, X, y_2 : B \vdash Y, y_1 : B, y_2 : A}}{\dfrac{\mathcal{N}, xN_e(y_1, y_2), X, x : \Box A, y_2 : B \vdash Y, y_1 : B}{\mathcal{N}, X, x : \Box A \vdash Y, x : \Box B} \; \mathbb{TL}(\Box_R^{en})} \; \mathbb{TL}(\Box_L^{en})$$

is transformed into the labelled derivation

$$\frac{\dfrac{y_1 : A, X \vdash Y, y_1 : B}{y_1 \in a, a \Vdash^\forall A, X \vdash Y, y_1 : B} \Vdash^\forall \quad \dfrac{y_2 : B, X \vdash Y, y_2 : A \quad \overline{y_2 \in a, X \vdash Y, y_2 \in a}^{\text{init}}}{\dfrac{y_2 : B, A \lhd a, X \vdash Y, y_2 \in a}{} \; \Box_R^{et}} \lhd}{\dfrac{a \in N(x), a \Vdash^\forall A, A \lhd a, X \vdash Y, x : \Box B}{X, x : \Box A \vdash Y, x : \Box B} \; \Box_L^{et}}$$

Observe that, in π_1, the label y_2 will no longer be active, hence the formula $y_2 : B$ can be weakened. The same with y_1 in π_2. Hence π_1/π_2 corresponds to π_1'/π_2' and the "only if" part holds. The "if" part uses similar proof theoretical arguments as in the intuitionistic or normal modal case, observing that applications of the forcing rules can be restricted so to be applied immediately after the modal rules. $\square$

$$\frac{N, xN_e(y_1, y_2), X, y_2 : B \vdash y_1 : B, Y}{N, X \vdash Y, x : \Box B} \; \mathsf{TL}(\Box_R^{en})$$

$$\frac{N, xN_e(y_1, y_2), X, y_2 :\perp \vdash Y}{N, xN_e(y_1, y_2), X \vdash Y} \; \mathsf{TL}(M^n)$$

$$\frac{N, xN y_1, y_1 : A, X \vdash Y \quad N, xN y_2, X \vdash Y, y_2 : A}{N, xN_e(y_1, y_2), x : \Box A, X \vdash Y} \; \mathsf{TL}(\Box_L^{en})$$

Fig. 13. Modal rules for $\mathsf{LbNS_E}$ and $\mathsf{LbNS_M}$ with y_1, y_2 fresh in $\Box_R^e$.

$$\frac{x \in a, x : A, a \Vdash^\vee A, X \vdash Y}{x \in a, a \Vdash^\vee A, X \vdash Y} \; \Vdash^\vee$$

$$\frac{A \lhd a, X \vdash Y, z : A \quad z \in a, A \lhd a, X \vdash Y}{A \lhd a, X \vdash Y} \; \lhd$$

$$\frac{}{x \in a, X \vdash Y, x \in a} \; \mathsf{init}^f$$

Fig. 14. Forcing rules, with z arbitrary in $\lhd_L$.

$$\frac{a \in N(x), a \Vdash^\vee A, A \lhd a, X \vdash Y}{x : \Box A, X \vdash Y} \; \Box_L^{ef}$$

$$\frac{z \in a, a \in N(x), X \vdash Y, x : \Box B, z : B \quad y : B, a \in N(x), X \vdash Y, x : \Box B, y \in a}{a \in N(x), X \vdash Y, x : \Box B} \; \Box_R^{ef}$$

$$\frac{a \in N(x), y \in a, X \vdash Y, x : \Box B, y : B}{a \in N(x), X \vdash Y, x : \Box B} \; \Box_R^{mf}$$

$$\frac{a \in N(x), a \Vdash^\vee A, X \vdash Y}{x : \Box A, X \vdash Y} \; \Box_L^{mf}$$

Fig. 15. Labelled systems G3E and G3M. a fresh in $\Box_L^e, \Box_L^m$ and y, z fresh in $\Box_R^e, \Box_R^m$.

Observe that the neighbourhood information is "hidden" in the nested approach. That is, creating of a nesting has a two-step interpretation: one of creating a neighbourhood and another of creating a world in it. These steps are separated in Negri's labelled systems, but the nesting information becomes superfluous after the creation of the related world. This indicates that the nesting approach is more efficient proof theoretically speaking when compared to labelled systems. Also, it is curious to note that this "two-step" interpretation makes the nested system *external*, in the sense that nestings cannot be interpreted inside the syntax of the logic. In fact, it makes use of the $\langle\,]$ modality [2].

8   Conclusion and Future Work

In this work we showed a semantical characterisation of intuitionistic, normal and non-normal modal systems, via a case-by-case translation between labelled nested to labelled sequent systems. In this way, we closed the cycle of syntax/semantic characterisation for a class of logical systems.

While some of the presented results are expected (or even not new as the semantical interpretation of nestings in intuitionistic logic), our approach is, as far as we know, the first done entirely using proof theoretical arguments. Indeed, the soundness and completeness results are left to the case of labelled systems, that carry within the syntax the semantic information explicitly. Using the results in [11], we were able to extend all the semantic discussion to the sequent case.

This work can be extended in a number of ways. First, it seems possible to propose nested and labelled systems for some paraconsistent systems and systems with negative modalities [7]. Our approach, both for sequentialising nested systems and for relating the so-called internal and external calculi could then be applied in such cases.

Another natural direction to follow is to complete this syntactical/semantical analysis for the classical cube [10]. This is specially interesting since MNC = K, that is, we should be able to smoothly collapse the neighbourhood approach into the relational one. We observe that nestings play an important role in these transformations, since it enables to modularly building proof systems.

Finally, it would be interesting to analyse to what extent the methodology of the present paper might be applied to shed light on the problem of finding intuitive semantics for substructural logics, like linear logic and its modal extensions [9].

References

1. Brünnler, K.: Deep sequent systems for modal logic. Arch. Math. Log. **48**, 551–577 (2009)
2. Chellas, B.F.: Modal Logic. Cambridge University Press, Cambridge (1980)
3. Dyckhoff, R., Negri, S.: Proof analysis in intermediate logics. Arch. Math. Log. **51**(1–2), 71–92 (2012)
4. Fitting, M.: Nested sequents for intuitionistic logics. Notre Dame J. Formal Log. **55**(1), 41–61 (2014)
5. Gentzen, G.: Investigations into logical deduction. In: The Collected Papers of Gerhard Gentzen, pp. 68–131 (1969)
6. Goré, R., Ramanayake, R.: Labelled tree sequents, tree hypersequents and nested (deep) sequents. In: AiML, vol. 9, pp. 279–299 (2012)
7. Lahav, O., Marcos, J., Zohar, Y.: Sequent systems for negative modalities. Logica Universalis **11**(3), 345–382 (2017). https://doi.org/10.1007/s11787-017-0175-2
8. Lellmann, B.: Linear nested sequents, 2-sequents and hypersequents. In: De Nivelle, H. (ed.) TABLEAUX 2015. LNCS (LNAI), vol. 9323, pp. 135–150. Springer, Cham (2015). https://doi.org/10.1007/978-3-319-24312-2_10
9. Lellmann, B., Olarte, C., Pimentel, E.: A uniform framework for substructural logics with modalities. In: LPAR-21, pp. 435–455 (2017)
10. Lellmann, B., Pimentel, E.: Modularisation of sequent calculi for normal and non-normal modalities. CoRR abs/1702.08193 (2017)
11. Lellmann, B., Pimentel, E., Ramanayake, R.: Sequentialising nested systems (2018, submitted). https://sites.google.com/site/elainepimentel/home/publications---elaine-pimentel
12. Maehara, S.: Eine darstellung der intuitionistischen logik in der klassischen. Nagoya Math. J. **7**, 45–64 (1954)
13. Mints, G.: Indexed systems of sequents and cut-elimination. J. Phil. Log. **26**(6), 671–696 (1997)
14. Negri, S.: Proof analysis in modal logic. J. Phil. Log. **34**, 507–544 (2005)
15. Negri, S.: Proof theory for non-normal modal logics: the neighbourhood formalism and basic results. IfCoLog J. Log. Appl. **4**(4), 1241–1286 (2017)

16. Poggiolesi, F.: The method of tree-hypersequents for modal propositional logic. In: Makinson, D., Malinowski, J., Wansing, H. (eds.) Towards Mathematical Philosophy. TL, vol. 28, pp. 31–51. Springer, Dordrecht (2009). https://doi.org/10.1007/978-1-4020-9084-4_3
17. Straßburger, L.: Cut elimination in nested sequents for intuitionistic modal logics. In: Pfenning, F. (ed.) FoSSaCS 2013. LNCS, vol. 7794, pp. 209–224. Springer, Heidelberg (2013). https://doi.org/10.1007/978-3-642-37075-5_14
18. Viganò, L.: Labelled Non-classical Logics. Kluwer, Alphen aan den Rijn (2000)

A Formalization of Brouwer's Argument
for Bar Induction

Ryota Akiyoshi[1,2]([⊠])

[1] WIAS, Waseda University, 1-6-1 Nishi Waseda, Shinjuku-ku, Tokyo, Japan
georg.logic@gmail.com
[2] Keio University, 2-15-45 Mita, Minato-ku, Tokyo, Japan

Abstract. Brouwer was a founder of intuitionism and he developed intuitionistic mathematics in 1920's. In particular, he proved the uniform continuity theorem using the fan theorem in 1927, which was derived from a stronger theorem called bar induction. For this principle Brouwer gave a justification which was an important source of BHK-interpretation, but it depends on an assumption which we call "the fundamental assumption" (FA). Since FA was neither explained or justified, many people have thought that Brouwer's argument is highly controversial. In this paper, we propose a way of formalizing Brouwer's argument using a method in infinitary proof theory. Also, based on our formalization, we give an explanation and justification of FA from the viewpoint of the practice of intuitionistic mathematics.

Keywords: Brouwer's bar induction · Intuitionism
Infinitary proof-theory

1 Introduction

Brouwer was a founder of intuitionism and he developed intuitionistic mathematics. In particular, he proved the uniform continuity theorem using the fan theorem in 1927, which was derived from a stronger theorem called bar induction. This argument is quite important because (1) it is an essential origin of BHK-interpretation for implication, and (2) it is very proof-theoretic in the sense that it involves transformation of proofs. However, since his argument depends on an assumption which we call "the fundamental assumption", it has been highly controversial.

This work is partially supported by JSPS KAKENHI 16K16690, 17H02265, and Core-to-Core Program (A. Advanced Research Networks). We would like to thank Yasuo Deguchi, Takashi Iida, Mitsuhiro Okada, and Yuta Takahashi for their helpful comments. Special thanks go to Sam Sanders for his encouragement and organizing an opportunity of presenting this paper at Munich Center for Mathematical Philosophy. Moreover, important feedbacks from people working on constructive mathematics and proof-theory in the department of mathematics of LMU München were very helpful for finishing this paper.

L. S. Moss et al. (Eds.): WoLLIC 2018, LNCS 10944, pp. 77–90, 2018.
https://doi.org/10.1007/978-3-662-57669-4_4

First, let us state Brouwer's bar induction BI in modern notations and explain his argument [10, 15, 20].

$BI :\equiv ((a) \land (b) \land (c) \land (d)) \rightarrow A(\langle\rangle)$ where

a. $\forall f \exists x B(f, x)$,
b. $\forall f \forall y (B(f, y) \rightarrow B(f, y + 1))$,
c. $\forall f \forall x (B(f, x) \rightarrow A(\overline{f}(x)))$,
d. $\forall n (\forall y (A(n * \langle y \rangle)) \rightarrow A(n))$.

Here, n ranges over finite sequences of natural numbers, f ranges over functions from natural numbers to the set of natural numbers, and $\overline{f}(m)$ in (c) denotes the m-tuple ordered pair $\langle f(0), \ldots, f(m-1) \rangle$. Therefore, if we consider a tree T whose root is $\langle\rangle$ and the k-th successor node of n is $n * \langle k \rangle$ for each $k \in \omega$, then f determines a path in T. For example, $f(m) = 0$ for each $m \in \omega$ means the leftmost path in T. The intuition behind BI is as follows. (a) states that the tree considered is well-founded (relative to B) in the sense that every path in T eventually hits with a node in B. (b) states that B is monotonic, (c) states that all nodes in B have a property (or predicate) A. (d) states that A is inductive; if all successor node of n has A, then n also has A. Therefore, BI is a kind of "backward" induction on T.

Brouwer's argument proceeds in the following way [6]: (i) we consider a proof d of (a). (ii) The fundamental assumption (FA) states that we may assume that d is canonical. (iii) Under other assumptions (b)–(d), we transform d into another proof of the conclusion $A(\langle\rangle)$. Finally, he concludes $((a) \land (b) \land (c) \land (d)) \rightarrow A(\langle\rangle)$.

The problematic part is the step (ii). Brouwer writes "Now, if the relations employed in any given proof can be decomposed into basic relations, its "canonical" form (...) employs only basic relations..." [22, p. 460]. Since this assumption was neither explained or justified, many people have thought that the argument is controversial. For example, Iemhoff writes "...the proof Brouwer gave is by many considered to be no proof at all since it uses an assumption on the form of proofs for which no rigorous argument is provided" [12].

In the present paper, we propose a formalization of Brouwer's argument of BI via a tool called the Ω-rule in proof-theory. The Ω-rule was introduced by Buchholz and used for ordinal analysis of the theories of inductive definition and subsystems of second-order arithmetic ([2–4, 7, 8]). The basic idea comes from BHK-interpretation of $\forall X A \rightarrow A(T)$ (which corresponds to the parameter-free Π_1^1-comprehension axiom) where $A(X)$ is arithmetical. The Ω-rule says the following: if there is a uniform transformation of *all* proofs of $\forall X A$ into another ones of $A(T)$, then we can infer $\forall X A \rightarrow A(T)$. We must be careful about the range of "all" since if it had contained the Ω-rule itself, then this definition would break down because of the circularity of the definition. This is solved by considering all *normalized arithmetical* proofs of $\forall X$ instead of all proofs of it.

Here, a reader might already notice the following analogy informally sketched in [1]. The formula $\forall X A$ is very close to $\forall f \exists x B(f, x)$ in BI since the notion of function defined on natural numbers is definable by $\forall X$. Hence, the formula $\forall X A \rightarrow A(T)$ is analogous to $\forall f \exists x B(f, x) \rightarrow A(\langle\rangle)$ in BI. The restriction to cut-free arithmetical proofs in the Ω-rule is quite similar to Brouwer's FA. Indeed,

Parsons already pointed out this similarity ([22, p. 451]) without mentioning the Ω-rule. Our main idea is to give an analysis of Brouwer's argument by formalizing it by a concrete mathematical tool (the Ω-rule) and analyzing the result.

Before explaining our basic results, we explain our philosophical standpoint about Brouwer. As is well known, he claims that mathematics is essentially independent of language and developed the theory of creative subject. This thesis makes almost all attempts to examine Brouwer's argument for BI using a formal tool very difficult. In this paper, we focus on only the objective side of intuitionistic mathematics though it would be meaningful to find a connection between our proposal and the theory of creative subject in the future work.

Now we explain our results and its importance. By the Ω-rule adopted into our setting, we prove BI in an infinitary natural deduction $ELBI^{\Omega}$. This theorem is called Embedding in proof-theory (Theorem 1) and it is proposed that the embedding theorem corresponds to Brouwer's argument of BI:

First Claim: Brouwer's argument corresponds to the embedding theorem in proof-theory.

We remark that our meta-theory is not classical, that is, it is based on intuitionistic logic and the exact strength of it is discussed in Sect. 5.

Secondly, we explain the role of FA in Brouwer's argument for BI. As suggested before, the restriction to normalized arithmetical proofs in the Ω-rule is quite close to FA. This restriction is necessary for defining the Ω-rule without committing to a vicious circle. Hence, the requirement of FA is reasonable from this point of view.

Second Claim: Brouwer's FA is needed for avoiding the vicious circle in the definition of his infinitary rule.

However, one might ask whether FA demands too much because it claims that all proofs of $\forall f \exists x B$ must be normalized. In proof-theory of a stronger theory beyond arithmetic, the normalization theorem holds only for a restricted class of formulas (Collapsing Theorem). To answer this question, we prove that if the conclusion of a given derivation is $\forall f \exists x B$ with Π_1^0-formula B, then it is normalized (Theorem 2). This is a partial result, but supports our interpretation of Brouwer's argument because we need only BI with such a formula B to prove his uniform continuity theorem (cf. [5]). Also, this restricted BI is enough to interpret the theory of inductive definition for Brouwer's ordinal called $ID_1(\mathcal{O})$ (cf. [16]). Therefore, we claim the following:

Third Claim: Brouwer's FA does not demand too much from the viewpoint of our normalization theorem and the practice of intuitionistic mathematics.

The structure of this paper is as follows. In Sect. 2, we introduce the basic finitary system called $ELBI$ and infinitary systems $ELBI^{\omega}$ and $ELBI^{\Omega}$. To support the second claim above, we prove BI in $ELBI^{\Omega}$ in Sect. 3. For our third claim, we prove the normalization theorem of derivations of $\forall f \exists x B$ with Π_1^0-formula B in Sect. 4. Finally, we give our conclusion and remarks.

2 Definitions of Formal Systems

We introduce the system $ELBI$, which was first introduced and investigated in [13]. The basic system EL is conservative over Heyting arithmetic (HA). EL has variables and quantifiers for functions on natural numbers. Informally, function symbols represent choice sequences. We introduce EL in a natural deduction style for our purpose although it is introduced in Hilbert-style in [13].

The language of EL is based on a two-sorted intuitionistic predicate logic. In particular, it has variables f, g and quantifiers $\forall f, \exists f$ for functions. The constant 0, function (relation) symbols for primitive recursive functions (relations) , and the predicate symbol $\perp$ for absurdity are also included. Function variables and symbols for primitive recursive functions are called *function terms* (f.t for short). A term of the form $S \ldots S(0)$ is called a *numeral*. The numerical terms (n.t. for short) are numerical variables, numerals, and $h(t)$ where h is a f.t. and t is a n.t.

An atomic formula is $R t_1 \ldots t_n$, where R is an n-ary primitive recursive predicate. Formulas are obtained from atomic formulas by $\wedge, \vee, \rightarrow, \exists x, \forall x, \exists f, \forall f$. If A is a formula, then its *negation* $\neg A$ is defined as $A \rightarrow \perp$. We write AX for the set of arithmetical axioms of EL except for the induction axiom. The notation $!y!$ (or $!g!$) means that y (or g) is an eigenvariable. In the inference rules presented below, the standard proviso for an eigenvariable is assumed.

Definition 1. *The axioms and inference rules of EL are*

$$(Ax_A)\ \frac{}{A}\ \text{with } A \in AX \qquad (\wedge I)\ \frac{A_0 \quad A_1}{A_0 \wedge A_1} \qquad (\wedge E)\ \frac{A_0 \wedge A_1}{A_k}\ k \in \{0,1\}$$

$$(\vee I)\ \frac{A_k}{A_0 \vee A_1}\ k \in \{0,1\} \qquad (\vee E, i)\ \frac{A_0 \vee A_1 \quad \overset{[A_0]^i}{\underset{C}{\vdots}} \quad \overset{[A_1]^i}{\underset{C}{\vdots}}}{C} \qquad (\rightarrow I, i)\ \frac{\overset{[A]^i}{\underset{B}{\vdots}}}{A \rightarrow B}$$

$$(\rightarrow E)\ \frac{A \quad A \rightarrow B}{B} \qquad (\forall I^0)\ \frac{A(y/x)}{\forall x A}\ !y! \qquad (\forall E^0)\ \frac{\forall x A}{A(t/x)}\ \text{where } t \text{ is a n.t.}$$

$$(\forall I^1)\ \frac{A(g/f)}{\forall f A}\ !g! \qquad (\forall E^1)\ \frac{\forall f A}{A(h/f)}\ \text{where } h \text{ is a f.t.} \qquad (\exists I^0)\ \frac{A(t/x)}{\exists x A}\ \text{where } t \text{ is a n.t.}$$

$$(\exists E^0, i)\ \frac{\exists x A \quad \overset{[A(y/x)]^i}{\underset{C}{\vdots}}}{C}\ !y! \qquad (\exists I^1)\ \frac{A(h/f)}{\exists f A}\ \text{where } h \text{ is a f.t.} \qquad (\exists E^1, i)\ \frac{\exists f A \quad \overset{[A(g/f)]^i}{\underset{C}{\vdots}}}{C}\ !g!$$

$$(\perp)\ \frac{\perp}{A} \qquad (\text{Ind}, i)\ \frac{A(0) \quad \overset{[A(y)]^i}{\underset{A(y+1)}{\vdots}}}{A(t/y)}\ !y!$$

Remark 1. The set AX of axioms includes the arithmetical axioms. We assume that all axioms in AX are formulated without any free number variable by taking their universal closures. For the sake of simplicity, we do not include the axiom of choice for quantifier-free formulas.

The standard coding of finite sequences of natural numbers can be done in *EL*. We adopt the following standard notations:

1. $seq(x)$: x is the code of a finite sequence of natural numbers such as $\langle a_0, \ldots, a_n \rangle$,
2. $lh(x) = m$: the length of x, which is the code of a sequence $\langle n_0, \ldots, n_{m-1} \rangle$,
3. $(x)_i$: the i-th member of x if $x = \langle n_0, \ldots, n_{m-1} \rangle$ and $i < m$,
4. $a * b = \langle a_0, \ldots, a_n, b_0, \ldots, b_m \rangle$: the concatenation of a and b if $a = \langle a_0, \ldots, a_n \rangle$ and $b = \langle b_0, \ldots, b_m \rangle$,
5. If f is a function term, then $\overline{f}(n) := \langle f(0), \ldots, f(n-1) \rangle$.
6. If $x = \langle n_0, \ldots, n_m \rangle$, then $[x](i) := (x)_i$ for $i < m$.

In the following, let $\forall n, \exists m$ range over finite sequences of natural numbers. This abbreviation is possible because $\forall n A$ (or $\exists m A$) reads as $\forall x(seq(x) \rightarrow A(x))$ (or $\exists x(seq(x) \land A(x))$). Likewise, we use introduction and elimination rules for quantifiers over finite sequences of natural numbers.

Definition 2. *ELBI is obtained by adding the following axiom to EL:*

$$(BI) \quad \overline{(\mathbf{a} \land \mathbf{b} \land \mathbf{c} \land \mathbf{d}) \rightarrow A(\langle \rangle)}$$

where:

a. $\forall f \exists x B(f, x)$,
b. $\forall f \forall y(B(f, y) \rightarrow B(f, y + 1))$,
c. $\forall f \forall y(B(f, y) \rightarrow A(\overline{f}(y)))$,
d. $\forall n(\forall y(A(n * \langle y \rangle) \rightarrow A(n)))$.

Next we introduce the infinitary systems $ELBI^\omega$ and $ELBI^\Omega$ with the B^Ω-rule. Informally, the system $ELBI^\omega$ is an infinitary system obtained from EL by replacing an induction axiom with the ω-rule and allowing the infinitely long definition of a function constant h. The B^Ω-rule is defined by quantifying over proofs in $ELBI^\omega$, and $ELBI^\Omega$ is obtained by adding the B^Ω-rule to $ELBI^\omega$.

A term without any free number and functional variables is called *closed*. When no confusion likely occurs, we identify a closed term with its value. Moreover, we introduce all function constants h defined by $h(0) := a_0, h(1) := a_1, \ldots$ where $a_n \in \omega$ for all $n < \omega$. The computational axioms for such h are also assumed via infinitary λ-terms which is originally due to [19]. In particular, a paring function and a projection function for an infinitary long term are included.

Definition 3. *The axioms and inference rules of $ELBI^\omega$ are:*

$$(Ax_A) \ \frac{}{A} \ \text{with } A \in AX \qquad (\land I) \ \frac{\overset{\vdots}{A_0} \quad \overset{\vdots}{A_1}}{A_0 \land A_1} \qquad (\land E) \ \frac{\overset{\vdots}{A_0 \land A_1}}{A_k} \ k \in \{0, 1\}$$

$$(\vee I)\ \frac{A_k}{A_0 \vee A_1}\ k \in \{0,1\} \qquad (\vee E, i)\ \frac{A_0 \vee A_1 \quad \overset{[A_0]^i}{\underset{\vdots}{C}} \quad \overset{[A_1]^i}{\underset{\vdots}{C}}}{C} \qquad (\to I, i)\ \frac{\overset{[A]^i}{\underset{\vdots}{B}}}{A \to B}$$

$$(\to E)\ \frac{A \quad A \to B}{B} \qquad (\forall I^0)\ \frac{\ldots A(n/x) \ldots}{\forall x A} \qquad (\forall E^0)\ \frac{\forall x A}{A(t/x)}$$

$$(\exists I^0)\ \frac{A(t/x)}{\exists x A} \qquad (\exists E^0, i)\ \frac{\exists x A \quad \overset{[A(n/x)]^i}{\underset{\vdots}{\ldots C \ldots}}}{C} \qquad (\forall I^1)\ \frac{A(g/f)}{\forall f A}\ !g!$$

$$(\forall E^1)\ \frac{\forall f A}{A(h/f)} \qquad (\exists I^1)\ \frac{A(h/f)}{\exists f A} \qquad (\exists E^1, i)\ \frac{\exists f A \quad \overset{[A(g/f)]^i}{\underset{\vdots}{C}}}{C}\ !g!$$

$$(\bot)\ \frac{\bot}{A}\ \text{where } A \text{ is atomic.}$$

In an elimination rule, the premise containing the occurrence of the logical symbol eliminated is called its *major premise*.

Next, we define $ELBI^\Omega$ with the B^Ω-rule. To define this rule, we need the notion of redex for deductions in $ELBI^\omega$.

Definition 4. *We define redex and the reduction relation between deductions called "contr" as:*

$$\frac{\frac{A_0 \quad A_1}{A_0 \wedge A_1}\wedge I}{A_i}\wedge E\ \ contr\ \ A_i \qquad \frac{\frac{A_i}{A_0 \vee A_1}\vee I \quad \overset{[A_0]^i}{\underset{\vdots}{C}} \quad \overset{[A_1]^i}{\underset{\vdots}{C}}}{C}\vee E, i\ \ contr\ \ \overset{A_i}{\underset{\vdots}{C}}$$

$$\frac{A \quad \frac{\overset{[A]^i}{\underset{\vdots}{B}}}{A \to B}\to I, i}{B}\to E\ \ contr\ \ \overset{A}{\underset{\vdots}{B}} \qquad \frac{\frac{\ldots A(n/x)\ldots}{\forall x A}\forall I^0}{A(t/x)}\forall E^0\ \ contr\ \ A(m/x)$$

where m is a natural number obtained by computing a function constant in t if there is.

$$\frac{A(t/x)}{\exists x A}\exists I^1 \qquad \frac{\overset{\vdots}{\exists x A} \quad \overset{[A(n/x)]^i}{\underset{\vdots}{C}}}{C}\exists E^0,i \qquad contr \qquad \frac{A(m)}{\overset{\vdots}{C}}$$

where m is a natural number obtained by computing a function constant in t if there is.

$$\frac{\dfrac{A(g/f)}{\forall f A}\forall I^1}{A(h/f)}\forall E^1 \quad contr \quad A(h/f) \qquad \frac{\dfrac{A(h/f)}{\exists f A}\exists I^1 \quad \overset{[A(g/f)]^i}{\underset{\vdots}{C}}}{C}\exists E^1,i \quad contr \quad \frac{A(h/g)}{\overset{\vdots}{C}}$$

In the following two rules, Q is an elimination rule and C is its major premise.

$$\frac{A_0 \vee A_1 \quad \overset{[A_0]^i}{\underset{\vdots}{C}} \quad \overset{[A_1]^i}{\underset{\vdots}{C}}}{\dfrac{C}{D}Q}\vee E,i \qquad contr \qquad \frac{A_0 \vee A_1 \quad \overset{[A_0]^i}{\underset{\vdots}{\dfrac{C}{D}Q}} \quad \overset{[A_1]^i}{\underset{\vdots}{\dfrac{C}{D}Q}}}{D}\vee E,i$$

$$\frac{\exists x A \quad \overset{[A(n)/x]^i}{\underset{\vdots}{C}}}{\dfrac{C}{D}Q}\exists E^0,i \qquad contr \qquad \frac{\exists x A \quad \overset{[A(n/x)]^i}{\underset{\vdots}{\dfrac{C}{D}Q}}}{D}\exists E^0,i$$

$$\frac{\exists f A \quad \overset{[A(g/f)]^i}{\underset{\vdots}{C}}}{\dfrac{C}{D}Q}\exists E^1,i \qquad contr \qquad \frac{\exists f A \quad \overset{[A(g/f)]^i}{\underset{\vdots}{\dfrac{C}{D}Q}}}{D}\exists E^1,i$$

A deduction in $ELBI^\omega$ without any redex is called *normal*. As usual, a deduction is closed if it does not contain any open assumption.

Definition 5. $ELBI^\Omega$ is obtained by adding the following rule to $ELBI^\omega$:

$$\frac{\{q:\forall f \exists x B(f,x)\}}{\overset{\vdots}{\underset{\vdots}{A}}}$$

$$(B^\Omega,i)\,\frac{\dots A \dots}{\forall f \exists x B(f,x)) \to A}$$

where:

1. $q \in |B^\Omega|$,
2. $|B^\Omega| := \{d | d$ is a closed normal deduction in $ELBI^\omega$ of $\forall f \exists x B(f,x)\}$.

3   Embedding of Brouwer's Bar Induction

In this section, we derive BI via the B^Ω-rule. We write $T \ni d \vdash A$ to mean that d is a deduction of A in a formal system T.

Lemma 1. *Let $ELBI^\omega \ni d \vdash \forall f \exists x B(f, x)$ be a closed normal deduction. Then, we may find a deduction d' of d and a term $t(f)$ so that $ELBI^\omega \ni d' \vdash B(f, t(f))$.*

Proof. By the standard argument of analyzing the structure of a given normal and closed deduction (Cf. 6.2.7D in [18, p. 189]). Note that $t(f)$ could contain an occurrence of f because of our formulation of $\forall I^1$ and the fact that a numerical term could contain an occurrence of a function variable.   □

Lemma 2. *Any closed instance of BI is derivable in $ELBI^\Omega$:*

$$ELBI^\Omega \vdash (\mathbf{a} \wedge \mathbf{b} \wedge \mathbf{c} \wedge \mathbf{d}) \to A(\langle\rangle).$$

where:

a. $\forall f \exists x B(f, x)$,
b. $\forall f \forall y (B(f, y) \to B(f, y+1))$,
c. $\forall f \forall y (B(f, y) \to A(\bar{f}(y)))$,
d. $\forall n (\forall y (A(n * \langle y \rangle) \to A(n)))$.

Proof. To prove the theorem, it suffices to show that there exists a deduction h in $ELBI^\Omega$ such that

$$\mathbf{b}, \mathbf{c}, \mathbf{d} \vdash \mathbf{a} \to A(\langle\rangle). \tag{1}$$

Then, the theorem is proved by propositional logic.

We use the B^Ω-rule to show (1). Suppose that a normal closed deduction d of $\forall f \exists x B(f, x)$ in $ELBI^\omega$ is given. We have to transform this derivation into another deduction $\mathfrak{g}(d)$ of $A(\langle\rangle)$. By Lemma 1, we have a closed normal deduction d' of $B(f, t(f))$. To prove the theorem, we show the proposition

$$\mathfrak{g}(d) \text{ is a deduction in } ELBI^\omega \text{ of } A(\langle\rangle) \text{ with } As(\mathfrak{g}(d)) \subseteq \{\mathbf{b}, \mathbf{c}, \mathbf{d}\}, \tag{2}$$

where $As(\mathfrak{g}(d))$ is the set of open assumption of $\mathfrak{g}(d)$.

We consider the tree $\mathcal{T}$ of finite sequences of natural numbers in the following way:

1. the root of $\mathcal{T}$ is $\langle\rangle$,
2. if a is a node of $\mathcal{T}$, then its n-th successor of $\mathcal{T}$ is $a * \langle n \rangle$.

Consider any infinite sequence of natural numbers, say $g = \langle g_0, g_1, \dots \rangle$, which is a function constant. Then g can be regarded as a "path" of $\mathcal{T}$.

We appeal BI of specific form for $\mathcal{T}$ to prove (2) in meta-level. (We discuss the status of this meta BI in Sect. 5) Let h be a function constant. Define

- $P(h, m) :\Leftrightarrow ELBI^\omega \vdash B(h, n)$ from $\mathbf{b}, \mathbf{c}, \mathbf{d}$ & $n \leq m$ for some n.
- $Q(\langle a_0, \dots, a_l \rangle) :\Leftrightarrow ELBI^\omega \vdash A(\langle a_0, \dots, a_l \rangle)$ from $\mathbf{b}, \mathbf{c}, \mathbf{d}$,

where $0 \le l$ and $\mathbf{b}, \mathbf{c}, \mathbf{d}$ are the assumptions of BI to be proved.

Let us state the meta-BI. Suppose that

  i. $\forall f \exists x P(f, x)$,
  ii. $\forall f \forall y (P(f, y) \to P(f, y + 1))$,
  iii. $\forall f \forall y (P(f, y) \to Q(\overline{f}(y)))$,
  iv. $\forall n (\forall y (Q(n * \langle y \rangle)) \to Q(n))$.

Then $Q(\langle \rangle)$, that is, $ELBI^\omega \vdash A(\langle \rangle)$ from $\mathbf{b}, \mathbf{c}, \mathbf{d}$.

  i. $\forall f \exists x P(f, x)$

First, there is a closed deduction d' of $B(f, t(f))$ in $ELBI^\omega$ as shown before. Then, if a function constant h is given, we get a closed normal derivation in $ELBI^\omega$ of $B(h, t(h))$ and compute $t(h)$ by the computation rule for h.

  ii. $\forall f \forall y (P(f, y) \to P(f, y + 1))$

Suppose that $P(h, m)$ for any given h, m, that is, $ELBI^\omega \vdash B(h, m)$ from $\mathbf{b}$, $\mathbf{c}, \mathbf{d}$ for some $n \le m$. We can get the desired conclusion because $n \le m < m + 1$.

  iii. $\forall f \forall y (P(f, y) \to Q(\overline{f}(y)))$

Assume that $P(g, m)$ for any given g, m. We have to prove $Q(\overline{g}(m))$, that is, $ELBI^\Omega \vdash A(\langle g(0), \dots, g(m-1) \rangle)$ from $\mathbf{b}, \mathbf{c}, \mathbf{d}$.

From the assumption, we have $ELBI^\omega \vdash B(g, n)$ from $\mathbf{b}$, $\mathbf{c}$, $\mathbf{d}$ for some $n \le m$. Using the assumption $\mathbf{b}$ (monotonicity) and the rules $\forall E^0, \to E$, we have $ELBI^\omega \vdash B(g, m)$ from $\mathbf{b}$, $\mathbf{c}$, $\mathbf{d}$. Therefore, applying the assumption $\mathbf{c}$, we get the desired conclusion $Q(\overline{g}(m))$. In a traditional notation of proof-figure, this transformation is as follows:

$$
\cfrac{
\cfrac{
\cfrac{B(g,n)}{\quad}\;
\cfrac{b : \forall f \forall y (B(f, y) \to B(f, y+1))}{B(g,n) \to B(g, n+1)}\ \forall E^0
}{B(g, n+1)}\ {\to}E
}{\cfrac{\vdots}{B(g,m)} \qquad \cfrac{c : \forall f \forall n (B(f, n) \to A(\overline{f}(n)))}{\cfrac{B(g,m) \to A(\overline{g}(m))}{}}\ \forall E^0, {\to}E}
$$
$$
\cfrac{\qquad\qquad\qquad}{A(\overline{g}(m))}\ {\to}E
$$

  iv. $\forall n (\forall y (Q(n * \langle y \rangle)) \to Q(n))$

Assume that $Q(n * \langle w \rangle)$ for any $w \in \omega$ and a sequence $\mathbf{n} = \langle n_0, \dots, n_l \rangle$, that is, $ELBI^\Omega \vdash A(\langle n_0, \dots, n_l \rangle * \langle w \rangle)$ from $\mathbf{b}, \mathbf{c}, \mathbf{d}$. Using the assumption $\mathbf{d}$ ("A is inductive"), we obtain a deduction of $A(n)$ from $\mathbf{b}$, $\mathbf{c}$, $\mathbf{d}$. This transformation is presented in the following way:

$$
\cfrac{
\cfrac{A(\mathbf{n} * \langle w_0 \rangle) \quad A(\mathbf{n} * \langle w_1 \rangle), \dots}{\forall w (A(\mathbf{n} * \langle w \rangle))}\ \forall I^0 \qquad d : \forall w (A(\mathbf{n} * \langle w \rangle)) \to A(\mathbf{n})
}{A(\mathbf{n})}\ {\to}E
$$

Since the assumptions (i)–(iv) in the meta BI are proved, we conclude $Q(\langle\rangle)$:

$$ELBI^{\Omega} \vdash A(\langle\rangle) \text{ from } \mathbf{b}, \mathbf{c}, \mathbf{d}.$$

Therefore (2) is proved, hence the theorem is proved.   □

In proof-figure, these transformations in the argument above are

$$
\cfrac{
\cfrac{
\cfrac{\{d : \forall f \exists x B(f,x)\}}{\dots A(\langle\rangle) \dots}\; \mathfrak{g}(d)
}{\forall f \exists x B(f,x) \to A(\langle\rangle)}\; B^{\Omega}
}{\mathbf{a} \wedge \mathbf{b} \wedge \mathbf{c} \wedge \mathbf{d} \to A(\langle\rangle)}
$$

If Γ is a set of formulas, then we define $\Gamma' := \{A' | A \in \Gamma\}$.

Theorem 1. *If $ELBI \ni d : \Gamma \vdash A$, then $ELBI^{\Omega} \ni d^{\infty} : \Gamma' \vdash A'$ for any closed instance A' of A.*

Proof. By induction on d. We treat only three cases. We want to define the derivation d^{∞} such that $ELBI^{\Omega} \ni d^{\infty} : \Gamma' \vdash A'$.

Let $last(d)$ be the last inference rule of d. If $last(d)$ is BI, then the claim follows from Lemma 2. If $last(d) = \forall I^0$, then there is the immediate sub-deduction d_0 of $A(x)$. Then, d^{∞} is obtained by IH and $\forall I^0$ in $ELBI^{\omega}$:

$$
\cfrac{
\begin{array}{ccc}
\vdots\; d_0^{\infty}(0/x) & \vdots\; d_0^{\infty}(1/x) & \\
A'(0/x) & A'(1/x) & \dots
\end{array}
}{\forall x A'}
$$

If $last(d) = \mathrm{Ind}$, then d is of the form

$$
\cfrac{
\begin{array}{cc}
 & [A(y)]^i \\
\vdots\; d_0 & \vdots\; d_1 \\
A(0) & A(y+1)
\end{array}
}{A(t/y)}
$$

We identify t' with some numeral, say n. The argument proceeds by induction on n. If $n = 0$, then d^{∞} is just $d_{\langle 0 \rangle}^{\infty}$. Otherwise, $n = m + 1$. By IH, we have the derivation of $A'(m)$. Then, we get the required derivation:

$$
\begin{array}{c}
\vdots \\
A'(m) \\
\vdots\; d_1^{\infty}(m/y) \\
A'(m+1)
\end{array}
$$

This completes the proof of this theorem.   □

4 Normalization Theorem

In this section, we prove the normalization theorem for derivations in $ELBI^\Omega$ with the following restriction; the B^Ω-rule is restricted into $\forall f \exists x \forall y B(f, x, y)$ where $B(f, x, y)$ is without any numerical and functional quantifier. Moreover, we assume that if $D_0 \to D_1$ occurs in $B(f, x, y)$ and $\to$ is the outermost implication symbol in it, then f does not occur in D_0 at all. This fragment is enough to prove the uniform continuity theorem [5] and transfinite induction on Brouwer's ordinals [16].

For the normalization theorem, we need to extend the notion of redex:

Definition 6. *Redex in Definition 4 is also redex in $ELBI^\Omega$. The conversion rules for them are defined in the standard way. Moreover, we add the following to redex and conversion rules in $ELBI^\Omega$.*

$$
\cfrac{\forall f\exists x B(f,x) \quad \cfrac{\{q:\forall f\exists x B(f,x)\} \\ \vdots \\ \dots C \dots}{\forall f\exists x B(f,x) \to C}\,B^\Omega}{\cfrac{C}{D}\,Q}\to E
\qquad contr \qquad
\cfrac{\forall f\exists x B(f,x) \quad \cfrac{\{q:\forall f\exists x B(f,x)\} \\ \vdots \\ \cfrac{C}{\dots D \dots}\,Q}{\forall f\exists x B(f,x)\to D}\,B^\Omega}{D}\to E
$$

where Q is an elimination rule and C is its major premise.

By the standard reduction method (cf. [18, pp. 178–182]) with suitable modifications, we can prove the following (Weak Predicative Normalization):

Lemma 3. *If $ELBI^\Omega \ni d \vdash \forall f \exists x \forall y B(f, x, y)$, then we can construct $ELBI^\Omega \ni d'$ without any redex in Definition 6.*

When D is provable without any redex in Definition 6, we write $ELBI^\Omega \vdash_0 D$. Note that such a derivation could contain the following "detour" (Fig. 1):

$$
\cfrac{\forall f\exists x B(f,x) \quad \cfrac{\{q:\forall f\exists x B(f,x)\} \\ \vdots \\ \dots D \dots}{\forall f\exists x B(f,x)\to D}\,B^\Omega}{D}\to E
$$

Theorem 2 (Collapsing Theorem). *Suppose that $ELBI^\Omega \ni d \vdash_0 D$ such that $As(d) = \emptyset$ and D is $\forall f \exists x \forall y B(f, x, y)$ or its subformula, then there is an operator $\mathcal{D}(\cdot)$ on deductions such that $ELBI^\omega \ni \mathcal{D}(d) \vdash_0 D$ with $As(\mathcal{D}(d)) = \emptyset$.*

Proof. By induction on d. The case in which the last inference rule is B^Ω is excluded by the assumption. Consider the crucial case of Fig. 1. Let d_0 be the left subdeduction of $\forall f \exists x B(f, x)$ and d_1 the right deduction of $\forall f \exists x B(f, x) \to D$. By IH, $ELBI^\omega \ni \mathcal{D}(d_0) \vdash_0 \forall f \exists x B(f, x)$ with $As(\mathcal{D}(d_0)) = \emptyset$.

There is the subdeduction of d_1 indexed by $\mathcal{D}(d_0)$. We write this as $d_{\mathcal{D}(d_0)}$. Note that the conclusion of $d_{\mathcal{D}(d_0)}$ is D. Observe that $As(d_{\mathcal{D}(d_0)}) \subseteq As(d_1) \cup$

$As(\mathcal{D}(d_0)) = \emptyset \cup \emptyset = \emptyset$. Therefore, applying IH to $d_{\mathcal{D}(d_0)}$, we get the required derivation $ELBI^\omega \ni \mathcal{D}(d_{\mathcal{D}(d_0)}) \vdash_0 D$ with $As(\mathcal{D}(d_{\mathcal{D}(d_0)})) = \emptyset$.

Let $last(d)$ be the last inference of d. If $last(d)$ is $\forall I^1, \exists E^0, \forall I^0$ or other cases except $\rightarrow I$, then we apply IH and the same inference symbol. If the last inference is $\rightarrow I$, then $D \equiv D_0 \rightarrow D_1$. By the assumption on the form of D, it is quantifier-free. Now, we consider cases according to whether D_0 is true or not. If D_0 is false, then D is provable. Otherwise, we replace D_0 as an open assumption by D_0 as axiom. Note that D_0 is true and quantifier-free. Then, D_1 is provable without any free assumption, hence we apply IH to this and $\rightarrow I$.   □

If we restrict the form of BI into $\forall f \exists x \forall y B(f, x, y)$ as we did for the B^Ω-rule in $ELBI^\Omega$, we prove the following by combining Theorems 1 and 2:

Corollary 1. *Suppose that $ELBI \ni d \vdash \forall f \exists x \forall y B(f, x, y)$ with $As(d) = \emptyset$. Then, we can construct d_0 such that $ELBI^\omega \ni d_0 \vdash \forall f \exists x \forall y B(f, x, y)$.*

5   Conclusion and Remarks

We proposed that the embedding theorem (Theorem 1) gives a formalization of Brouwer's argument of BI in [6], or at least gives a plausible prototype of it. As explained in Introduction, Theorem 2 gives an evidence of Brouwer's FA to be reasonable from the viewpoint of intuitionistic mathematics because we need only BI for such a formula to prove the uniform continuity theorem (cf. [5]).

Going back to our second claim, we explain the role of FA in our setting. The B^Ω-rule says that if there is a uniform transformation of *all normalized arithmetical* proofs of $\forall f \exists x B(f, x)$ into another ones of $A(\langle\rangle)$, then we can infer $\forall f \exists x B(f, x) \rightarrow A(\langle\rangle)$. If a given proof had contained the B^Ω-rule, then this definition would be circular; it is defined by *quantifying* over a set of proofs, so this assumes the existence of it in advance. Now the role of FA is quite clear. Brouwer supposed FA to avoid the vicious circle in the definition of the B^Ω-rule. Here, we may refer to Dummett's criterion for impredicativity; When he is defining a domain, he cannot use a quantifier over that domain ([9, p. 531]).

Let us describe two problems to be investigated before closing this paper. First, the exact strength of our meta-theory remains open. At first sight, P, Q used in the meta-BI in Lemma 2 refer to the provability predicate of $ELBI^\omega$. Hence, the complexity of these predicates should be Π^0_∞ because it is essentially ω-arithmetic. However, the language contains uncountably many infinitely long terms. The standard coding method for them does not work here. Moreover, we cannot restrict choice sequences to recursive ones in the presence of BI ([11]). Additionally, the use of meta-BI in Lemma 2 seems unavoidable.

Secondly, it would be interesting to formulate the B^Ω-rule in Sambin's Minimalist Foundation. The notion of choice sequence and a formulation of BI are given in their setting (cf. [14,17]). It does not use infinitely long terms to represent choice sequences, hence we could reduce the strength of the meta-BI.

Finally, according to van Atten's recent work [21], Brouwer's notion of implication is parametric. Since this point seems very similar to our approach via the Ω-rule, we will investigate the relationship between them in the future work.

References

1. Akiyoshi, R.: An interpretation of Brouwer's argument for bar theorem via infinitary proof theory. In: Abstract Volume of XXIII World Congress of Philosophy, p. 18 (2013)
2. Akiyoshi, R.: An ordinal-free proof of the complete cut-elimination theorem for $\Pi_1^1\text{-}CA + BI$ with the ω-rule. In: Gabbay, D., Prosorov, O., (eds.) The Mints' Memorial Issue of IfCoLog Journal of Logics and Their Applications, pp. 867–884 (2017)
3. Akiyoshi, R., Terui, K.: Strong normalization for the parameter-free polymorphic lambda calculus based on the omega-rule. In: First International Conference on Formal Structures for Computation and Deduction (FSCD), Leibniz International Proceedings in Informatics, vol. 52 (2016)
4. Akiyoshi, R., Mints, G.: An extension of the omega-rule. Arch. Math. Log. **55**(3), 593–603 (2016)
5. Berger, J.: The logical strength of the uniform continuity theorem. In: Beckmann, A., Berger, U., Löwe, B., Tucker, J.V. (eds.) CiE 2006. LNCS, vol. 3988, pp. 35–39. Springer, Heidelberg (2006). https://doi.org/10.1007/11780342_4
6. Brouwer, L.E.J.: Über definitionsbereiche von funktionen. Math. Ann. **97**, 60–75 (1927). English translation with introduction by Charles Parsons in [22]
7. Buchholz, W.: The $\Omega_{\mu+1}$-rule. In: Buchholz, W., Feferman, S., Pohlers, W., Sieg, W. (eds.) Iterated Inductive Definitions and Subsystems of Analysis: Recent Proof-Theoretical Studies. Lecture Notes in Mathematics, vol. 897, pp. 188–233. Springer, Heidelberg (1981). https://doi.org/10.1007/BFb0091894
8. Buchholz, W.: Explaining the Gentzen-Takeuti reduction steps. Arch. Math. Log. **40**, 255–272 (2001)
9. Dummett, M.A.E.: Frege: Philosophy of Language, 2nd edn. Duckworth, London (1993)
10. Dummett, M.A.E.: Elements of Intuitionism, 2nd edn. Oxford University Press, Oxford (2000). Volume 39 of Oxford Logic Guides
11. Kreisel, G., Howard, W.: Transfinite induction and bar induction of types zero and one, and the role of continuity in intuitionistic analysis. J. Symb. Log. **31**, 325–358 (1966)
12. Iemhoff, R.: Intuitionism in the philosophy of mathematics. In: Zalta, E.N. (ed.) The Stanford Encyclopedia of Philosophy. Summer 2013 Edition. https://plato.stanford.edu/entries/intuitionism/
13. Kreisel, G., Troelstra, A.S.: Formal systems for some branches of intuitionistic analysis. Ann. Math. Log. **1**, 229–387 (1970)
14. Maietti, M., Sambin, G.: Why topology in the minimalist foundation must be pointfree. Log. Log. Philos. **22**, 167–199 (2013)
15. Martino, E., Giaretta, P.: Brouwer, Dummett and the bar theorem. In: Atti del Congresso nazionale di logica, Montecatini Terme, 1–5 ottobre 1979, a cura di Sergio Bernini, Bibliopolis, Napoli, pp. 541–558 (1981)
16. de Lavalette, G.R.R.: Extended bar induction in applicative theories. Ann. Pure Appl. Log. **50**, 139–189 (1990)
17. Sambin, G.: The Basic Picture: Structures for Constructive Topology. Oxford University Press, to be published
18. Schwichtenberg, H., Troelstra, A.S.: Basic Proof Theory, 2nd edn. Cambridge University Press, Cambridge (2000)

19. Tait, W.W.: Infinitely long terms of transfinite type. In: Crossley, J.N., Dummet, M. (eds.) Formal Systems and Recursive Functions, pp. 176–185, North-Holland (1965)
20. van Atten, M.: The development of intuitionistic logic. In: Zalta, E.N. (ed.) The Stanford Encyclopedia of Philosophy. Autumn 2017 Edition. https://plato.stanford.edu/entries/intuitionistic-logic-development/
21. van Atten, M.: Predicativity and parametric polymorphism of Brouwerian implication. Preprint in Arxiv (2017). https://arxiv.org/abs/1710.07704
22. van Heijenoort, J. (ed.): From Frege to Gödel: A Source Book in Mathematical Logic, pp. 1879–1931. Harvard University Press, Cambridge (1967)

Deciding Open Definability
via Subisomorphisms

Carlos Areces[1,2]($\boxtimes$), Miguel Campercholi[1,2], and Pablo Ventura[1,2]

[1] FaMAF, Universidad Nacional de Córdoba, Córdoba, Argentina
carlos.areces@gmail.com
[2] CONICET, Buenos Aires, Argentina

Abstract. Given a logic $\mathcal{L}$, the $\mathcal{L}$-Definability Problem for finite structures takes as input a finite structure A and a target relation T over the domain of A, and determines whether there is a formula of $\mathcal{L}$ whose interpretation in A coincides with T. In this note we present an algorithm that solves the definability problem for quantifier-free first order formulas. Our algorithm takes advantage of a semantic characterization of open definability via subisomorphisms, which is sound and complete. We also provide an empirical evaluation of its performance.

1 Introduction

Providing a logic $\mathcal{L}$ with a formal semantics usually starts with the definition of a function that, given a suitable structure A for $\mathcal{L}$ and a formula φ in $\mathcal{L}$, returns the *extension* of φ in A. Often, this extension is a set of tuples built from elements in A. Extensions, also called *definable sets*, are the elements that will be referred by the formulas of $\mathcal{L}$ in a given structure, and in that sense, define the expressivity of $\mathcal{L}$. Definable sets are one of the central objects studied by Model Theory, and it is usually an interesting question to investigate, given a logic $\mathcal{L}$, which are the definable sets of $\mathcal{L}$ over a given structure A, or, more concretely, whether a particular set of tuples is a definable set of $\mathcal{L}$ over A. This is what we call the *Definability Problem for $\mathcal{L}$ over A*.

One of the goals of Computational Logic is to understand the computational complexity of different problems for different logics and define efficient algorithms to solve them. Among the most investigated inference problems is *Satisfiability*: given a formula φ from a given logic $\mathcal{L}$ decide whether there exists a structure that makes φ true. In recent years, and motivated by concrete applications, other reasoning problems have sparkled interest. A well known example is the *Model Checking Problem* used in software verification to check that a given property P (expressed as a formula in the verification language) holds in a given formal representation S of the system (see, e.g., [1,2]). From a more general perspective, model checking can be defined as follows: given a structure A, and a formula φ decide which is the extension T of φ in A. From that perspective, the definability problem can be understood as the *inverse* problem of model checking: given a structure A and a target set T it asks whether there is a formula φ whose

© Springer-Verlag GmbH Germany, part of Springer Nature 2018
L. S. Moss et al. (Eds.): WoLLIC 2018, LNCS 10944, pp. 91–105, 2018.
https://doi.org/10.1007/978-3-662-57669-4_5

extension is T. A further example of a reasoning task related to definability comes from computational linguistics. The *Generation of Referring Expressions* (GRE) problem can be defined as follows: given a context C and an target object t in C, generate a grammatically correct description (in some natural language) that represents t, differentiating it from other objects in C (see [3] for a survey). Most of the work in this area is focused on the content determination problem (i.e., finding the properties that single out the target object) and leaves the actual realization (i.e., expressing this content as a grammatically correct expression) to standard techniques. As it is discussed in [4,5] the content realization part of the GRE problem can be understood as the task that, given a structure A that represents the context C, and an object t in the domain of A returns a formula φ in a suitable logic $\mathcal{L}$ whose extension in A coincides with t. Of course, this will be possible only if t is definable for $\mathcal{L}$ over A.

The complexity of the definability problem for a number of logics has already been investigated. Let **FO** be first-order logic with equality in a vocabulary without constant symbols. The computational complexity of **FO**-definability was already discussed in 1978 [6,7], when a semantic characterization of the problem, based on automorphisms, placed **FO**-definability within coNP. Much more recently, in [8], a polynomial-time algorithm for **FO**-definability was given, which uses calls to a graph-isomorphism subroutine as an oracle. As a consequence, **FO**-definability is shown to be inside GI (defined as the set of all languages that are polynomial-time Turing reducible to the graph isomorphism problem). The authors also show that the problem is GI-hard and, hence, GI-complete. Interestingly, Willard showed in [9], that the complexity of the definability problem for the fragment of **FO** restricted to conjunctive queries (**CQ**) –i.e., formulas of the form $\exists \bar{x} \bigwedge_i C_i$, where each conjunct C_i is atomic– was coNEXPTIME-complete. The complexity upper bound followed from a semantic characterization of **CQ**-definability in terms of polymorphisms given in [10], while the lower bound is proved by an encoding of a suitable tiling problem. The complexity of definability has been investigated also for some modal languages: [4] shows that for the basic modal logic K, the definability problem is in P; in [5] the result is extended to some fragments of K where the use of negation is limited.

In [11] we prove that the definability problem for open formulas is coNP-complete. We also show that if the size and the arity of the target relation are taken as parameters the definability problem for open formulas is coW[1]-complete for every vocabulary with at least one, at least binary, relation. Given its complexity, algorithms to decide open decidability should be designed with care. In this article we introduce an algorithm that takes advantage of a semantic characterization of open definability via subisomorphisms presented in [12], which is sound and complete. We present an implementation of this algorithm and provide a preliminary empirical evaluation of its performance.

2   Preliminaries

In this section we provide some basic definitions and fix notation. We assume basic knowledge of first-order logic. For a detailed account see, e.g., [13].

We focus on definability by open first-order formulas in a purely relational first-order vocabulary, i.e., without function or constant symbols. For a relation symbol R in a vocabulary τ, let $ar(R)$ denote the arity of R. In what follows, all vocabularies are assumed to be finite and purely relational. We assume that the language contains variables from a countable, infinite set $\mathsf{VAR} = \{x_1, x_2, \ldots, x_n, \ldots\}$. Variables from VAR are the only *terms* in the language. *Atomic formulas* are either of the form $v_i = v_j$ or $R(\bar{v})$, where $v_i, v_j \in \mathsf{VAR}$, $\bar{v}$ is a sequence of variables in VAR of length k and R is a relation symbol of arity k. *Open formulas* are Boolean combinations of atomic formulas (or, equivalently, quantifier-free first order formulas). We shall often write just formula instead of open formula. We write $\varphi(v_1, \ldots, v_k)$ for an open formula φ whose variables are included in $\{v_1, \ldots, v_k\}$.

Let τ be a vocabulary. A τ-*structure* (or *model*) is a pair $\boldsymbol{A} = \langle A, \cdot^{\boldsymbol{A}} \rangle$ where A is a non-empty set (the *domain* or *universe*), and $\cdot^{\boldsymbol{A}}$ is an *interpretation function* that assigns to each k-ary relation symbol R in τ a subset $R^{\boldsymbol{A}}$ of A^k. If $\boldsymbol{A}$ is a structure we write A for its domain and $\cdot^{\boldsymbol{A}}$ for its interpretation function. Given a formula $\varphi(v_1, \ldots, v_k)$, and a sequence of elements $\bar{a} = \langle a_1, \ldots, a_k \rangle \in A^k$ we write $\boldsymbol{A} \models \varphi[\bar{a}]$ if φ is true in $\boldsymbol{A}$ under an assignment that maps v_i to a_i.

We say that a subset $T \subseteq A^k$ is *open-definable* in $\boldsymbol{A}$ if there is an open first-order formula $\varphi(x_1, \ldots, x_k)$ in the vocabulary of $\boldsymbol{A}$ such that $T = \{\bar{a} \in A^k : \boldsymbol{A} \models \varphi[\bar{a}]\}$.

In this article we study the following computational decision problem:

OpenDef

Instance: A finite relational structure $\boldsymbol{A}$ and a relation T over the domain of $\boldsymbol{A}$.

Question: Is T open-definable in $\boldsymbol{A}$?

We shall refer to the relations $R^{\boldsymbol{A}}$ for $R \in \boldsymbol{\mathcal{L}}$ as *base relations* and to T as the *target relation*. Let $f : \mathrm{dom}\, f \subseteq A \to A$ be a function. Given $S \subseteq A^m$, we say that f *preserves* S if for all $\langle s_1, \ldots, s_m \rangle \in S \cap (\mathrm{dom}\, f)^m$ we have $\langle f s_1, \ldots, f s_m \rangle \in S$. The function f is a *subisomorphism* (subiso for short) of $\boldsymbol{A}$ provided that f is injective, and both f and f^{-1} preserve $R^{\boldsymbol{A}}$ for each $R \in \tau$. (Note that a subiso of $\boldsymbol{A}$ is exactly an isomorphism between two substructures of $\boldsymbol{A}$.) We denote the set of all subisomorphisms of $\boldsymbol{A}$ by $\mathrm{subIso}\,\boldsymbol{A}$.

The following characterization of open-definability is key to our study.

Theorem 1 ([12, Theorem 3.1]). *Let $\boldsymbol{A}$ be a finite relational structure and $T \subseteq A^m$. The following are equivalent:*

1. *T is open-definable in $\boldsymbol{A}$.*
2. *T is preserved by all subisomorphisms γ of $\boldsymbol{A}$.*
3. *T is preserved by all subisomorphisms γ of $\boldsymbol{A}$ with $|\mathrm{dom}\,\gamma| \leq m$.*

Proof. The equivalence of (1) and (2) is proved in [12, Theorem 3.1]. Certainly (2) implies (3), so we show that (3) implies (2). Let γ be a subiso of $\boldsymbol{A}$ and let $\langle a_1, \ldots, a_m \rangle \in T \cap (\mathrm{dom}\,\gamma)^m$. Note that the restriction $\gamma|_{\{a_1,\ldots,a_m\}}$ is a subiso of $\boldsymbol{A}$. Thus, $\gamma(\bar{a}) = \gamma|_{\{a_1,\ldots,a_m\}}(\bar{a}) \in T$.

3   Computing OpenDef Efficiently

In view of Theorem 1 a possible strategy to decide OpenDef(A, T) is to check that every subiso of A preserves T. The following results show that in many cases it is not necessary to check every member of subIso A. Let $\mathbb{S}(A)$ denote the set of substructures of A.

Theorem 2. *Let A be a finite relational structure. Suppose $S \subseteq \mathbb{S}(A)$ and $\mathcal{F} \subseteq$ subIso A are such that:*

(i) *no two members of S are isomorphic,*
(ii) *aut $B \subseteq \mathcal{F}$ for all $B \in S$, and*
(iii) *for every $C \in \mathbb{S}(A) \setminus S$ there are $C' \in S$ and $\rho, \rho^{-1} \in \mathcal{F}$ such that $\rho : C \to C'$ is an isomorphism.*

Then every $\gamma \in$ subIso A is a composition of functions in $\mathcal{F}$.

Proof. Let $\gamma : C \to C'$ be an isomorphism with $C, C' \in \mathbb{S}(A)$. If both C and C' are in S, then $C = C'$ and $\gamma \in$ aut $C \subseteq \mathcal{F}$. Suppose next that $C \notin S$ and $C' \in S$. Then there are $\rho, \rho^{-1} \in \mathcal{F}$ such that $\rho : C \to C'$ is an isomorphism. Let $\alpha := \gamma\rho^{-1}$, and observe that $\alpha \in$ aut $C' \subseteq \mathcal{F}$. So $\gamma = \alpha\rho$ is a composition of functions in $\mathcal{F}$. To conclude assume that neither C nor C' are in S. There are $B \in S$ and $\rho, \rho^{-1}, \delta, \delta^{-1} \in \mathcal{F}$ such that $\rho : C \to B$ and $\delta : C' \to B$ are isomorphisms. Note that $\alpha := \delta\gamma\rho^{-1}$ is an automorphism of B, and thus $\alpha \in \mathcal{F}$. So $\gamma = \delta^{-1}\alpha\rho$ is a composition of functions in $\mathcal{F}$.

Let T be a set of tuples, we define the *spectrum* of T as

$$\operatorname{spec} T := \{|\{a_1, \ldots, a_n\}| : \langle a_1, \ldots, a_n \rangle \in T\}.$$

Combining Theorems 1 and 2 we obtain the following characterization of open-definability which is essential to our proposed algorithm.

Corollary 1. *Let A be a finite relational structure and $T \subseteq A^n$. Suppose S and $\mathcal{F}$ are as in Theorem 2, and let $\mathcal{F}' := \{\gamma \in \mathcal{F} : |\operatorname{dom}\gamma| \in \operatorname{spec}(T)\}$. Then T is open-definable in A if and only if T is preserved by all subisos in $\mathcal{F}'$.*

Proof. If T is open-definable in A then it must be preserved by every subiso of A; so the right-to-left direction is clear. Suppose T is preserved by all $\gamma \in \mathcal{F}'$. Fix $\delta \in$ subIso A and let $\langle a_1, \ldots, a_n \rangle \in T$ such that $\{a_1, \ldots, a_n\} \subseteq \operatorname{dom}\delta$. Note that $|\{a_1, \ldots, a_n\}| \in \operatorname{spec} T$. Set $\delta' := \delta|_{\{a_1, \ldots, a_n\}}$; by Theorem 2 we know that there are $\delta_1, \ldots, \delta_m \in \mathcal{F}$ such that $\delta' = \delta_1 \circ \cdots \circ \delta_m$. Since each δ_j is a bijection we have $|\operatorname{dom}\delta_j| = |\operatorname{dom}\delta| \in \operatorname{spec} T$, and hence $\delta_j \in \mathcal{F}'$ for $j \in \{1, \ldots, m\}$. So, as δ' is a composition of functions that preserve T, we have that δ' preserves T. Finally observe that $\langle \delta a_1, \ldots, \delta a_n \rangle = \langle \delta' a_1, \ldots, \delta' a_n \rangle \in T$.

Next, we show that it is possible to delete redundancies in the relations of the input structure without affecting definability.

Let $\bar{a} = \langle a_1, \ldots, a_n \rangle$ be a tuple. We define the *pattern* of $\bar{a}$, denoted by pattern $\bar{a}$, to be the partition of $\{1, \ldots, n\}$ such that i, j are in the same block if and only if $a_i = a_j$. For example, pattern $\langle a, a, b, c, b, c \rangle = \{\{1, 2\}, \{3, 5\}, \{4, 6\}\}$. We write $\lfloor \bar{a} \rfloor$ to denote the tuple obtained from $\bar{a}$ by deleting every entry equal to a prior entry. E.g., $\lfloor \langle a, a, b, c, b, c \rangle \rfloor = \langle a, b, c \rangle$. Given a set of tuples S and a pattern θ let $\lfloor S \rfloor := \{\lfloor \bar{s} \rfloor : \bar{s} \in S\}$ and $S_\theta := \{\bar{s} \in S : \text{pattern } \bar{s} = \theta\}$. For a structure $\mathbf{A}$ and an integer $k \geq 1$.

- let $\mathbf{A}_{\leq k}$ be the structure obtained from $\mathbf{A}$ by deleting every relation of arity greater than k, and
- let $\lfloor \mathbf{A} \rfloor$ be the structure obtained from $\mathbf{A}$ by replacing each relation R of $\mathbf{A}$ by the relations $\{\lfloor R_{\text{pattern } \bar{a}} \rfloor : \bar{a} \in R\}$.

Lemma 1. *Let $\mathbf{A}$ be a structure $T \subseteq A^m$, and take $k := \max(\text{spec}(T))$. Then, T is open-definable in $\mathbf{A}$ if and only if T is open-definable in $\lfloor \mathbf{A} \rfloor_{\leq k}$.*

Proof. First observe that if $\gamma : D \subseteq A \to A$ is injective, then for each relation R on A we have that: γ preserves R if and only if γ preserves $\lfloor R_{\text{pattern } \bar{a}} \rfloor$ for every $\bar{a} \in R$. It follows that subIso $\mathbf{A} = $ subIso $\lfloor \mathbf{A} \rfloor$. Hence, by Theorem 1, we have that T is open-definable in $\mathbf{A}$ iff T is open-definable in $\lfloor \mathbf{A} \rfloor$. Next, notice that Corollary 1 implies that T is open-definable in $\lfloor \mathbf{A} \rfloor$ iff T is preserved by every subisomorphism with domain of size at most $k = \max(\text{spec}(T))$. Now observe that if S is a relation of $\lfloor \mathbf{A} \rfloor$ with arity greater than k and $D \subseteq A$ has size at most k, then the restriction of S to D is empty (because tuples in S have no repetitions). In consequence, the relations of $\lfloor \mathbf{A} \rfloor$ with arity at most k determine the subisos of $\lfloor \mathbf{A} \rfloor$ that have to preserve T for T to be open-definable.

4   Implementation

In this section we present an algorithm for deciding definability, a pseudo-code of the implementation is shown in Algorithm 1.

Algorithm 1 decides whether T is open-definable in model $\mathbf{A}$. We assume implemented the following functions:

modelThinning ($\mathbf{A}$, k): Where $\mathbf{A}$ is a model and $k > 0$, returns $\lfloor \mathbf{A} \rfloor_{\leq k}$ as in Lemma 1.

computeSpectrum (T): Where T is a set of tuples, returns a reverse sorted list containing spec T.

cardinalityCheck ($\mathbf{A}$, S): Where $\mathbf{A}$ and S are models, returns True if for each $R \in \mathcal{L}$, $|R^{\mathbf{A}}| = |R^S|$ (see Lemma 2 below).

submodels ($\mathbf{A}$, n): Where $\mathbf{A}$ is a model and n is a natural number, generates all submodels of $\mathbf{A}$ with cardinality n.

To better describe the way our algorithm works let us define a tree $\text{Tr}^{\mathbf{A}}$. The root of $\text{Tr}^{\mathbf{A}}$ is $\mathbf{A}$. Given a node $\mathbf{B}$ of $\text{Tr}^{\mathbf{A}}$, its children are the submodels of $\mathbf{B}$ with l elements, where l is the greatest number in spec T strictly less than $|\mathbf{B}|$. So, if the numbers in $spec(T)$ are $l_1 > l_2 > \cdots > l_r$, the level j of $\text{Tr}^{\mathbf{A}}$ (for

Algorithm 1. Open Definability Algorithm

1: **function** ISOPENDEF(A, T)
2:     $spectrum$ = computeSpectrum(T)
3:     A = modelThinning(A, max($spectrum$))
        {* $spectrum$ is a reverse sorted list of natural numbers *}
4:     **global** $S = \emptyset$
5:     **if** $spectrum = [\,]$ **then**
6:         **return** True
7:     **for** $B \in$ submodels(A, **head**($spectrum$)) **do**
8:         **if** not isOpenDefR(B, T, **tail**($spectrum$)) **then**
9:             **return** False
10:    **return** True

11: **function** ISOPENDEFR(A, T, $spectrum$)
12:    **for** $S \in S$ **do**
13:        **if** cardinalityCheck(A,S) **then**
14:            **if** there is an iso $\gamma : A \to S$ **then**
15:                **if** γ and γ^{-1} preserve T **then**
16:                    **return** True
17:                **else**
18:                    **return** False
19:    **for** $\gamma \in$ aut(A) **do**
20:        **if** γ does not preserve T **then**
21:            **return** False
22:    $S = S \cup \{A\}$
23:    **if** $spectrum = [\,]$ **then**
24:        **return** True
25:    **for** $B \in$ submodels(A, **head**($spectrum$)) **do**
26:        **if** not isOpenDefR(B, **tail**($spectrum$)) **then**
27:            **return** False
28:    **return** True

$j \geq 1$) contains all submodels of A with l_j elements. The level 0 contains only A. As Algorithm 1 runs it traverses Tr^A depth-first. It starts from level 1, unless $|A| \in \mathrm{spec}\,T$ in which case it starts from level 0. For every node B it performs the following tasks:

– Check if there is an isomorphism $\gamma : B \to S$ for some $S \in S$;
  1. if there is no such isomorphism, inspect the automorphisms of B. If any member of aut B fails to preserve T, return False. Otherwise, add B to S and proceed to process the children of B.
  2. if there is such a γ, check if γ and γ^{-1} preserve T. If either preservation fails return False. Otherwise, move on to an unprocessed sibling of B.

When the algorithm runs out of nodes to process it returns True. An example of an execution of the algorithm is shown in the Appendix.

In our implementation of Algorithm 1 isomorphisms are checked as constraint satisfaction problems which are solved using MINION (version 1.8) [14]. When checking for the existence of an isomorphism between structures A and B we first verify that their domains have the same size and that $|R^A| = |R^B|$ for each R in the vocabulary. After this, it is enough to verify if there is an injective homomorphism between the structures. This is done on Line 14.

Lemma 2. *Let A and B be relational finite structures such that $|R^A| = |R^B|$. Then, any bijective R-homomorphism from A to B is an R-isomorphism.*

Proof. Suppose γ^{-1} is not a homomorphism. Then there is $\bar{b}$ such that $\bar{b} \in R^B$ but $\gamma^{-1}(\bar{b}) \notin R^A$, thus $|R^A| \neq |R^B|$.

Knowing that the constraint satisfaction problem is hard, it pays to try to limit the number of calls to MINION. In particular the cardinality tests described above are cheap ways of discovering non isomorphisms. Note that it is more likely for the tests to fail in large vocabularies. In this regard, observe that trading A for $\lfloor A \rfloor$ (Lemma 1) increases the size of the underlying vocabulary without increasing the size of the domain.

5   Empirical Tests

In this section we present a preliminary empirical evaluation of our implementation of Algorithm 1, written in Python 3 under GPL 2 license, source code available at https://github.com/pablogventura/OpenDef. All tests were performed using an Intel Xeon E5-2620v3 processor with 12 cores (however, our algorithm does not make use of parallelization) at 2.40 GHz, 128 GiB DDR4 RAM 2133 MHz. Memory was never an issue in the tests we ran. As there is no standard test set for definability problems, we tested on random instances. The main purpose of these experiments is to convey an idea of the times our tool requires to compute different kinds of instances, and how the parameters of the instances impact these running times. In view of the strategies our algorithm implements, we decided to consider the following parameters:

1. Size of the universe of the input model.
2. Number of relations in the input model.
3. Size of the relations in the input model (see the paragraph below).
4. Arity of the target relation.

Parameters (2) and (3) aims at testing our tool on models with different kinds of underlying "structure". When considering how the number of tuples in base relations affects the structure, it is easy to see that this depends on the size of the universe (adding an edge to a graph with 3 nodes has a much greater impact than adding and edge to a graph with 50 nodes). Thus, our measure for a base relation R of A is its *density*, defined as follows:

$$\text{density}(R) = \frac{|R|}{|A|^{\text{ar}(R)}}.$$

Since exchanging a base relation for its complement does not affect open-definability, it makes sense to consider only base relations with density at most 0.5.

For each test we compute median wall-time (in seconds), over 300 problem instances in each configuration.

5.1   On the Number of Isomorphism Types of Submodels

A key feature of our algorithm is that it takes advantage of models with a small number of isomorphism types among its substructures (because in that case the set S in Algorithm 1 remains small throughout the computation). One can think of a model with few isomorphism types among its substructures as having a *regular* or *symmetrical* structure. On the other hand, if there is a high diversity of isomorphism types of submodels the model has less regularities our strategy can exploit. Define the k-*diversity* of a structure A as the number of isomorphism types among its substructures of size k. At this point it is important to note that the k-diversity that is prone to have the greatest impact on the running times is for $k = \max(\operatorname{spec} T)$. This is due to the fact that submodels of size greater than k are not processed and the number of submodels of size less than k is substantially smaller (assuming $k \leq |A|/2$, and these are much easier to test for isomorphisms. It is worth pointing out here that when thinking of the k-diversity as great or small, this should be done by comparing it to the total number of subsets of size k of A.

To be able to evaluate the impact of the diversity in the running times we need to generate instances of varying diversities; this is accomplished by noting two things. First, models with sparse (low density) relations have low diversity since most submodels do not contain any edges at all. Second, a small number of base relations of small arity entails a low number of possible isomorphism types (e.g., there are only 10 directed graphs with two elements). Thus, by varying the density and number of base relations in our test configurations we are able to obtain a range of diversities. Since diversity grows rapidly as a function of the density of the base relations we use a logarithmic increments for densities.

In all but one of our test configurations (Fig. 2) we use only binary base relations. This is due to a couple of reasons:

- All target relations have arity at most three and thus the function **modelThinning** will erase base relations of arity greater than 3.
- The point of varying base relations is to generate a range of diversities (as explained above) and this can be accomplished adding either ternary or binary base relations. However, using binary relations we have finer control over the increments in diversity.

5.2   Tests with Definable Target

When the target relation is definable, our algorithm has to find and check the greatest possible number of subisos of A. Hence, to get a sense of the worst case times we ran most of the tests for the definable case. The instances are generated by first fixing the universe, number and density of base relations, and then each base relation is built by randomly picking its tuples. In all instances the target relation contains all tuples of the given length, thus being obviously definable.

In the tests shown in the first line of Fig. 1 the target is binary. The size of the domains are 30, 40 and 50, and the number of base relations are 1, 5 and 10. All base relations have arity 2, and their densities go through $\frac{0.5}{2^4}, \frac{0.5}{2^3}, \frac{0.5}{2^2}, \frac{0.5}{2^1}, \frac{0.5}{2^0}$. For the tests displayed in the second line the arity of the target is 3.

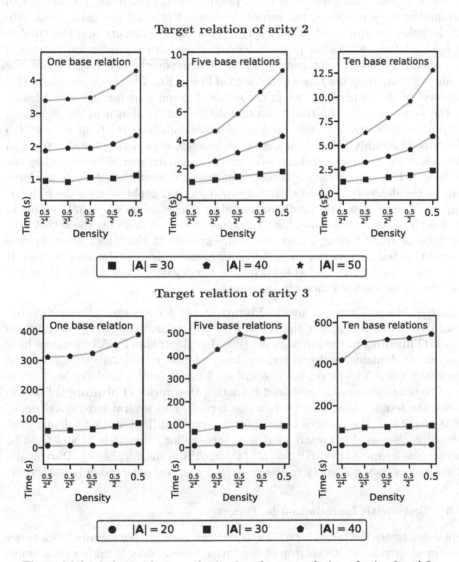

Fig. 1. Multiple base relations of arity 2 and target relation of arity 2 and 3.

The first thing that we notice is the considerable difference in times between the instances with binary and ternary targets. This is easily understood when considering that every one of the $\binom{|A|}{k}$ submodels of size $k = \max \operatorname{spec}(T)$ has to be processed, and in these examples k coincides with the arity of the target. Regarding the impact of universe sizes, we can clearly observe a polynomial behavior, quadratic for the binary target and cubic for the ternary target. Again, this is in direct correspondence to the number of submodels processed in each case. Finally, let us analyze the effect of increasing the number and density of the base relations. Note that these parameters do not affect the number of submodels to be processed but rather the amount of work processing each submodel takes (as discussed in Sect. 5.1). Increasing the density and the number of base relations makes the problem harder, since both represent an increase in diversity. However, it is interesting to note that the overhead is rather small. For example, comparing the times in the second line of Fig. 1 between instances with universe of 40 elements, we see in the leftmost graph that for one base relation of the lowest density the time is around 300 s, and, as shown in the rightmost graph, the time for the same universe size and ten base relations of arity 0.5 the time is roughly 550 s. A modest time increase considering the difference in the *size* of the compared instances (here size means number of bits encoding the instance). This behavior is explained by noting that, even though there is a large gap in the diversity of the two instances, most isomorphism tests done in the higher diversity case are rapidly answered by the function **cardinalityCheck**. Although many more isomorphism tests have to be carried out, the actual calls to the subroutine testing isomorphisms are avoided in the higher diversity case (recall that testing relational structures for isomorphism is GI-complete [15]). It should be noted that this behavior may not carry over to situations where the base relations are not randomly generated.

Effect of Model Thinning. **modelThinning** can have a great impact in running times. Figure 2 charts the times with and without applying the function **modelThinning** to the input model (line 3 in Algorithm 1). All instances have 20-element domains, and one ternary base relation whose density varies as in the above tests. The target is the set of all 3-tuples, and thus definable.

The time difference is explained by noting that **modelThinning** is likely to break the ternary base relation into one ternary and several binary and unary relations. Thus, the chances of the function **cardinalityCheck** to detect non-isomorphic submodels greatly increases, because for submodels B and C to be isomorphic we must have $|R^B| = |R^C|$ for *each* R in the vocabulary. Once again, this analysis may not apply to base relations that are not randomly generated.

5.3   Tests with Non-definable Target

The cases where the target relation T is *not* definable are potentially much easier for our algorithm, as it can stop computation the moment it finds a subisomorphism that does not preserve T. On the left of Fig. 3 we can see the times of the non-definable case, where the input structures have the same parameters as in the rightmost graph in the second line of Fig. 1 (which we repeat on the

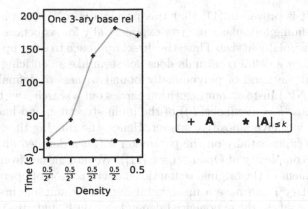

Fig. 2. $|A| = 20$, one base ternary relation and $T = A^3$.

right of Fig. 3). To generate a non-definable targets for each instance, we randomly generated ternary relations of density 0.5 until a non-definable one was encountered.

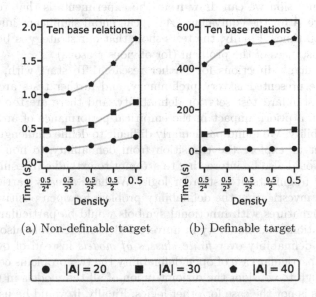

(a) Non-definable target   (b) Definable target

| ● |A| = 20 | ■ |A| = 30 | ♣ |A| = 40 |

Fig. 3. Comparing definable and non-definable cases.

6   Conclusions and Future Work

We designed and implemented an algorithm that decides the definability problem for open first-order formulas with equality over a purely relational vocabulary. The algorithm is based on the semantic characterization of open definability given in [12]. It is proved in [11] that this problem is coNP-complete and that the shortest defining formulas can grow exponentially for sequences of instances that grow polynomially in size. Thus, the direct approach to solve open definability by a search for a defining formula does not seem like an enticing alternative. (Also note that any kind of polynomially bound witness of definability would imply coNP $\subseteq$ NP.) Instead, our algorithm carries out a search over the submodels of size at most $k = \max(\mathrm{spec}(T))$ of the input structure, and has to exhaust the search space to give a positive answer. Hence, the running time of our algorithm depends exponentially on the parameter k (for a detailed analysis of the parameterized complexity of OpenDef see [11]). We also gather from these facts that the refinement of the original semantic characterization of open definability [12] into Corollary 1 can have a meaningful impact for suitable instances. The empirical tests confirm the exponential dependency on k, and give us an insight into the impact of the other parameters considered. Notably, we learned from the tests that, for randomly generated input structures, varying the number and density of the base relations does not have a great impact in running times. As noted in the discussion in Sect. 5.2, this is due to the fact that, as diversity grows and more isomorphism tests have to be calculated, the number of these tests that are easily decided by comparing cardinalities also grows. Another interesting conclusion we can draw from the experiments is that replacing an input structure $\boldsymbol{A}$ for the equivalent $\lfloor \boldsymbol{A} \rfloor_{\leq k}$ can entail significant improvements in computation times. Finally, the tests show that our strategy is better suited for negative instances of the problem (for obvious reasons).

There are many directions for further research. To start with, the empirical testing we presented is very preliminary, and further tests are necessary. There are no standard test sets for definability, and there are too many variables that can, a priori, impact in the empirical performance of any algorithm testing definability. We found particularly difficult to define challenging test sets that would let us explore the transition from definability to non-definability smoothly. It would also be interesting to attempt to extend our results (and our tool) to other fragments of first-order logic. We have already carried out preliminary work investigating the definability problem for open positive formulas. Considering signatures with functional symbols would be particularly interesting for applications of definability in universal algebra. It would also be natural to investigate definability over *finite classes of models* instead of over a single model. In the particular case of open definability the two problems coincide as it would be enough to consider the disjoint union of all the models in the class as input, but this is not the case for other logics. Finally, it would be useful to find classes of modes in which the problem is well conditioned, and where polynomial algorithms exist.

Acknowledgments. This work was partially supported by grant ANPCyT-PICT-2013-2011, STIC-AmSud "Foundations of Graph Structured Data (FoG)", and the Laboratoire International Associé "INFINIS". We thank also the reviewers for useful comments and minor corrections.

A   An Execution of OpenDef

To illustrate how Algorithm 1 works, we run it for the following instance. The input structure is the graph $G = (\{0,1,2,3\}, E)$

and the target relation is

$$T = \{(a,b,c,d) : E(b,c) \wedge (a = b \vee c = d)\}.$$

A brute force algorithm based on Theorem 1 would have to check that every sub-isomorphisms of G preserves T. That is, for each subuniverse S of G compute all sub-isomorphisms with domain S, and check them for preservation. This means going trough every subuniverse listed in the tree below, where the automorphisms of all subuniverses will be calculated and checked for preservation.

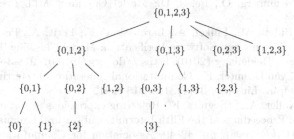

The first improvement of Algorithm 1 allows us to prune tree levels whose nodes do not have cardinality in spec(T) (this is correct due to Lemma 1). In the current example, it only processes subuniverses with size in spec(T) = $\{2,3\}$. This amounts to process only the following subuniverses. (The node *Root* is not a subuniverse to be processed; we add it to better visualize the tree traversed by the DFS.)

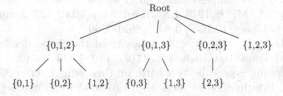

The second improvement is due to the order in which Algorithm 1 computes the subisomorphisms with a given domain S. It first tries to find an isomorphism from S to an already processed subuniverse. In the case where such an isomorphism exists, no other isomorphisms from S are computed, and the subuniverses contained in S are pruned (this is correct by Corollary 1). The subuniverses of G actually processed by Algorithm 1 are shown below. The bold nodes are in S, and therefore are the only ones that are checked for automorphisms and subuniverses.

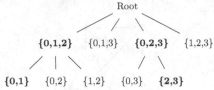

In view of the strategies Algorithm 1 employs we get a good idea of what makes for a well conditioned instances A, T:

- Target relations with a small spectrum compared to the power set of A.
- Structures with few isomorphisms types for substructures with cardinality in the spectrum.

References

1. Clarke, E., Grumberg, O., Peled, D.: Model Checking. MIT Press, Cambridge (1999)
2. Berard, B., Bidoit, M., Finkel, A., Laroussinie, F., Petit, A., Petrucci, L., Schnoebelen, P.: Systems and Software Verification: Model-Checking Techniques and Tools. Springer, Heidelberg (2010). https://doi.org/10.1007/978-3-662-04558-9
3. Krahmer, E., van Deemter, K.: Computational generation of referring expressions: a survey. Comput. Linguist. **38**(1), 173–218 (2012)
4. Areces, C., Koller, A., Striegnitz, K.: Referring expressions as formulas of description logic. In: Proceedings of the Fifth International Natural Language Generation Conference (INLG 2008), pp. 42–49. Association for Computational Linguistics (2008)
5. Areces, C., Figueira, S., Gorín, D.: Using logic in the generation of referring expressions. In: Pogodalla, S., Prost, J.-P. (eds.) LACL 2011. LNCS (LNAI), vol. 6736, pp. 17–32. Springer, Heidelberg (2011). https://doi.org/10.1007/978-3-642-22221-4_2
6. Paredaens, J.: On the expressive power of the relational algebra. Inf. Process. Lett. **7**(2), 107–111 (1978)
7. Bancilhon, F.: On the completeness of query languages for relational data bases. In: Winkowski, J. (ed.) MFCS 1978. LNCS, vol. 64, pp. 112–123. Springer, Heidelberg (1978). https://doi.org/10.1007/3-540-08921-7_60
8. Arenas, M., Diaz, G.: The exact complexity of the first-order logic definability problem. ACM Trans. Database Syst. **41**(2), 13:1–13:14 (2016)
9. Willard, R.: Testing expressibility is hard. In: Proceedings of the 16th International Conference on Principles and Practice of Constraint Programming (CP 2010), pp. 9–23 (2010)
10. Jeavons, P., Cohen, D., Gyssens, M.: How to determine the expressive power of constraints. Constraints **4**(2), 113–131 (1999)
11. Areces, C., Campercholi, M., Penazzi, D., Ventura, P.: The complexity of definability by open first-order formulas. Logic J. IGPL (2017, submitted)

12. Campercholi, M., Vaggione, D.: Semantical conditions for the definability of functions and relations. Algebra Univers. **76**(1), 71–98 (2016)
13. Ebbinghaus, H., Flum, J., Thomas, W.: Mathematical Logic. Springer, Heidelberg (1994). https://doi.org/10.1007/978-1-4757-2355-7
14. Gent, I., Jefferson, C., Miguel, I.: MINION: a fast, scalable, constraint solver. In: Proceedings of the 17th European Conference on Artificial Intelligence (ECAI 2006), pp. 98–102. IOS Press (2006)
15. Zemlyachenko, V.N., Korneenko, N.M., Tyshkevich, R.I.: Graph isomorphism problem. J. Sov. Math. **29**(4), 1426–1481 (1985)

APAL with Memory Is Better

Alexandru Baltag, Aybüke Özgün[(⊠)], and Ana Lucia Vargas Sandoval

ILLC, University of Amsterdam, Amsterdam, The Netherlands
ozgunaybuke@gmail.com, ana.varsa@gmail.com

Abstract. We introduce Arbitrary Public Announcement Logic with Memory (APALM), obtained by adding to the models a 'memory' of the initial states, representing the information before any communication took place ("the prior"), and adding to the syntax operators that can access this memory. We show that APALM is recursively axiomatizable (in contrast to the original Arbitrary Public Announcement Logic, for which the corresponding question is still open). We present a complete recursive axiomatization, that uses a natural finitary rule, we study this logic's expressivity and the appropriate notion of bisimulation.

1 Introduction

Arbitrary Public Announcement Logic (APAL) and its relatives are natural extensions of Public Announcements Logic (PAL), involving the addition of operators $\Box\varphi$ and $\Diamond\varphi$, quantifying over public announcements $[\theta]\varphi$ of some given type. These logics are of great interest both philosophically and from the point of view of applications. Motivations range from supporting an analysis of Fitch's paradox [16] by modeling notions of 'knowability' (expressible as $\Diamond K\varphi$), to determining the existence of communication protocols that achieve certain goals (cf. the famous Russian Card problem, given at a mathematical Olympiad [17]), and more generally to epistemic planning [9], and to inductive learnability in empirical science [5]. Many such extensions have been investigated, starting with the original APAL [2], and continuing with its variants GAL (Group Announcement Logic) [1], Future Event Logic [20], FAPAL (Fully Arbitrary Public Announcement Logic) [24], APAL$^+$ (Positive Arbitrary Announcement Logic) [19], BAPAL (Boolean Arbitrary Public Announcement Logic) [18], etc.

One problem with the above formalisms, with the exception of BAPAL[1], is that they all use *infinitary* axiomatizations. It is therefore not guaranteed that the validities of these logics are recursively enumerable.[2] The seminal paper

---

[1] BAPAL is a very weak version, allowing $\Box\varphi$ to quantify over only purely propositional announcements - no epistemic formulas.

[2] APAL$^+$ is known to be decidable, hence its validities must be r.e., but no recursive axiomatization is known. Also, note that APAL$^+$ is still very weak, in that it quantifies only over positive epistemic announcements, thus not allowing public announcements of ignorance, which are precisely the ones driving the solution process in puzzles such as the Muddy Children.

© Springer-Verlag GmbH Germany, part of Springer Nature 2018
L. S. Moss et al. (Eds.): WoLLIC 2018, LNCS 10944, pp. 106–129, 2018.
https://doi.org/10.1007/978-3-662-57669-4_6

on APAL [2] proved completeness using an infinitary rule, and then went on to claim that in theorem-proving[3] this rule can be replaced by the following finitary rule: from $\chi \rightarrow [\theta][p]\varphi$, infer $\chi \rightarrow [\theta]\Box\varphi$, as long as the propositional variable p is "fresh". This is a natural $\Box$-introduction rule, similar to the introduction rule for universal quantifier in FOL, and it is based on the intuition that variables that do not occur in a formula are irrelevant for its truth value, and thus can be taken to stand for any "arbitrary" formula (via some appropriate change of valuation). However, the soundness of this rule was later disproved via a counterexample by Kuijer [11]. Thus, a long-standing open question concerns *finding a 'strong' version of APAL for which there exists a recursive axiomatization.*[4]

In this paper, we solve this open question. Our diagnosis of Kuijer's counterexample is that it makes an essential use of a known undesirable feature of PAL and APAL, namely their *lack of memory*: the updated models "forget" the initial states. As a consequence, the expressivity of the APAL $\Box$-modality reduces after any update. This is what invalidates the above rule. We fix this problem by adding to the models a memory of the *initial* epistemic situation W^0, representing the information *before any non-trivial communication took place* ("the prior"). Since communication - gaining more information - deletes possibilities, the set W of currently possible states is a (possibly proper) subset of the set W^0 of initial states. On the syntactic side, we add an operator φ^0 saying that "φ was initially the case" (before all communication). To mark the initial states, we also need a constant 0, stating that "no non-trivial communication has taken place yet". Therefore, 0 will be true only in the initial epistemic situation. It is convenient, though maybe not absolutely necessary, to add a universal modality $U\varphi$ that quantifies over all currently possible states.[5] In the resulting Arbitrary Public Announcement Logic with Memory (APALM), the arbitrary announcement operator $\Box$ quantifies over updates (not only of epistemic formulas but) of arbitrary formulas that do not contain the operator $\Box$ itself.[6] As a result, the range of $\Box$ is wider than in standard APAL, covering announcements that may

---

[3] This means that from any proof of a theorem from the axioms that uses the infinitary rule we can obtain a finitary proof of the same theorem, by using the finitary rule instead.

[4] Here, by 'strong' version we mean one that allows quantification over a sufficiently wide range of announcements (sufficiently wide to avoid Liar-like circles) as intended by a similar restriction in the original APAL or in its group-restricted version GAL.

[5] From an epistemic point of view, it would be more natural to replace U by an operator Ck that describes current common knowledge and quantifies only over currently possible states that are accessible by epistemic chains from the actual state. We chose to stick with U for simplicity and leave the addition of Ck to APAL for future work.

[6] This restriction is necessary to produce a well-defined semantics that avoids Liar-like vicious circles. In standard APAL, the restriction is w.r.t. inductive construct $\Diamond\varphi$. Thus, formulas of the form $\langle\Diamond p\rangle\varphi$ are allowed in original APAL. APAL and APALM expressivities seem to be incomparable, and that would still be the case if we dropped the above restriction.

refer to the initial situation (by the use of the operators 0 and φ^0) or to all currently possible states (by the use of $U\varphi$).

We show that the original finitary rule proposed in [2] *is* sound for APALM and, moreover, it forms the basis of a *complete recursive axiomatization*.[7] Besides its technical advantages, APALM is valuable in its own respect. Maintaining a record of the initial situation in our models helps us to formalize updates that refer to the 'epistemic past' such as "what you said, I knew already" [15]. This may be useful in treating certain epistemic puzzles involving reference to past information states, e.g. "What you said did not surprise me" [12]. The more recent *Cheryl's Birthday* problem also contains an announcement of the form "At first I didn't know when Cheryl's birthday is, but now I know" (although in this particular puzzle the past-tense announcement is redundant and plays no role in the solution).[8] See [15] for more examples.

Note though that the 'memory' of APALM is of a very specific kind. Our models do not remember the whole history of communication, but only the initial epistemic situation (before any communication). Correspondingly, in the syntax we do *not* have a 'yesterday' operator $Y\varphi$, referring to the previous state just before the last announcement as in [14], but only the operator φ^0 referring to the initial state. We think of this 'economy' of memory as *a (positive) "feature, not a bug"* of our logic: a detailed record of all history is simply not necessary for solving the problem at hand. In fact, keeping all the history and adding a $Y\varphi$ operator would greatly complicate our task by invalidating some of the standard nice properties of PAL (e.g. the composition axiom, stating that any sequence of announcements is equivalent to a single announcement).[9]

So we opt for simplicity, enriching the models and language just enough to recover the full expressivity of $\Box$ after updates, and thus establish the soundness of the $\Box$-introduction rule. The minimalistic-memory character of our semantics is rather natural and it is similar to the one encountered in Bayesian models[10],

---

[7] We use a slightly different version of this rule, which is easily seen to be equivalent to the original version in the presence of the usual PAL reduction axioms.

[8] Cheryl's Birthday problem was part of the 2015 Singapore and Asian Schools Math Olympiad, and became viral after it was posted on Facebook by Singapore TV presenter Kenneth Kong.

[9] As a consequence, having such a rich memory of history would destroy some of the appealing features of the APAL operator (e.g. its $S4$ character: $\Box\varphi \to \Box\Box\varphi$), and would force us to distinguish between "knowability via one communication step" $\Diamond K$ versus "knowability via a finite communication sequence" $\Diamond^* K$, leading to an unnecessarily complex logic.

[10] In such models, only the 'prior' and the 'posterior' information states are taken to be relevant, while all the intermediary steps are forgotten. As a consequence, all the evidence gathered in between the initial and the current state can be compressed into one set E, called "the evidence" (rather than keeping a growing tail-sequence of all past evidence sets). Similarly, in our logic, all the past communication is compressed in its end-result, namely in the set W of current possibilities, which plays the same role as the evidence set E in Bayesian models.

with their distinction between 'prior' and 'posterior' (aka current) probabilities.[11]

On the technical side, our completeness proof involves an essential detour into an alternative semantics for APALM ('pseudo-models'), in the style of Subset Space Logics (SSL) [10,13]. This reveals deep connections between apparently very different formalisms. Moreover, this alternative semantics is of independent interest, giving us a more general setting for modeling knowability and learn-ability (see, e.g., [5,7,8]). Various combinations of PAL or APAL with subset space semantics have been investigated in the literature [4–7,21,22,25,26]. Following SSL-style, our pseudo-models come with a given family of *admissible sets* of worlds, which in our context represent "publicly announceable" (or commu-nicable) propositions.[12] We interpret □ in pseudo-models as the so-called 'effort' modality of SSL, which quantifies over updates with announceable propositions (regardless of whether they are syntactically definable or not). The finitary □-introduction rule is obviously *sound for the effort modality*, because of its more 'semantic' character. This observation, together with the fact that APALM mod-els (unlike original APAL models) can be seen as a special case of pseudo-models, lie at the core of our soundness and completeness proof.

Due to the page limit, we skipped some of the proofs. Readers interested to see all the relevant proofs, as well as a other related results, may consult the extended online version of this paper at https://analuciavsblog.wordpress.com/page/.

2   Syntax, Semantics, and Axiomatization

Let *Prop* be a countable set of propositional variables and $\mathcal{AG} = \{1, \ldots, n\}$ be a finite set of agents. The language $\mathcal{L}$ of APALM (Arbitrary Public Announcement Logic with Memory) is recursively defined by the grammar:

$$\varphi ::= \top \mid p \mid 0 \mid \varphi^0 \mid \neg\varphi \mid \varphi \wedge \varphi \mid K_i\varphi \mid U\varphi \mid \langle\theta\rangle\varphi \mid \Diamond\varphi,$$

with $p \in Prop$, $i \in \mathcal{AG}$, and $\theta \in \mathcal{L}_{-\Diamond}$ is a formula in the sublanguage $\mathcal{L}_{-\Diamond}$ obtained from $\mathcal{L}$ by removing the $\Diamond$ operator. Given a formula $\varphi \in \mathcal{L}$, we

---

[11] Moreover, this limited form of memory is enough to 'simulate' the yesterday operator $Y\varphi$ by using context-dependent formulas. For instance, the dialogue in Cheryl's birthday puzzle (Albert: "I don't know when Cheryl's birthday is, but I know that Bernard doesn't know it either"; Bernard: "At first I didn't know when Cheryl's birthday is, but I know now"; Albert: "Now I also know"), can be simulated by the following sequence of announcements: first, the formula $0 \wedge \neg K_a c \wedge K_a \neg K_b c$ is announced (where 0 marks the fact that this is the first announcement), then $(\neg K_b c)^0 \wedge K_b c$ is announced, and finally $K_a c$ is announced. (Here, we encode e.g. 'Albert knows Cheryl's birthday' as $K_a c = \bigvee\{K_a(d \wedge m) : d \in D, m \in M\}$, where D is the set of possible days and M is the set of possible months.).

[12] In SSL, the set of admissible sets is sometimes, but not always, taken to be a topology. Here, it will be a Boolean algebra with epistemic operators.

denote by P_φ the set of all propositional variables occurring in φ. We employ the usual abbreviations for $\bot$ and the propositional connectives $\lor, \to, \leftrightarrow$. The dual modalities are defined as $\hat{K}_i\varphi := \neg K_i \neg\varphi$, $E\varphi := \neg U \neg\varphi$, $\Box\varphi := \neg\Diamond\neg\varphi$, and $[\theta]\varphi := \neg\langle\theta\rangle\neg\varphi$.[13]

We read $K_i\varphi$ as "φ is *known* by agent i"; $\langle\theta\rangle\varphi$ as "θ can be *truthfully announced*, and *after* this public announcement φ is true". U and E are the global universal and existential modalities quantifying over all *current* possibilities: $U\varphi$ says that "φ is true in *all current alternatives* of the actual state". $\Diamond\varphi$ and $\Box\varphi$ are the (existential and universal) arbitrary announcement operators, quantifying over updates with formulas in $\mathcal{L}_{-\Diamond}$. We can read $\Box\varphi$ as "φ is *stably true* (under public announcements)": i.e., φ stays true no matter what (true) announcements are made. The constant 0, meaning that "*no (non-trivial) announcements* took place yet", holds only at the *initial time*. Similarly, the formula φ^0 means that "*initially* (prior to all communication), φ was true".

Definition 1 (Model, Initial Model, and Relativized Model)

- A model *is a tuple* $\mathcal{M} = (W^0, W, \sim_1, \ldots, \sim_n, \|\cdot\|)$, *where* $W \subseteq W^0$ *are non-empty sets of states,* $\sim_i \subseteq W^0 \times W^0$ *are equivalence relations labeled by 'agents'* $i \in \mathcal{AG}$, *and* $\|\cdot\| : Prop \to \mathcal{P}(W^0)$ *is a valuation function that maps every propositional atom* $p \in Prop$ *to a set of states* $\|p\| \subseteq W^0$. W^0 *is the* initial domain, *representing the initial informational situation before any communication took place; its elements are called* initial states. *In contrast,* W *is the* current domain, *representing the current information, and its elements are called* current states.
- *For every model* $\mathcal{M} = (W^0, W, \sim_1, \ldots, \sim_n, \|\cdot\|)$, *we define its* initial model $\mathcal{M}^0 = (W^0, W^0, \sim_1, \ldots, \sim_n, \|\cdot\|)$, *whose current and initial domains are both given by the initial domain of the original model* $\mathcal{M}$.
- *Given a model* $\mathcal{M} = (W^0, W, \sim_1, \ldots, \sim_n, \|\cdot\|)$ *and a set* $A \subseteq W$, *we define the* relativized model $\mathcal{M}|A = (W^0, A, \sim_1, \ldots, \sim_n, \|\cdot\|)$.

For states $w \in W$ and agents i, we will use the notation $w_i := \{s \in W : w \sim_i s\}$ to denote the restriction to W of w's equivalence class modulo $\sim_i$.

Definition 2 (Semantics). *Given a model* $\mathcal{M} = (W^0, W, \sim_1, \ldots, \sim_n, \|\cdot\|)$, *we define a truth set* $[\![\varphi]\!]_\mathcal{M}$ *for every formula* φ. *When the current model* $\mathcal{M}$ *is understood, we skip the subscript and simply write* $[\![\varphi]\!]$. *The definition of* $[\![\varphi]\!]$ *is by recursion on* φ:

---

[13] The update operator $\langle\theta\rangle\varphi$ is often denoted by $\langle!\theta\rangle\varphi$ in Public Announcement Logic literature; we skip the exclamation sign, but we will use the notation $\langle!\rangle$ for this modality and $[!]$ for its dual when we do not want to specify the announcement formula θ (so that ! functions as a placeholder for the content of the announcement).

$$[\![\top]\!] = W \qquad\qquad [\![\varphi \wedge \psi]\!] = [\![\varphi]\!] \cap [\![\psi]\!]$$

$$[\![p]\!] = \|p\| \cap W \qquad\qquad [\![K_i\varphi]\!] = \{w \in W : w_i \subseteq [\![\varphi]\!]\}$$

$$[\![0]\!] = \begin{cases} W^0 & \text{if } W = W^0 \\ \emptyset & \text{otherwise} \end{cases} \qquad [\![U\varphi]\!] = \begin{cases} W & \text{if } [\![\varphi]\!] = W \\ \emptyset & \text{otherwise} \end{cases}$$

$$[\![\varphi^0]\!] = [\![\varphi]\!]_{\mathcal{M}^0} \cap W \qquad\qquad [\![\langle\theta\rangle\varphi]\!] = [\![\varphi]\!]_{\mathcal{M}|[\theta]}$$

$$[\![\neg\varphi]\!] = W - [\![\varphi]\!] \qquad\qquad [\![\Diamond\varphi]\!] = \bigcup\{[\![\langle\theta\rangle\varphi]\!] : \theta \in \mathcal{L}_{-\Diamond}\}$$

Observation 1. *Note that we have*

$$w \in [\![\Box\varphi]\!] \quad \text{iff} \quad w \in [\![\langle\theta\rangle\varphi]\!] \text{ for every } \theta \in \mathcal{L}_{-\Diamond}.$$

In some of our inductive proofs, we will need a complexity measure on formulas different from the standard one[14]:

Lemma 1. *There exists a well-founded strict partial order $<$ on $\mathcal{L}$, such that:*

1. *if φ is a subformula of ψ,*   2. *$\langle\theta\rangle\varphi < \Diamond\varphi$, for all $\theta \in \mathcal{L}_{-\Diamond}$.*
   then $\varphi < \psi$,

Proposition 1. *We have $[\![\varphi]\!] \subseteq W$, for all formulas $\varphi \in \mathcal{L}$.*

(The proof, in the online version, is by induction on the complexity $<$ from Lemma 1.)

Definition 3 (APALM Models and Validity). *An APALM model is a tuple $\mathcal{M} = (W^0, W, \sim_1, \ldots, \sim_n, \|\cdot\|)$ such that $W = [\![\theta]\!]_{\mathcal{M}^0}$ for some $\theta \in \mathcal{L}_{-\Diamond}$; i.e. $\mathcal{M}$ can be obtained by updating its initial model $\mathcal{M}^0$ with some formula in $\mathcal{L}_{-\Diamond}$. A formula φ is valid on APALM models if it is true in all current states $s \in W$ (i.e. $[\![\varphi]\!]_{\mathcal{M}} = W$) for every APALM model $\mathcal{M} = (W^0, W, \sim_1, \ldots, \sim_n, \|\cdot\|)$.*

APALM models are our *intended* models, in which the current information range comes from updating the initial range with some public announcement.

We arrive now at the main result of our paper.

Theorem 1. *(Soundness and Completeness) APALM validities are recursively enumerable. Indeed, the following axiom system **APALM** (given in Table 1, where recall that P_φ is the set of propositional variables in φ) is sound and complete wrt APALM models:*

Intuitive Reading of the Axioms. Parts (I) and (II) should be obvious. The axiom R[$\top$] says that *updating with tautologies is redundant*. The reduction laws

---

[14] The standard notion requires only that formulas are more complex than their subformulas, while we also need that $\Diamond\varphi$ is more complex than $\langle\theta\rangle\varphi$. To the best of our knowledge, such a complexity measure was first introduced in [3] for the original APAL language from [2]. Similar measures have later been introduced for topological versions of APAL in [22,23].

Table 1. The axiomatization **APALM**. (Here, $\varphi, \psi, \chi \in \mathcal{L}$, while $\theta, \rho \in \mathcal{L}_{-\diamond}$.)

	(I) Basic Axioms of system APALM:
(CPL)	all classical propositional tautologies and Modus Ponens
$(S5_{K_i})$	all $S5$ axioms and rules for knowledge operator K_i
$(S5_U)$	all $S5$ axioms and rules for U operator
$(U\text{-}K_i)$	$U\varphi \to K_i\varphi$
	(II) Axioms and rules for dynamic modalities [!]:
$(K_!)$	*Kripke's axiom for* [!]: $[\theta](\psi \to \varphi) \to ([\theta]\psi \to [\theta]\varphi)$
$(Nec_!)$	*Necessitation for* [!]: from φ, infer $[\theta]\varphi$.
(RE)	*Replacement of Equivalents* [!]: from $\theta \leftrightarrow \rho$, infer $[\theta]\varphi \leftrightarrow [\rho]\varphi$.
Reduction laws:	
$(R[\top])$	$[\top]\varphi \leftrightarrow \varphi$
(R_p)	$[\theta]p \leftrightarrow (\theta \to p)$
$(R_\neg)$	$[\theta]\neg\psi \leftrightarrow (\theta \to \neg[\theta]\psi)$
(R_{K_i})	$[\theta]K_i\psi \leftrightarrow (\theta \to K_i[\theta]\psi)$
$(R_{[!]})$	$[\theta][\rho]\chi \leftrightarrow [\langle\theta\rangle\rho]\chi$
(R^0)	$[\theta]\varphi^0 \leftrightarrow (\theta \to \varphi^0)$
(R_U)	$[\theta]U\varphi \leftrightarrow (\theta \to U[\theta]\varphi)$
(R_0)	$[\theta]0 \leftrightarrow (\theta \to (U\theta \wedge 0))$
	(III) Axioms and rules for 0 and *initial* operator ^{0}:
(Ax_0)	0^0
$(0\text{-}U)$	$0 \to U0$
$(0\text{-}eq)$	$0 \to (\varphi \leftrightarrow \varphi^0)$
(Nec^0)	*Necessitation for* ^{0}: from φ, infer φ^0
Equivalences with ^{0}:	
(Eq_p^0)	$p^0 \leftrightarrow p$
$(Eq_\neg^0)$	$(\neg\varphi)^0 \leftrightarrow \neg\varphi^0$
$(Eq_\wedge^0)$	$(\varphi \wedge \psi)^0 \leftrightarrow (\varphi^0 \wedge \psi^0)$
Implications with ^{0}:	
(Imp_U^0)	$(U\varphi)^0 \to U\varphi^0$
(Imp_i^0)	$(K_i\varphi)^0 \to K_i\varphi^0$
$(Imp_\square^0)$	$(\square\varphi)^0 \to \varphi$
	(III) Elim-axiom and Intro-rule for $\square$:
$([!]\square\text{-elim})$	$[\theta]\square\varphi \to [\theta \wedge \rho]\varphi$
$([!]\square\text{-intro})$	from $\chi \to [\theta \wedge p]\varphi$, infer $\chi \to [\theta]\square\varphi$ (for $p \notin P_\chi \cup P_\theta \cup P_\varphi$).

that do not contain ^{0}, U or 0 are well-known PAL axioms. R_U is the natural reduction law for the universal modality. The axiom R^0 says that the truth value of φ^0 formulas stays the same in time (because the superscript ^{0} serves as a time stamp), so they can be treated similarly to atoms. Ax_0 says that 0 was initially the case, and R_0 says that at any later stage (after any update) 0 can only be true if it was already true before the update and the update was trivial (universally true). Together, these two say that the constant 0 characterizes states where no non-trivial communication has occurred. Axiom $0\text{-}U$ is a sychronicity constraint: *if no non-trivial communication has taken place yet, then this is the case in all the currently possible states.* Axiom 0-*eq* says that *initially, φ is equivalent to its initial correspondent φ^0.* The Equivalences with ^{0} express that ^{0} distributes over negation and over conjunction. Imp$_\square^0$ says that *if initially φ was stably true (under any further announcements), then φ is the case now.* Taken together, the elimination axiom [!]□-elim and introduction rule [!]□-intro say that *φ is a stable truth after an announcement θ iff φ stays true after any more informative announcement* (of the form $\theta \wedge \rho$).[15]

Proposition 2. *The following schemas and inference rules are derivable in* **APALM**, *where $\varphi, \psi, \chi \in \mathcal{L}$ and $\theta \in \mathcal{L}_{-\Diamond}$:*

1. *from $\vdash \varphi \leftrightarrow \psi$, infer $\vdash [\theta]\varphi \leftrightarrow [\theta]\psi$*
2. $\vdash \langle\theta\rangle 0 \leftrightarrow (0 \wedge U\theta)$
3. $\vdash \langle\theta\rangle\psi \leftrightarrow (\theta \wedge [\theta]\psi)$
4. $\vdash \square\varphi \rightarrow [\theta]\varphi$
5. *from $\vdash \chi \rightarrow [p]\varphi$, infer $\vdash \chi \rightarrow \square\varphi$*
   $(p \notin P_\chi \cup P_\varphi)$
6. $\vdash (\varphi \rightarrow \psi)^0 \leftrightarrow (\varphi^0 \rightarrow \psi^0)$
7. $\vdash \varphi^{00} \leftrightarrow \varphi^0$
8. $\vdash \square\varphi^0 \leftrightarrow \varphi^0$, *and* $\vdash \Diamond\varphi^0 \leftrightarrow \varphi^0$

9. $\vdash (\square\varphi)^0 \rightarrow \square\varphi^0$
10. $\vdash (0 \wedge \Diamond\varphi^0) \rightarrow \varphi$
11. $\vdash \varphi \rightarrow (0 \wedge \Diamond\varphi)^0$
12. $\vdash \varphi \rightarrow \psi^0$ *if and only if*
    $\vdash (0 \wedge \Diamond\varphi) \rightarrow \psi$
13. $\vdash [\theta](\psi \wedge \varphi) \leftrightarrow ([\theta]\psi \wedge [\theta]\varphi)$
14. $\vdash [\theta][p]\psi \leftrightarrow [\theta \wedge p]\psi$
15. $\vdash [\theta]\bot \leftrightarrow \neg\theta$

Proposition 3. *All S4 axioms and inference rules for $\square$ are derivable in* **APALM**.

3   An Analysis of Kuijer's Counterexample

To understand Kuijer's counterexample [11] to the soundness of the finitary □-elimination rule for the original APAL, recall that in the APAL □ quantifies only over updates with *epistemic* formulas. More precisely, the APAL semantics of □ is given by

$$w \in [\![\square\varphi]\!] \quad \text{iff} \quad w \in [\![[\theta]\varphi]\!] \text{ for every } \theta \in \mathcal{L}_{epi},$$

where $\mathcal{L}_{epi}$ is the sublanguage generated from propositional atoms $p \in Prop$ using only the Boolean connectives $\neg\varphi$ and $\varphi \wedge \psi$ and the epistemic operators $K_i\varphi$.

Kuijer takes the formula $\gamma := p \wedge \hat{K}_b\neg p \wedge \hat{K}_a K_b p$, and shows that

$$[\hat{K}_b p]\square\neg\gamma \rightarrow [q]\neg\gamma.$$

is valid on APAL models. (In fact, it is also valid on APALM models!) But then, by the [!]□-intro rule (or rather, by its weaker consequence in Proposition 2(5)), the formula

$$[\hat{K}_b p]\square\neg\gamma \rightarrow \square\neg\gamma$$

---

[15] The "freshness" of the variable $p \in P$ in the rule [!]□-intro ensures that it represents any 'generic' announcement.

should also be valid. But this is contradicted by the model $\mathcal{M}$ in Fig. 1. The premise $[\hat{K}_b p]\Box\neg\gamma$ is true at w in $\mathcal{M}$, since $\Box\neg\gamma$ holds at w in $\mathcal{M}|[\![\hat{K}_b p]\!]$: indeed, the only way to falsify it would be by deleting the lower-right node from Fig. 2a while keeping (all other nodes, and in particular) the upper-right node. But in $\mathcal{M}|[\![\hat{K}_b p]\!]$ the upper and lower right nodes can't be separated by epistemic sentences: they are bisimilar.

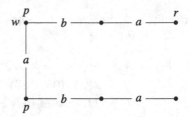

Fig. 1. An initial model $\mathcal{M}$. Worlds are nodes in the graph, valuation is given by labeling the node with the true atoms, and epistemic relations are given by labeled arrows.

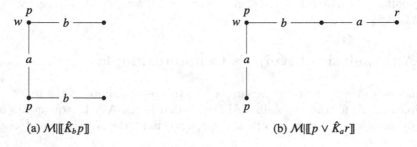

(a) $\mathcal{M}|[\![\hat{K}_b p]\!]$ (b) $\mathcal{M}|[\![p \vee \hat{K}_a r]\!]$

Fig. 2. Two updates of $\mathcal{M}$.

In contrast, the conclusion $\Box\neg\gamma$ is *false* at w in $\mathcal{M}$, since in that original model the two mentioned nodes *could be separated*. Indeed, we could perform an alternative update with the formula $p \vee \hat{K}_a r$, yielding the epistemic model $\mathcal{M}|[\![p \vee \hat{K}_a r]\!]$ shown in Fig. 2b, in which γ *is true* at w (contrary to the assertion that $\Box\neg\gamma$ was true in $\mathcal{M}$).

To see that the counterexample does not apply to APALM, notice that APALM models keep track of the initial states, so technically speaking the updated model $\mathcal{M}|[\![\hat{K}_b p]\!]$ consists now of the initial structure in Fig. 1 together with current set of worlds W in Fig. 2a. But in this model, $\Box\neg\gamma$ is *no longer true* at w (and so the premise $[\hat{K}_b p]\Box\neg\gamma$ was *not* true in $\mathcal{M}$ when considered as an APALM model!). Indeed, we *can* perform a new update of $\mathcal{M}|[\![\hat{K}_b p]\!]$ with the formula $(p \vee \hat{K}_a r)^0$, which yields an updated model whose current set of worlds is given in Fig. 3:

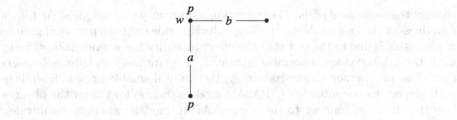

Fig. 3. The current worlds resulting from updating $\mathcal{M}|[\![\hat{K}_bp]\!]$ with $(p \vee \hat{K}_ar)^0$.

Note that, in this new model, γ *is* the case at w (-thus showing that $\Box\neg\gamma$ was *not* true at w in $\mathcal{M}|[\![\hat{K}_bp]\!]$). So the counterexample simply does not work for APALM.

Moreover, we can see that the un-soundness of $[!]\Box$-intro rule for APAL has to do with its lack of memory, which leads to information loss after updates: while initially (in $\mathcal{M}$) there were epistemic sentences (e.g. $p \vee \hat{K}_ar$) that could separate the two nodes mentioned above, there are no such sentences after the update.

APALM solves this by keeping track of the initial states, and referring back to them, as in $(p \vee \hat{K}_ar)^0$.

4   Soundness, via Pseudo-model Semantics

To start with, note that the even the soundness of our axiomatic system is not a trivial matter. As we saw from Kuijer's counterexample, the analogue of our finitary $\Box$-introduction rule was *not* sound for APAL. To prove its soundness on APALM models, we need a detour into an equivalent semantics, in the style of Subset Space Logics (SSL) [10,13]: *pseudo-models*.[16]

We first introduce an auxiliary notion: 'pre-models' are just SSL models, coming with a given family $\mathcal{A}$ of "admissible sets" of worlds (which can be thought of as the *communicable* propositions). We interpret $\Box$ in these structures as the so-called "effort modality" of SSL, which quantifies over updates with admissible propositions in $\mathcal{A}$. Our 'pseudo-models' are pre-models with additional closure conditions (saying that the family of admissible sets includes the valuations and is closed under complement, intersection and epistemic operators). These conditions imply that *every set definable by $\Diamond$-free formulas is admissible*, and this

---

[16] A more direct soundness proof on APALM models is in principle possible, but would require at least as much work as our detour. Unlike in standard EL, PAL or DEL, the meaning of an APALM formula depends, not only on the valuation of the atoms occurring in it, but also on the family $\mathcal{A}$ of all sets definable by $\mathcal{L}_{-\Diamond}$-formulas. The move from models to pseudo-models makes explicit this dependence on the family $\mathcal{A}$, while also relaxing the demands on $\mathcal{A}$ (which is no longer required to be exactly the family of $\mathcal{L}_{-\Diamond}$-definable sets), and thus makes the soundness proof both simpler and more transparent. Since we will need pseudo-models for our completeness proof anyway, we see no added value in trying to give a more direct proof.

ensures the soundness of our □-elimination axiom on pseudo-models. As for the soundness of the long-problematic □-introduction rule on (both pre- and) pseudo-models, this is due to the fact that the effort modality has a more 'robust' range than the arbitrary announcement operator: it quantifies over admissible sets, regardless of whether these sets are syntactically definable or not. Soundness with respect to our intended (APALM) models then follows from the observation that they (in contrast to the original APAL models) are in fact equivalent to a *special case of pseudo-models*: the "standard" ones (in which the admissible sets in $\mathcal{A}$ are *exactly* the sets definable by $\Diamond$-free formulas).

Definition 4 (Pre-model). *A* pre-model *is a tuple* $\mathcal{M} = (W^0, \mathcal{A}, \sim_1, \ldots, \sim_n , \| \cdot \|)$, *where* W^0 *is the prior domain,* $\sim_i$ *are equivalence relations on* W^0, $\| \cdot \| : Prop \to \mathcal{P}(W^0)$ *is a valuation map, and* $\mathcal{A} \subseteq \mathcal{P}(W^0)$ *is a family of subsets of the initial domain, called* admissible sets *(representing the propositions that can be publicly announced).*

Given a set $A \subseteq W^0$ and a state $w \in A$, we use the notation $w_i^A := \{s \in A : w \sim_i s\}$ to denote the restriction to A of w's equivalence class modulo $\sim_i$.

Definition 5 (Pre-model Semantics). *Given a pre-model* $\mathcal{M} = (W^0, \mathcal{A}, \sim_1 , \ldots, \sim_n, \| \cdot \|)$, *we recursively define a* truth set $[\![\varphi]\!]_A$ *for every formula* φ *and subset* $A \subseteq W^0$:

$$[\![\top]\!]_A = A \qquad\qquad [\![\varphi \wedge \psi]\!]_A = [\![\varphi]\!]_A \cap [\![\psi]\!]_A$$

$$[\![p]\!]_A = \|p\| \cap A \qquad\qquad [\![K_i\varphi]\!]_A = \{w \in A : w_i^A \subseteq [\![\varphi]\!]_A\}$$

$$[\![0]\!]_A = \begin{cases} A & \text{if } A = W^0 \\ \emptyset & \text{otherwise} \end{cases} \qquad [\![U\varphi]\!]_A = \begin{cases} A & \text{if } [\![\varphi]\!]_A = A \\ \emptyset & \text{otherwise} \end{cases}$$

$$[\![\varphi^0]\!]_A = [\![\varphi]\!]_{W^0} \cap A \qquad\qquad [\![\langle\theta\rangle\varphi]\!]_A = [\![\varphi]\!]_{[\theta]_A}$$

$$[\![\neg\varphi]\!]_A = A - [\![\varphi]\!]_A \qquad\qquad [\![\Diamond\varphi]\!]_A = \bigcup\{[\![\varphi]\!]_B : B \in \mathcal{A}, B \subseteq A\}$$

Observation 2. *Note that, for all* $w \in A$, *we have*

$$w \in [\![\Box\varphi]\!]_A \quad \text{iff} \quad \forall B \in \mathcal{A}(w \in B \subseteq A \Rightarrow w \in [\![\varphi]\!]_B),$$

which fits with the semantics of the 'effort' modality in SSL [10,13]. *Moreover, it is easy to see that* $[\![\varphi]\!]_A \subseteq A$ *for all* $A \in \mathcal{A}$ *and* $\varphi \in \mathcal{L}$.

Definition 6 (Pseudo-models and Validity). *A* pseudo-model *is a pre-model* $\mathcal{M} = (W^0, \mathcal{A}, \sim_1, \ldots, \sim_n, \| \cdot \|)$, *satisfying the following closure conditions:*

1. $\|p\| \in \mathcal{A}$, *for all* $p \in Prop$,
2. $W^0 \in \mathcal{A}$,
3. *if* $A \in \mathcal{A}$ *then* $(W^0 - A) \in \mathcal{A}$,
4. *if* $A, B \in \mathcal{A}$ *then* $(A \cap B) \in \mathcal{A}$,
5. *if* $A \in \mathcal{A}$ *then* $K_i A \in \mathcal{A}$, *where* $K_i A := \{w \in W^0 : \forall s \in W^0(w \sim_i s \Rightarrow s \in A)\}$.

    A formula $\varphi \in \mathcal{L}$ *is* valid on pseudo-models *if it is true in all admissible sets* $A \in \mathcal{A}$ *of every pseudo-model* $\mathcal{M}$, *i.e,* $[\![\varphi]\!]_A = A$ *for all* $A \in \mathcal{A}$ *and all* $\mathcal{M}$.

Proposition 4. *Given a pseudo-model* $(W^0, \mathcal{A}, \sim_1, \ldots, \sim_n, \| \cdot \|)$, $A \in \mathcal{A}$ *and* $\theta \in \mathcal{L}_{-\Diamond}$, *we have* $[\![\theta]\!]_A \in \mathcal{A}$.

Proof. The proof is by subformula induction on θ, where we denote by IH the induction hypothesis. The base cases and the inductive cases for the Booleans are immediate (using the conditions in Definition 6).

Case $\theta := \psi^0$. By the semantics, $[\![\psi^0]\!]_A = [\![\psi]\!]_{W^0} \cap A \in \mathcal{A}$, since $[\![\psi]\!]_{W^0} \in \mathcal{A}$ (by the fact that $W^0 \in \mathcal{A}$ and IH), $A \in \mathcal{A}$ (by assumption), and $\mathcal{A}$ is closed under intersection.

Case $\theta := K_i \psi$.

$$
\begin{aligned}
[\![K_i\psi]\!]_A &= \{w \in A : w_i^A \subseteq [\![\varphi]\!]_A\} \\
&= A \cap \{w \in W^0 : w_i^A \subseteq [\![\varphi]\!]_A\} \\
&= A \cap \{w \in W^0 : \forall s \in W^0((s \in A \text{ and } w \sim_i s) \Rightarrow s \in [\![\varphi]\!]_A)\} \\
&= A \cap \{w \in W^0 : \forall s \in W^0(w \sim_i s \Rightarrow (s \in A \Rightarrow s \in [\![\varphi]\!]_A))\} \\
&= A \cap \{w \in W^0 : \forall s \in W^0(w \sim_i s \Rightarrow (s \in (X - A) \text{ or } s \in [\![\varphi]\!]_A))\} \\
&= A \cap \{w \in W^0 : \forall s \in W^0(w \sim_i s \Rightarrow s \in ((X - A) \cup [\![\varphi]\!]_A)\} \\
&= A \cap K_i((X - A) \cup [\![\varphi]\!]_A)
\end{aligned}
$$

(by Definition 6(3–5), since $A \in \mathcal{A}$ and $[\![\varphi]\!]_A \in \mathcal{A}$(by IH).)

Case $\theta := U\psi$. By definition, $[\![U\psi]\!]_A \in \{\emptyset, A\} \subseteq \mathcal{A}$.

Case $\theta := \langle \varphi \rangle \psi$. Since $A \in \mathcal{A}$, we have $[\![\varphi]\!]_A \in \mathcal{A}$ (by IH), and hence $[\![\langle \varphi \rangle \psi]\!]_A = [\![\psi]\!]_{[\![\varphi]\!]_A} \in \mathcal{A}$ (by the semantics and IH again).

Proposition 5. *The system* **APALM** *is sound wrt pseudo-models.*

To prove Proposition 5, we need the following:

Lemma 2. *Let* $\mathcal{M} = (W^0, \mathcal{A}, \sim_1, \ldots, \sim_n, \| \cdot \|)$ *and* $\mathcal{M}' = (W^0, \mathcal{A}, \sim_1, \ldots, \sim_n,$ $\| \cdot \|')$ *be two pseudo-models and* $\varphi \in \mathcal{L}$ *such that* $\mathcal{M}$ *and* $\mathcal{M}'$ *differ only in the valuation of some* $p \notin P_\varphi$. *Then, for all* $A \in \mathcal{A}$, *we have* $[\![\varphi]\!]_A^{\mathcal{M}} = [\![\varphi]\!]_A^{\mathcal{M}'}$.

Proof. The proof follows by subformula induction on φ. Let $\mathcal{M} = (W^0, \mathcal{A}, \sim_1$ $, \ldots, \sim_n, \| \cdot \|)$ and $\mathcal{M}' = (W^0, \mathcal{A}, \sim_1, \ldots, \sim_n, \| \cdot \|')$ be two pseudo-models such that $\mathcal{M}$ and $\mathcal{M}'$ differ only in the valuation of some $p \notin P_\varphi$ and let $A \in \mathcal{A}$. We want to show that $[\![\varphi]\!]_A^{\mathcal{M}} = [\![\varphi]\!]_A^{\mathcal{M}'}$. The base cases $\varphi := q$, $\varphi := \top$, $\varphi := 0$, and the inductive cases for Booleans are immediate.

Case $\varphi := \psi^0$. Note that $P_{\psi^0} = P_\psi$. Then, by IH, we have that $[\![\psi]\!]_A^{\mathcal{M}'} = [\![\psi]\!]_A^{\mathcal{M}}$ for every $A \in \mathcal{A}$, in particular for $W^0 \in \mathcal{A}$. Thus $[\![\psi]\!]_{W^0}^{\mathcal{M}'} = [\![\psi]\!]_{W^0}^{\mathcal{M}}$. Then, $[\![\psi]\!]_{W^0}^{\mathcal{M}'} \cap A = [\![\psi]\!]_{W^0}^{\mathcal{M}} \cap A$ for all $A \in \mathcal{A}$. By the semantics of the initial operator on pseudo-models, we obtain $[\![\psi^0]\!]_A^{\mathcal{M}'} = [\![\psi^0]\!]_A^{\mathcal{M}}$.

Case $\varphi := K_i \psi$. Note that $P_{K_i\psi} = P_\psi$. Then, by IH, we have that $[\![\psi]\!]_A^{\mathcal{M}} = [\![\psi]\!]_A^{\mathcal{M}'}$. Observe that $[\![K_i\psi]\!]_A^{\mathcal{M}} = \{w \in A : w_i^A \subseteq [\![\psi]\!]_A^{\mathcal{M}}\}$ and, similarly, $[\![K_i\psi]\!]_A^{\mathcal{M}'} = \{w \in A : w_i^A \subseteq [\![\psi]\!]_A^{\mathcal{M}'}\}$. Then, since $[\![\psi]\!]_A^{\mathcal{M}} = [\![\psi]\!]_A^{\mathcal{M}'}$, we obtain $[\![K_i\psi]\!]_A^{\mathcal{M}} = [\![K_i\psi]\!]_A^{\mathcal{M}'}$.

Case $\varphi := U\psi$. Note that $P_{U\psi} = P_\psi$. Then, by IH, we have that $[\![\psi]\!]_A^{\mathcal{M}'} = [\![\psi]\!]_A^{\mathcal{M}}$ for every $A \in \mathcal{A}$. We have two case: (1) If $[\![\psi]\!]_A^{\mathcal{M}'} = [\![\psi]\!]_A^{\mathcal{M}} = A$, then $[\![U\psi]\!]_A^{\mathcal{M}'} = A = [\![U\psi]\!]_A^{\mathcal{M}}$. (2) If $[\![\psi]\!]_A^{\mathcal{M}'} = [\![\psi]\!]_A^{\mathcal{M}} \neq A$, then $[\![U\psi]\!]_A^{\mathcal{M}'} = [\![U\psi]\!]_A^{\mathcal{M}} = \emptyset$.

Case $\varphi := \langle\theta\rangle\psi$. Note that $P_{\langle\theta\rangle\psi} = P_\theta \cup P_\psi$. By IH, we have $[\![\theta]\!]_A^{\mathcal{M}'} = [\![\theta]\!]_A^{\mathcal{M}}$ and $[\![\psi]\!]_A^{\mathcal{M}'} = [\![\psi]\!]_A^{\mathcal{M}}$ for every $A \in \mathcal{A}$. By Proposition 4, we know that $[\![\theta]\!]_A^{\mathcal{M}} = [\![\theta]\!]_A^{\mathcal{M}'} \in \mathcal{A}$. Therefore, in particular, we have $[\![\psi]\!]_{[\![\theta]\!]_A^{\mathcal{M}'}}^{\mathcal{M}'} = [\![\psi]\!]_{[\![\theta]\!]_A^{\mathcal{M}}}^{\mathcal{M}}$. Therefore, by the semantics of $\langle!\rangle$ on pseudo-models, we obtain $[\![\langle\theta\rangle\psi]\!]_A^{\mathcal{M}'} = [\![\langle\theta\rangle\psi]\!]_A^{\mathcal{M}}$.

Case $\varphi := \Diamond\psi$. Note that $P_{\Diamond\psi} = P_\psi$. Since *the same* family of sets $\mathcal{A}$ is carried by both models $\mathcal{M}$ and $\mathcal{M}'$ and since (by IH) $[\![\psi]\!]_A^{\mathcal{M}'} = [\![\psi]\!]_A^{\mathcal{M}}$ for all $A \in \mathcal{A}$, we get:

$$[\![\Diamond\psi]\!]_A^{\mathcal{M}'} = \bigcup\{[\![\psi]\!]_B^{\mathcal{M}'} : B \in \mathcal{A}, B \subseteq A\} = \bigcup\{[\![\psi]\!]_B^{\mathcal{M}} : B \in \mathcal{A}, B \subseteq A\} = [\![\Diamond\psi]\!]_A^{\mathcal{M}}.$$

Proof of Proposition 5. The soundness of most of the axioms follows simply by spelling out the semantics. We present here only the soundness of the rule $[!]\square$-intro:

Suppose $\models \chi \rightarrow [\theta \wedge p]\varphi$ and $\not\models \chi \rightarrow [\theta]\square\varphi$, where $p \notin P_\chi \cup P_\theta \cup P_\varphi$. The latter means that there exists a pseudo model $\mathcal{M} = (W^0, \mathcal{A}, \sim_1, \ldots, \sim_n, \|\cdot\|)$ such that for some $A \in \mathcal{A}$ and some $w \in A$, $w \notin [\![\chi \rightarrow [\theta]\square\varphi]\!]_A^{\mathcal{M}}$. Therefore $w \in [\![\chi \wedge \neg[\theta]\square\varphi]\!]_A^{\mathcal{M}}$. Thus we have (1) $w \in [\![\chi]\!]_A^{\mathcal{M}}$ and (2) $w \in [\![\neg[\theta]\square\varphi]\!]_A^{\mathcal{M}}$. Because of (2), $w \in [\![\langle\theta\rangle\Diamond\neg\varphi]\!]_A^{\mathcal{M}}$, and, by the semantics, $w \in [\![\Diamond\neg\varphi]\!]_{[\![\theta]\!]_A^{\mathcal{M}}}^{\mathcal{M}}$. Therefore, applying the semantics, we obtain (3) there exists $B \in \mathcal{A}$ s.t. $w \in B \subseteq [\![\theta]\!]_A^{\mathcal{M}} \subseteq A$ and $w \in [\![\neg\varphi]\!]_B^{\mathcal{M}}$.

Now consider the pre-model $\mathcal{M}' = (W^0, \mathcal{A}, \sim_1, \ldots, \sim_n, \|\cdot\|')$ such that $\|p\|' := B$ and $\|q\|' = \|q\|$ for any $q \neq p \in Prop$. In order to use Lemma 2 we must show that $\mathcal{M}'$ is a pseudo-model. For this we only need to verify that $\mathcal{M}'$ satisfies the closure conditions given in Definition 6. First note that $\|p\|' := B \in \mathcal{A}$ by the construction of $\mathcal{M}'$, so $\|p\|' \in \mathcal{A}$. For every $q \neq p$, since $\|q\|' = \|q\|$ and $\|q\| \in \mathcal{A}$ we have $\|q\|' \in \mathcal{A}$. Since $\mathcal{A}$ is the same for both $\mathcal{M}$ and $\mathcal{M}'$, and $\mathcal{M}$ is a pseudo model, the rest of the closure conditions are already satisfied for $\mathcal{M}'$. Therefore $\mathcal{M}'$ is a pseudo model. Now continuing with our soundness proof, note that by Lemma 2 and since $p \notin P_\chi \cup P_\theta \cup P_\varphi$ we obtain $[\![\chi]\!]_A^{\mathcal{M}'} = [\![\chi]\!]_A^{\mathcal{M}}$, $[\![\theta]\!]_A^{\mathcal{M}'} = [\![\theta]\!]_A^{\mathcal{M}}$ and $[\![\neg\varphi]\!]_A^{\mathcal{M}'} = [\![\neg\varphi]\!]_A^{\mathcal{M}}$. Since $\|p\|' = B \subseteq [\![\theta]\!]_A^{\mathcal{M}'} \subseteq A$ we have $\|p\|' = [\![p]\!]_A^{\mathcal{M}'}$. Because of (3) we have that $w \in [\![\theta]\!]_A^{\mathcal{M}'}$ and $w \in [\![\neg\varphi]\!]_B^{\mathcal{M}'} = [\![\neg\varphi]\!]_{[\![p]\!]_A^{\mathcal{M}'}}^{\mathcal{M}'} = [\![\langle p\rangle\neg\varphi]\!]_A^{\mathcal{M}'}$. Thus $w \in [\![p]\!]_A^{\mathcal{M}'}$, so $w \in [\![\theta \wedge p]\!]_A^{\mathcal{M}'} = [\![\theta]\!]_A^{\mathcal{M}'} \cap [\![p]\!]_A^{\mathcal{M}'} = [\![p]\!]_A^{\mathcal{M}'}$ simply because $[\![p]\!]_A^{\mathcal{M}'} \subseteq [\![\theta]\!]_A^{\mathcal{M}'}$. Since $w \in [\![\neg\varphi]\!]_{[\![p]\!]_A^{\mathcal{M}'}}^{\mathcal{M}'}$ we obtain $w \in [\![\neg\varphi]\!]_{[\![\theta \wedge p]\!]_A^{\mathcal{M}'}}^{\mathcal{M}'}$. Putting everything together, $w \in [\![\theta \wedge p]\!]_A^{\mathcal{M}'}$ and $w \in [\![\neg\varphi]\!]_{[\![\theta \wedge p]\!]_A^{\mathcal{M}'}}^{\mathcal{M}'}$, we obtain that $w \in [\![\langle\theta \wedge p\rangle\neg\varphi]\!]_A^{\mathcal{M}'}$ and $w \in [\![\chi]\!]_A^{\mathcal{M}'}$. Therefore $\mathcal{M}', w \models \chi \wedge \langle\theta \wedge p\rangle\neg\varphi$, which contradicts the validity of $\chi \rightarrow [\theta \wedge p]\varphi$.

Definition 7 (Standard Pre-model). *A pre-model* $\mathcal{M} = (W^0, \mathcal{A}, \sim_1, \ldots, \sim_n, \|\cdot\|)$ *is* standard *if and only if* $\mathcal{A} = \{[\![\theta]\!]_{W^0} : \theta \in \mathcal{L}_{-\Diamond}\}$.

Proposition 6. *Every standard pre-model is a pseudo-model.*

Proof. Let $\mathcal{M} = (W^0, \mathcal{A}, \sim_1, \ldots, \sim_n, \|\cdot\|)$ be a standard pre-model. This implies that $\mathcal{A} = \{[\![\theta]\!]_{W^0} : \theta \in \mathcal{L}_{-\Diamond}\}$. We need to show that $\mathcal{M}$ satisfies the closure conditions given in Definition 6. Conditions (1) and (2) are immediate.

For (3): let $A \in \mathcal{A}$. Since $\mathcal{M}$ is a standard pre-model, we know that $A = [\![\theta]\!]_{W^0}$ for some $\theta \in \mathcal{L}_{-\Diamond}$. Since $\theta \in \mathcal{L}_{-\Diamond}$, we have $\neg\theta \in \mathcal{L}_{-\Diamond}$, thus, $[\![\neg\theta]\!]_{W^0} \in \mathcal{A}$. Observe that $[\![\neg\theta]\!]_{W^0} = W^0 - [\![\theta]\!]_{W^0}$, thus, we obtain $W^0 - A \in \mathcal{A}$.

For (4): let $A, B \in \mathcal{A}$. Since $\mathcal{M}$ is a standard pre-model, $A = [\![\theta_1]\!]_{W^0}$ and $B = [\![\theta_2]\!]_{W^0}$ for some $\theta_1, \theta_2 \in \mathcal{L}_{-\Diamond}$. Since $\theta_1, \theta_2 \in \mathcal{L}_{-\Diamond}$, we have $\theta_1 \wedge \theta_2 \in \mathcal{L}_{-\Diamond}$, thus, $[\![\theta_1 \wedge \theta_2]\!]_{W^0} \in \mathcal{A}$. Observe that $[\![\theta_1 \wedge \theta_2]\!]_{W^0} = [\![\theta_1]\!]_{W^0} \cap [\![\theta_2]\!]_{W^0} = A \cap B$, thus, we obtain $A \cap B \in \mathcal{A}$.

For (5): let $A \in \mathcal{A}$. Since $\mathcal{M}$ is a standard pre-model, $A = [\![\theta]\!]_{W^0}$ for some $\theta \in \mathcal{L}_{-\Diamond}$. Since $\theta \in \mathcal{L}_{-\Diamond}$, we have $K_i\theta \in \mathcal{L}_{-\Diamond}$, thus, $[\![K_i\theta]\!]_{W^0} \in \mathcal{A}$. Observe that $[\![K_i\theta]\!]_{W^0} = \{w \in W^0 : \forall s \in W^0(w \sim_i s \Rightarrow s \in [\![\theta]\!]_{W^0})\} = K_i[\![\theta]\!]_{W^0}$, thus, we obtain $K_iA \in \mathcal{A}$.

Equivalence Between the Standard Pseudo-models and APALM Models. For Proposition 7 only, we use the notation $[\![\varphi]\!]_A^{PS}$ to refer to pseudo-model semantics (as in Definition 5) and use $[\![\varphi]\!]_{\mathcal{M}}$ to refer to APALM semantics (as in Definition 1).

Proposition 7

1. *For every standard pseudo-model $\mathcal{M} = (W^0, \mathcal{A}, \sim_1, \ldots, \sim_n, \|\cdot\|)$ and every set $A \in \mathcal{A}$, we denote by $\mathcal{M}_A$ the model $\mathcal{M}_A = (W^0, A, \sim_1, \ldots, \sim_n, \|\cdot\|)$. Then:*
   (a) For every $\varphi \in \mathcal{L}$, we have $[\![\varphi]\!]_{\mathcal{M}_A} = [\![\varphi]\!]_A^{PS}$ for all $A \in \mathcal{A}$.
   (b) $\mathcal{M}_A$ is an APALM model, for all $A \in \mathcal{A}$.
2. *For every APALM model $\mathcal{M} = (W^0, W, \sim_1, \ldots, \sim_n, \|\cdot\|)$, we denote by $\mathcal{M}'$ the pre-model $\mathcal{M}' = (W^0, \mathcal{A}, \sim_1, \ldots, \sim_n, \|\cdot\|)$, where $\mathcal{A} = \{[\![\theta]\!]_{\mathcal{M}^0} : \theta \in \mathcal{L}_{-\Diamond}\}$. Then*
   (a) $\mathcal{M}'$ is a standard pseudo-model.
   (b) For every $\varphi \in \mathcal{L}$, we have $[\![\varphi]\!]_{\mathcal{M}} = [\![\varphi]\!]_W^{PS}$.

The proof of Proposition 7 needs the following lemma.

Lemma 3. *Let $\mathcal{M} = (W^0, \mathcal{A}, \sim_1, \ldots, \sim_n, \|\cdot\|)$ a standard pseudo-model, $A \in \mathcal{A}$ and $\varphi \in \mathcal{L}$, then the following holds: $[\![\Diamond\varphi]\!]_A = \bigcup\{[\![\langle\theta\rangle\varphi]\!]_A : \theta \in \mathcal{L}_{-\Diamond}\}$.*

Proof. For ($\subseteq$): Let $w \in [\![\Diamond\varphi]\!]_A$. Then, by the semantics of $\Diamond$ in pseudo-models. there exists some $B \in \mathcal{A}$ such that $w \in B \subseteq A$ and $w \in [\![\varphi]\!]_B$. Since $\mathcal{M}$ is standard, we know that $A = [\![\psi]\!]_{W^0}$ and $B = [\![\chi]\!]_{W^0}$ for some $\psi, \chi \in \mathcal{L}_{-\Diamond}$. Moreover, since $B = [\![\chi]\!]_{W^0} \subseteq A = [\![\psi]\!]_{W^0}$, we have $B = [\![\chi]\!]_{W^0} \cap [\![\psi]\!]_{W^0} = [\![\chi^0]\!]_{[\![\psi]\!]_{W^0}} = [\![\chi^0]\!]_A$, and so $w \in [\![\varphi]\!]_B = [\![\varphi]\!]_{[\![\chi^0]\!]_A} = [\![\langle\chi^0\rangle\varphi]\!]_A \subseteq \bigcup\{[\![\langle\theta\rangle\varphi]\!]_A : \theta \in \mathcal{L}_{-\Diamond}\}$. For ($\supseteq$): Let $w \in \bigcup\{[\![\langle\theta\rangle\varphi]\!]_A : \theta \in \mathcal{L}_{-\Diamond}\}$. Then we have $w \in [\![\langle\theta\rangle\varphi]\!]_A = [\![\varphi]\!]_{[\![\theta]\!]_A}$, for some $\theta \in \mathcal{L}_{-\Diamond}$, and since $[\![\theta]\!]_A \in \mathcal{A}$ (by Proposition 4) and $[\![\theta]\!]_A \subseteq A$ (by Observation 2), it follows that $w \in [\![\Diamond\varphi]\!]_A$ (by the semantics of $\Diamond$ in pseudo-models).

Proof of Proposition 7

1. Let $\mathcal{M} = (W^0, \mathcal{A}, \sim_1, \ldots, \sim_n, \|\cdot\|)$ be a standard pseudo-model. $A \in \mathcal{A}$ implies $A = [\![\theta]\!]_{W^0}^{PS} \subseteq W$ for some θ, hence $\mathcal{M}_A = (W^0, A, \sim_1, \ldots, \sim_n, \|\cdot\|)$ is a model.
   (a) The proof is by induction on the complexity measure $<$ from Lemma 1. The base cases and the inductive cases for Booleans are straightforward.
   Case $\theta := \psi^0$. We have $[\![\psi]\!]_A^{PS} = [\![\psi]\!]_{W^0}^{PS} \cap A = [\![\psi]\!]_{\mathcal{M}_A^0} \cap A = [\![\psi^0]\!]_{\mathcal{M}_A}$ (by Definition 5, IH, and Definition 2).
   Case $\theta := K_i\psi$. We have $[\![K_i\psi]\!]_A^{PS} = \{w \in A : w_i^A \subseteq [\![\psi]\!]_A^{PS} = \{w \in A : w_i \subseteq [\![\psi]\!]_{\mathcal{M}_A}\} = [\![K_i\psi]\!]_{\mathcal{M}_A}$ (by Definition 5, IH, and Definition 2).
   Case $\theta := U\psi$. By Definitions 2 and 5, we have:

$$[\![U\psi]\!]_{\mathcal{M}_A} = \begin{cases} W & \text{if } [\![\psi]\!]_{\mathcal{M}_A} = A \\ \emptyset & \text{otherwise} \end{cases} \qquad [\![U\psi]\!]_A^{PS} = \begin{cases} A & \text{if } [\![\psi]\!]_A^{PS} = A \\ \emptyset & \text{otherwise} \end{cases}$$

By IH, $[\![\psi]\!]_A^{PS} = [\![\psi]\!]_{\mathcal{M}_A}$, therefore, $[\![U\psi]\!]_A^{PS} = [\![U\psi]\!]_{\mathcal{M}_A}$.
   Case $\varphi := \langle\psi\rangle\chi$. By Definition 2, we know that $[\![\langle\psi\rangle\chi]\!]_{\mathcal{M}_A} = [\![\chi]\!]_{\mathcal{M}_A|[\![\psi]\!]_{\mathcal{M}_A}}$. Now consider the relativized model $\mathcal{M}_A|[\![\psi]\!]_{\mathcal{M}_A} = (W^0, [\![\psi]\!]_{\mathcal{M}_A}, \sim_1, \ldots, \sim_n, \|\cdot\|)$. By Lemma 1(1) and IH, we have $[\![\psi]\!]_{\mathcal{M}_A} = [\![\psi]\!]_A^{PS}$. Moreover, by the definition of standard pseudo-models, we know that $A = [\![\theta]\!]_{W^0}^{PS}$ for some $\theta \in \mathcal{L}_{-\diamond}$. Therefore, $[\![\psi]\!]_{\mathcal{M}_A} = [\![\psi]\!]_A^{PS} = [\![\psi]\!]_{[\![\theta]\!]_{W^0}^{PS}}^{PS} = [\![\langle\theta\rangle\psi]\!]_{W^0}^{PS}$. Therefore, $[\![\psi]\!]_{\mathcal{M}_A} \in \mathcal{A}$. We then have

$$[\![\langle\psi\rangle\chi]\!]_{\mathcal{M}_A} = [\![\chi]\!]_{\mathcal{M}_{[\![\psi]\!]_{\mathcal{M}_A}}} = [\![\chi]\!]_{\mathcal{M}_{[\![\psi]\!]_A^{PS}}} = [\![\chi]\!]_{[\![\psi]\!]_A^{PS}}^{PS} = [\![\langle\psi\rangle\chi]\!]_A^{PS},$$

by the semantics and IH on ψ and on χ (since $[\![\psi]\!]_A^{PS} \in \mathcal{A}$).
   Case $\varphi := \Diamond\psi$. We have:

$$[\![\Diamond\psi]\!]_{\mathcal{M}_A} = \bigcup\{[\![\langle\chi\rangle\psi]\!]_{\mathcal{M}_A} : \chi \in \mathcal{L}_{-\diamond}\} \qquad (\text{Defn.2})$$
$$= \bigcup\{[\![\langle\chi\rangle\psi]\!]_W^{PS} : \chi \in \mathcal{L}_{-\diamond}\} \qquad (\text{Lemma 1.2, IH})$$
$$= [\![\Diamond\psi]\!]_A^{PS} \qquad (\text{Lemma 3, since } \mathcal{M} \text{ is a standard pseudo model})$$

   (b) By part (a), $[\![\varphi]\!]_{\mathcal{M}_A^0} = [\![\varphi]\!]_{\mathcal{M}_{W^0}} = [\![\varphi]\!]_{W^0}^{PS}$ for all φ. Since $\mathcal{M}$ is standard, we have $A = [\![\theta]\!]_{W^0}^{PS} = [\![\theta]\!]_{\mathcal{M}_A^0}$ for some $\theta \in \mathcal{L}_{-\diamond}$, so $\mathcal{M}_A$ is an APALM model.
2. Let $\mathcal{M} = (W^0, W, \sim_1, \ldots, \sim_n, \|\cdot\|)$ be an APALM model. Since $\mathcal{A} = \{[\![\theta]\!]_{\mathcal{M}^0} : \theta \in \mathcal{L}_{-\diamond}\} \subseteq \mathcal{P}(W^0)$, the model $\mathcal{M}' = (W^0, \mathcal{A}, \sim_1, \ldots, \sim_n, \|\cdot\|)$ is a pre-model. Therefore, the semantics given in Definition 5 is defined on $\mathcal{M}' = (W^0, \mathcal{A}, \sim_1, \ldots, \sim_n, \|\cdot\|)$.
   (a) By Proposition 6, it suffices to prove that the pre-model $\mathcal{M}' = (W^0, \mathcal{A}, \sim_1, \ldots, \sim_n, \|\cdot\|)$ is standard, i.e. that $\{[\![\theta]\!]_{\mathcal{M}^0} : \theta \in \mathcal{L}_{-\diamond}\} = \{[\![\theta]\!]_{W^0}^{PS} : \theta \in \mathcal{L}_{-\diamond}\}$. For this, we need to show that for every APALM model $\mathcal{M} = (W^0, W, \sim_1, \ldots, \sim_n, \|\cdot\|)$, we have $[\![\theta]\!]_{\mathcal{M}} = [\![\theta]\!]_W^{PS}$ for all $\theta \in \mathcal{L}_{-\diamond}$.

We prove this by subformula induction on θ. The base cases and the inductive cases for Booleans are straightforward.

Case $\theta := \psi^0$. Then $[\![\psi^0]\!]_{\mathcal{M}} = [\![\psi]\!]_{\mathcal{M}^0} \cap W = [\![\psi]\!]_{W^0}^{PS} \cap W = [\![\psi]\!]_W^{PS}$ (by Definition 2, IH, and Definition 5).

Case $\theta := K_i\psi$. We have $[\![K_i\psi]\!]_{\mathcal{M}} = \{w \in W : w_i \subseteq [\![\psi]\!]_{\mathcal{M}}\} = \{w \in W : w_i^W \subseteq [\![\psi]\!]_W^{PS}\} = [\![K_i\psi]\!]_W^{PS}$ (by Definition 2, IH, and Definition 5).

Case $\theta := U\psi$. By Definitions 2 and 5, we have:

$$[\![U\psi]\!]_{\mathcal{M}} = \begin{cases} W & \text{if } [\![\psi]\!]_{\mathcal{M}} = W \\ \emptyset & \text{otherwise} \end{cases} \qquad [\![U\psi]\!]_W^{PS} = \begin{cases} W & \text{if } [\![\psi]\!]_W^{PS} = W \\ \emptyset & \text{otherwise} \end{cases}$$

By IH, $[\![\psi]\!]_W^{PS} = [\![\psi]\!]_{\mathcal{M}}$, therefore, $[\![U\psi]\!]_W^{PS} = [\![U\psi]\!]_{\mathcal{M}}$.

Case $\theta := \langle\psi\rangle\chi$. By Definition 2, we know that $[\![\langle\psi\rangle\chi]\!]_{\mathcal{M}} = [\![\chi]\!]_{\mathcal{M}|[\![\psi]\!]_{\mathcal{M}}}$. Now consider the relativized model $\mathcal{M}|[\![\psi]\!]_{\mathcal{M}} = (W^0, [\![\psi]\!]_{\mathcal{M}_A}, \sim_1, \ldots, \sim_n, \|\cdot\|)$. By Lemma 1(1) and IH on ψ, we have $[\![\psi]\!]_{\mathcal{M}} = [\![\psi]\!]_W^{PS}$. Moreover, by the definition of APALM models, we know that $W = [\![\theta]\!]_{\mathcal{M}^0}$ for some $\theta \in \mathcal{L}_{-\diamond}$. Therefore, $[\![\psi]\!]_{\mathcal{M}} = [\![\psi]\!]_{\mathcal{M}^0|[\theta]_{\mathcal{M}^0}} = [\![\langle\theta\rangle\psi]\!]_{\mathcal{M}^0}$. Therefore, since $\langle\theta\rangle\psi \in \mathcal{L}_{-\diamond}$, the model $\mathcal{M}|[\![\psi]\!]_{\mathcal{M}}$ is also an APALM model obtained by updating the initial model $\mathcal{M}^0$. We then have

$$[\![\langle\psi\rangle\chi]\!]_{\mathcal{M}} = [\![\chi]\!]_{\mathcal{M}|[\![\psi]\!]_{\mathcal{M}}} \hfill \text{(Defn. 2)}$$

$$= [\![\chi]\!]_{\mathcal{M}|[\![\psi]\!]_W^{PS}} \hfill \text{(IH on } \psi)$$

$$= [\![\chi]\!]_{[\![\psi]\!]_W^{PS}}^{PS} \qquad \text{(IH on } \chi \text{ since } \mathcal{M}|[\![\psi]\!]_{\mathcal{M}} \text{ is an APALM model)}$$

$$= [\![\langle\psi\rangle\chi]\!]_W^{PS} \hfill \text{(Defn. 5)}$$

(b) The proof of this part follows by $<$-induction on φ (where $<$ is as in Lemma 1). All the inductive cases are similar to ones in the above proof, except for the case $\varphi := \Diamond\psi$, shown below:

$$[\![\Diamond\psi]\!]_{\mathcal{M}} = \bigcup\{[\![\langle\chi\rangle\psi]\!]_{\mathcal{M}} : \chi \in \mathcal{L}_{-\diamond}\} \hfill \text{(Defn. 2)}$$

$$= \bigcup\{[\![\langle\chi\rangle\psi]\!]_W^{PS} : \chi \in \mathcal{L}_{-\diamond}\} \hfill \text{(Lemma 1.2, IH)}$$

$$= [\![\Diamond\psi]\!]_W^{PS} \qquad \text{(Lemma 3, since } \mathcal{M}' \text{ is a standard pseudo model)}$$

Corollary 1. *Validity on standard pseudo-models coincides with APALM validity.*

Proof. This is a straightforward consequence of Proposition 7.

Corollary 2. *The system* **APALM** *is sound wrt APALM models.*

This follows immediately from Proposition 5 and Corollary 1.

It is important to note that the equivalence between standard pseudo-models and APALM models (given by Proposition 7 above, and underlying our soundness result) is *not* trivial. It relies in particular on the equivalence between the effort modality and the arbitrary announcement operator $\Box$ on standard pseudo-models, which holds *only because our models and language retain the memory of the initial situation* (see Lemma 3). Hence, a similar equivalence fails for the original APAL.

5 Completeness

In this section we prove the completeness of **APALM**. First, we show completeness with respect to pseudo-models, via an innovative modification of the standard canonical model construction. This is based on a method previously used in [6], that makes an essential use of the finitary □-introduction rule, by requiring our canonical theories T to be (not only maximally consistent, but also) "witnessed". Roughly speaking, a theory T is witnessed if every $\Diamond\varphi$ occurring in every "existential context" in T is witnessed by some atomic formula p, meaning that $\langle p \rangle \varphi$ occurs in the same existential context in T. Our canonical pre-model will consist of all *initial, maximally consistent, witnessed theories* (where a theory is 'initial' if it contains the formula 0). A Truth Lemma is proved, as usual. Completeness for (both pseudo-models and) APALM models follows from the observation that our canonical pre-model *is standard*, hence it is (a standard pseudo-model, and thus) equivalent to a genuine APALM model.

We now proceed with the details. The appropriate notion of "existential context" is represented by *possibility forms*, in the following sense.

Definition 8 (Necessity forms and possibility forms). *For any finite string $s \in (\{\bullet^0\} \cup \{\varphi\to \mid \varphi \in \mathcal{L}\} \cup \{K_i : i \in \mathcal{A}\} \cup \{U\} \cup \{\rho \mid \rho \in \mathcal{L}_{-\Diamond}\})^* = NF_{\mathcal{L}}$, we define pseudo-modalities $[s]$ and $\langle s \rangle$. These pseudo-modalities are functions mapping any formula $\varphi \in \mathcal{L}$ to another formula $[s]\varphi \in \mathcal{L}$ (necessity form), respectively $\langle s \rangle\varphi \in \mathcal{L}$ (possibility form). The necessity forms are defined recursively as $[\epsilon]\varphi = \varphi$, $[s, \bullet^0]\varphi = [s]\varphi^0$, $[s, \varphi\to]\varphi = [s](\varphi \to \varphi)$, $[s, K_i]\varphi = [s]K_i\varphi$, $[s, U]\varphi = [s]U\varphi$, $[s, \rho]\varphi = [s][\rho]\varphi$, where ϵ is the empty string. For possibility forms, we set $\langle s \rangle\varphi := \neg[s]\neg\varphi$.*

Example: $[K_i, \bullet^0, \Diamond p\to, 0, U]$ *is a necessity form s.t.* $[K_i, \bullet^0, \Diamond p\to, 0, U]\varphi = K_i(\Diamond p \to [0]U\varphi)^0$.

Definition 9 (Theories: witnessed, initial, maximal). *Let $\mathcal{L}^P$ be the language of APALM based only on some countable set P of propositional variables. Similarly, let NF^P denote the corresponding set of strings defined based on $\mathcal{L}^P$. A P-theory is a consistent set of formulas in $\mathcal{L}^P$ (where "consistent" means consistent with respect to the axiomatization of APALM formulated for $\mathcal{L}^P$). A maximal P-theory is a P-theory Γ that is maximal with respect to $\subseteq$ among all P-theories; in other words, Γ cannot be extended to another P-theory. A P-witnessed theory is a P-theory Γ such that, for every $s \in NF^P$ and $\varphi \in \mathcal{L}^P$, if $\langle s \rangle\Diamond\varphi$ is consistent with Γ then there is $p \in P$ such that $\langle s \rangle\langle p \rangle\varphi$ is consistent with Γ (or equivalently: if $\Gamma \vdash [s][p]\neg\varphi$ for all $p \in P$, then $\Gamma \vdash [s]\Box\neg\varphi$). A P-theory Γ is called initial if $0 \in \Gamma$. A maximal P-witnessed theory Γ is a P-witnessed theory that is not a proper subset of any P-witnessed theory. A maximal P-witnessed initial theory Γ is a maximal P-witnessed theory such that $0 \in \Gamma$.*

The proofs of the following lemmas are in the online version.

Lemma 4. *For every necessity form $[s]$, there exist formulas $\theta \in \mathcal{L}_{-\Diamond}$ and $\psi \in \mathcal{L}$ such that for all $\varphi \in \mathcal{L}$, we have*

$$\vdash [s]\varphi \text{ iff } \vdash \psi \to [\theta]\varphi.$$

Lemma 5. *The following rule is admissible in* **APALM**:

$$\text{if } \vdash [s][p]\varphi \text{ then } \vdash [s]\Box\varphi, \text{ where } p \notin P_s \cup P_\varphi.$$

Lemma 6. *For every maximal P-witnessed theory Γ, and every formula $\varphi, \psi \in \mathcal{L}^P$:*

1. $\Gamma \vdash \varphi$ iff $\varphi \in \Gamma$
2. $\varphi \notin \Gamma$ iff $\neg\varphi \in \Gamma$,
3. $\varphi \wedge \psi \in \Gamma$ iff $\varphi \in \Gamma$ and $\psi \in \Gamma$,
4. $\varphi \in \Gamma$ and $\varphi \to \psi \in \Gamma$ implies $\psi \in \Gamma$.
5. $\mathbf{APALM}_P \subseteq \Gamma$, where $\mathbf{APALM}_P$ is **APALM** formulated for $\mathcal{L}^P$.

Lemma 7. *For every $\Gamma \subseteq \mathcal{L}^P$, if Γ is a P-theory and $\Gamma \nvdash \neg\varphi$ for some $\varphi \in \mathcal{L}^P$, then $\Gamma \cup \{\varphi\}$ is a P-theory. Moreover, if Γ is P-witnessed, then $\Gamma \cup \{\varphi\}$ is also P-witnessed.*

Lemma 8. *If $\{\Gamma_i\}_{i\in\mathbb{N}}$ is an increasing chain of P-theories such that $\Gamma_i \subseteq \Gamma_{i+1}$, then $\bigcup_{n\in\mathbb{N}} \Gamma_n$ is a P-theory.*

Lemma 9. *For every maximal P-witnessed theory T, both $\{\theta \in \mathcal{L}^P : K_i\theta \in T\}$ and $\{\theta \in \mathcal{L}^P : U\theta \in T\}$ are P-witnessed theories.*

Lemma 10 (Lindenbaum's Lemma). *Every P-witnessed theory Γ can be extended to a maximal P-witnessed theory T_Γ.*

Lemma 11 (Extension Lemma). *Let P be a countable set of propositional variables and P' be a countable set of fresh propositional variables, i.e., $P \cap P' = \emptyset$. Let $\widetilde{P} = P \cup P'$. Then, every initial P-theory Γ can be extended to an initial $\widetilde{P}$-witnessed theory $\widetilde{\Gamma} \supseteq \Gamma$, and hence to a maximal $\widetilde{P}$-witnessed initial theory $T_\Gamma \supseteq \Gamma$.*

To define our canonical pseudo-model, we first put, for all maximal P-witnessed theories T, S:

$$T \sim_U S \text{ iff } \forall\varphi \in \mathcal{L}^P (U\varphi \in T \text{ implies } \varphi \in S).$$

Definition 10 (Canonical Pre-Model). *Given a maximal P-witnessed initial theory T_0, the canonical pre-model for T_0 is a tuple $\mathcal{M}^c = (W^c, \mathcal{A}^c, \sim_1^c, \ldots, \sim_n^c, \|\cdot\|^c)$ such that:*

- $W^c = \{T : T$ is a maximal P-witnessed theory such that $T_0 \sim_U T\}$,
- $\mathcal{A}^c = \{\widehat{\theta} : \theta \in \mathcal{L}^P_{-\Diamond}\}$ where $\widehat{\varphi} = \{T \in W^c : \varphi \in T\}$ for any $\varphi \in \mathcal{L}^P$,

– *for every* $T, S \in W^c$ *and* $i \in \mathcal{AG}$ *we define:*

$$T \sim_i^c S \quad iff \quad \forall \varphi \in \mathcal{L}^P \, (K_i \varphi \in T \text{ implies } \varphi \in S).$$

– $\|p\|^c = \{T \in W^c \; : \; p \in T\} = \widehat{p}.$

As usual, it is easy to see (given the $S5$ axioms for K_i and for U) that $\sim_U$ and $\sim_i^c$ are equivalence relations.

The proofs of the first three results below are online.

Lemma 12 (Existence Lemma for K_i). *Let T be a maximal P-witnessed theory, $\alpha \in \mathcal{L}_{-\Diamond}^P$, and $\varphi \in \mathcal{L}^P$ such that $\alpha \in T$ and $K_i[\alpha]\varphi \notin T$. Then, there is a maximal P-witnessed theory S such that $T \sim_i S$, $\alpha \in S$ and $[\alpha]\varphi \notin S$.*

Lemma 13 (Existence Lemma for U). *Let T be a maximal P-witnessed theory, $\alpha \in \mathcal{L}_{-\Diamond}^P$, and $\varphi \in \mathcal{L}^P$ such that $\alpha \in T$ and $U[\alpha]\varphi \notin T$. Then, there is a maximal P-witnessed theory S such that $T \sim_U S$, $\alpha \in S$ and $[\alpha]\varphi \notin S$.*

Corollary 3. *For $\varphi \in \mathcal{L}$, we have $\widehat{U\varphi} = W^c$ if $\widehat{\varphi} = W^c$, and $\widehat{U\varphi} = \emptyset$ otherwise.*

Lemma 14. *Every element $T \in W^c$ is an initial theory (i.e. $0 \in T$).*

Proof. Let $T \in W^c$. By the construction of W^c, we have $T_0 \sim_U T$. Since $0 \to U0$ is an axiom and T_0 is maximal, $(0 \to U0) \in T_0$. Thus, since $0 \in T_0$, we obtain $U0 \in T_0$ (by Lemma 6(4)). Therefore, by the definition of $\sim_U$ and since $T_0 \sim_U T$, we have that $0 \in T$.

Corollary 4. *For all $\varphi \in \mathcal{L}^P$, we have $\widehat{\varphi} = \widehat{\varphi^0}$.*

Proof. Since $0 \in T$ for all $T \in W^c$, we obtain by axiom (0-eq) that $\varphi \leftrightarrow \varphi^0 \in T$ for all $T \in W^c$. Therefore, $\widehat{\varphi} = \widehat{\varphi^0}$.

Lemma 15 (Truth Lemma). *Let $\mathcal{M}^c = (W^c, \mathcal{A}^c, \sim_1^c, \ldots, \sim_n^c, V^c)$ be the canonical pre-model for some T_0 and $\varphi \in \mathcal{L}^P$. Then, for all $\alpha \in \mathcal{L}_{-\Diamond}^P$, we have $[\![\varphi]\!]_{\widehat{\alpha}} = \widehat{\langle \alpha \rangle \varphi}.$*

Proof. The proof is by $<$-induction on φ, using the following *induction hypothesis* (IH): for all $\psi < \varphi$, we have $[\![\psi]\!]_{\widehat{\alpha}} = \widehat{\langle \alpha \rangle \psi}$ for all $\alpha \in \mathcal{L}_{-\Diamond}$. The cases for the Boolean connectives are straightforward. The cases for K_i and U are standard, using $\vdash \langle \alpha \rangle K_i \psi \leftrightarrow \alpha \wedge K_i [\alpha] \psi$ and Lemma 12 for K_i, and $\vdash \langle \alpha \rangle U \psi \leftrightarrow \alpha \wedge U [\alpha] \psi$ and Lemma 13 for U.

Base case $\varphi := \top$. Then $[\![\top]\!]_{\widehat{\alpha}} = \widehat{\alpha} = \widehat{\langle \alpha \rangle \top}$, by Definition 5 and the fact that $\vdash \alpha \leftrightarrow \langle \alpha \rangle \top$.

Base case $\varphi := p$. Then $[\![p]\!]_{\widehat{\alpha}} = \|p\|^c \cap \widehat{\alpha} = \widehat{p} \cap \widehat{\alpha} = \widehat{p \wedge \alpha} = \widehat{\langle \alpha \rangle p}$, by Definition 5, the defn. of $\| \cdot \|^c$, R_p, and Proposition 2(3).

Base case $\varphi := 0$. Then $[\![0]\!]_{\widehat{\alpha}} = W^c$ if $\widehat{\alpha} = W^c$, and $[\![0]\!]_{\widehat{\alpha}} = \emptyset$ otherwise. Also, $\widehat{\langle \alpha \rangle 0} = \widehat{0 \wedge U\alpha} = \widehat{0} \cap \widehat{U\alpha} = \widehat{0} \cap \widehat{U\alpha} = \widehat{U\alpha}$ (by Propositions 2(2) and Lemma 14). By Corollary 3, $\widehat{U\alpha} = W^c$ if $\widehat{\alpha} = W^c$, and $\widehat{U\alpha} = \emptyset$ otherwise. So $[\![0]\!]_{\widehat{\alpha}} = \widehat{\langle \alpha \rangle 0}$.

Case $\varphi := \psi^0$. Follows easily from $\widehat{\top} = W^c$ and R[$\top$], Corollary 4, and R^0.

Case $\varphi := \langle\chi\rangle\psi$. Straightforward, using the fact that $\vdash \langle\alpha\rangle\langle\chi\rangle\psi \leftrightarrow \langle\langle\alpha\rangle\chi\rangle\psi$ (by $R_{[!]}$).

Case $\varphi := \Diamond\psi$.

($\Rightarrow$) Suppose $T \in [\![\Diamond\psi]\!]_{\widehat{\alpha}}$. This means, by Definition 5, that $\alpha \in T$ and there exists $B \in \mathcal{A}^c$ such that $T \in B \subseteq \widehat{\alpha}$ and $T \in [\![\psi]\!]_B$ (see Observation 2). By the construction of $\mathcal{A}^c$, we know that $B = \widehat{\theta}$ for some $\theta \in \mathcal{L}^P_{-\Diamond}$. Therefore, $T \in [\![\psi]\!]_B$ means that $T \in [\![\psi]\!]_{\widehat{\theta}}$. Moreover, since $\widehat{\theta} \subseteq \widehat{\alpha}$ and, thus, $\widehat{\theta} = \widehat{\alpha} \cap \widehat{\theta} = \widehat{\alpha \wedge \theta}$, we obtain $T \in [\![\psi]\!]_{\widehat{\alpha \wedge \theta}}$. By Lemma 1(1), we have $\psi < \Box\psi$. Therefore, by IH, we obtain $T \in \widehat{\langle\alpha \wedge \theta\rangle\psi}$. Then, by axiom ([!]$\Box$-elim) and the fact that T is maximal, we conclude that $T \in \widehat{\langle\alpha\rangle\Diamond\psi}$.

($\Leftarrow$) Suppose $T \in \widehat{\langle\alpha\rangle\Diamond\psi}$, i.e., $\langle\alpha\rangle\Diamond\psi \in T$. Then, since T is a maximal P-witnessed theory, there is $p \in P$ such that $\langle\alpha\rangle\langle p\rangle\psi \in T$. By Lemma 1(2), we know that $\langle p\rangle\psi < \Diamond\psi$. Thus, by IH on $\langle p\rangle\psi$, we obtain that $T \in [\![\langle p\rangle\psi]\!]_{\widehat{\alpha}}$. This means, by Definition 5 and Observation 2, that $T \in [\![\psi]\!]_{[\![p]\!]_{\widehat{\alpha}}} \subseteq [\![p]\!]_{\widehat{\alpha}}$. Since $p < \Diamond\psi$, by IH on p, we obtain that $[\![p]\!]_{\widehat{\alpha}} = \widehat{\langle\alpha\rangle p} \subseteq \widehat{\alpha}$. By the construction of $\mathcal{A}^c$, we moreover have $\widehat{\langle\alpha\rangle p} \in \mathcal{A}^c$. Therefore, as $T \in [\![\psi]\!]_{\widehat{\langle\alpha\rangle p}}$ and $\widehat{\langle\alpha\rangle p} \subseteq \widehat{\alpha}$, by Definition 5, we conclude that $T \in [\![\Diamond\psi]\!]_{\widehat{\alpha}}$.

Corollary 5. *The canonical pre-model $\mathcal{M}^c$ is standard (and hence a pseudo-model).*

Proof. $\mathcal{A}^c = \{\widehat{\theta} : \theta \in \mathcal{L}^P_{-\Diamond}\} = \{\widehat{\langle\top\rangle\theta} : \theta \in \mathcal{L}^P_{-\Diamond}\} = \{[\![\theta]\!]_{\widehat{\top}} : \theta \in \mathcal{L}^P_{-\Diamond}\} = \{[\![\theta]\!]_{W^c} : \theta \in \mathcal{L}^P_{-\Diamond}\}$.

Lemma 16. *For every $\varphi \in \mathcal{L}^P$, if φ is consistent then $\{0, \Diamond\varphi\}$ is an initial P_φ-theory.*

Proof. Let $\varphi \in \mathcal{L}^P$ s.t. $\varphi \not\vdash \bot$. By the *Equivalences with* ^{0} in Table 1, we have $\vdash \bot^0 \leftrightarrow (p \wedge \neg p)^0 \leftrightarrow (p^0 \wedge \neg p^0) \leftrightarrow (p \wedge \neg p) \leftrightarrow \bot$. Therefore, $\vdash \psi \to \bot^0$ iff $\vdash \psi \to \bot$ for all $\psi \in \mathcal{L}^P$. Then, by Proposition 2(12), we obtain $\vdash \varphi \to \bot$ iff $\vdash (0 \wedge \Diamond\varphi) \to \bot$. Since $\varphi \not\vdash \bot$, we have $0 \wedge \Diamond\varphi \not\vdash \bot$, i.e., $\{0, \Diamond\varphi\}$ is a P_φ-theory. By definition, it is an initial one.

Corollary 6. *APALM is complete with respect to standard pseudo models.*

Proof. Let φ be a consistent formula. By Lemma 16, $\{0, \Diamond\varphi\}$ is an initial P_φ-theory. By Extension and Lindenbaum Lemmas, we can extend P_φ to some $P \supseteq P_\varphi$, and extend $\{0, \Diamond\varphi\}$ to some maximal P-witnessed theory T_0 such that $(0 \wedge \Diamond\varphi) \in T_0$. So T_0 is initial, and we can construct the canonical pseudo-model $\mathcal{M}^c$ for T_0. Since $\Diamond\varphi \in T_0$ and T_0 is P-witnessed, there exists $p \in P$ such that $\langle p\rangle\varphi \in T_0$. By Truth Lemma (applied to $\alpha := p$), we get $T_0 \in [\![\varphi]\!]_{\widehat{p}}$. Hence, φ is satisfied at T_0 in the set $\widehat{p} \in \mathcal{A}^c$.

Corollary 7. *APALM is complete with respect to APALM models.*

This follows immediately from Corollaries 1 and 6.

6 Expressivity

To compare APALM and its fragments with basic epistemic logic (and its extension with the universal modality), consider the *static fragment* $\mathcal{L}_{-\Diamond,\langle!\rangle}$, obtained from $\mathcal{L}$ by removing both the $\Diamond$ operator and the dynamic modality $\langle\varphi\rangle\psi$; as well as the *present-only fragment* $\mathcal{L}_{-\Diamond,\langle!\rangle,0,\varphi^0}$, obtained by removing the operators 0 and φ^0 from $\mathcal{L}_{-\Diamond,\langle!\rangle}$; and finally the *epistemic fragment* $\mathcal{L}_{epi}$, obtained by further removing the universal modality U from $\mathcal{L}_{-\Diamond,\langle!\rangle,0,\varphi^0}$. For every APALM model $\mathcal{M} = (W^0, W, \sim_1, \ldots, \sim_n, \|\cdot\|)$, consider its *initial epistemic model* $\mathcal{M}^{initial} = (W^0, \sim_1, \ldots, \sim_n, \|\cdot\|)$ and its *current epistemic model* $\mathcal{M}^{current} = (W, \sim_1 \cap W \times W, \ldots, \sim_n \cap W \times W, \|\cdot\| \cap W)$.

Proposition 8. *The fragment $\mathcal{L}_{-\Diamond}$ is co-expressive with the static fragment $\mathcal{L}_{-\Diamond,\langle!\rangle}$. In fact, every formula $\varphi \in \mathcal{L}_{-\Diamond}$ is provably equivalent with some formula $\psi \in \mathcal{L}_{-\Diamond,\langle!\rangle}$ (by using **APALM** reduction laws to eliminate dynamic modalities, as in standard PAL).*

Proposition 9. *The static fragment $\mathcal{L}_{-\Diamond,\langle!\rangle}$ (and hence, also $\mathcal{L}_{-\Diamond}$) is strictly more expressive than the present-only fragment $\mathcal{L}_{-\Diamond,\langle!\rangle,0,\varphi^0}$, which in turn is more expressive than the epistemic fragment $\mathcal{L}_{epi}$. In fact, each of the operators 0 and φ^0 independently increase the logic's expressivity.*

Kuijer's counterexample shows that the standard epistemic bisimulation is not appropriate for APALM, so we now define a new such notion:

Definition 11 (APALM Bisimulation). *An APALM bisimulation between APALM models $\mathcal{M}_1 = (W_1^0, W_1, \sim_1, \ldots, \sim_n, \|\cdot\|)$ and $\mathcal{M}_2 = (W_2^0, W_2, \sim_1, \ldots, \sim_n, \|\cdot\|)$ is a total bisimulation B (in the usual sense)[17] between the corresponding initial epistemic models $\mathcal{M}_1^{initial}$ and $\mathcal{M}_2^{initial}$, with the property that: if $s_1 B s_2$, then $s_1 \in W_1$ holds iff $s_2 \in W_2$ holds. Two current states $s_1 \in W_1$ and $s_2 \in W_2$ are APALM-bisimilar if there exists an APAL bisimulation B between the underlying APALM models such that $s_1 B s_2$.*

Since APALM models are always of the form $\mathcal{M} = (\mathcal{M}^0)|\theta$ for some $\theta \in \mathcal{L}_{-\Diamond}$, we have a characterization of APALM-bisimulation only in terms of the initial models:

Proposition 10. *Let $\mathcal{M}_1 = (W_1^0, W_1, \sim_1, \ldots, \sim_n, \|\cdot\|)$ and $\mathcal{M}_2 = (W_2^0, W_2, \sim_1, \ldots, \sim_n, \|\cdot\|)$ be APALM models, and let $B \subseteq W_1^0 \times W_2^0$. The following are equivalent:*

---

[17] A *total bisimulation* between epistemic models $(W, \sim_1, \ldots, \sim_n, \|\cdot\|)$ and $(W', \sim_1', \ldots, \sim_n', \|\cdot\|')$ is an epistemic bisimulation relation (satisfying the usual valuation and back-and-forth conditions from Modal Logic) $B \subseteq W \times W'$, such that: for every $s \in W$ there exists some $s' \in W'$ with sBs'; and dually, for every $s' \in W'$ there exists some $s \in W$ with sBs'.

1. B is an APALM bisimulation between $\mathcal{M}_1$ and $\mathcal{M}_2$;
2. B is a total bisimulation between $\mathcal{M}_1^{initial}$ and $\mathcal{M}_2^{initial}$ (or equivalently, an APALM bisimulation between $\mathcal{M}_1^0$ and $\mathcal{M}_2^0$), and $\mathcal{M}_1 = (\mathcal{M}_1^0)|\theta$, $\mathcal{M}_2 = (\mathcal{M}_2^0)|\theta$ for some common formula $\theta \in \mathcal{L}_{-\Diamond}$.

So, to check for APALM-bisimilarity, it is enough to check for total bisimilarity between the initial models and for both models being updates with the same formula.

Proposition 11. APALM formulas are invariant under APALM-bisimulation: if $s_1 B s_2$ for some APALM-bisimulation relation between APALM models $\mathcal{M}_1 = (W_1^0, W_1, \sim_1, \ldots, \sim_n, \|\cdot\|)$ and $\mathcal{M}_2 = (W_2^0, W_2, \sim_1, \ldots, \sim_n, \|\cdot\|)$, then: $s_1 \in [\![\varphi]\!]_{\mathcal{M}_1}$ iff $s_2 \in [\![\varphi]\!]_{\mathcal{M}_2}$. The (Hennessy-Milner) converse holds for finite models: if $\mathcal{M}_1 = (W_1^0, W_1, \sim_1, \ldots, \sim_n, \|\cdot\|)$ and $\mathcal{M}_2 = (W_2^0, W_2, \sim_1, \ldots, \sim_n, \|\cdot\|)$ are APALM models with W_1^0, W_2^0 finite, then $s_1 \in W_1$ and $s_2 \in W_2$ satisfy the same APALM formulas iff they are APALM-bisimilar.

7   Conclusions and Future Work

This paper solves the open question of finding a strong variant of APAL that is recursively axiomatizable. Our system APALM is inspired by our analysis of Kuijer's counterexample [11], which lead us to add to APAL a 'memory' of the initial situation. The soundness and completeness proofs crucially rely on a Subset Space-like semantics and on the equivalence between the effort modality and the arbitrary announcement modality, thus revealing the strong link between these two formalisms.

It seems clear that our method works for other versions of APAL, and in ongoing work we are looking at a recursive axiomatization GALM for a memory-enhanced variant of GAL (Group Announcement Logic) [1]. As in GAL, the $\Box$ operator quantifies only over announcements that are known by some of the agents, so GALM seems better fit than APALM for treating puzzles involving epistemic dialogues.

Acknowledgments. We thank the anonymous WOLLIC referees for their extremely valuable comments on a previous draft of this paper.

References

1. Ågotnes, T., Balbiani, P., van Ditmarsch, H., Seban, P.: Group announcement logic. J. Appl. Log. **8**, 62–81 (2010)
2. Balbiani, P., Baltag, A., van Ditmarsch, H., Herzig, A., Hoshi, T., de Lima, T.: 'Knowable' as 'known after an announcement'. Rev. Symb. Log. **1**, 305–334 (2008)
3. Balbiani, P., van Ditmarsch, H.: A simple proof of the completeness of APAL. Stud. Log. **8**(1), 65–78 (2015)

4. Balbiani, P., van Ditmarsch, H., Kudinov, A.: Subset space logic with arbitrary announcements. In: Lodaya, K. (ed.) ICLA 2013. LNCS, vol. 7750, pp. 233–244. Springer, Heidelberg (2013). https://doi.org/10.1007/978-3-642-36039-8_21
5. Baltag, A., Gierasimczuk, N., Özgün, A., Vargas Sandoval, A.L., Smets, S.: A dynamic logic for learning theory. In: Madeira, A., Benevides, M. (eds.) DALI 2017. LNCS, vol. 10669, pp. 35–54. Springer, Cham (2018). https://doi.org/10.1007/978-3-319-73579-5_3
6. Baltag, A., Özgün, A., Vargas Sandoval, A.L.: Topo-logic as a dynamic-epistemic logic. In: Baltag, A., Seligman, J., Yamada, T. (eds.) LORI 2017. LNCS, vol. 10455, pp. 330–346. Springer, Heidelberg (2017). https://doi.org/10.1007/978-3-662-55665-8_23
7. Bjorndahl, A.: Topological subset space models for public announcements. In: van Ditmarsch, H., Sandu, G. (eds.) Jaakko Hintikka on Knowledge and Game-Theoretical Semantics. OCL, vol. 12, pp. 165–186. Springer, Cham (2018). https://doi.org/10.1007/978-3-319-62864-6_6
8. Bjorndahl, A., Özgün, A.: Topological subset space models for belief. In: Proceedings of the 16th TARK. EPTCS, vol. 251, pp. 88–101. Open Publishing Association (2017)
9. Bolander, T., Andersen, M.B.: Epistemic planning for single-and multi-agent systems. J. Appl. Non-Class. Log. **21**, 9–34 (2011)
10. Dabrowski, A., Moss, L.S., Parikh, R.: Topological reasoning and the logic of knowledge. Ann. Pure Appl. Log. **78**, 73–110 (1996)
11. Kuijer, L.B.: Unsoundness of $R(\Box)$ (2015). http://personal.us.es/hvd/APAL_counterexample.pdf
12. McCarthy, J.: Formalization of two puzzles involving knowledge. In: Formalizing Common Sense: Papers by John McCarthy, (Original 1978–81). Ablex Publishing Corporation (1990)
13. Moss, L.S., Parikh, R.: Topological reasoning and the logic of knowledge. In: Proceedings of the 4th TARK, pp. 95–105. Morgan Kaufmann (1992)
14. Renne, B., Sack, J., Yap, A.: Dynamic epistemic temporal logic. In: He, X., Horty, J., Pacuit, E. (eds.) LORI 2009. LNCS (LNAI), vol. 5834, pp. 263–277. Springer, Heidelberg (2009). https://doi.org/10.1007/978-3-642-04893-7_21
15. van Benthem, J.: One is a Lonely Number: On the Logic of Communication. Lecture Notes in Logic, pp. 96–129. Cambridge University Press, Cambridge (2002)
16. van Benthem, J.: What one may come to know. Analysis **64**, 95–105 (2004)
17. van Ditmarsch, H.: The russian cards problem. Studia Logica **75**, 31–62 (2003)
18. van Ditmarsch, H., French, T.: Quantifying over Boolean announcements (2017). https://arxiv.org/abs/1712.05310
19. van Ditmarsch, H., French, T., Hales, J.: Positive announcements (2018). https://arxiv.org/abs/1803.01696
20. van Ditmarsch, H., French, T., Pinchinat, S.: Future event logic-axioms and complexity. Adv. Modal Log. **8**, 77–99 (2010)
21. van Ditmarsch, H., Knight, S., Özgün, A.: Arbitrary announcements on topological subset spaces. In: Bulling, N. (ed.) EUMAS 2014. LNCS (LNAI), vol. 8953, pp. 252–266. Springer, Cham (2015). https://doi.org/10.1007/978-3-319-17130-2_17
22. van Ditmarsch, H., Knight, S., Özgün, A.: Announcement as effort on topological spaces. In: Proceedings of the 15th TARK, pp. 95–102 (2015)
23. van Ditmarsch, H., Knight, S., Özgün, A.: Announcement as effort on topological spaces–Extended version. Synthese, 1–43 (2015)
24. van Ditmarsch, H., van der Hoek, W., Kuijer, L.B.: Fully arbitrary public announcements. Adv. Modal Log. **11**, 252–267 (2016)

25. Wang, Y.N., Ågotnes, T.: Multi-agent subset space logic. In: Proceedings of the 23rd IJCAI, pp. 1155–1161. IJCAI/AAAI (2013)
26. Wáng, Y.N., Ågotnes, T.: Subset space public announcément logic. In: Lodaya, K. (ed.) ICLA 2013. LNCS, vol. 7750, pp. 245–257. Springer, Heidelberg (2013). https://doi.org/10.1007/978-3-642-36039-8_22

Lindenbaum and Pair Extension Lemma in Infinitary Logics

Marta Bílková[2,3], Petr Cintula[2(✉)], and Tomáš Lávička[1,3]

[1] Institute of Information Theory and Automation, Czech Academy of Sciences,
Pod vodárenskou věží 4, 182 08 Prague, Czech Republic
`lavicka.thomas@gmail.com`
[2] Institute of Computer Science, Czech Academy of Sciences,
Pod vodárenskou věží 2, 182 07 Prague, Czech Republic
`marta.bilkova@ff.cuni.cz, cintula@cs.cas.cz`
[3] Faculty of Arts, Charles University,
Jana Palacha 1/2, 116 38 Prague, Czech Republic

Abstract. The abstract Lindenbaum lemma is a crucial result in algebraic logic saying that the prime theories form a basis of the closure systems of all theories of an arbitrary given logic. Its usual formulation is however limited to finitary logics, i.e., logics with Hilbert-style axiomatization using finitary rules only. In this contribution, we extend its scope to all logics with a countable axiomatization and a well-behaved disjunction connective. We also relate Lindenbaum lemma to the Pair extension lemma, other well-known result with many applications mainly in the theory of non-classical modal logics. While a restricted form of this lemma (to pairs with finite right-hand side) is, in our context, equivalent to Lindenbaum lemma, we show a perhaps surprising result that in full strength it holds for finitary logics only. Finally we provide examples demonstrating both limitations and applications of our results.

Keywords: Lindenbaum lemma · Pair extension lemma
Infinitary logic · Infinitary deduction rule · Strong disjunction
Prime theory

1 Introduction

Lindenbaum's lemma, originally proved in 20s and published and attributed to Lindenbaum by Tarski in 1930, states that any consistent theory of classical predicate logic can be extended to a maximal consistent one. It is a crucial lemma used to prove strong completeness not only for classical logic, but for many non-classical logics as well. However, in some non-classical logics only a weaker variant of the lemma is true, and there are two lines of research generalizing this lemma.

P. Cintula and T. Lávička are supported by the project GA17-04630S of the Czech Science Foundation; M. Bílková is supported by the project SEGA GAČR-DFG 16-07954J; P. Cintula also acknowledges the support of RVO 67985807 and JSPS-16-08.

© Springer-Verlag GmbH Germany, part of Springer Nature 2018
L. S. Moss et al. (Eds.): WoLLIC 2018, LNCS 10944, pp. 130–144, 2018.
https://doi.org/10.1007/978-3-662-57669-4_7

The first one, which still usually bears the name 'Lindenbaum lemma', was proved in the framework of Abstract Algebraic Logic and says that any theory not proving a certain formula can be extended into a prime theory not proving it as well.

The other extension is known as Pair extension lemma and was originally proved by Belnap in the early 70s. It works with a notion of pair (two sets of formulas such that the former proves no finite disjunction of elements of the later) and has been used to prove strong completeness of various logics where the classical negation is not available, including intuitionistic logic, relevant and other substructural logics [9, 20]. It is especially useful to prove canonical completeness w.r.t. Kripke semantics of these logics and their modal extensions.

As originally proven, both Lindenbaum and Pair extension lemmata rely on the fact that a union of a chain of consistent theories is a consistent theory, and thus apply to finitary logics only; in which context it is easy to see that both lemmata are actually equivalent. Generalizations of Lindenbaum lemma to logics with infinitary axiomatizations have been considered both in algebraic and modal logic tradition.

- In 1977, Sundholm proved strong completeness of Von Wright's temporal logic [25]. In 1984, Goldblatt proved a general result about the existence of maximally consistent theories satisfying certain prescribed closure conditions [12], and later in his 1993 book [13] outlined a general approach to prove Lindenbaum lemma in an infinitary setting. In 1994, Segerberg [24] offered a general method of strong completeness proofs for non-compact modal logics, using saturated sets of formulas, which in many cases coincide with maximally consistent theories. More recently, de Lavalette et al. [17] used a similar proof of Lindenbaum lemma to prove completeness, and investigate limitations of canonicity, of infinitary axiomatization of PDL and some related non-compact modal logics (such as epistemic logics with a common knowledge modality). These results only concern modal logics and assume the background propositional logic to be classical.
- In the algebraic logic tradition let us mention the seminal work of Franco Montagna [19] who used an infinitary rule and a storage operator to axiomatize infinitary versions of prominent fuzzy logics; followed by the work of Vidal et al. [26] who managed to avoid the storage operator and proved analogous results for logics with countably many truth constants and Baaz–Moneiro Δ connective; and finally the work of Kułacka [16] who proved a particular form of the infinitary Lindenbaum lemma directly for the infinitary version of the basic fuzzy logic BL and its prominent axiomatic extensions such as the infinitary product and Łukasiewicz logics. There is also a series of papers on fuzzy logics with truth constants [10, 11, 15] where (among others) the authors use infinitary rules to establish the Pavelka-style completeness for these logics; however all those papers contain a gap in the proof of the appropriate Lindenbaum lemma. This gap was later corrected by Cintula [3] who proved the Lindenbaum lemma for logics involving two special kinds of infinitary rules necessary for the completeness proof.

In this paper, we offer a general abstract algebraic perspective at both lemmata for infinitary logics, widening the area of their applicability. The paper is organized as follows: after a short preliminary section, where we fix the notation and recall the needed notions, we generalize the results mentioned above in two directions.

- First, in Sect. 3 we prove a Lindenbaum lemma (Theorem 1) for infinitary logics in a general setting, making two assumptions about the logic at hand only: a well-behaved disjunction connective being present (or definable) in the language, and the logic having a countable axiomatization.[1] We show that neither of these assumptions can be easily dropped and that infinitary Łukasiewicz logic $Ł_\infty$ (Example 1) is a prominent example of logic satisfying both of them, and we use the Lindenbaum lemma to obtain a new and simple proof of its standard completeness (Proposition 2).
- Second, in Sect. 4 we formulate a general Pair extension lemma, show that in our context Lindenbaum lemma is equivalent to its restricted form (for pairs with a finite right-hand side, Theorem 2) whereas, surprisingly, it holds in the full strength for finitary logics only (Theorem 3).

2 Preliminaries

By $Fm_\mathcal{L}$ we denote the set of formulas in a given propositional language $\mathcal{L}$. An $\mathcal{L}$-consecution is a pair $\Gamma \rhd \varphi$, where Γ is a set of formulas. Given a set of $\mathcal{L}$-consecutions L, we write $\Gamma \vdash_L \varphi$ rather than $\Gamma \rhd \varphi \in L$. A logic in the language $\mathcal{L}$ is a set of $\mathcal{L}$-consecutions such that $\vdash_L$ is a structural consequence relation, i.e., for each $\Gamma \cup \Delta \cup \Pi \cup \{\varphi\} \subseteq Fm_\mathcal{L}$ we have:

- $\Gamma, \varphi \vdash_L \varphi$.   (Reflexivity)
- If $\Delta \vdash_L \psi$ for each $\psi \in \Pi$ and $\Gamma, \Pi \vdash_L \varphi$, then $\Gamma, \Delta \vdash_L \varphi$.   (Cut)
- If $\Gamma \vdash_L \varphi$, then $\sigma[\Gamma] \vdash_L \sigma(\varphi)$ for each $\mathcal{L}$-substitution σ.   (Structurality)

A logic is said to be *finitary* if it satisfies the following additional condition for each $\Gamma \cup \{\varphi\} \subseteq Fm_\mathcal{L}$:

- If $\Gamma \vdash_L \varphi$, then there is a *finite* $\Gamma' \subseteq \Gamma$ such that $\Gamma' \vdash_L \varphi$.   (Finitarity)

On the left-hand sides of consequence relations we often write Γ, Δ and Γ, φ for $\Gamma \cup \Delta$ and $\Gamma \cup \{\varphi\}$ respectively.

A *theory* of a logic L is a set of formulas closed under the consequence relation. Note that the set Th(L) all theories of L contains $Fm_\mathcal{L}$ and is closed under arbitrary intersections and therefore it can be seen as a *closure system* and a domain of a *complete lattice*. Therefore we can use notions such as the theory generated by a set Γ (the smallest element of Th(L) containing Γ), a basis of Th(L) (a subset of Th(L) separating theories and formulas), or (finitely) meet-irreducible elements of Th(L).

---

[1] cf. [24], where it is explicitly assumed that there are only countably many instances of infinitary rules for the proof to work.

Lemma 1 (Lindenbaum). *Let* L *be a finitary logic. Then the (finitely) meet-irreducible theories form a basis of* Th(L).

Let us briefly recall a proof of this well-known claim to illustrate the crucial role of finitarity for the Lindenbaum lemma: Take $T \in$ Th(L) such that $\varphi \notin T$ and define $\mathcal{T} = \{S \in \text{Th(L)} \mid T \subseteq S \text{ and } \varphi \notin S\}$; due to the maximality principle (Zorn's lemma) $\mathcal{T}$ has a maximal element T' (w.r.t. $\subseteq$) which is clearly meet-irreducible (because the intersection of *all* strictly bigger theories contains φ). Note that without the finitarity assumption we would not be able to use the Zorn's lemma; indeed without it the union of a chain of theories from $\mathcal{T}$ need not belong to $\mathcal{T}$.

An *axiomatic system* $\mathcal{AS}$ in the language $\mathcal{L}$ is a set of $\mathcal{L}$-consecutions, closed under arbitrary substitutions. The elements of $\mathcal{AS}$ of the form $\Gamma \rhd \varphi$ are called *axioms* if $\Gamma = \emptyset$, *finitary deduction rules* if Γ is finite, and *infinitary deduction rules* otherwise.

A *proof* of a formula φ from a set of formulas Γ in $\mathcal{AS}$ is a well-founded tree (with no infinitely-long branch) labeled by formulas in such a way that

- its root is labeled by φ, and leaves by axioms of $\mathcal{AS}$ or elements of Γ,
- if a node is labeled by ψ and $\Delta \neq \emptyset$ is the set of labels of its preceding nodes, then $\Delta \rhd \psi \in \mathcal{AS}$.

We write $\Gamma \vdash_{\mathcal{AS}} \varphi$ if there is a proof of φ from Γ. The relation $\vdash_{\mathcal{AS}}$ is the least logic containing $\mathcal{AS}$.

An axiomatic system $\mathcal{AS}$ is called a presentation of a logic L if $\vdash_L = \vdash_{\mathcal{AS}}$. Recall that a logic is finitary iff it has a presentation without infinitary rules.

We say that a logic is *countably axiomatizable* if it has a countable presentation. Clearly, any finitary logic is countably axiomatizable (provided, of course, it has countable language). Let us present another example of such a logic.

Example 1. Consider the language $\mathcal{L}_L$ with primitive connectives $\rightarrow$ and $\neg$ and two additional defined connectives:

$$\varphi \,\&\, \psi = \neg(\varphi \rightarrow \neg\psi) \qquad\qquad \varphi \lor \psi = (\varphi \rightarrow \psi) \rightarrow \psi.$$

The infinite-valued Łukasiewicz logic Ł in the language $\mathcal{L}_L$ can be axiomatized using four axioms and the rule of *modus ponens,* and therefore it is a finitary logic [4,18].

The *infinitary Łukasiewicz logic* Ł$_\infty$ is an extension of Ł by the following infinitary rule (we define $\varphi^1 = \varphi$ and $\varphi^{n+1} = \varphi^n \,\&\, \varphi$):

$$\{\neg\varphi \rightarrow \varphi^n \mid n > 0\} \rhd \varphi. \qquad\qquad (\text{Ł}_\infty)$$

Clearly, Ł$_\infty$ is countably axiomatizable. Let $[0,1]_L$ be the standard MV-algebra, that is an algebra with the domain $[0,1]$ of reals and with the operations defined as follows:

$$a \rightarrow^{[0,1]_L} b = \min\{1, 1 - a + b\} \qquad\qquad \neg^{[0,1]_L} a = 1 - a.$$

It is easy to compute the semantics of additional connectives:

$$a \,\&^{[0,1]_L}\, b = \max\{0, a + b - 1\} \qquad a \vee^{[0,1]_L} b = \max\{a, b\}.$$

It is well known (see e.g. [14, 16], or our Proposition 2 for an alternative proof) that $[0, 1]_L$ is a complete semantics for L_∞: i.e. for each $\Gamma \cup \{\varphi\} \subseteq Fm_{\mathcal{L}_L}$ we have:[2]

$$\Gamma \vdash_{L_\infty} \varphi \iff \text{for every homomorphism } e \colon \boldsymbol{Fm}_{\mathcal{L}_L} \to [0, 1]_L$$
$$e(\varphi) = 1 \text{ whenever } e[\Gamma] \subseteq \{1\}.$$

Using the completeness we can prove that L_∞ is not finitary. Indeed for each m we can show that:

$$\{\neg\varphi \to \varphi^n \mid 0 < n \le m\} \nvdash_{L_\infty} \varphi$$

by setting $e(\varphi) = \frac{m}{m+1}$ which gives us for each $n \le m$: $e(\neg\varphi) = \frac{1}{m+1} = e(\varphi^m) \le e(\varphi^n)$ and so $e(\neg\varphi \to \varphi^n) = 1$.

Definition 1. *A binary connective $\vee$ is a* strong disjunction *in a logic L whenever $\varphi \vdash_L \varphi \vee \psi$, $\psi \vdash_L \varphi \vee \psi$, and for all sets of formulas Γ, Φ, Ψ and a formula χ we have (by $\Phi \vee \Psi$ we denote the set $\{\varphi \vee \psi \mid \varphi \in \Phi, \psi \in \Psi\}$):*

$$\frac{\Gamma, \Phi \vdash \chi \qquad \Gamma, \Psi \vdash \chi}{\Gamma, \Phi \vee \Psi \vdash \chi}. \qquad\qquad \text{(sPCP)}$$

If the metarule is valid for Φ and Ψ being singletons we omit the prefix 'strong'.

The metarule is known as the *strong proof by cases property* (or just the proof by cases property PCP for Φ and Ψ being singletons). In finitary logics, any disjunction is a strong disjunction, but in general they may differ [6].

It is well known (see e.g. [6]) that if $\vee$ is a disjunction, then *finitely meet-irreducible* theories coincide with *prime* theories, i.e., those theories where for each φ and ψ, if $\varphi \vee \psi \in T$, then $\varphi \in T$ or $\psi \in T$. Therefore in such contexts we use these two terms interchangeably.

It is easy to observe that any disjunction satisfies the following properties:

(PD) $\quad \varphi \vdash_L \varphi \vee \psi \qquad\qquad\qquad\qquad \psi \vdash_L \varphi \vee \psi$

(I) $\quad\; \varphi \vee \varphi \vdash_L \varphi$

(C) $\quad \varphi \vee \psi \vdash_L \psi \vee \varphi$

(A) $\quad \varphi \vee (\psi \vee \chi) \vdash_L (\varphi \vee \psi) \vee \chi \qquad (\varphi \vee \psi) \vee \chi \vdash_L \varphi \vee (\psi \vee \chi)$

The following characterization of strong disjunction is often useful.

Proposition 1 ([6, Theorem 4.5]). *Let L be a logic with a presentation $\mathcal{AS}$. Then a connective $\vee$ is a strong disjunction in L if and only if it satisfies (PD), (I), (C), and for every rule $\Gamma \rhd \varphi$ in $\mathcal{AS}$ it is the case that its $\vee$-form is provable in L, that is for every formula χ we have: $\Gamma \vee \chi \vdash_L \varphi \vee \chi$.*

---

[2] By $\boldsymbol{Fm}_{\mathcal{L}_L}$ we denote the absolutely free algebra of type $\mathcal{L}_L$.

Lemma 2. *The connective $\vee$ is a strong disjunction in L_∞.*

Proof. It is known that $\vee$ is a (strong) disjunction in Ł (see e.g. [5]). Therefore, thanks to the right-to-left direction of the previous theorem, Ł and also L_∞ prove (PD), (I), (C), and $\vee$-form of *modus ponens*. Then, thanks to the left-to-right direction of the previous theorem, it suffices to prove the $\vee$-form of (L_∞) i.e.,

$$\{(\neg\varphi \to \varphi^n) \vee \chi \mid n \in \omega\} \vdash_{L_\infty} \varphi \vee \chi.$$

To do so we prove (in Łukasiewicz logic Ł):[3]

$$(\neg\varphi \to \varphi^n) \vee \chi \vdash_L \neg(\varphi \vee \chi) \to (\varphi \vee \chi)^n$$

and observe that a simple use of (L_∞) completes the proof.   □

3   Lindenbaum Lemma for Some Infinitary Logics

Before we prove the lemma, we introduce a useful technical tool which is interesting on its own (see the next section). First observe that if $\vee$ is a disjunction in logic L and $\Delta = \{\varphi_1, \ldots, \varphi_n\}$ is a finite non-empty set of formulas we can define $\bigvee \Delta = \varphi_1 \vee (\varphi_2 \vee \cdots \vee \varphi_n)\ldots)$ and, thanks to (C) and (A), the bracketing does not matter if the derivability is all we are interested in.

Therefore for each such logic we can define a relation $\Vdash_L$ between sets of formulas as:[4]

$$\Gamma \Vdash_L \Delta \quad \text{iff} \quad \text{there is a finite non-empty } \Delta' \subseteq \Delta \text{ and } \Gamma \vdash_L \bigvee \Delta'.$$

It is known that if L is a finitary logic, then $\Vdash_L$ is the so-called symmetric consequence relation as defined e.g. in [9]. In general this relation need not satisfy the proper cut rule of those consequence relations (see Example 5) but we can show it satisfies a particular form of cut, which will be instrumental to prove Lindenbaum lemma.

Lemma 3. *Let L be a logic with strong disjunction $\vee$. Then the relation $\Vdash_L$ has the so called* finite strong cut property, *i.e., for each sets Γ_1, Γ_2, Φ of formulas and each finite sets Δ_1, Δ_2 of formulas we have:*

$$\frac{\{\Gamma_1 \Vdash_L \Delta_1 \cup \{\varphi\} \mid \varphi \in \Phi\} \qquad \Gamma_2 \cup \Phi \Vdash_L \Delta_2}{\Gamma_1 \cup \Gamma_2 \Vdash_L \Delta_1 \cup \Delta_2}.$$

Proof. Let us by χ denote the formula $\bigvee(\Delta_1 \cup \Delta_2)$. From the assumption and properties of $\vee$ we obtain $\Gamma_1, \Gamma_2, \Phi \vdash_L \chi$ and $\Gamma_1, \Gamma_2 \vdash_L \varphi \vee \chi$ for each $\varphi \in \Phi$. As clearly $\Gamma_1, \Gamma_2, \chi \vdash_L \chi$ we can use sPCP to obtain $\Gamma_1, \Gamma_2, \Phi \vee \chi \vdash_L \chi$ and the regular cut of L to get the claim.   □

---

[3] Since we know that $\vee$ has sPCP in Ł it suffices to show that $\neg\varphi \to \varphi^n \vdash_L \neg(\varphi \vee \chi) \to (\varphi \vee \chi)^n$ and $\chi \vdash_L \neg(\varphi \vee \chi) \to (\varphi \vee \chi)^n$. The first one is provable because $\to$ is antitone in the first argument and monotone in the second. The second one holds by $\chi \vdash_L (\varphi \vee \chi)^n$ and $\psi \vdash_L \delta \to \psi$.

[4] By way of convention we say that $\Gamma \Vdash_L \emptyset$ iff $\Gamma \vdash_L \varphi$ for each formula φ.

The final ingredient used in the proof of the Lindenbaum lemma below is the notion of a pair: we say that $\langle \Gamma, \Delta \rangle$ is a *pair* in L if $\Gamma \nvdash_L \Delta$. Furthermore we say that a pair $\langle \Gamma, \Delta \rangle$ is *full* if $\Gamma \cup \Delta = Fm_{\mathcal{L}}$; note in such case Γ has to be a prime theory.

Theorem 1 (Lindenbaum lemma for infinitary logics). *Let* L *be a countably axiomatizable logic with a strong disjunction. Then prime theories form a basis of* Th(L).

Proof. Assume that a theory T and $\chi \notin T$ are given. We construct a prime theory $T' \supseteq T$ such that $\chi \notin T'$. We first enumerate all rules $\Lambda_i \rhd \varphi_i$ in the existing countable axiomatic system, and then define two increasing sequences of sets of formulas Γ_i and Δ_i such that each $\langle \Gamma_i, \Delta_i \rangle$ is a pair with Δ_i finite. We start by putting $\Gamma_0 = T$ and $\Delta_0 = \{\chi\}$. In the induction step, we distinguish two cases:

- If $\langle \Gamma_i \cup \{\varphi_i\}, \Delta_i \rangle$ is a pair, then we define $\Delta_{i+1} = \Delta_i$ and $\Gamma_{i+1} = \Gamma_i \cup \{\varphi_i\}$.
- If $\langle \Gamma_i \cup \{\varphi_i\}, \Delta_i \rangle$ is not a pair, then there has to be $\chi_i \in \Lambda_i$ such that $\langle \Gamma_i, \Delta_i \cup \{\chi_i\} \rangle$ is a pair; indeed, otherwise we would have:

$$\frac{\dfrac{\{\Gamma_i \Vdash \Delta_i \cup \{\chi_i\} \mid \chi_i \in \Lambda_i\} \qquad \Lambda_i \Vdash \varphi_i}{\Gamma_i \Vdash \Delta_i \cup \{\varphi_i\}} \qquad \Gamma_i \cup \{\varphi_i\} \Vdash \Delta_i}{\Gamma_i \Vdash \Delta_i}.$$

Thus we can define $\Gamma_{i+i} = \Gamma_i$ and $\Delta_{i+i} = \Delta_i \cup \{\chi_i\}$.

Finally, define $T' = \bigcup \Gamma_i$ and $\Delta = \bigcup \Delta_i$. We can assume w.l.o.g. that our axiomatic system contains a 'dummy' rule $\varphi \rhd \varphi$ for each formula φ, so that, due to the construction, $T' \cup \Delta = Fm_{\mathcal{L}}$. So when we show that $\langle T', \Delta \rangle$ is a pair, we have that T' is a prime theory and the proof is done.

First we show that for each φ we have: if $T' \vdash \varphi$, then $\varphi \in \Gamma_j$ for some j. Let us fix a proof of φ from T'; we prove the claim for each formula which is a label of some of its nodes. If the node is a leaf the claim is trivial. Consider a node obtained using rule $\Lambda_i \rhd \varphi_i$. If we have proceeded by the first case of the induction step we have $\varphi_i \in \Gamma_{i+1}$. Let us see that we couldn't have proceeded by the second case: consider $\chi_i \in \Lambda_i$. We know that $T' \vdash \chi_i$ (it is a label of a node preceding φ_i) and so the induction assumption gives us j such that $\Gamma_j \vdash \chi_i$ and so $\langle \Gamma_{\max\{i+1,j\}}, \Delta_{\max\{i+1,j\}} \rangle$ is not a pair, yielding a contradiction.

Now we can conclude the proof that $\langle T', \Delta \rangle$ is a pair. Assume otherwise, then we have $T' \vdash \bigvee \Delta_0$ for some finite Δ_0. Then $\Delta_0 \subseteq \Delta_i$ for some i and so by the previous claim there is j such that $\langle \Gamma_{\max\{i,j\}}, \Delta_{\max\{i,j\}} \rangle$ is not a pair, a contradiction. $\qquad \square$

It is to be noted that neither of the two assumptions of Theorem 1 (of the countability of the axiomatic system and of having a strong disjunction) can be omitted. We present two examples of logics satisfying only one of these conditions and failing the Lindenbaum lemma (and thus also the other condition).

Example 2 (A logic with a strong disjunction and no countable presentation).
Consider a language $\mathcal{L}$ consisting of a binary connective $\vee$, a unary connective s, and two constants $\mathbf{0}$ and $\boldsymbol{\omega}$. Let us by $\mathbf{n}$ denote the formula defined inductively as $(\mathbf{n}+1) = s(\mathbf{n})$.

Let L be a logic in $\mathcal{L}$ axiomatized by the rules (PD), (C), (I), and (A), $\vee$-forms of these rules, and the following rules for each infinite set $C \subseteq \omega$:

$$\{\mathbf{i} \vee \psi \mid i \in C\} \rhd \psi. \tag{Inf$_C$}$$

First we use Proposition 1 to show that $\vee$ is a strong disjunction in L. Indeed $\vee$-forms of some rules are directly part of its presentation and for the remaining ones we use (A), e.g. we know that $(\mathbf{i} \vee \psi) \vee \chi \vdash_\mathrm{L} \mathbf{i} \vee (\psi \vee \chi)$ and so obviously $\{(\mathbf{i} \vee \psi) \vee \chi \mid i \in C\} \vdash_\mathrm{L} \psi \vee \chi$.

We prove that Lindenbaum lemma fails in L. Consider a subset A of 2^ω:

$$A = \{\omega\} \cup \{X \subseteq \omega \mid X \text{ finite and for each } i \in \omega \colon 2i \notin X \text{ or } 2i+1 \notin X\}.$$

Note that A is closed under arbitrary intersections and so $\langle A, \vee \rangle$ with

$$X \vee Y = \bigcap_{Z \in A, X \cup Y \subseteq Z} Z$$

is a complete join-semilattice and observe that

$$X \vee Y = \begin{cases} X \cup Y & \text{if } X \cup Y \in A \\ \omega & \text{otherwise (i.e., when } \{2i, 2i+1\} \subseteq X \cup Y). \end{cases}$$

Consider an algebra $\boldsymbol{A} = \langle A, \vee, s, \mathbf{0}, \boldsymbol{\omega} \rangle$ where $\mathbf{0} = \{0\}$, $\boldsymbol{\omega} = \omega$, and $s(\{i\}) = \{i+1\}$ and $s(X) = \emptyset$ otherwise (note that $\mathbf{n} = \{n\}$). We show that $\boldsymbol{A}$ is an algebraic model of L, i.e.,

$$\Gamma \vdash_\mathrm{L} \varphi \implies \text{ for every homomorphism } e \colon \boldsymbol{Fm}_\mathcal{L} \to \boldsymbol{A}$$
$$e(\varphi) = \omega \text{ whenever } e[\Gamma] \subseteq \{\omega\}.$$

Obviously it suffices to check the rules.

- Soundness of the rules (C), (I), and (A) and their $\vee$-forms is straightforward (recall that the join of any sets from A is ω if and only if its union contains $\{2i, 2i+1\}$ for some i)
- Next consider a rule (Inf$_C$) and an evaluation e such that $e(\psi) \neq \omega$. We know that $e(\psi)$ is a finite set and so for $m = \max(e(\psi))$ we have $e(\mathbf{m}+\mathbf{2} \vee \psi) = e(\psi) \cup \{m+2\} \neq \omega$.

To conclude the proof we show that for a theory T generated by the set of formulas $\{2\mathbf{i} \vee 2\mathbf{i}+\mathbf{1} \mid i \in \omega\}$ we have $T \nvdash_\mathrm{L} \mathbf{0}$ while $T' \vdash_\mathrm{L} \mathbf{0}$ for each prime theory $T' \supseteq T$. To show the first claim just observe that for an arbitrary homomorphism $e \colon \boldsymbol{Fm}_\mathcal{L} \to \boldsymbol{A}$ we have $e(2\mathbf{i} \vee 2\mathbf{i}+\mathbf{1}) = \omega$. As regards the second claim, since T' is prime, we obtain for each i that $2\mathbf{i} \in T'$ or $2\mathbf{i}+\mathbf{1} \in T'$. Thus there is an infinite set C such that $\{\mathbf{i} \vee \mathbf{0} \mid i \in C\} \subseteq T'$ and so by (Inf$_C$) we obtain $T \vdash_\mathrm{L} \mathbf{0}$.

Example 3 (A countably axiomatizable logic without a strong disjunction). Consider a language with one unary operation box $\Box$, we write $\Box^n$ as a shortcut for the n-fold application of $\Box$. In this example we consider a logic L axiomatized by the infinitary rules (Inf_n) for each $n \in \omega$:

$$\{\Box^m(\varphi) \mid m > n\} \rhd \varphi. \tag{Inf$_n$}$$

Clearly this logic is countably axiomatizable. We show that Lindenbaum lemma fails in L. First we show that if $\Gamma, \varphi \vdash_L \chi$, then $\chi = \varphi$ or $\Gamma \vdash_L \chi$. We prove it by induction for each δ in the proof of χ from $\Gamma \cup \{\varphi\}$. The only non-trivial case is if δ follows by the application of an infinitary rule $\{\Box^m(\delta) \mid m > n\} \rhd \delta$. Let us set $n' = k$ if $\varphi = \Box^k(\delta)$ for some $k > n$ and $n' = n$ otherwise. Due to the induction assumption we have $\Gamma \vdash_L \Box^m(\delta)$ for each $m > n'$. Thus $\Gamma \vdash_L \delta$.

Therefore if T is a theory so is $T \cup \{\psi\}$ and so the only finitely meet-irreducible theory is $Fm_{\mathcal{L}}$. Now, since obviously our logic has non-trivial theories (e.g., the $\emptyset$), the finitely meet-irreducible theories do not form a basis of $\text{Th}(L)$.

In the previous section we showed that infinitary Łukasiewicz logic $Ł_\infty$ has both a strong disjunction and a countable axiomatization, thus we know that the Lindenbaum lemma works in $Ł_\infty$. Let us conclude this section by giving one more example of such a logic, and by showing how we can use Lindenbaum lemma to obtain an easy proof of completeness for $Ł_\infty$ with respect to the standard MV-algebra $[0,1]_Ł$ (a folklore result which is implicit e.g. in [14], or could be obtained from a more complicated axiomatization of this algebra from [16]).

Example 4. In Definition 4 of [17], an infinitary logic PDL_ω is given as the smallest consequence relation induced by the axioms of PDL, containing *modus ponens,* the infinitary rule:

$$\{[\alpha; \beta^n]\varphi \mid n \in \mathbb{N}\} \rhd [\alpha; \beta^*]\varphi, \tag{Inf*}$$

and closed under Cut, Weakening, Deduction, and Strong Necessitation metarules. To present a countable axiomatization in our sense, consider the axioms of PDL, rules *modus ponens* and the (Inf^*) rule, plus all the box-forms of these rules $([\alpha]\Gamma \rhd [\alpha]\varphi$, for each α and $\Gamma \rhd \varphi$). As the classical disjunction of PDL is a strong disjunction—not immediate but not hard to see either (cf. Example 6)—Theorem 1 applies, and we obtain Lindenbaum Lemma for PDL_ω. This suffices to prove strong completeness of PDL_ω, where Lindenbaum lemma is used also in the construction of a canonical model to prove the Truth lemma (Lemma 2 of [17]).

Proposition 2. *The infinitary Łukasiewicz logic $Ł_\infty$ is sound and complete w.r.t. the standard MV-algebra $[0,1]_Ł$.*

Proof. It is easy to check that the semantics is sound. For completeness assume that $\Gamma \nvdash_{Ł_\infty} \varphi$ and that T is a prime theory of $Ł_\infty$ separating Γ from φ (its existence follows form the Lindenbaum lemma). Clearly T is a consistent theory of L, we show that it is a maximal such theory: consider $\psi \notin T$ then, thanks to the infinitary rule ($Ł_\infty$), there has to be some $n \in \omega$ such that $\neg\psi \to \psi^n \notin T$. Due to the prelinearity theorem $\vdash_Ł (\varphi \to \chi) \lor (\chi \to \varphi)$ and primeness of T we obtain $\psi^n \to \neg\psi \in T$. Thus $T, \psi \vdash_Ł \neg\psi$ (using *modus ponens* and the fact that $\psi \vdash_Ł \psi^m$) and so $T \cup \{\psi\}$ is inconsistent (because we have $\psi, \neg\psi \vdash_Ł \chi$).

By the standard techniques in algebraic logic we can find a counterexample to $\Gamma \nvdash_{Ł_\infty} \varphi$ over the Lindenbaum–Tarski algebra $Ł_T$. MV-algebras are the equivalent algebraic semantics of Ł (see e.g. [2]) and so $Ł_T$ is an MV-algebra. There is a lattice isomorphism between theories of Ł extending T and congruences on $Ł_T$ (an instance of an abstract result of algebraic logic [7]) and because T is a maximally consistent theory of Ł, we conclude that $Ł_T$ is a simple MV-algebra. However, e.g. from [2] we know that simple MV-algebras are up to isomorphism subalgebras of $[0,1]_Ł$ and so the rest of the proof is straightforward.   □

4   Pair Extension Lemma

As we have seen in the introduction, the Lindenbaum lemma is related to the *Pair extension lemma*, which says that each pair can be extended into a full pair (a pair $\langle \Gamma', \Delta' \rangle$ is an *extension* of $\langle \Gamma, \Delta \rangle$ when $\Gamma' \supseteq \Gamma$ and $\Delta' \supseteq \Delta$).

The Pair extension lemma is known to be equivalent to soundness of the form of the cut rule used in the definition of symmetric consequence relations, see [9]. The next example shows that, while $Ł_\infty$ enjoys the Lindenbaum lemma, it does not enjoy the Pair extension lemma and so $\Vdash_{Ł_\infty}$ is not a symmetric consequence relation.

Example 5. Let us consider the infinitary Łukasiewicz logic $Ł_\infty$ and a set $\Delta = \{\neg\varphi^n \mid n > 0\} \cup \{\varphi\}$. Obviously $\langle \emptyset, \Delta \rangle$ is a pair: just use Proposition 2 and evaluate $e(\varphi) = \frac{m}{m+1}$ and observe that $e(\varphi \lor \bigvee_{0 < i \le m} \neg\varphi^i) = \frac{m}{m+1} \ne 1$. But it cannot be extended to a full pair: indeed, if $\langle \Gamma, \Delta' \rangle$ would be such a full pair, we could argue as in Proposition 2 to infer that Γ is a maximally consistent theory, and, as shown in the proof of that proposition, every such theory contains at least one formula from Δ—a contradiction with $\langle \Gamma, \Delta' \rangle$ being a pair.

On the other hand, as our proof of Theorem 1 suggests, it should be very easy to see that the restriction of Pair extension lemma to pairs with *finite* right-hand sides is equivalent with the Lindenbaum lemma. In fact, already in a weaker setting (assuming that $\lor$ is merely a disjunction) we can prove a stronger claim which also illuminates the role of the finite strong cut rule and strong proof by cases property.

Theorem 2. *Let L be a logic with a countable axiomatization $\mathcal{AS}$ and a disjunction $\lor$. Then the following are equivalent:*

1. *$\Vdash_L$ enjoys the Pair extension lemma for pairs with finite right-hand sides, i.e., each pair $\langle \Gamma, \Delta \rangle$ where Δ is finite can be extended into a full one.*
2. *$\Vdash_L$ enjoys the finite strong cut rule, i.e., for each sets Γ_1, Γ_2, Φ of formulas and each finite sets Δ_1, Δ_2 of formulas:*

$$\frac{\{\Gamma_1 \Vdash_L \Delta_1 \cup \{\varphi\} \mid \varphi \in \Phi\} \qquad \Gamma_2 \cup \Phi \Vdash_L \Delta_2}{\Gamma_1 \cup \Gamma_2 \Vdash_L \Delta_1 \cup \Delta_2}.$$

3. *For each rule $\Gamma \rhd \varphi$ in $\mathcal{AS}$ and each formula χ we have*

$$\{\gamma \vee \chi \mid \gamma \in \Gamma\} \vdash_{\mathrm{L}} \varphi \vee \chi,$$

i.e., $\vee$ is a strong disjunction.

4. *L enjoys the Lindenbaum lemma, i.e., prime theories form a basis of* Th(L).

Proof. 1→2: Assume (i) $\Gamma_1 \Vdash_{\mathrm{L}} \Delta_1 \cup \{\varphi\}$ for each $\varphi \in \Phi$ and (ii) $\Gamma_2 \cup \Phi \Vdash_{\mathrm{L}} \Delta_2$. To obtain contradiction suppose that $\Gamma_1 \cup \Gamma_2 \nVdash_{\mathrm{L}} \Delta_1 \cup \Delta_2$. Thus there is a full pair $\langle \Gamma', \Delta' \rangle$ extending $\langle \Gamma_1 \cup \Gamma_2, \Delta_1 \cup \Delta_2 \rangle$. Due to (i) there can be no $\varphi \in \Phi \cap \Delta'$. Therefore $\Phi \subseteq \Gamma'$ and so by (ii) we get $\Gamma' \Vdash_{\mathrm{L}} \Delta'$, a contradiction.

2→3: A simple application of the strong cut rule:

$$\frac{\{\{\gamma \vee \chi \mid \gamma \in \Gamma\} \Vdash_{\mathrm{L}} \{\chi, \gamma\} \mid \gamma \in \Gamma\} \qquad \Gamma \Vdash_{\mathrm{L}} \{\varphi\}}{\{\gamma \vee \chi \mid \gamma \in \Gamma\} \Vdash_{\mathrm{L}} \{\varphi, \chi\}}.$$

3→4: This follows by Lindenbaum lemma (Theorem 1).

4→1: The proof is obvious.   □

Example 6. Consider again PDL_ω from Example 4. It is a simple exercise to see that, due to the presence of the classical negation (and using Cut, W, Ded), the logic PDL_ω, as defined in [17], induces a consequence relation $\Vdash_{\mathrm{PDL}_\omega}$ which satisfies the finite strong cut rule. This is another alternative proof that Lindenbaum lemma is available for PDL_ω, and that the classical disjunction of PDL_ω is a strong disjunction.

Let us turn our attention to the full Pair extension lemma for pairs $\langle \Gamma, \Delta \rangle$ with an arbitrary Δ. We can still prove a similar theorem. It moreover illuminates some interesting limitations of the Pair extension lemma: namely, the full Pair extension lemma, in presence of infinitely many propositional variables in the language, entails the finitarity of the logic in question. Thus in particular $\Vdash_{\mathrm{L}}$ is a symmetric consequence relation, if and only if, L is a finitary logic.

Theorem 3. *Let L be a logic, whose language contains countably many propositional variables, with a countable axiomatization $\mathcal{AS}$ and a disjunction $\vee$. Then the following are equivalent:*

1. $\Vdash_{\mathrm{L}}$ *enjoys the Pair extension lemma, i.e., each pair $\langle \Gamma, \Delta \rangle$ can be extended into a full one.*
2. $\Vdash_{\mathrm{L}}$ *enjoys the strong cut rule, i.e., for each sets $\Gamma_1, \Gamma_2, \Phi, \Delta_1, \Delta_2$ of formulas:*

$$\frac{\{\Gamma_1 \Vdash_{\mathrm{L}} \Delta_1 \cup \{\varphi\} \mid \varphi \in \Phi\} \qquad \Gamma_2 \cup \Phi \Vdash_{\mathrm{L}} \Delta_2}{\Gamma_1 \cup \Gamma_2 \Vdash_{\mathrm{L}} \Delta_1 \cup \Delta_2}.$$

3. *For each rule $\Gamma \rhd \varphi$ in $\mathcal{AS}$, each set of formulas Δ and each surjective function $f \colon \Gamma \to \Delta$ we have*

$$\{\gamma \vee f(\gamma) \mid \gamma \in \Gamma\} \Vdash_{\mathrm{L}} \Delta \cup \{\varphi\}.$$

4. *L is finitary.*

Proof. The implication 4→1 is a well-known fact valid in general, for a proof see e.g. [9,20];[5] the proof of the implication 1→2 is analogous to the proof of the corresponding claim in the previous theorem; and the implication 2→3 is a simple application of the strong cut rule:

$$\frac{\{\{\gamma \vee f(\gamma) \mid \gamma \in \Gamma\} \Vdash_L \Delta \cup \{\gamma\} \mid \gamma \in \Gamma\} \qquad \Gamma \Vdash_L \{\varphi\}}{\{\gamma \vee f(\gamma) \mid \gamma \in \Gamma\} \Vdash_L \Delta \cup \{\varphi\}}.$$

To prove the remaining implication 3→4 assume that L is not finitary. There has to be a proper infinitary rule $\Gamma \rhd \varphi$ in $\mathcal{AS}$ (i.e., for no finite $\Gamma' \subseteq \Gamma$ we have $\Gamma' \vdash_L \varphi$) and, since $\mathcal{AS}$ is countable and closed under substitutions, there are only finitely many variables occurring in Γ. Assume that $\Gamma = \{\gamma_1, \gamma_2, \dots\}$ and let $\Delta = \{p_1, p_2, \dots\}$ be the infinite (!) set of all variables not occurring in $\Gamma \cup \{\varphi\}$.

We define a function $f \colon \Gamma \to \Delta$ as follows: $f(\gamma_i) = p_i$. Clearly f is surjective and so from our assumption we know that

$$\{\gamma_i \vee p_i \mid i > 0\} \Vdash_L \Delta \cup \{\varphi\}.$$

Therefore there is a non-empty finite set of variables $\Delta' \subseteq \Delta$ such that

$$\{\gamma_i \vee p_i \mid i > 0\} \vdash_L \varphi \vee \bigvee \Delta'.$$

Pick any $p_n \in \Delta'$ and define a substitution σ in the following way: $\sigma(p) = p$ if p occurs in $\Gamma \cup \{\varphi\}$, $\sigma(p_i) = \varphi$ for $p_i \in \Delta'$, and $\sigma(p_i) = \gamma_n$ otherwise. Then by structurality we obtain

$$\{\gamma_i \vee \varphi \mid p_i \in \Delta'\} \cup \{\gamma_i \vee \gamma_n \mid p_i \notin \Delta'\} \vdash_L \varphi \vee \varphi \vee \cdots \vee \varphi.$$

Thus by the properties of disjunction we obtain

$$\{\gamma_i \mid p_i \in \Delta'\} \vdash_L \varphi,$$

which is a contradiction with the fact that $\Gamma \vdash_L \varphi$ is a proper infinitary rule. □

Remark 1. In the proof of 3→4, we have substantially used the fact that there are infinitely many variables in the language. In Example 7 below we demonstrate that there is indeed an infinitary logic with finitely many variables which still has full Pair extension lemma.

To do so, we need to show that the properties 1–3 are equivalent even without assuming existence of infinitely many variables. Clearly, the proofs of the implications from top to bottom do not use the assumption and the implication 3→1 can be proven by the following simple modification of the proof of Lindenbaum lemma (Theorem 1).

---

[5] The simplest proof would be to enumerate all formulas, then in each step add the processed formula either to the left or to the right side of the pair (which can always be done thanks to the cut rule), finitarity will then guarantee that the union of the pair will in the end still be a pair.

Let $\langle \Gamma, \Delta \rangle$ be the pair we want to extend; we start by Γ_0 being the theory generated by Γ and $\Delta_0 = \Delta$. We only need to show that the second case of the induction procedure can still be carried out, the rest of the proof remains the same. We have a pair $\langle \Gamma_i, \Delta_i \rangle$ and need to process the rule $\Lambda_i \rhd \varphi_i$. We assume that $\langle \Gamma_i \cup \{\varphi_i\}, \Delta_i \rangle$ is not a pair (i.e., $\Gamma_i, \varphi_i \vdash_L \bigvee \Delta_i'$ for some finite $\Delta_i' \subseteq \Delta_i$) and $\langle \Gamma_i, \Delta_i \cup \{\chi\} \rangle$ is not a pair for any $\chi \in \Lambda_i$. Thus we can define $f(\chi) = \bigvee \Delta_\chi$, where $\Delta_\chi \subseteq \Delta_i$ is a finite set such that $\Gamma_i \vdash_L \chi \vee \bigvee \Delta_\chi$. By the point 3 we obtain

$$\{\chi \vee \bigvee \Delta_\chi \mid \chi \in \Lambda_i\} \Vdash_L \{\bigvee \Delta_\chi \mid \chi \in \Gamma\} \cup \{\varphi_i\}.$$

Thus $\Gamma_i \vdash \varphi_i \vee \bigvee \Delta_i''$ for some $\Delta_i'' \subseteq \Delta_i$. Using PCP we obtain $\Gamma_i \vdash \bigvee(\Delta_i' \cup \Delta_i'')$ (cf. the proof of Lemma 3), i.e., $\langle \Gamma_i, \Delta_i \rangle$ is not a pair, a contradiction.

Example 7 (An infinitary logic with finitely many variables and full Pair extension lemma). We consider a language with only finitely many variables $Var = \{p_1, \ldots, p_n\}$, a disjunction connective $\vee$, and a constant $\mathbf{n}$ for every natural number n. A logic L in this language is presented by an axiomatic system consisting of rules (PD), (I), (A), (C), an infinitary rule

$$\{\mathbf{n} \mid n \in \omega\} \rhd p_1, \tag{Inf}$$

and $\vee$-forms of all these rules.

It is easy to see that L is infinitary—it is enough to consider models of size 2. As it is clearly countably axiomatizable and $\vee$ is a strong disjunction, if we show that it satisfies condition 3 of Theorem 3 we obtain full Pair extension lemma due to the previous remark. For the rules (PD), (I), (A), and (C) and their $\vee$-forms it is a consequence of the fact that they have finitely many premises and are closed under its $\vee$-forms. Thus it remains to be shown that the rule (Inf) satisfies it too (the argument for its $\vee$-form is basically the same).

Consider a set Δ and a surjective function $f \colon \{\mathbf{n} \mid n \in \omega\} \to \Delta$. Note that $f(\mathbf{n})$ is a disjunction of constants and variables (in the limit case with just one disjunct).

First assume that there are $m, l \in \omega$ such that $\mathbf{l}$ is one of the disjunct of $f(\mathbf{m})$. Then $\mathbf{l} \vdash f(\mathbf{m})$ and so $\mathbf{l} \vdash f(\mathbf{m}) \vee f(\mathbf{l}) \vee p_1$, both by (PD). By (PD) we also obtain $f(\mathbf{l}) \vdash f(\mathbf{m}) \vee f(\mathbf{l}) \vee p_1$, and so by PCP

$$\{\mathbf{n} \vee f(\mathbf{n}) \mid n \in \omega\} \vdash f(\mathbf{m}) \vee f(\mathbf{l}) \vee p_1.$$

If there are no such $m, l \in \omega$, then each formula from Δ is a disjunction of variables. Since there are only finitely many variables, there is a finite set $\Delta' \subseteq \Delta$ such that each variable from some formula from Δ occurs in some formula from Δ', and so, by the properties of disjunction, for each $\chi \in \Delta$ we have $\chi \vdash \bigvee \Delta'$. Thus also for each n we have $\mathbf{n} \vee f(\mathbf{n}) \vdash \mathbf{n} \vee \bigvee \Delta'$, and therefore by $\{\mathbf{n} \vee \bigvee \Delta' \mid n \in \omega\} \vdash p_1 \vee \bigvee \Delta'$ (which is the $\vee$-form of our infinitary rule) we obtain

$$\{\mathbf{n} \vee f(\mathbf{n}) \mid n \in \omega\} \vdash p_1 \vee \bigvee \Delta'.$$

5   Conclusion and Future Work

In this paper we have proved a general form of Lindenbaum lemma for a wide class of non-finitary logics and explored its relation with the Pair extension lemma. We have seen how it can be used to obtain some known completeness results. Thus not only we can subsume numerous ad hoc proofs of similar results scattered in the literature, but more importantly, we can easily prove them for newly defined logics (here especially the characterization of strong disjunction comes in handy). Our first further goal here is to prove strong completeness and (limited) canonicity of not yet considered infinitary proof theory for non-classical variants of PDL [8,22,23], and of infinitary proof theory for substructural epistemic logics of [1,21], when extended with a common knowledge modality.

On the theoretical side we plan to explore the relation with symmetrical consequence relations of [9] to overcome the limitations implied by Theorem 3. The problem here is that our symmetrization of a logic L, the relation $\Vdash_L$, has build-in finitarity on the righthand side. It would be indeed interesting to explore other ways how to symmetrize L to avoid this problem, solution of which could also help to overcome the restriction to finitely meet-irreducible theories in Theorem 1. Here a notion of ω-prime theory and perhaps infinitary disjunction will be needed.

Finally, Condition 3 of Theorem 3 defines a new notion of super-strong disjunction. In the finitary context it coincides with simple disjunction, but studying it as a notion of its own right could lead to interesting results.

References

1. Bílková, M., Majer, O., Peliš, M.: Epistemic logics for sceptical agents. J. Log. Comput. **26**(6), 1815–1841 (2016)
2. Cignoli, R., D'Ottaviano, I.M., Mundici, D.: Algebraic Foundations of Many-Valued Reasoning. Trends in Logic, vol. 7. Kluwer, Dordrecht (1999)
3. Cintula, P.: A note on axiomatizations of Pavelka-style complete fuzzy logics. Fuzzy Sets Syst. **292**, 160–174 (2016)
4. Cintula, P., Hájek, P., Noguera, C. (eds.): Handbook of Mathematical Fuzzy Logic (in 2 Volumes). Studies in Logic, Mathematical Logic and Foundations, vols. 37, 38. College Publications, London (2011)
5. Cintula, P., Noguera, C.: A general framework for mathematical fuzzy logic. In: Cintula, P., Hájek, P., Noguera, C. (eds.) Handbook of Mathematical Fuzzy Logic - Volume 1. Studies in Logic, Mathematical Logic and Foundations, vol. 37, pp. 103–207. College Publications, London (2011)
6. Cintula, P., Noguera, C.: The proof by cases property and its variants in structural consequence relations. Studia Logica **101**(4), 713–747 (2013)
7. Czelakowski, J.: Protoalgebraic Logics. Trends in Logic, vol. 10. Kluwer, Dordrecht (2001)
8. Degen, J., Werner, J.: Towards intuitionistic dynamic logic. Log. Log. Philos. **15**(4), 305–324 (2006)
9. Dunn, J.M., Hardegree, G.M.: Algebraic Methods in Philosophical Logic. Oxford Logic Guides, vol. 41. Oxford University Press, Oxford (2001)

10. Esteva, F., Godo, L., Hájek, P., Navara, M.: Residuated fuzzy logics with an involutive negation. Arch. Math. Log. **39**(2), 103–124 (2000)
11. Esteva, F., Godo, L., Montagna, F.: The LΠ and LΠ$\frac{1}{2}$ logics: two complete fuzzy systems joining Łukasiewicz and product logics. Arch. Math. Log. **40**(1), 39–67 (2001)
12. Goldblatt, R.: An abstract setting for Henkin proofs. Topoi **3**(1), 37–41 (1984)
13. Goldblatt, R.: Mathematics of Modality. CSLI Lecture Notes, vol. 43. CSLI Publications Stanford University (1993)
14. Hay, L.S.: Axiomatization of the infinite-valued predicate calculus. J. Symb. Log. **28**(1), 77–86 (1963)
15. Horčík, R., Cintula, P.: Product Łukasiewicz logic. Arch. Math. Logic **43**(4), 477–503 (2004)
16. Kułacka, A.: Strong standard completeness for continuous t-norms. Fuzzy Sets Syst. https://doi.org/10.1016/j.fss.2018.01.001
17. de Lavalette, G.R., Kooi, B., Verbrugge, R.: Strong completeness and limited canonicity for PDL. J. Log. Lang. Inf. **7**(1), 69–87 (2008)
18. Łukasiewicz, J., Tarski, A.: Untersuchungen über den Aussagenkalkül. Comptes Rendus des Séances de la Société des Sciences et des Lettres de Varsovie, cl. III **23**(iii), 30–50 (1930)
19. Montagna, F.: Notes on strong completeness in Łukasiewicz, product and *BL* logics and in their first-order extensions. In: Aguzzoli, S., Ciabattoni, A., Gerla, B., Manara, C., Marra, V. (eds.) Algebraic and Proof-theoretic Aspects of Non-classical Logics. LNCS (LNAI), vol. 4460, pp. 247–274. Springer, Heidelberg (2007). https://doi.org/10.1007/978-3-540-75939-3_15
20. Restall, G.: An Introduction to Substructural Logics. Routledge, New York (2000)
21. Sedlár, I.: Epistemic extensions of modal distributive substructural logics. J. Log. Comput. **26**(6), 1787–1813 (2016)
22. Sedlár, I.: Propositional dynamic logic with Belnapian truth values. In: Beklemishev, L., Demri, S., Máté, A. (eds.) Advances in Modal Logic, vol. 11, pp. 530–519. College Publications (2016)
23. Sedlár, I.: Non-classical PDL on the cheap. In: Arazim, P., Lávička, T. (eds.) The Logica Yearbook 2016, pp. 239–256. College Publications (2017)
24. Segerberg, K.: A model existence theorem in infinitary propositional modal logic. J. Philos. Log. **23**, 337–367 (1994)
25. Sundholm, G.: A completeness proof for an infinitary tense-logic. Theoria **43**(1), 47–51 (1977)
26. Vidal, A., Bou, F., Esteva, F., Godo, L.: On strong standard completeness in some MTL$_\Delta$ expansions. Soft. Comput. **21**(1), 125–147 (2017)

The Epistemology of Nondeterminism

Adam Bjorndahl[(⊠)]

Carnegie Mellon University, Pittsburgh, USA
abjorn@andrew.cmu.edu

Abstract. This paper proposes new semantics for propositional dynamic logic (PDL), replacing the standard relational semantics. Under these new semantics, program execution is represented as fundamentally deterministic (i.e., functional), while nondeterminism emerges as an epistemic relationship between the agent and the system: intuitively, the nondeterministic outcomes of a given process are precisely those that cannot be ruled out in advance. We formalize these notions using topology and the framework of dynamic topological logic (DTL) [1]. We show that DTL can be used to interpret the language of PDL in a manner that captures the intuition above, and moreover that continuous functions in this setting correspond exactly to deterministic processes. We also prove that certain axiomatizations of PDL remain sound and complete with respect to the corresponding classes of dynamic topological models. Finally, we extend the framework to incorporate knowledge using the machinery of subset space logic [2], and show that the topological interpretation of public announcements as given in [3] coincides exactly with a natural interpretation of test programs.

1 Introduction

Propositional dynamic logic (PDL) is a framework for reasoning about the effects of *nondeterministic programs* (or, more generally, *nondeterministic actions*).[1] The standard models for PDL are relational structures interpreted as state transition diagrams: each program π is associated with a binary relation R_π on the state space, and $xR_\pi y$ means that state y is one possible result of executing π in x.

What is the sense of "possibility" at play here? This paper explores an epistemic account. The standard models for PDL treat nondeterminism as a primitive, unanalyzed notion: effectively, for each state x, π is interpreted as nondeterministic at x just in case $|\{y : xR_\pi y\}| > 1$. But one might hope for a logic that provides some insight into the *nature* and *source* of nondeterminism, rather than simply stipulating its existence.

We investigate a richer class of models for nondeterministic program execution which differ from the standard models in two key respects: (1) states

---

[1] Mathematical treatments of nondeterminism have a long history in computer science (see, e.g., [4–7]), though this paper focuses specifically on the semantics of PDL. See [8] for an overview and brief history of this branch of modal logic.

© Springer-Verlag GmbH Germany, part of Springer Nature 2018
L. S. Moss et al. (Eds.): WoLLIC 2018, LNCS 10944, pp. 145–162, 2018.
https://doi.org/10.1007/978-3-662-57669-4_8

completely determine the effects of actions, and (2) nondeterminism emerges, loosely speaking, as a kind of epistemic relationship between a given agent (or collection of agents) and the program (or action) in question. As we argue in the next section, to make this relationship precise we need structures rich enough to represent *potential observations*; for this we make use of *topology*. The resulting framework is very closely related to *dynamic topological logic* (DTL) as developed by Kremer and Mints [1]; roughly speaking, we show that DTL embeds a faithful interpretation of PDL. Furthermore, we demonstrate that *continuity* in this setting coincides exactly with the notion of determinism: that is, *determinism is continuity in the observation topology*.

The rest of the paper is organized as follows. In Sect. 2 we review the basics of PDL and present the intuitions that motivate the development of our new models and the importance of "potential observations" in the epistemic interpretation of nondeterminism. In Sect. 3 we motivate and review the use of topology for this purpose, and connect it to dynamic topological logic. This provides the tools we need to formalize our epistemic conception of nondeterminism and establish the correspondence between determinism and continuity mentioned above. In Sect. 4 we transform PDL models into DTL models in a manner that preserves the truth value of all PDL formulas, and use this to prove that certain standard PDL axiomatizations are also sound and complete with respect to corresponding classes of DTL models. In Sect. 5 we enrich our semantics using the machinery of *subset space logic* [2] in order to reason simultaneously about both knowledge and knowability in the context of nondeterministic program execution; furthermore, we show how *public announcements* [9], appropriately generalized to the topological setting [3], can be captured using *test programs*. Section 6 concludes with a brief discussion of ongoing work. Proofs and other details are collected in Appendix A.

2   Review and Motivation

Fix a countable set of *primitive propositions* PROP and a countable set of *programs* Π. The language of PDL, denoted $\mathcal{L}_{PDL}$, is given by

$$\varphi ::= p \mid \neg\varphi \mid \varphi \wedge \psi \mid \langle\pi\rangle\varphi,$$

where $p \in$ PROP, $\pi \in \Pi$, and $\langle\pi\rangle\varphi$ is read, "after some execution of π, φ is true". Often Π is constructed from a set of more "basic" programs by closing under certain operations, but for the moment we will take it for granted as a structureless set. A **(standard) PDL model** is a relational frame $(X, (R_\pi)_{\pi\in\Pi})$ together with a valuation $v :$ PROP $\to 2^X$; Boolean formulas are interpreted in the usual way, while $\langle\pi\rangle$ is interpreted as existential quantification over the R_π-accessible states:

$$x \models p \text{ iff } x \in v(p)$$
$$x \models \neg\varphi \text{ iff } x \not\models \varphi$$
$$x \models \varphi \wedge \psi \text{ iff } x \models \varphi \text{ and } x \models \psi$$
$$x \models \langle\pi\rangle\varphi \text{ iff } (\exists y)(xR_\pi y \text{ and } y \models \varphi).$$

Thus, $\langle\pi\rangle\varphi$ is true at a state x just in case some possible execution of π at x results in a φ-state.

Following standard conventions, we write $[\pi]$ as an abbreviation for $\neg\langle\pi\rangle\neg$, so we have

$$x \models [\pi]\varphi \text{ iff } (\forall y)(xR_\pi y \text{ implies } y \models \varphi).$$

We also treat R_π as a set-valued function when convenient, with $R_\pi(x) = \{y \in X : xR_\pi y\}$.

It is easy to adjust the standard models for PDL so that each state completely determines the outcome of each action: simply replace the relations R_π with functions $f_\pi : X \to X$. To emphasize this shift we introduce modalities $\bigcirc_\pi$ into the language, reading $\bigcirc_\pi\varphi$ as "after execution of π, φ holds"; these modalities are interpreted using the functions f_π in the natural way:

$$x \models \bigcirc_\pi\varphi \text{ iff } f_\pi(x) \models \varphi.$$

Perhaps the most direct attempt to formalize nondeterminism as an epistemic notion in this setting is to interpret the "nondeterministic outcomes" of π to be precisely those outcomes that the agent considers possible.

Somewhat more formally, supposing we have access to a knowledge modality K with corresponding dual $\hat{K}$ (so $\hat{K}\varphi$ is read "the agent considers φ possible"), we might define

$$\langle\pi\rangle\varphi \equiv \hat{K}\bigcirc_\pi\varphi.$$

Crucially, however, this seems to miss the essence of nondeterminism. For instance, according to this definition, when the agent in question happens to be very uncertain about π, we are forced to interpret π as having a great many possible nondeterministic outcomes. But there is a clear conceptual distinction between those outcomes of π that are possible as far as some agent knows—perhaps an agent with very poor information—as opposed to those outcomes that would remain possible *even with good information*. And it seems to be the latter concept that aligns more closely with our intuitions regarding nondeterminism.

For a simple example, imagine running a random number generator. This seems like a canonical example of a nondeterministic process. Note that what is important here is not merely that you do not, in fact, know what number will be generated in advance, but also that you are unable *in principle* to determine this in advance.[2] By contrast, imagine running a program that queries a

---

[2] To be sure, if you had access to a more advanced set of tools than are standardly available, perhaps you *could* make such a determination. And in this case, thinking of the random number generator as a nondeterministic process loses much of its intuitive appeal. Indeed, *any* nondeterministic process whatsoever might be viewed as deterministic relative to a sufficiently powerful set of tools (e.g., from God's perspective). Thus, nondeterminism can be naturally construed as a *relative* notion that depends on a fixed background set of "feasible measurements". We make this precise below.

given database and prints the result; we would not want to call this program nondeterministic even if you happened to be ignorant about the contents of the database.

This is a distinction we want to respect. The relevant epistemic notion, then, is not what any given agent currently happens to know, but what they *could come to know*. This is where topology comes in: the notion of "potential knowledge" or "knowability" is naturally represented in topological spaces.

3   Topology, Dynamics, and Determinism

3.1   Topological Spaces and Models

A **topological space** is a set X together with a collection $\mathcal{T} \subseteq 2^X$ of subsets of X such that $\emptyset, X \in \mathcal{T}$ and $\mathcal{T}$ is closed under unions and finite intersections. Elements of $\mathcal{T}$ are called *open* and $\mathcal{T}$ is called the *topology*.

There are various intuitions that help to make sense of this definition, most of which tap into the notion of topology as the mathematics of physical space and proximity.[3] Here, though, we focus instead on epistemic intuitions, through which topology is naturally interpreted as a formalization of evidence and potential observations. In fact, these two intuitions overlap in cases where the relevant observations are measurements about locations in space.

Informally, if we think of X as a set of possible worlds encoding the potential uncertainties one may have, then we can think of open sets $U \in \mathcal{T}$ as the results of measurements or observations. More precisely, we can understand U to represent the observation that rules out precisely those worlds $x \notin U$. On this view, each $U \in \mathcal{T}$ corresponds to a possible state of knowledge, and the topology $\mathcal{T}$ itself can be conceptualized as the set of available observations.[4]

A core notion in topology is that of the *interior* of a set $A \subseteq X$, defined by:

$$int(A) = \{x \in A : (\exists U \in \mathcal{T})(x \in U \subseteq A)\}.$$

The interior of A therefore consists of those points x that are "robustly" inside A, in the sense that there is some "witness" $U \in \mathcal{T}$ to x's membership in A. When we interpret elements of $\mathcal{T}$ as the results of possible measurements, the notion of interior takes on a natural epistemic interpretation: x lies in the interior of A just in case there is *some* measurement one could potentially take that would

---

[3] For a standard introduction to topological notions, we refer the reader to [10].

[4] Suppose, for a simple example, that you measure your height and obtain a reading of 180 ± 2 cm. If we represent the space of your possible heights using the positive real numbers, $\mathbb{R}^+$, then it is natural to identify this measurement with the open interval $(178, 182)$. And with this measurement in hand, you can safely deduce that you are, for instance, less than 183 cm tall, while remaining uncertain about whether you are, say, taller than 179 cm.

entail A. In other words, the worlds in the interior of A are precisely the worlds where A *could come to be known*.[5]

The dual of the interior operator is called *closure*:

$$cl(A) = X \setminus int(X \setminus A)$$
$$= \{x \in X \ : \ (\forall U \in \mathcal{T})(x \in U \Rightarrow U \cap A \neq \emptyset)\}.$$

Thus, epistemically speaking, worlds in the closure of A are precisely those worlds in which A is compatible with *every possible measurement*. The closure operator therefore offers a mathematical realization of our intuition about nondeterminism: namely, that a nondeterministic outcome of a program is one that remains possible no matter how good the agent's state of information.

A **topological model** M is a topological space $(X, \mathcal{T})$ together with a *valuation* $v : \text{PROP} \to 2^X$. In such models we interpret the basic modal language $\mathcal{L}_\square$ defined by

$$\varphi ::= p \,|\, \neg\varphi \,|\, \varphi \wedge \psi \,|\, \square\varphi,$$

where $p \in \text{PROP}$, via the usual recursive clauses for the Boolean connectives together with the following addition:

$$x \models \square\varphi \text{ iff } x \in int(\llbracket\varphi\rrbracket),$$

where $\llbracket\varphi\rrbracket = \{x \in X : x \models \varphi\}$. We also make use of the dual modality $\Diamond$, defined by

$$x \models \Diamond\varphi \text{ iff } x \in cl(\llbracket\varphi\rrbracket).$$

Following the discussion above, we read $\square\varphi$ as "φ is knowable" or "φ can be ascertained" and $\Diamond\varphi$ as "φ is unfalsifiable" or "φ cannot be ruled out".

3.2 Dynamic Topological Models

Kremer and Mints [1] introduce the notion of a *dynamic topological model*, which is simply a topological model equipped with a continuous function $f : X \to X$. Since we wish to capture the execution of a multitude of programs, we generalize this notion slightly to topological models equipped with a family of functions, one for each program $\pi \in \Pi$. Moreover, continuity is not something we will want to take for granted; we therefore drop this requirement as well.

---

[5] One might wonder about the closure conditions on topologies. Finite intersections can perhaps be accounted for by identifying them with sequences of measurements, but what about unions? One intuition comes by observing that for any set A, $int(A) = \bigcup\{U \in \mathcal{T} : U \subseteq A\}$, so $int(A)$ is the information state that arises from learning *that* A is true without learning *what particular measurement* was taken to ascertain this fact. This idea is formalized in [3] using public announcements; we direct the reader to this work for a more detailed discussion of this point.

A **dynamic topological model** is a tuple $(X, \mathfrak{T}, \{f_\pi\}_{\pi \in \Pi}, v)$ where $(X, \mathfrak{T}, v)$ is a topological model and each f_π is a function (not necessarily continuous) from X to X. In such models we can interpret the language $\mathcal{L}_{\square, \bigcirc}$ defined by

$$\varphi ::= p \mid \neg \varphi \mid \varphi \wedge \psi \mid \square \varphi \mid \bigcirc_\pi \varphi,$$

where $p \in \text{PROP}$ and $\pi \in \Pi$, via the additional semantic clause:

$$x \models \bigcirc_\pi \varphi \text{ iff } f_\pi(x) \models \varphi.$$

This provides the final tool we need to formalize the re-interpretation of nondeterministic program execution sketched in Sect. 2:

$$\langle \pi \rangle \varphi \equiv \Diamond \bigcirc_\pi \varphi.$$

Semantically:

$$x \models \langle \pi \rangle \varphi \text{ iff } x \models \Diamond \bigcirc_\pi \varphi$$
$$\text{iff } x \in cl(f_\pi^{-1}(\llbracket \varphi \rrbracket)).$$

So: φ is a nondeterministic outcome of π (at x) just in case it cannot be ruled out (at x) that φ will hold after π is executed. Topologically: every measurement at x (i.e., every open neighbourhood of x) is compatible with a state where φ results from executing π. Call this the *epistemic interpretation of nondeterminism*.

3.3   Determinism as Continuity

The epistemic interpretation of nondeterminism accords with our earlier intuitions about random number generators and database queries. Consider a process *rand* that randomly displays either a 0 or a 1, and an agent who (we presume) is unable to measure in advance the relevant quantities that determine the output of this process. This means that both $\bigcirc_{rand}0$ and $\bigcirc_{rand}1$ are compatible with every measurement the agent can take (in advance of running the process), so $\Diamond \bigcirc_{rand}0$ and $\Diamond \bigcirc_{rand}1$ both hold, i.e., $\langle rand \rangle 0 \wedge \langle rand \rangle 1$.[6] By contrast, consider a process *query* that outputs the next entry in a given database and an agent who can look up that entry in advance (which is, say, 0). This means there is a measurement that guarantees $\bigcirc_{query}0$, so $\square \bigcirc_{query}0$ holds, which yields $[query]0$.

What exactly is (non)determinism in this setting? It is tempting to describe a (non)deterministic process as one in which the output state can(not) be determined in advance. But this is far too liberal: in principle, states may encode many details that are far beyond the ability of the agent to measure precisely,

---

[6] Of course, once the process *rand* has been called, presumably the agent *can* ascertain its output (e.g., by looking at the screen). In particular, if (say) the number displayed is 1, then this knowable, and therefore it is not the case that $\bigcirc_{rand}\Diamond 0$ holds. This shows that the order of the two modalities is crucial in establishing the proper correspondence with nondeterminism. Thanks to an anonymous reviewer for suggesting this clarification.

which would then trivially render every process nondeterministic. For instance, if the state description encodes the current temperature of some internal components of the system (e.g., as a seed value for the *rand* process), then even *after* executing a simple program like *query* the user will not be in a position to know the true state.

The correct notion is more subtle. Consider again the *rand* process: the crucial feature of this program is not that it produces a *state* that cannot be determined in advance, but that it produces a *measurable quantity*—namely, the number displayed—that cannot be determined in advance.

To make the same point slightly more formally: it may be that no measurement at x rules out *all* the other states (indeed, this will be the case whenever $\{x\}$ is not open). This is necessary but *not sufficient* for nondeterminism because it may still be possible to learn (in advance) everything there is to know about the effects of executing π (as describable in the language).

This suggests the following refined account of determinism: a deterministic process is one in which everything *one could learn* about the state of the system *after the program is executed* one can also determine in advance of the program execution.

This account aligns perfectly with the topological notion of *continuity*. Intuitively: a function is continuous if small changes in the input produce small changes in the output. Topologically: f is **continuous at** x if, for every open neighbourhood V of $f(x)$, there exists an open neighbourhood U of x such that $f(U) \subseteq V$. And, finally, epistemically: f is continuous at x if every measurement V compatible with the output state $f(x)$ can be guaranteed in advance by some measurement U compatible with the input state x. So the definition of continuity corresponds exactly to our refined account of determinism. In other words: *determinism is continuity in the observation topology*.

Continuity of f_π can be *defined* in the object language $\mathcal{L}_{\square,\bigcirc}$ by the formula[7]

$$\bigcirc_\pi \square \varphi \rightarrow \square \bigcirc_\pi \varphi.$$

Unsurprisingly, this is precisely the scheme that Kremer and Mints call "the axiom of continuity" in their axiomatization of the class of (continuous) dynamic topological models [1]. It reads: "If, *after* executing π, φ is (not only true, but also) measurably true, then it is possible to take a measurement *before* executing π that guarantees that φ will be true after executing π." So this scheme expresses the idea that one need not actually execute π in order to determine whatever could be determined after its execution: all the measurable effects of π can be determined in advance. Again, this is determinism. Continuity is determinism.

4   Axiomatization and Model Transformation

We restrict our attention in this section to *serial* PDL models, in which each R_π is serial: $(\forall x)(\exists y)(x R_\pi y)$. Thus, we rule out the possibility of a program π producing no output at all in some states (intuitively, "crashing"), which corresponds

---

[7] This claim is made precise and proved in Appendix A.1.

to the fact that the functions in dynamic topological models are assumed to be total (i.e., everywhere defined). This allows for a cleaner translation between the two paradigms; in Sect. 5 we consider a framework that drops this assumption.

Table 1. Axioms and rules of inference for SPDL$_0$

(CPL)	All propositional tautologies	Classical propositional logic
(K$_\pi$)	$[\pi](\varphi \to \psi) \to ([\pi]\varphi \to [\pi]\psi)$	Distribution
(D$_\pi$)	$[\pi]\varphi \to \langle\pi\rangle\varphi$	Seriality
(MP)	From φ and $\varphi \to \psi$ deduce ψ	Modus ponens
(Nec$_\pi$)	From φ deduce $[\pi]\varphi$	Necessitation

The most basic version of serial PDL (without any operations on programs) is axiomatized by the axioms and rules of inference given in Table 1. Call this system SPDL$_0$.

Theorem 1. *SPDL$_0$ is a sound and complete axiomatization of the language $\mathcal{L}_{PDL}$ with respect to the class of all serial PDL models.*[8]

Using the epistemic interpretation of nondeterminism given in Sect. 3.2, we can also interpret the language $\mathcal{L}_{PDL}$ directly in dynamic topological models:

$$x \models \langle\pi\rangle\varphi \text{ iff } x \in cl(f_\pi^{-1}(\llbracket\varphi\rrbracket)).$$

And, dually:

$$x \models [\pi]\varphi \text{ iff } x \in int(f_\pi^{-1}(\llbracket\varphi\rrbracket)).$$

This puts us in a position to evaluate our re-interpretation in a precise way. Namely, we can ask: are all the properties of nondeterministic program execution that are captured by standard (serial) PDL models preserved under this new interpretation? And we can answer in the affirmative:

Theorem 2. *SPDL$_0$ is a sound and complete axiomatization of the language $\mathcal{L}_{PDL}$ with respect to the class of all dynamic topological models.*

Proof. Soundness of (CPL) and (MP) is immediate. Soundness of (Nec$_\pi$) follows from the fact that $f_\pi^{-1}(X) = X$ and $int(X) = X$, and soundness of (D$_\pi$) follows from the fact that, for all $A \subseteq X$, $int(A) \subseteq cl(A)$. Finally, to see that (K$_\pi$) is sound, observe that

$$int(f_\pi^{-1}(\llbracket\varphi \to \psi\rrbracket)) \cap int(f_\pi^{-1}(\llbracket\varphi\rrbracket)) = int(f_\pi^{-1}(\llbracket\varphi \to \psi\rrbracket) \cap f_\pi^{-1}(\llbracket\varphi\rrbracket))$$
$$\subseteq int(f_\pi^{-1}(\llbracket\psi\rrbracket)).$$

---

[8] This can be proved using very standard techniques; see, e.g., [11].

The proof of completeness proceeds by way of a model-transformation construction we provide in Appendix A.2: specifically, we show that every serial PDL model can be transformed into a dynamic topological model in a manner that preserves the truth of all formulas in $\mathcal{L}_{PDL}$ (Proposition 2). By Theorem 1, every non-theorem of SPDL_0 is refuted on some serial PDL model, so our transformation produces a dynamic topological model that refutes the same formula, thereby establishing completeness.   □

Typically one works with richer versions of PDL in which the set of programs Π is equipped with one or more operations corresponding, intuitively, to ways of building new programs from old programs. Standard examples include:

- *Sequencing*: $\pi_1; \pi_2$ executes π_1 followed immediately by π_2.
- *Nondeterministic union*: $\pi_1 \cup \pi_2$ nondeterministically chooses to execute either π_1 or π_2.
- *Iteration*: π^* repeatedly executes π some nondeterministic finite number of times.

Can we make sense of these operations in our enriched epistemic setting? The latter two transform deterministic programs into nondeterministic programs, and for this reason they are difficult to interpret in a setting where program execution is fundamentally deterministic (i.e., interpreted by functions). We return to discuss this point in Sect. 6. Sequencing, on the other hand, does not have this issue; one might guess that it is straightforwardly captured by the condition

$$f_{\pi_1;\pi_2} = f_{\pi_2} \circ f_{\pi_1}.$$

While function composition certainly seems like the natural way to interpret sequential program execution, there is a wrinkle in the axiomatization. PDL with sequencing is axiomatized by including the following axiom scheme:

$$(\text{Seq}) \qquad \langle \pi_1; \pi_2 \rangle \varphi \leftrightarrow \langle \pi_1 \rangle \langle \pi_2 \rangle \varphi.$$

Interestingly, this scheme is *not* valid in arbitrary dynamic topological models. This is because

$$[\![\langle \pi_1; \pi_2 \rangle \varphi]\!] = cl(f_{\pi_1;\pi_2}^{-1}([\![\varphi]\!])) = cl(f_{\pi_1}^{-1}(f_{\pi_2}^{-1}([\![\varphi]\!]))),$$

whereas

$$[\![\langle \pi_1 \rangle \langle \pi_2 \rangle \varphi]\!] = cl(f_{\pi_1}^{-1}(cl(f_{\pi_2}^{-1}([\![\varphi]\!]))));$$

the extra closure operator means we have

$$[\![\langle \pi_1; \pi_2 \rangle \varphi]\!] \subseteq [\![\langle \pi_1 \rangle \langle \pi_2 \rangle \varphi]\!]$$

but not, in general, equality.[9]

---

[9] For instance, consider the set $X = \{a, b\}$ equipped with the topology $\mathcal{T} = \{\emptyset, \{b\}, X\}$, and the function f_π defined by $f_\pi(a) = b$ and $f_\pi(b) = a$. Let $v(p) = \{a\}$. Then, since $f_\pi \circ f_\pi = id$, we have $cl(f_{\pi;\pi}^{-1}([\![p]\!])) = cl(\{a\}) = \{a\}$, whereas $cl(f_\pi^{-1}(cl(f_\pi^{-1}([\![p]\!])))) = cl(f_\pi^{-1}(cl(\{b\}))) = cl(f_\pi^{-1}(X)) = X$.

A function $f : X \to Y$ is called **open** if for every open subset $U \subseteq X$, the set $f(U)$ is open in Y. It turns out that when the function f_{π_1} is open, the mismatch above vanishes (all of the following claims are proved in Appendix A.3):

Lemma 1. *Let* $(X, \mathcal{T}, \{f_\pi\}_{\pi \in \Pi}, v)$ *be a dynamic topological model. If* f_{π_1} *is open, then*

$$[\![\langle \pi_1; \pi_2 \rangle \varphi]\!] = [\![\langle \pi_1 \rangle \langle \pi_2 \rangle \varphi]\!].$$

Say that a dynamic topological model is *open* if each f_π is open.

Theorem 3. *SPDL$_0$ + (Seq) is a sound and complete axiomatization of the language* $\mathcal{L}_{PDL}$ *with respect to the class of all open dynamic topological models.*

Like continuity, openness of the function f_π can be defined in the object language; in fact, it is defined by the converse of the scheme defining continuity:

$$\Box \bigcirc_\pi \varphi \to \bigcirc_\pi \Box \varphi.$$

Roughly speaking, this says that whatever you can (in principle) predict about executing π beforehand you could also come to know afterward. This has a "perfect recall" type flavour, except the relevant epistemic notion is not what is actually known but what *could come to be known*. Besides serving to validate the standard sequencing axiom, this principle also plays a crucial role in the next section, where we extend the present framework to incorporate knowledge.

5   Knowledge and Learning

To study the epistemology of nondeterministic program execution, we want to be able to reason not only about what *can be* known, but also about what *is* known. To do so we need a richer semantic setting, for which we turn to *topological subset models* [2]; essentially, these use an additional parameter to keep track of the current state of information, through which a standard knowledge modality can be interpreted.

Topological subset models have experienced renewed interest in recent years [3,12–14], beginning with the work in [3] studying public announcements in the topological setting. Standard semantics for public announcement logic take the *precondition* of an announcement of φ to be the truth of φ; in the topological setting, this precondition is strengthened to the *knowability* of φ. As we will see, this interpretation of public announcements is recapitulated in the present framework via a natural interpretation of *test programs*.

5.1   Incorporating Knowledge

A **topological subset model** just *is* a topological model $(X, \mathcal{T}, v)$; the difference lies in the semantic clauses for truth, which are defined with respect to *pairs* of the form (x, U), where $x \in U \in \mathcal{T}$; such pairs are called *epistemic scenarios*. Intuitively, x represents the actual world, while U captures the agent's current information and thus what they know. Formally, we interpret the language $\mathcal{L}_{K,\square}$ given by

$$\varphi ::= p \mid \neg\varphi \mid \varphi \wedge \psi \mid K\varphi \mid \square\varphi,$$

where $p \in \text{PROP}$, as follows:

$$(x, U) \models p \text{ iff } x \in v(p)$$
$$(x, U) \models \neg\varphi \text{ iff } (x, U) \not\models \varphi$$
$$(x, U) \models \varphi \wedge \psi \text{ iff } (x, U) \models \varphi \text{and} (x, U) \models \psi$$
$$(x, U) \models K\varphi \text{ iff } U \subseteq \llbracket\varphi\rrbracket^U$$
$$(x, U) \models \square\varphi \text{ iff } x \in int(\llbracket\varphi\rrbracket^U),$$

where $\llbracket\varphi\rrbracket^U = \{x \in U : (x, U) \models \varphi\}$. So the agent knows φ in the epistemic scenario (x, U) just in case it is guaranteed by their current information U.

We next define *dynamic topological subset models* by incorporating functions f_π as above. Of course, we need subset-style semantics for the dynamic modalities. Perhaps the most natural way to define the updated epistemic scenario is as follows:

$$(x, U) \models \bigcirc_\pi\varphi \text{ iff } (f_\pi(x), f_\pi(U)) \models \varphi.$$

This definition raises two issues, one technical and the other conceptual. First, as a technical matter, the definition only makes sense if $f_\pi(U)$ is open— otherwise $(f_\pi(x), f_\pi(U))$ is not an epistemic scenario. So we have here another reason to restrict our attention to *open* functions f_π.

Second, conceptually, in a sense this framework does not permit *learning*. True, an agent's state of knowledge changes in accordance with program execution, but every "live" possibility $y \in U$ is preserved as the corresponding state $f_\pi(y)$ in the updated information set $f_\pi(U)$. Intuitively, then, the agent can never truly eliminate possibilities.

Dynamic epistemic logic [15] is a modern and vibrant area of research concerned exactly with this issue of how to capture the dynamics of information update. But rather than explicitly importing machinery from this paradigm (e.g., announcements) to represent learning in the present context, we can take advantage of a mechanic that PDL already has available: crashing. In Sect. 4 we restricted attention to standard PDL models that were serial, corresponding in our framework to total functions. We now drop this assumption to allow *partial* functions $f_\pi : X \rightharpoonup X$ that are undefined at some points in X. This allows the corresponding updates to effectively delete states and thus capture information update in much the same way that, e.g., public announcements do.

A **dynamic topological subset model (over Π)** is a topological subset model together with a family of partial, open functions[10] $f_\pi : X \rightharpoonup X$, $\pi \in \Pi$.

---

[10] Typically the concept of openness is applied to total functions, but the definition makes sense for partial functions as well: f is open provided, for all open U, $f(U) = \{y \in X : (\exists x \in U)(f(x) = y)\}$ is open.

Formulas of the language $\mathcal{L}_{K,\Box,\bigcirc}$ given by

$$\varphi ::= p \,|\, \neg\varphi \,|\, \varphi \wedge \psi \,|\, K\varphi \,|\, \Box\varphi \,|\, \bigcirc_\pi \varphi,$$

are interpreted at epistemic scenarios via the semantic clauses introduced above, taking care to note that the righthand side of the clause defining $\bigcirc_\pi$ now carries the implicit assumption that $f_\pi(x)$ is actually defined:

$$(x, U) \models \bigcirc_\pi \varphi \text{ iff } f_\pi(x) \text{ is defined and } (f_\pi(x), f_\pi(U)) \models \varphi.$$

We provide a sound and complete axiomatization of this logic in Appendix A.4.

5.2 Test Programs and Public Announcements

A standard enrichment of PDL expands the set of programs Π to include *test programs*. Unlike other program constructions, test programs are not built from existing programs but instead from formulas in the language: given a formula φ, the program $\varphi?$ is introduced to be interpreted by the relation $R_{\varphi?}$ defined by

$$x R_{\varphi?} y \text{ iff } x = y \text{ and } x \models \varphi.$$

Intuitively, the program $\varphi?$ crashes at states where φ is false, and otherwise does nothing.

Test programs are deterministic, so can be represented just as easily by functions:

$$f_{\varphi?}(x) = \begin{cases} x & \text{if } x \models \varphi \\ \text{undefined} & \text{otherwise.} \end{cases}$$

But to make sense of this definition in dynamic topological subset models, two issues must be addressed. First, the relation $x \models \varphi$ is not actually defined in subset models—formulas are evaluated with respect to epistemic scenarios, not individual states. However, it is easy to see that when φ belongs to the fragment $\mathcal{L}_{\Box,\bigcirc}$, its truth in an epistemic scenario (x, U) is independent of U; in this case, we can simply declare that $x \models \varphi$ iff $(x, U) \models \varphi$ for some (equivalently, all) open sets U containing x. We therefore restrict the formation of test programs to $\varphi \in \mathcal{L}_{\Box,\bigcirc}$.

Second, $f_{\varphi?}$ may not be open. Indeed, $f_{\varphi?}$ is open just in case $[\![\varphi]\!] = \{x : x \models \varphi\}$ is open. If $[\![\varphi]\!]$ is not open then it contains at least one state $x \in [\![\varphi]\!] \setminus int([\![\varphi]\!])$, that is, a state at which φ is true but not *measurably* true. Intuitively, at such states the test program $\varphi?$ should crash, since it must fail to determine that φ is true. This motivates the following revised definition of $f_{\varphi?}$:

$$f_{\varphi?}(x) = \begin{cases} x & \text{if } x \in int([\![\varphi]\!]) \\ \text{undefined} & \text{otherwise.} \end{cases}$$

These functions *are* open. Moreover, under these revised semantics, we have:

$$(x, U) \models \bigcirc_{\varphi?}\psi \text{ iff } f_{\varphi?}(x) \text{ is defined and } (f_{\varphi?}(x), f_{\varphi?}(U)) \models \psi$$
$$\text{iff } x \in int([\![\varphi]\!]) \text{ and } (x, U \cap int([\![\varphi]\!])) \models \psi,$$

which coincides exactly with the topological definition of a public announcement of φ as given in [3].[11]

6   Future Work

We have formalized a relatively simple idea: namely, that the nondeterministic outcomes of a process are precisely those that the agent cannot rule out in advance. Using the tools of topology to represent potential observations, we have demonstrated a striking connection between deterministic processes and continuous functions, proved that certain axiomatizations of PDL remain sound and complete when reinterpreted in this enriched setting, and established a natural identity between test programs and public announcements.

Many questions remain, both conceptual and technical. What is the relationship between the (probabilistic) notion of *chance* and the topological construal of nondeterminism presented here? Is there a way to import "nondeterministic" operations on programs, such as nondeterministic union or iteration, into this setting? Or is it perhaps better to focus on deterministic analogues of these program constructions, such as, "If φ do π_1, else do π_2", or, "Do π until φ"? How much of dynamic epistemic logic can be recovered as program execution in dynamic topological subset models? For instance, can we make sense of test programs based on epistemic formulas (i.e., formulas that include the K modality), as we can with public announcements? And how can we extend the axiomatization of dynamic topological subset models given in Appendix A.4 to include test programs? These questions and more are the subject of ongoing research.

A   Proofs and Details

A.1   Characterizing Continuity

A **dynamic topological frame (over** Π**)** is a tuple $F = (X, \mathcal{T}, \{f_\pi\}_{\pi \in \Pi})$ where $(X, \mathcal{T})$ is a topological space and each $f_\pi : X \to X$. In other words, a frame is simply a dynamic topological model without a valuation function. A frame F is said to *validate* a formula φ just in case φ is true at every point of every model of the form (F, v).

Proposition 1. *The formula scheme* $\bigcirc_\pi \square \varphi \to \square \bigcirc_\pi \varphi$ *defines the class of dynamic topological frames in which* f_π *is continuous: that is, for every dynamic topological frame F, F validates every instance of* $\bigcirc_\pi \square \varphi \to \square \bigcirc_\pi \varphi$ *iff* f_π *is continuous.*

Proof. First suppose that M is a dynamic topological model in which f_π is continuous, and let x be a point in this model satisfying $\bigcirc_\pi \square \varphi$. Then $f_\pi(x) \in$

---
[11] Or, rather, it coincides with the dual of the definition given in [3], but this is not an important difference.

$int([\![\varphi]\!])$. By continuity, the set $U = f_\pi^{-1}(int([\![\varphi]\!]))$ is open. Moreover, it is easy to see that $x \in U$ and $U \subseteq [\![\bigcirc_\pi\varphi]\!]$, from which it follows that $x \models \Box\bigcirc_\pi\varphi$.

Conversely, suppose that F is a dynamic topological frame in which f_π is not continuous. Let U be an open subset of X such that $A = f_\pi^{-1}(U)$ is not open, and let $x \in A \setminus int(A)$; consider a valuation v such that $v(p) = U$. In the resulting model, since $f_\pi(x) \in U = int(U)$, we have $x \models \bigcirc_\pi\Box p$. On the other hand, since by definition $x \notin int(f_\pi^{-1}(U))$, we have $x \not\models \Box\bigcirc_\pi p$.   □

A.2   Model Transformation

Our task in this section is to transform an arbitrary serial PDL model into a dynamic topological model in a truth-preserving manner. The intuition for this transformation is fairly straightforward: in a serial PDL model, each state may be nondeterministically compatible with many possible execution paths corresponding to all the possible ways of successively traversing R_π-edges. In a dynamic topological model, by contrast, all such execution paths must be differentiated by state—roughly speaking, this means we need to create a new state for each possible execution path in the standard model. Then, to preserve the original notion of nondeterminism, we overlay a topological structure that "remembers" which new states originated from the same state in the standard model by rendering them topologically indistinguishable.

Let $M = (X, (R_\pi)_{\pi\in\Pi}, v)$ be a serial PDL model. Let Π^* denote the set of all finite sequences from Π. A map $\alpha : \Pi^* \to X$ is called a **network (through M)** provided $(\forall \boldsymbol{\pi} \in \Pi^*)(\forall \pi \in \Pi)(\alpha(\boldsymbol{\pi})R_\pi\alpha(\boldsymbol{\pi}, \pi))$. In other words, a network α must respect R_π-edges in the sense that it associates with each sequence $(\pi_1, \ldots, \pi_n)$ a path $(x_1, \ldots, x_{n+1})$ through X such that, for each i, $x_i R_{\pi_i} x_{i+1}$.[12] Networks through M constitute the points of the dynamic topological model we are building:
$$\tilde{X} = \{\alpha : \alpha \text{ is a network through } M.\}.$$

The topology we equip $\tilde{X}$ with is particularly simple: for each $x \in X$, let $U_x = \{\alpha \in \tilde{X} : \alpha(\emptyset) = x\}$. Clearly the sets U_x partition X and so form a topological basis; let $\mathcal{T}$ be the topology they generate.

Next we define the functions $f_\pi : \tilde{X} \to \tilde{X}$. Intuitively, $\alpha \in \tilde{X}$ is a complete record of what paths will be traversed in the original state space X for every sequence of program executions. Therefore, after executing π, the updated record $f_\pi(\alpha)$ should simply consist in those paths determined by α that start with an execution of π:
$$f_\pi(\alpha)(\boldsymbol{\pi}) = \alpha(\pi, \boldsymbol{\pi}).$$

Finally, define $\tilde{v} : \text{PROP} \to 2^{\tilde{X}}$ by
$$\tilde{v}(p) = \{\alpha \in \tilde{X} : \alpha(\emptyset) \in v(p)\}.$$

Let $\tilde{M} = (\tilde{X}, \mathcal{T}, (f_\pi)_{\pi\in\Pi}, \tilde{v})$.

---

[12] Specifically, $x_1 = \alpha(\emptyset)$ and $(\forall i \geq 2)(x_i = \alpha(x_1, \ldots, x_{i-1}))$.

Proposition 2. *For every* $\varphi \in \mathcal{L}_{PDL}$, *for every* $\alpha \in \tilde{X}$, $(\tilde{M}, \alpha) \models \varphi$ *iff* $(M, \alpha(\emptyset)) \models \varphi$.

Proof. We proceed by induction on the structure of φ. The base case when $\varphi = p \in \text{PROP}$ follows directly from the definition of $\tilde{v}$:

$$(\tilde{M}, \alpha) \models p \text{ iff } \alpha \in \tilde{v}(p)$$
$$\text{iff } \alpha(\emptyset) \in v(p)$$
$$\text{iff } (M, \alpha(\emptyset)) \models p.$$

The inductive steps for the Boolean connectives are straightforward. So suppose inductively that the result holds for φ; we wish to show it holds for $\langle \pi \rangle \varphi$.

Let $\alpha \in \tilde{X}$ and $x = \alpha(\emptyset)$. By definition, $(\tilde{M}, \alpha) \models \langle \pi \rangle \varphi$ iff $\alpha \in cl(f_\pi^{-1}(\llbracket \varphi \rrbracket_{\tilde{M}}))$. Since the topology is generated by a partition, we know that U_x is a minimal neighbourhood of α, and therefore the preceding condition is equivalent to:

$$U_x \cap f_\pi^{-1}(\llbracket \varphi \rrbracket_{\tilde{M}}) \neq \emptyset.$$

This intersection is nonempty just in case there exists an $\alpha' \in \tilde{X}$ such that $\alpha'(\emptyset) = x$ and $f_\pi(\alpha') \in \llbracket \varphi \rrbracket_{\tilde{M}}$. By the induction hypothesis,

$$f_\pi(\alpha') \in \llbracket \varphi \rrbracket_{\tilde{M}} \text{ iff } (\tilde{M}, f_\pi(\alpha')) \models \varphi$$
$$\text{iff } (M, f_\pi(\alpha')(\emptyset)) \models \varphi$$
$$\text{iff } (M, \alpha'(\pi)) \models \varphi.$$

So to summarize, we have shown that $(\tilde{M}, \alpha) \models \langle \pi \rangle \varphi$ iff there exists an $\alpha' \in \tilde{X}$ such that $\alpha'(\emptyset) = x$ and $(M, \alpha'(\pi)) \models \varphi$. Since we know that $\alpha'(\emptyset) R_\pi \alpha'(\pi)$, this is in turn equivalent to $(M, x) \models \langle \pi \rangle \varphi$, which completes the induction.   □

A.3   Sequencing

Lemma 1. *Let* $(X, \mathcal{T}, \{f_\pi\}_{\pi \in \Pi}, v)$ *be a dynamic topological model. If* f_{π_1} *is open, then*

$$\langle \pi_1; \pi_2 \rangle \varphi = \llbracket \langle \pi_1 \rangle \langle \pi_2 \rangle \varphi \rrbracket.$$

Proof. It suffices to show that $\llbracket \langle \pi_1; \pi_2 \rangle \varphi \rrbracket \supseteq \llbracket \langle \pi_1 \rangle \langle \pi_2 \rangle \varphi \rrbracket$. So let

$$x \in \llbracket \langle \pi_1 \rangle \langle \pi_2 \rangle \varphi \rrbracket = cl(f_{\pi_1}^{-1}(cl(f_{\pi_2}^{-1}(\llbracket \varphi \rrbracket))));$$

then for every open neighbourhood U containing x, we know that $U \cap f_{\pi_1}^{-1}(cl(f_{\pi_2}^{-1}(\llbracket \varphi \rrbracket))) \neq \emptyset$. This implies that $f_{\pi_1}(U) \cap cl(f_{\pi_2}^{-1}(\llbracket \varphi \rrbracket)) \neq \emptyset$; since $f_{\pi_1}(U)$ is open, it follows that $f_{\pi_1}(U) \cap f_{\pi_2}^{-1}(\llbracket \varphi \rrbracket) \neq \emptyset$ as well. This then implies that $U \cap f_{\pi_1}^{-1}(f_{\pi_2}^{-1}(\llbracket \varphi \rrbracket)) \neq \emptyset$, and therefore

$$x \in cl(f_{\pi_1}^{-1}(f_{\pi_2}^{-1}(\llbracket \varphi \rrbracket))) = \llbracket \langle \pi_1; \pi_2 \rangle \varphi \rrbracket,$$

as desired.   □

Say that a dynamic topological model is *open* if each f_π is open.

Theorem 3. $SPDL_0 + (Seq)$ *is a sound and complete axiomatization of the language* $\mathcal{L}_{PDL}$ *with respect to the class of all open dynamic topological models.*

Proof. Lemma 1 shows that (Seq) is valid in the class of all open dynamic topological models. For completeness, it suffices to observe that the dynamic topological model $\tilde{M}$ constructed in Appendix A.2 is itself open: indeed, for each basic open U_x, we have

$$f_\pi(U_x) = \{\alpha \in \tilde{X} : x R_\pi \alpha(\emptyset)\}$$
$$= \bigcup_{y \in R(x)} U_y,$$

which of course is open.   □

Proposition 3. *The formula scheme* $\Box \bigcirc_\pi \varphi \rightarrow \bigcirc_\pi \Box \varphi$ *defines the class of dynamic topological frames in which* f_π *is open: that is, for every dynamic topological frame* F, F *validates every instance of* $\Box \bigcirc_\pi \varphi \rightarrow \bigcirc_\pi \Box \varphi$ *iff* f_π *is open.*

Proof. First suppose that M is a dynamic topological model in which f_π is open, and let x be a point in this model satisfying $\Box \bigcirc_\pi \varphi$. Then $x \in int(f_\pi^{-1}(\llbracket \varphi \rrbracket))$. By openness, the set $V = f(int(f_\pi^{-1}(\llbracket \varphi \rrbracket)))$ is open. Moreover, it is easy to see that $f_\pi(x) \in V$ and $V \subseteq \llbracket \varphi \rrbracket$, from which it follows that $x \models \bigcirc_\pi \Box \varphi$.

Conversely, suppose that F is a dynamic topological frame in which f_π is not open. Let U be an open subset of X such that $A = f_\pi(U)$ is not open, and let $x \in U$ be such that $f_\pi(x) \in A \setminus int(A)$; consider a valuation v such that $v(p) = A$. In the resulting model, since $x \in U$ and $f_\pi(U) \subseteq \llbracket p \rrbracket$, we have $x \models \Box \bigcirc_\pi p$. On the other hand, since by definition $f_\pi(x) \notin int(\llbracket p \rrbracket)$, we have $x \not\models \bigcirc_\pi \Box p$.   □

A.4   Dynamic Topological Epistemic Logic

We provide a sound and complete axiomatization of the language $\mathcal{L}_{K,\Box,\bigcirc}$ with respect to the class of all dynamic topological subset models.

Let CPL denote the axioms and rules of classical propositional logic, let $S5_K$ denote the S5 axioms and rules for the K modality, and let $S4_\Box$ denote the S4 axioms and rules for the $\Box$ modality (see, e.g., [11]). Let (KI) denote the axiom scheme $K\varphi \rightarrow \Box \varphi$, and set

$$EL_{K,\Box} := CPL + S5_K + S4_\Box + (KI).$$

Theorem 4 ([3, Theorem 1]). $EL_{K,\Box}$ *is a sound and complete axiomatization of* $\mathcal{L}_{K,\Box}$ *with respect to the class of all dynamic topological subset models.*

Let DTEL denote the axiom system $EL_{K,\Box}$ together with the axiom schemes and rules of inference given in Table 2.

Theorem 5. *DTEL is a sound and complete axiomatization of* $\mathcal{L}_{K,\Box,\bigcirc}$ *with respect to the class of all dynamic topological subset models.*

Table 2. Additional axioms and rules of inference for DTEL

$(\neg\text{-PC}_\pi)$	$\bigcirc_\pi\neg\varphi \leftrightarrow (\neg\bigcirc_\pi\varphi \wedge \bigcirc_\pi\top)$	Partial commutativity of $\neg$
$(\wedge\text{-C}_\pi)$	$\bigcirc_\pi(\varphi \wedge \psi) \leftrightarrow (\bigcirc_\pi\varphi \wedge \bigcirc_\pi\psi)$	Commutativity of $\wedge$
$(K\text{-PC}_\pi)$	$\bigcirc_\pi\top \rightarrow (\bigcirc_\pi K\varphi \leftrightarrow K(\bigcirc_\pi\top \rightarrow \bigcirc_\pi\varphi))$	Partial commutativity of K
(O_π)	$(\Box\neg\bigcirc_\pi\varphi \wedge \bigcirc_\pi\top) \rightarrow \bigcirc_\pi\Box\neg\varphi$	Openness
(Mon_π)	From $\varphi \rightarrow \psi$ deduce $\bigcirc_\pi\varphi \rightarrow \bigcirc_\pi\psi$	Monotonicity

Proof. Soundness of $\mathsf{EL}_{K,\Box}$ follows as in the proof given in [3, Theorem 1], while soundness of the additions presented in Table 2 is easy to check. The presence of $\bigcirc_\pi\top$ in $(\neg\text{-PC}_\pi)$ accounts for the fact that f_π can be partial (since both $\neg\bigcirc_\pi\varphi$ and $\neg\bigcirc_\pi\neg\varphi$ are true at states where f_π is undefined), and plays an analogous role in $(K\text{-PC}_\pi)$ and (O_π). Similarly, the usual "necessitation" rule for $\bigcirc_\pi$ is not valid, since even if φ is a theorem, $\bigcirc_\pi\varphi$ still fails at states where f_π is undefined.

Completeness is proved by a canonical model construction. Let X denote the set of all maximal DTEL-consistent subsets of $\mathcal{L}_{K,\Box,\bigcirc}$. Define a binary relation $\sim$ on X by

$$x \sim y \Leftrightarrow (\forall\varphi \in \mathcal{L}_{K,\Box,\bigcirc})(K\varphi \in x \Leftrightarrow K\varphi \in y).$$

Clearly $\sim$ is an equivalence relation; let $[x]$ denote the equivalence class of x under $\sim$. For each $x \in X$, define

$$R(x) = \{y \in X : (\forall\varphi \in \mathcal{L}_{K,\Box,\bigcirc})(\Box\varphi \in x \Rightarrow \varphi \in y)\}.$$

Let $\mathcal{B} = \{R(x) : x \in X\}$, and let $\mathcal{T}$ be the topology generated by $\mathcal{B}$. It is easy to check that $\mathcal{B}$ is a basis for $\mathcal{T}$, and each $R(x)$ is a minimal neighbourhood about x (see, e.g., [16]). Given $x \in X$, define

$$f_\pi(x) = \begin{cases} \{\varphi : \bigcirc_\pi\varphi \in x\} & \text{if } \bigcirc_\pi\top \in x \\ \text{undefined} & \text{otherwise.} \end{cases}$$

Lemma 2. *Each f_π is a partial, open function $X \rightharpoonup X$.*

For each $p \in \text{PROP}$, set $v(p) := \{x \in X : p \in x\}$. Let $\mathcal{X} = (X, \mathcal{T}, \{f_\pi\}_{\pi\in\Pi}, v)$. Clearly $\mathcal{X}$ is a dynamic topological subset model.

Lemma 3 (Truth Lemma). *For every $\varphi \in \mathcal{L}_{K,\Box,\bigcirc}$, for all $x \in X$, $\varphi \in x$ iff $(\mathcal{X}, x, [x]) \models \varphi$.*

Completeness is an easy consequence: if φ is not a theorem of DTEL, then $\{\neg\varphi\}$ is consistent and so can be extended by Lindenbaum's lemma to some $x \in X$; by Lemma 3, we have $(\mathcal{X}, x, [x]) \not\models \varphi$. $\qquad\Box$

References

1. Kremer, P., Mints, G.: Dynamic topological logic. Ann. Pure Appl. Logic **131**, 133–158 (2005)
2. Dabrowski, A., Moss, L., Parikh, R.: Topological reasoning and the logic of knowledge. Ann. Pure Appl. Logic **78**, 73–110 (1996)
3. Bjorndahl, A.: Topological subset space models for public announcements. In: van Ditmarsch, H., Sandu, G. (eds.) Jaakko Hintikka on Knowledge and Game-Theoretical Semantics. OCL, vol. 12, pp. 165–186. Springer, Cham (2018). https://doi.org/10.1007/978-3-319-62864-6_6
4. Rabin, M.O., Scott, D.: Finite automata and their decision problems. IBM J. Res. **3**(2), 115–125 (1959)
5. Dijkstra, E.W.: A Discipline of Programming. Prentice-Hall, Englewood Cliffs (1976)
6. Francez, N., Hoare, C.A.R., Lehmann, D.J., de Roever, W.P.: Semantics of non-determinism, concurrency, and communication. J. Comput. Syst. Sci. **19**, 290–308 (1979)
7. Søndergaard, H., Sestoft, P.: Non-determinism in functional langauges. Comput. J. **35**(5), 514–523 (1992)
8. Troquard, N., Balbiani, P.: Propositional dynamic logic. The Stanford Encyclopedia of Philosophy ((Spring 2015 Edition)) Zalta, E.N. (ed.). https://plato.stanford.edu/archives/spr2015/entries/logic-dynamic/
9. Plaza, J.: Logics of public communications. Synthese **158**, 165–179 (2007)
10. Munkres, J.: Topology, 2nd edn. Prentice-Hall, Englewood Cliffs (2000)
11. Blackburn, P., de Rijke, M., Venema, Y.: Modal Logic. Cambridge Tracts in Theoretical Computer Science, vol. 53. Cambridge University Press, Cambridge (2001)
12. van Ditmarsch, H., Knight, S., Özgün, A.: Announcement as effort on topological spaces. In: Proceedings of the 15th conference on Theoretical Aspects of Rationality and Knowledge (TARK), pp. 95–102 (2015)
13. Baltag, A., Özgün, A., Vargas Sandoval, A.L.: Topo-logic as a dynamic-epistemic logic. In: Baltag, A., Seligman, J., Yamada, T. (eds.) LORI 2017. LNCS, vol. 10455, pp. 330–346. Springer, Heidelberg (2017). https://doi.org/10.1007/978-3-662-55665-8_23
14. Bjorndahl, A., Özgün, A.: Logic and topology for knowledge, knowability, and belief. In: Lang, J. (ed.) Proceedings of the 16th conference on Theoretical Aspects of Rationality and Knowledge (TARK) (2017)
15. van Ditmarsch, H., van der Hoek, W., Kooi, B.: Dynamic Epistemic Logic. Springer, Heidelberg (2008). https://doi.org/10.1007/978-1-4020-5839-4
16. Aiello, M., van Benthem, J., Bezhanishvili, G.: Reasoning about space: the modal way. J. Logic Comput. **13**(6), 889–920 (2003)

Parameterized Complexity of Some Prefix-Vocabulary Fragments of First-Order Logic

Luis Henrique Bustamante[1]([✉]), Ana Teresa Martins[1], and Francicleber Ferreira Martins[2]

[1] Department of Computing, Federal University of Ceará, Fortaleza, Brazil
{lhbusta,ana}@lia.ufc.br
[2] Federal University of Ceará, Campus of Quixadá, Quixadá, Brazil
fran@lia.ufc.br

Abstract. We analyze the parameterized complexity of the satisfiability problem for some prefix-vocabulary fragments of First-order Logic with the finite model property. Here we examine three natural parameters: the quantifier rank, the vocabulary size and the maximum arity of relation symbols. Following the classical classification of decidable prefix-vocabulary fragments, we will see that, for all relational *classes of modest complexity* and some *classical classes*, fixed-parameter tractability is achieved by using the above cited parameters.

Keywords: Satisfiability
Prefix-vocabulary fragments of first-order logic
Fixed-parameter tractability

1 Introduction

Parameterized complexity theory has been applied extensively to many logical problems [1,16,22,24,29]. Moreover, the model checking problem for first-order logic (FO), the problem of verifying whether a first-order sentence φ holds in a given structure $\mathfrak{A}$, plays a central role in the characterization of the parameterized intractability [7,8,17]. In this vein, we address the problem of the analysis of parameterized complexity for the satisfiability problem of some decidable fragments of FO. We are mainly interested in the syntactic defined prefix-vocabulary classes well explored in the textbook "The Classical Decision Problem" of Börger, Grädel, and Gurevich [4, Chaps. (6–7)]. Up to our knowledge, it is the first time that the parameterized satisfiability of prefix-vocabulary classes of first-order logic is systematically studied.

The parameterized complexity theory [5,8] is a branch of computational complexity in which the analysis of complexity is expressed in terms of an additional term, not only the input size, but some structural parameter obtained from the problem. The central notion of parameterized complexity theory is what we call

L. S. Moss et al. (Eds.): WoLLIC 2018, LNCS 10944, pp. 163–178, 2018.
https://doi.org/10.1007/978-3-662-57669-4_9

fixed-parameter tractability that corresponds to a relaxed version of classical tractability. A parameterized problem is said to be fixed-parameter tractable, if there exists an algorithm that runs in time $f(k) \cdot |x|^{O(1)}$, where $|x|$ is the input size, k is the parameter, and f is some arbitrary computable function. FPT is the class of all fixed-parameter tractable parameterized problems. A canonical example is the propositional satisfiability parameterized by the number of propositional variables. We denote this problem by p-SAT, and the brute force algorithm solves it in time $2^k \cdot n$ where k is the number of propositional variables and n is the size of the propositional formula.

We consider the satisfiability problem of some prefix-vocabulary classes of first-order logic (FO). A prefix-vocabulary fragment $[\Pi, \bar{p}, \bar{f}]$ is a set of first-order logic formulas in the prenex normal form where Π is a string in $\{\exists, \forall, \exists^*, \forall^*\}$ that represents the quantifier pattern as a regular expression, $\bar{p}$ is a relation arity sequence $(p_1, p_2, \ldots)$ such that p_a is the number of relations in the vocabulary with arity a, and $\bar{f}$ is a function arity sequence $(f_1, f_2, \ldots)$ such that f_a is the number of functions in the vocabulary with arity a. The relational classes are the prefix-vocabulary classes without function symbols, i.e. $\bar{f} = (0)^1$, which we represent by $[\Pi, \bar{p}]$. When we use the equality symbol $=$, we represent the fragment by $[\Pi, \bar{p}, \bar{f}]_=$ (see Definition 1).

The parameterized version of the satisfiability problem p-κ-SAT(X) consider whether a first-order formula φ in a prefix-vocabulary class X has a model fixing some parametrization κ which is a function that assigns a natural number (the parameter) to each formula in X. Here we consider the quantifier rank $\mathrm{qr}(\varphi)$ of a sentence φ, the maximum number of nested quantifiers occurring in φ, the size of the vocabulary $\mathrm{vs}(\varphi)$ occurring in φ, and the maximum arity of relation symbols $\mathrm{ar}(\varphi)$ occurring in φ as parameters.

The classification of the satisfiability problem for prefix-vocabulary classes into solvable and unsolvable cases has a long tradition within mathematical logic. In [4, Chaps. (6–7)], the authors present a complete investigation of decidability and complexity for many prefix-vocabulary classes. Seven of these are *maximal* with respect decidability, and we list them in Table 1.

Table 1. Maximal prefix-vocabulary classes.

Prefix-vocabulary class	References
(1) $[\exists^*\forall^*, \mathrm{all}]_=,$	Bernays and Schönfinkel [3]
(2) $[\exists^*\forall^2\exists^*, \mathrm{all}],$	Gödel [11], Kalmár [18], Schütte [27]
(3) $[\mathrm{all}, (\omega), (\omega)],$	Löb [20], Gurevich [13]
(4) $[\exists^*\forall\exists^*, \mathrm{all}, \mathrm{all}],$	Gurevich [14]
(5) $[\exists^*, \mathrm{all}, \mathrm{all}]_=,$	Gurevich [15]
(6) $[\mathrm{all}, (\omega), (1)]_=,$	Rabin [25]
(7) $[\exists^*\forall\exists^*, \mathrm{all}, (1)]_=,$	Shelah [28]

---

^{1} $(0) = (0, 0, \ldots).$

The other classes are classified as *classical*, and *classes with modest complexity* [4, Sect. 6.4]. Almost all maximal decidable classes have high complexity; deterministic and nondeterministic exponential time. For the classes with modest complexity, their satisfiability problems lay in P, NP, CoNP, PSPACE, Σ_2^p and Π_2^p (see Appendix 1). The decidability of some of these classes is obtained via finite model property, the property that guarantees a finite model for every satisfiable formula in the class. In this paper, we develop the parameterized complexity analysis for the classical cases, and all relational classes with modest complexity.

The strategy used here is to define a set of fixed-parameter reductions from p-κ-SAT(X), for some decidable fragments X of FO, to the propositional parameterized satisfiability p-SAT. Due to the closure of the class FPT under this kind of reduction, these problems are in FPT. The results are strongly supported by the propositionalization procedure of first-order logic [10,26] combined with the finite model property. We summarize them in Tables 3 and 6. Furthermore, we observe that, for the satisfiability of [all, (ω)] and [$\exists^*\forall^*$, all], when parameterized by the quantifier rank only, the problem is hard for paraNP, the class of problems decidable by a nondeterministic algorithm in time $f(k) \cdot |x|^{O(1)}$, where $|x|$ is the input size, k is the parameter, and f is some arbitrary computable function.

We organize the text in the following way. In the next section, we provide the basic tools of first-order logic and parameterized complexity theory. In Sect. 3, we analyze the parameterized complexity of the classical prefix-vocabulary fragments. In Sect. 4, we present the fixed-parameter tractability for all relational classes with modest complexity. We conclude the paper with a consideration of the choice of the parameters and other prefix-vocabulary classes that should be investigated as future works. In Appendix 1, we exhibit all relational prefix-vocabulary classes of modest complexity with their parameterized complexity classification. In Appendix 2, we show the proofs of some theorems.

2   Preliminaries

In this section, we will present some concepts and theorems useful for the parameterized complexity analysis that we will exhibit on Sects. 3 and 4.

2.1   Decidable Prefix-Vocabulary Fragments

We assume basic knowledge of propositional and first-order logic [6], and of computational complexity [23].

Recall that a *vocabulary* τ is a finite set of relation, function and constant symbols. Each symbol is associated with a natural number, its *arity*. The arity of τ is the maximum arity of its symbols. A τ-*structure* $\mathfrak{A}$ is a tuple $(A, R_1^{\mathfrak{A}}, \ldots R_r^{\mathfrak{A}}, f_1^{\mathfrak{A}}, \ldots, f_s^{\mathfrak{A}}, c_1^{\mathfrak{A}}, \ldots c_t^{\mathfrak{A}})$ such that A is a non-empty set, called the *domain*, and each $R_i^{\mathfrak{A}}$ is a relation under $A^{\mathrm{arity}(R_i)}$ interpreting the symbol $R_i \in \tau$, each $f_i^{\mathfrak{A}}$ is a function from $A^{\mathrm{arity}(f_i)}$ to A interpreting the symbol $f_i \in \tau$, and each $c_i^{\mathfrak{A}}$ is an element of A interpreting the symbol c_i. Here, we assume structures with finite domain, and, without loss of generality, we use a domain of naturals $\{1, \ldots, n\}$ denoted by $[n]$.

A τ-term is a variable x, a constant c, or a function symbol applied to τ-terms $f(t_1, t_2, \ldots, t_m)$. If R is a relation symbol, and $t_1, t_2, \ldots t_m$ are τ-terms, then $R(t_1, t_2, \ldots, t_m)$, and $t_1 = t_2$ are τ-formulas, which we call *atomic formulas*. If φ and ψ are τ-formulas, then $(\varphi \wedge \psi)$, $(\varphi \vee \psi)$, $\neg\varphi$ are τ-formulas. If x is a variable, and φ is a τ-formula, then $\forall x\varphi$ and $\exists x\varphi$ are τ-formulas. A *sentence* is a formula in which every variable in a subformula is in the scope of a corresponding quantifier. A formula is in *prenex normal form* if it is of the form $Q_1 x_1 \ldots Q_\ell x_\ell \psi$, such that ψ is a quantifier-free formula and $Q_1 \ldots Q_\ell \in \{\exists, \forall\}^*$ is the *prefix*. We define the *quantifier rank* $\mathrm{qr}(\varphi)$ as the maximum number of nested quantifiers occurring in φ. If φ is an atomic formula, then $\mathrm{qr}(\varphi) = 0$. If $\varphi := \neg\varphi'$, then $\mathrm{qr}(\varphi) = \mathrm{qr}(\varphi')$. If $\varphi := (\psi \,\square\, \theta)$, where $\square \in \{\wedge, \vee\}$, then $\mathrm{qr}(\varphi) = \max\{\mathrm{qr}(\psi), \mathrm{qr}(\theta)\}$. If $\varphi := Qx\psi$, where $Q \in \{\exists, \forall\}$, then $\mathrm{qr}(\varphi) = \mathrm{qr}(\psi) + 1$. For a fixed formula φ, we define τ_φ as the set of symbols occurring in the formula φ, $\mathrm{vs}(\varphi)$, the *vocabulary size*, as the number of symbols in τ_φ, and $\mathrm{ar}(\varphi)$ as the *maximum arity* of a symbol in τ_φ.

Let $\mathfrak{A}$ be a τ-structure and $a_1, a_2, \ldots, a_m$ be elements of the domain. If $\varphi(x_1, x_2, \ldots x_m)$ is a τ-formula with free variables $x_1, x_2, \ldots x_m$, then we write $\mathfrak{A} \models \varphi(a_1, a_2, \ldots a_m)$ to denote that $\mathfrak{A}$ satisfies φ if $x_1, x_2, \ldots x_m$ are interpreted by $a_1, a_2, \ldots a_m$, respectively. If φ is a sentence, then we write $\mathfrak{A} \models \varphi$ to denote that $\mathfrak{A}$ satisfies φ, or that $\mathfrak{A}$ is a model of φ.

The *satisfiability problem* consists of deciding, given a formula φ, if there exists a model $\mathfrak{A}$ for the formula φ or not. For first-order logic in general, the satisfiability problem SAT(FO) in undecidable [6]. In this work, we are interested in decidable fragments definable by their prefix of quantifiers and the vocabulary.

Definition 1 (Prefix-Vocabulary Classes [4]). *A prefix-vocabulary fragment $[\Pi, \bar{p}, \bar{f}]$ is a set of first-order formulas in the prenex normal form, without equality symbol $=$, where Π is a quantifier pattern, i.e., a string on $\{\exists, \forall, \exists^*, \forall^*\}$ as a regular expression, $\bar{p}$ is a relation arity sequence $(p_1, p_2, \ldots)$, and $\bar{f}$ is a function arity sequence $(f_1, f_2, \ldots)$ where $p_a, f_a \in \mathbb{N} \cup \{\omega\}$ are the number of relations and functions of arity a, respectively. Occasionally, we use **all** to denote an arbitrary sequence of arities $(\omega, \omega, \cdots)$, or an arbitrary prefix. We denote an empty arity sequence $(0, 0, \ldots)$ by (0). In case $\bar{f} = (0)$, we may write $[\Pi, \bar{p}]$ instead of $[\Pi, \bar{p}, (0)]$. The prefix-vocabulary fragment $[\Pi, \bar{p}, \bar{f}]_=$ is defined in the same way, but allowing formulas with equality symbol $=$.*

For example, $[\exists^2 \forall^*, (1)]$ is the class of formulas in prenex normal form with prefix $\exists^2 \forall^*$, in the vocabulary with one monadic relation, and without function symbols.

The classification problem for the satisfiability of first-order logic consider then, for each subset $X \subseteq FO$, if SAT(X) is decidable, or undecidable. In Chap. 6 of [4], the decidable cases are divided into *maximal* (see Table 1), *classical* (see Table 2), and *modest complexity* classes (see Table 5).

In this work, we are mainly interested in the *classical* and *modest complexity* classes restricted to their relational cases while we observe that the propositionalization procedure handles the parameterized complexity classification. The *classical* classes are depicted in Table 2 and we will explore them in Sect. 3.

Table 2. Classical prefix-vocabulary classes.

Prefix-vocabulary class	References
(A) $[\text{all}, (\omega)]_{(=)}$	Löwenheim [21]
(B) $[\exists^*\forall^*, \text{all}]$	Bernays and Schönfinkel [3]
(C) $[\exists^*\forall\exists^*, \text{all}]$	Ackermann [2]
(D) $[\exists^*\forall^2\exists^*, \text{all}]$	Gödel [11], Kalmár [18], Schütte [27]

The strategy for decidability for most of these classes is carried out by the *finite model property*. For a class X with the finite model property, one can think of an algorithm that iterates over the structure size. For each possible structure $\mathfrak{A}$ with that fixed size, it verifies whether $\mathfrak{A} \models \varphi$, and, simultaneously, verifies if $\neg\varphi$ is a valid sentence. Moreover, it is possible to obtain an upper bound on the size of the structure. The following lemma specify the size of the model for formulas in the classical classes, and we will use them in most of the results of Sects. 3 and 4.

Lemma 1 ([4,27])

(i) *Let ψ be a satisfiable sentence in $[\text{all}, \omega]$. Then ψ has a model with at most 2^m elements where ψ has m monadic predicates.*

(ii) *Let ψ be a satisfiable sentence in $[\text{all}, \omega]_=$. Then ψ has a model with at most $q \cdot 2^m$ elements where ψ has quantifier rank q and m monadic predicates.*

(iii) *Let $\psi := \exists x_1 \ldots \exists x_p \forall y_1 \ldots \forall y_m \varphi$ be a satisfiable sentence in $[\exists^*\forall^*, \text{all}]_=$. Then ψ has a model with at most $\max(1, p)$ elements.*

(iv) *Let $\psi := \exists x_1 \ldots \exists x_p \forall y_1 \forall y_2 \exists z_1 \ldots \exists z_m \varphi$ be a satisfiable sentence in $\exists^p\forall^2\exists^m$ containing t predicates of maximal arity h. Then ψ has a model with cardinality at most*

$$4^{10tm^2 2^h (p+1)^{h+4}} + p.$$

In some cases, an upper bound on the running time of the satisfiability problem can be found. Using nondeterminism, we can guess a structure with a size less than or equal to the size provided by the finite model property, and then we evaluate the input formula on this structure. For example, the satisfiability problem for $[\text{all}, (\omega)]$ is in NTIME(2^n), where n is the size of the formula and, for the class $[\exists^*\forall^2\exists^*, \text{all}]$, the same problem is in NTIME($2^{n/\log n}$). The complexity of the satisfiability for most of these classes is addressed in [9,12,19].

The second group of prefix-vocabulary classes that we are interested in this work are those on which the satisfiability problem is in P, NP, Co-NP, Σ_2^p, Π_2^p, and PSPACE designated as *modest classes* [4, Sect. 6.4]. In Table 5 (Appendix 1), we summarize the description of modest complexity classes with their respective complexity result from [4].

We give an example of how this classification works for the monadic classes in Fig. 1. The class $[\text{all}, (\omega)]$ and $[\text{all}, (\omega)]_=$ called the *Löwenhein class* and the *Löwenhein class with equality*, or, alternatively, *relational monadic fragments*, and all classes below these classes are considered as classes of modest complexity, and, $[\text{all}, (\omega), (\omega)]$ the *full monadic class*, and $[\text{all}, (\omega), (1)]_=$ the *Rabin's class* are maximal with respect to decidability.

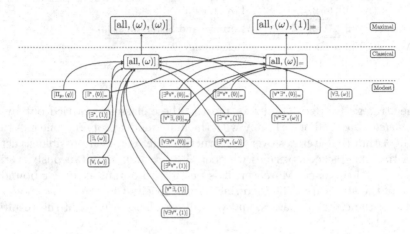

Fig. 1. The inclusion relation for monadic classes with modest complexity on Löwenheim's classes, and the maximal classes $[\text{all}, (\omega), (\omega)]$, $[\text{all}, (\omega), (1)]_=$

2.2   Parameterized Complexity

A *parameterized problem* is a pair (Q, κ) where $Q \subseteq \Sigma^*$ for some alphabet Σ is a decision problem and κ is a polynomial time computable function from Σ^* to natural numbers $\mathbb{N}$, called the *parameterization*. For an *instance* $x \in \Sigma^*$ of Q or (Q, κ), $\kappa(x) = k$ is the *parameter* of x. A *slice* of a parameterized problem (Q, κ) is the decision problem $(Q, \kappa)_\ell := \{x \in Q \mid \kappa(x) = \ell\}$. For a complete picture of parameterized complexity theory, we refer to the textbook [8].

We say that a problem (Q, κ) is *fixed-parameter tractable* if there is an algorithm that decides $x \in Q$ in time $f(\kappa(x)) \cdot p(|x|)$ for some computable function f, and a polynomial p. The class of all fixed-parameter tractable problems is called FPT.

For example, the propositional satisfiability problem p-SAT can be parameterized by the number of propositional variables. This is a trivial example of a fixed-parameter tractable problem that is witnessed by the brute-force algorithm with running time $2^k \cdot n$ where k is the number of propositional variables, and n the formula size.

Definition 2 (Parameterized Satisfiability). *Let* $X = [\Pi, \bar{p}, \bar{f}]_{(=)}$ *be a prefix-vocabulary class. We define* p-κ-SAT(X) *to be the parameterized version* *of the satisfiability problem for the class $X \subset FO$ with parameterization κ. For a given input formula $\varphi \in X$, is φ satisfiable considering $\kappa(\varphi)$ as the parameter?*

Three parameters emerge from the formalization of the prefix-vocabulary class X: the quantifier rank $\mathrm{qr}(\varphi)$, the vocabulary size $\mathrm{vs}(\varphi)$, and the maximum arity $\mathrm{ar}(\varphi)$. Each combination of these parameters leads to a different parameterized problem. For example, we have p-qr-SAT(X), p-vs-SAT(X), and p-(qr+vs)-SAT(X).

Given the parameterized problems (Q, κ) and (Q', κ'), an *fpt-reduction from* (Q, κ) *to* (Q', κ') is a mapping $R : \Sigma^* \to (\Sigma')^*$ such that: (i) For all $x \in \Sigma^*$ we have $(x \in Q \Leftrightarrow R(x) \in Q')$. (ii) R is computable by an *fpt-algorithm* (with respect to κ). That is, there is a computable function f and a polynomial p such that $R(x)$ is computable in time $f(\kappa(x)) \cdot p(|x|)$. (iii) There is a computable function $g : \mathbb{N} \to \mathbb{N}$ such that $\kappa'(R(x)) \leq g(\kappa(x))$ for all $x \in \Sigma^*$. Let C be a parameterized class. A parameterized problem (Q, κ) is *C-hard under fpt-reductions* if every problem in C is fpt-reducible to (Q, κ).

An important property of FPT is that this class is closed under fpt-reduction. Let $(Q, \kappa), (Q', \kappa')$ be parameterized problems and (Q', κ') in FPT. If there is an fpt-reduction from (Q, κ) to (Q', κ'), then (Q, κ) is in FPT too.

Let (Q, κ) be a parameterized problem. Then, (Q, κ) is in *paraNP*, if there is a nondeterministic algorithm that decides if $x \in Q$ in at most $f(\kappa(x)) \cdot p(|x|)$ steps, such that $f : \mathbb{N} \to \mathbb{N}$ is a computable function, and p is a polynomial. The class XP is the parameterized version of the exponential time class. A parameterized problem (Q, κ) is in XP, if there is an algorithm that decides if $x \in Q$ in at most $f(\kappa(x)) \cdot |x|^{g(\kappa(x))}$ steps, for some computable functions $f, g : \mathbb{N} \to \mathbb{N}$.

The following theorem says that, when we find a group of NP-hard slices of a parameterized problem, this problem is paraNP-hard. We will apply it to show that p-qr-SAT$([\mathrm{all}, \omega])$ is paraNP-hard when parameterized by the quantifier rank.

Theorem 1 [8]. *Let (Q, κ) be a parameterized problem, $\emptyset \subsetneq Q \subsetneq \Sigma^*$. Then (Q, κ) is paraNP-hard if, and only if, a union of finitely many slices of (Q, κ) is NP-hard.*

3   The Parameterized Complexity of Classical Decidable Classes

Our strategy to prove that p-κ-SAT(X), for some prefix-vocabulary class X and some parameterization κ, is fixed-parameter tractable is to give an fpt-reduction to the propositional satisfiability problem p-SAT, that is known to be in FPT. The essential tools for these results are the finite model property and the propositionalization procedure.

The finite model property provides a bound on the number of propositional variables introduced by the propositionalization procedure, and the process can be carried out in FPT when some proper parameter is chosen.

3.1   The Löwenheim Class and the Löwenheim Class with Equality

For the classical monadic fragments $[\text{all}, (\omega)]$ and $[\text{all}, (\omega)]_=$, we can show that the parameterized satisfiability problem is in FPT when parameterized by the vocabulary size and the quantifier rank. Moreover, based on Proposition 1 given below, we observe an evidence of intractability when parameterized by the quantifier rank only.

Theorem 2. *The satisfiability problem $p\text{-}(\text{qr}+\text{vs})\text{-SAT}([\text{all}, (\omega)])$ is in FPT.*

Proof. We will give an fpt-reduction to p-SAT. Let $\varphi \in [\text{all}, (\omega)]$ be a satisfiable formula with k monadic relation symbols and quantifier rank q. By Lemma 1 *(i)*, there is a model with at most 2^k elements. As we will transform φ into a propositional formula φ^*, we will represent each relation by 2^k propositional variables. More precisely, for each relation R_i with $1 \leq i \leq k$, and for each element $j \in [2^k]$, we use the variable p_{ij} to represent the truth value of $R_i(j)$. The translation works as follows. By structural induction on φ, apply the conversion of existential quantifiers into big disjunctions, and universal quantifiers into big conjunctions. We formally define this conversion as:

$$\varphi^* = \begin{cases} p_{ij} & \text{If } \varphi := R_i(j) \text{ is an atomic formula;} \\ \displaystyle\bigvee_{j \in [2^k]} (\psi[x/j])^* & \text{If } \varphi := \exists x \psi; \\ \displaystyle\bigwedge_{j \in [2^k]} (\psi[y/j])^* & \text{If } \varphi := \forall y \psi. \end{cases}$$

It is easy to see that φ has a model if and only if φ^* is a satisfiable formula. Each inductive step construct a formula of size $2^k \cdot |\varphi|$, and the whole process takes $O((2^k)^q \cdot n)$ where n is the size of φ. As the number of variables is bounded by $k \cdot 2^k$, this leads to the desired fpt-reduction. $\qquad\square$

The satisfiability problem for the Löwenheim class with equality $[\text{all}, (\omega)]_=$ is in FPT with respect to the vocabulary size and the quantifier rank.

Theorem 3. *The satisfiability problem $p\text{-}(\text{qr}+\text{vs})\text{-SAT}([\text{all}, (\omega)]_=)$ is in FPT.*

Proof. See Appendix 2. $\qquad\square$

However, when we choose the quantifier rank as the parameter, it is unlikely to obtain an fpt-algorithm for the satisfiability of the Löwenheim's class.

Proposition 1. *Unless $P = NP$, $p\text{-qr-SAT}([\text{all}, (\omega)])$ is not in XP.*

Proof. Assume, by contradiction, that p-qr-SAT$([\text{all}, (\omega)])$ is in XP. Then there is an algorithm that solves the problem in time $f(q) \cdot n^{g(q)}$, where n is the size of the formula, q is the quantifier rank of the input formula, and f and g are computable functions. Hence, for the first slice of the problem, $[\exists, (\omega)]$ and $[\forall, (\omega)]$ (see Table 5), there is an algorithm that runs in $f(1) \cdot n^{g(1)} \in O(n)$. This is a contradiction with the fact that these problems are NP-complete and with the reasonable assumption that $P \neq NP$. $\qquad\square$

As a consequence of Theorem 1 and that $[\exists, (\omega)]$ and $[\forall, (\omega)]$ are NP-complete problems, we have that p-qr-SAT[all, (ω)] is para-NP-hard under fpt-reductions.

Corollary 1. p-qr-SAT([all, (ω)]) *is paraNP-hard.*

Proof. The result follows directly from Theorem 1.

3.2   The Bernays-Schönfinkel-Ramsey Class

Taking the same idea from Theorem 2 for the monadic class, if we choose a parameter that bounds the number of propositional variables, and a propositionalization that can be conducted in FPT, we can provide an fpt-reduction for the parameterized satisfiability of prefix-vocabulary classes to p-SAT. This is the case of the Bernays-Schönfinkel-Ramsey class $[\exists^*\forall^*, \text{all}]_=$ when parameterized by the quantifier rank, the vocabulary size, and the maximum arity.

Theorem 4. p-(qr+vs+ar)-SAT([$\exists^*\forall^*$, all]) *and* p-(qr+vs+ar)-SAT([$\exists^*\forall^*$, all]$_=$) *are in FPT.*

Proof. See Appendix 2.

3.3   The Ackermann and Gödel-Kalmár-Shütte Classes

For the Ackermann class $[\exists^*\forall\exists^*, \text{all}]$ and Gödel-Kalmár-Shütte class $[\exists^*\forall^2\exists^*, \text{all}]$, by Lemma 1 *(iv)*, the finite model property provides a model of size related to the parameters of the class definition (the number of existential quantifiers, the number of relations, the maximum arity). If we consider the quantifier rank and the vocabulary size as the parameters, we can show it is in FPT.

Theorem 5. p-(qr+vs+ar)-SAT([$\exists^*\forall\exists^*$, all]) *and* p-(qr+vs+ar)-SAT([$\exists^*\forall^2\exists^*$, all]) *are in FPT.*

Proof. See Appendix 2.

Table 3. The parameterized complexity of the classical solvable cases.

Problem	Result
p-(qr+vs)-SAT([all, (ω)])	FPT (Theorem 2)
p-(qr)-SAT([all, (ω)])	paraNP-hard (Corollary 1)
p-(qr+vs)-SAT([all, (ω)]$_=$)	FPT (Theorem 3)
p-(qr+vs+ar)-SAT([$\exists^*\forall^*$, all]$_{(=)}$)	FPT (Theorem 4)
p-(qr+vs+ar)-SAT([$\exists^*\forall\exists^*$, all])	FPT (Theorem 5)
p-(qr+vs+ar)-SAT([$\exists^*\forall^2\exists^*$, all])	FPT (Theorem 5)

4   Modest Complexity Prefix-Vocabulary Classes

In this section, we analyze the parameterized complexity of prefix-vocabulary classes with modest complexity. We summarize our results on Table 6 (see Appendix 1). For these classes, it is possible to point out a parameter that put the problem in FPT. First, for some of these classes, the inclusion on the relational monadic class leads to an fpt result for the same parameters chosen. Then, we begin with a corollary of Theorem 2.

Corollary 2

(i) p-vs-SAT(X) *is in FPT for* $X \in \{[\exists, (\omega)], [\forall, (\omega)], [\forall\exists, (\omega)]\}$.
(ii) p-(qr+vs)-SAT(X) *is in FPT for* $X \in \{[\exists^2\forall^*, (\omega)], [\forall^*\exists^*, (\omega)]\}$.

Proof. (i) Let $\varphi \in X$ and $k = \text{vs}(\varphi)$. We already know that the finite model property gives us, for a satisfiable formula, a model of size at most 2^k. Then the same propositionalization procedure used in Theorem 2 leads to an fpt-reduction to p-SAT. In these cases, the quantifier rank is constant.

The claim *(ii)* follows directly from Theorem 2. □

Again, we can apply Lemma 1 *(i)* to give an fpt-reduction for the satisfiability problem of some classes with modest complexity.

Theorem 6. p-qr-SAT(X) *is in FPT for* $X \in \{[\forall^*\exists, (1)], [\forall\exists\forall^*, (1)], [\exists^*\forall^*, (1)]\}$.

Proof. Consider the satisfiability problem for the class $[\forall^*\exists, (1)]$. By Lemma 1(i), a satisfiable formula $\varphi := \forall x_1 \ldots \forall x_k \exists y \psi$ with one monadic relation and quantifier rank $q = \text{qr}(\varphi)$ has a model with size at most 2. Applying the same conversion of Theorem 2, the number of steps will be bounded by $2^q \cdot n$ where n is the size of φ. This will lead to an fpt-reduction to a propositional formula with two propositional variables.

The same reduction works for $[\forall\exists\forall^*, (1)]$ and $[\exists^*\forall^*, (1)]$. □

In the sequence, we can use Lemma 1 *(ii)* to obtain a reduction from the satisfiability of some classes with modest complexity to p-SAT, when we use the quantifier rank as a parameter.

Theorem 7. p-qr-SAT(X) *is in FPT for* $X \in \{ [\Pi_p, (m)]_=, [\Pi_p, (m)]_=, [\forall^*\exists, (0)]_=, [\forall\exists\forall^*, (0)]_=, [\forall^*\exists^*, (0)]_= \}$.

Proof. Consider the satisfiability problem for the class $[\Pi_p, (m)]_=$. Let φ be a satisfiable formula in $[\Pi_p, (m)]_=$ with at most m monadic relations and quantifier rank q. Then, by Lemma 1 *(ii)*, φ has a model with size at most $q \cdot 2^m$. Each monadic relation can be represented by $q \cdot 2^m$ propositional variables. So, in order to transform φ into a propositional formula, $q \cdot m \cdot 2^m$ propositional variables will be necessary to represent all relations. Applying the same conversion of Theorem 3, it will lead to an fpt-reduction to p-SAT.

The same reduction holds for $[\Pi_p, (q)]_=$ with at most p existential quantifiers. For $[\forall^*\exists, (0)]_=, [\forall\exists\forall^*, (0)]_=$, and $[\forall^*\exists^*, (0)]_=$, the size of the structure is bounded by $qr(\varphi)$ and the reduction follows in the same way. □

Formulas with a leading block of existential quantifiers can be handled with the finite model property of Lemma 1 *(iii)*.

Theorem 8. p-(qr+vs+ar)-SAT(X) *is in FPT for* $X \in \{[\exists^*\forall^*, all]_=, [\exists^*\forall^u, all]\}$.

Proof. Consider the satisfiability problem for the class $[\exists^*\forall^u, all]$. Let φ be a satisfiable formula in $[\exists^*\forall^u, all]$ with fixed natural u. So, $\varphi :=$ $\exists x_1 \ldots \exists x_k \forall y_1 \ldots \forall y_u \psi$. By Lemma 1 *(iii)*, φ has a model with size at most $k \leq \mathrm{qr}(\varphi)$. Let τ_φ the vocabulary of φ with maximum arity $\mathrm{ar}(\varphi)$ and size $\mathrm{vs}(\varphi)$. Applying the same conversion presented in Theorem 2, it will return a propositional formula with at most $\mathrm{vs}(\varphi) \cdot \mathrm{qr}(\varphi)^{\mathrm{ar}(\varphi)}$ propositional variables, and the whole process can be carried out by an fpt-algorithm.

For the class $[\exists^*\forall^*, all]_=$, the finite model property will provide a model of size 1, and each universally quantified variable could be handled as a dummy variable. $\qquad\square$

Theorem 9. p-qr-SAT(X) *is in FPT for* $X \in \{[\exists^*, (0)]_=, [\exists^*, (1)], [\exists^p\forall^*, \bar{s}]_=,$
$[\exists^2\forall^*, (0)]_=, [\exists^2\forall^*, (1)], [\exists^*\forall^*, (0)]_=\}$.

Proof. Consider the satisfiability problem for $[\exists^*\forall^*, (0)]_=$. By Lemma 1 *(iii)*, for a satisfiable formula $\varphi := \exists x_1 \ldots \exists x_k \forall y_1 \ldots \forall y_l \psi$ there is a model with size $k \leq \mathrm{qr}(\varphi)$. Applying the same translation of Theorem 2, and considering that only equality is in the vocabulary, the number of propositional variables obtained in the reduction is bounded by a function of $\mathrm{qr}(\varphi)$. This will lead to an fpt-reduction to p-SAT. $\qquad\square$

Theorem 10. p-qr-SAT(X) *is in FPT for* $X \in \{ [\exists^p\forall^2\exists^*, \bar{s}], [\exists^p\forall\exists^*, \bar{s}]_=,$
$[\exists^*\forall\exists, (0,1)]\}$.

Proof. We will consider the satisfiability problem of $[\exists^p\forall^2\exists^*, s]$ as an example. Let $\varphi := \exists x_1 \ldots \exists x_p \forall y_1 \forall y_2 \exists z_1 \ldots \exists z_k \psi$ be a first-order formula in a vocabulary with $\mathrm{vs}(\varphi)$ relation symbols of maximum arity (φ). By Lemma 1 *(iv)*, there is a model of size bounded by $\ell := 4^{10\mathrm{vs}(\varphi)k^2 2^{\mathrm{ar}(\varphi)}(p+1)^{\mathrm{ar}(\varphi)+4}} + p$ that satisfies φ. As $p, \mathrm{vs}(\varphi), \mathrm{ar}(\varphi)$ are constants, the size of this model is a function of k.

Then we can describe each relation with at most $\ell^{\mathrm{ar}(\varphi)}$ propositional variables. All relations are described by a binary string with length $\mathrm{vs}(\varphi) \cdot \ell^h$. By structural induction on φ, apply the same conversion of Theorem 5. Introduce one propositional variable to each possible assignment of a tuple to a relation symbol, which is a function of k. Clearly, this conversion is a function of k and n, the size of φ. This leads to an fpt reduction to p-SAT. $\qquad\square$

5   Conclusion

We saw that the finite model property and propositionalization procedure allow us to show the fixed-parameter tractability of the satisfiability problem for classical classes and relational classes of modest complexity. We investigated some natural parameters extracted from Definition 1: the quantifier rank, the vocabulary size, and the vocabulary arity. The fixed-parameter tractability considering one or a combination of these parameters can reveal the source of the intractability of the classical decision problem for the prefix-vocabulary classes investigated. For example, considering the relational monadic formulas, we can guarantee an fpt time execution for the satisfiability when the number of monadic relations and the quantifier rank are handled as the parameters. On the other hand, it is unlikely to obtain an fpt-algorithm for Lowenheim's class $[\text{all}, \omega]$ based on the quantifier rank as the parameter, as indicated by Proposition 1.

For $[\exists^*\forall^*, \text{all}]_=, [\exists^*\forall\exists^*, \text{all}]$, and $[\exists^*\forall^2\exists^*, \text{all}]$, the fixed-parameter tractability is obtained when we consider $qr(\varphi) + vs(\varphi) + ar(\varphi)$ as our parameterization function in Theorems 4 and 5. One question remains: What is the parameterized complexity of the satisfiability for these classes when some terms $(qr(\varphi), vs(\varphi), \text{and } ar(\varphi))$ are not considered in the parameterization function?

As a further work, we will investigate some modest complexity classes with unary functional symbols as the ones given on Table 4. In this case, other parameters may be considered like the maximum length of a term.

Table 4. Prefix-vocabulary classes of modest complexity with functions.

Prefix-vocabulary class	Complexity classification
$[\exists^*\forall^m, (0), (1)]_=$	NP-complete
$[\exists^*\forall\exists^*, (0), (1)]_=$	NP-complete
$[\exists^k\forall^*, (0), (1)]_=$	Co-NP-complete
$[\exists^*\forall^*, (0), (1)]_=$	Σ_2^p-complete

The methods applied here cannot be extended to classes without the finite model property like $[\text{all}, (\omega), (1)]_=$ and $[\exists^*\forall\exists^*, \text{all}, (1)]_=$. The parameterized complexity of these classes needs to be further investigated.

Appendix

Appendix 1

We reproduce the classification and computational complexity of the classes with modest complexity without function symbols from [4, Sect. 6.4].

We summarize the results of Sect. 4 here.

Table 5. Prefix-vocabulary classes of modest complexity

Prefix-vocabulary class	Complexity classification
$[\exists\forall^*, \text{all}]_=$	NP
$[\exists^*\forall^u, \text{all}]_=$ for $u \in \mathbb{N}$	
$[\exists^p\forall^2\exists^*, \bar{s}]$ for $p \in \mathbb{N}$ and $\bar{s}$ finite	
$[\exists^p\forall\exists^*, \bar{s}]_=$ for $p \in \mathbb{N}$ and $\bar{s}$ finite	
$[\Pi_p, (m)]_=$ $p, m \in \mathbb{N}$, and Π_p	
containing at most p universal quantifiers	
$[\exists^*, (0)]_=$	NP-complete
$[\exists^*, (1)]$	
$[\exists, (\omega)]$	
$[\forall, (\omega)]$	
$[\exists^p\forall^*, \bar{s}]_=$ for $p \in \mathbb{N}$ and $\bar{s}$ finite	Co-NP
$[\Pi_p, (m)]_=$ for $p, m \in \mathbb{N}$, and Π_p	
containing at most p existential quantifiers	
$[\exists^2\forall^*, (0)]_=$	Co-NP-complete
$[\exists^2\forall^*, (1)]$	
$[\forall^*\exists, (0)]_=$	
$[\forall^*\exists, (1)]$	
$[\forall\exists\forall^*, (0)]_=$	
$[\forall\exists\forall^*, (1)]$	
$[\exists^*\forall^*, (0)]_=$	Σ_2^p-complete
$[\exists^*\forall^*, (1)]$	
$[\exists^2\forall^*, (\omega)]$	
$[\forall^*\exists^*, (0)]_=$	Π_2^p-complete
$[\forall^*\exists^*, (\omega)]$	
$[\exists^*\forall\exists, (0, 1)]$	PSPACE-complete
$[\forall\exists, (\omega)]$	

Appendix 2

In this appendix, we reproduce the proofs of some theorems.

Proof. (Theorem 3) Using the same idea of Theorem 2, and the finite model property from Lemma 1 *(ii)*, we can describe an fpt-reduction from p-(qr+vs)-SAT($[\text{all}, (\omega)]_=$) to p-SAT.

For a satisfiable formula $\varphi \in [\text{all}, (\omega)]_=$ with at most $m = \text{vs}(\varphi)$ monadic relation symbols and quantifier rank $q = \text{qr}(\varphi)$, there is a model with at most $q \cdot 2^m$ elements by Lemma 1 *(ii)*. The number of steps on the conversion will be bounded by $O((q \cdot 2^m)^q \cdot n)$ where n is the formula size. Each atomic formula

Table 6. Prefix-vocabulary classes of modest complexity in which their satisfiability problem is in FPT with respect to some parameter.

Problem	Result
p-(qr+vs+ar)-SAT($[\exists\forall^*,\text{all}]_=$)	FPT(Theorem 8)
p-(qr+vs+ar)-SAT($[\exists^*\forall^u,\text{all}]_=$) for $u \in \mathbb{N}$	FPT (Theorem 8)
p-qr-SAT($[\exists^p\forall^2\exists^*,\bar{s}]$) for $p \in \mathbb{N}$ and $\bar{s}$ finite	FPT (Theorem 10)
p-qr-SAT($[\exists^p\forall\exists^*,\bar{s}]_=$) for $p \in \mathbb{N}$ and $\bar{s}$ finite	FPT (Theorem 10)
p-qr-SAT($[\Pi_p,(m)]_=$) $p,m \in \mathbb{N}$, Π_p containing at most p universal quantifiers	FPT (Theorem 7)
p-qr-SAT($[\exists^*,(0)]_=$)	FPT (Theorem 9)
p-qr-SAT($[\exists^*,(1)]$)	FPT (Theorem 9)
p-vs-SAT($[\exists,(\omega)]$)	FPT (Corollary 2)
p-vs-SAT($[\forall,(\omega)]$)	FPT (Corollary 2)
p-qr-SAT($[\exists^p\forall^*,\bar{s}]_=$) for $p \in \mathbb{N}$ and $\bar{s}$ finite $|\tau|$	FPT (Theorem 9)
p-qr-SAT($[\Pi_p,(m)]_=$ for $p,m \in \mathbb{N}$, and Π_p containing at most p existential quantifiers	FPT (Theorem 7)
p-qr-SAT($[\exists^2\forall^*,(0)]_=$)	FPT (Theorem 9)
p-qr-SAT($[\exists^2\forall^*,(1)]$)	FPT (Theorem 9)
p-qr-SAT($[\forall^*\exists,(0)]_=$)	FPT (Theorem 7)
p-qr-SAT($[\forall^*\exists,(1)]$)	FPT (Theorem 6)
p-qr-SAT($[\forall\exists\forall^*,(0)]_=$)	FPT (Theorem 7)
p-qr-SAT($[\forall\exists\forall^*,(1)]$)	FPT (Theorem 6)
p-qr-SAT($[\exists^*\forall^*,(0)]_=$)	FPT (Theorem 9)
p-qr-SAT($[\exists^*\forall^*,(1)]$)	FPT (Theorem 6)
p-(qr+vs)-SAT($[\exists^2\forall^*,(\omega)]$)	FPT (Corollary 2)
p-qr-SAT($[\forall^*\exists^*,(0)]_=$)	FPT (Theorem 7)
p-(qr+vs)-SAT($[\forall^*\exists^*,(\omega)]$)	FPT (Corollary 2)
p-qr-SAT($[\exists^*\forall\exists,(0,1)]$)	FPT (Theorem 10)
p-vs-SAT($[\forall\exists,(\omega)]$)	FPT (Corollary 2)

will be converted into a propositional variable, and this number is a function of q and m. Hence the whole process can be done in FPT.   □

Proof. (Theorem 4) Let φ be a satisfiable formula in $[\exists^*\forall^*,\text{all}]$ in the form $\exists x_1 \ldots \exists x_k \forall y_1 \ldots \forall y_\ell \psi$. By Lemma 1 *(iii)*, φ has a model of size at most $k \leq \text{qr}(\varphi)$. Then it will be necessary at most $\text{vs}(\varphi) \cdot k^{\text{ar}(\varphi)}$ propositional variables to represent the whole structure data. Applying the propositionalization described in Theorem 2, we will produce a satisfiable propositional formula with the number of variables bounded by $g(\text{qr}(\varphi),\text{vs}\varphi),\text{ar}_\varphi)$, for some computable function g. Clearly, this reduction can be done in FPT.

The same argument can be applied to the Ramsey's class $[\exists^*\forall^*,\text{all}]_=$.   □

Proof. (Theorem 5) We will consider the satisfiability problem of $[\exists^*\forall^2\exists^*, \text{all}]$. Let $\varphi := \exists x_1 \ldots \exists x_k \forall y_1 \forall y_2 \exists z_1 \ldots \exists z_\ell \psi$ be a first-order formula in a vocabulary with $\mathrm{vs}(\varphi)$ relation symbols of maximum arity $\mathrm{ar}(\varphi)$. By Lemma 1*(iv)*, there is a model of size bounded by $s := 4^{10 \cdot \mathrm{vs}(\varphi) \cdot \ell^2 \cdot 2^{\mathrm{ar}(\varphi)} \cdot (k+1)^{\mathrm{ar}(\varphi)+4}} + k$ that satisfies φ. Considering the propositionalization, the number of propositional variables will be bounded by $\mathrm{vs}(\varphi) \cdot s^{\mathrm{ar}(\varphi)}$. By structural induction on φ, apply the conversion of existential quantifiers into big disjunctions, and universal quantifiers into big conjunctions. Then it introduces one propositional variable to each possible assignment of tuples and relation symbols. This conversion is clearly a function of $s, \mathrm{vs}(\varphi), \mathrm{ar}(\varphi)$ and n, the size of φ. This lead to an fpt-reduction to p-SAT. The same argument can be applied to the Ackermann's class $[\exists^*\forall\exists^*, \text{all}]$.   □

References

1. Achilleos, A., Lampis, M., Mitsou, V.: Parameterized modal satisfiability. Algorithmica **64**(1), 38–55 (2012)
2. Ackermann, W.: Zum hilbertschen aufbau der reellen zahlen. Math. Ann. **99**(1), 118–133 (1928)
3. Bernays, P., Schönfinkel, M.: Zum entscheidungsproblem der mathematischen logik. Math. Ann. **99**(1), 342–372 (1928)
4. Börger, E., Grädel, E., Gurevich, Y.: The Classical Decision Problem. Springer, Heidelberg (1997). https://doi.org/10.1007/978-3-642-59207-2
5. Downey, R.G., Fellows, M.R.: Fundamentals of Parameterized Complexity. Springer, Heidelberg (2016). https://doi.org/10.1007/978-1-4471-5559-1
6. Ebbinghaus, H.D., Flum, J., Thomas, W.: Mathematical Logic. Springer, Heidelberg (2013). https://doi.org/10.1007/978-1-4757-2355-7
7. Flum, J., Grohe, M.: Fixed-parameter tractability, definability, and model-checking. SIAM J. Comput. **31**(1), 113–145 (2001)
8. Flum, J., Grohe, M.: Parameterized Complexity Theory. Springer, Heidelberg (2006). https://doi.org/10.1007/3-540-29953-X
9. Fürer, M.: Alternation and the ackermann case of the decision problem. L'Enseignement Math **27**, 137–162 (1981)
10. Gilmore, P.C.: A proof method for quantification theory: its justification and realization. IBM J. Res. Dev. **4**(1), 28–35 (1960)
11. Gödel, K.: Zum intuitionistischen aussagenkalkül. Anzeiger der Akademie der Wissenschaften in Wien **69**, 65–66 (1932)
12. Grädel, E.: Complexity of formula classes in first order logic with functions. In: Csirik, J., Demetrovics, J., Gécseg, F. (eds.) FCT 1989. LNCS, vol. 380, pp. 224–233. Springer, Heidelberg (1989). https://doi.org/10.1007/3-540-51498-8_21
13. Gurevich, Y.: The decision problem for the logic of predicates and of operations. Algebra Log. **8**(3), 160–174 (1969)
14. Gurevich, Y.: Formulas with one ∀. In: Selected Questions in Algebra and Logic; in memory of A. Mal'cev, pp. 97–110 (1973)
15. Gurevich, Y.: The decision problem for standard classes. J. Symb. Log. **41**(2), 460–464 (1976)
16. de Haan, R., Szeider, S.: Parameterized complexity results for symbolic model checking of temporal logics. In: Proceedings of the Fifteenth International Conference on Principles of Knowledge Representation and Reasoning, pp. 453–462. AAAI Press (2016)

17. de Haan, R., Szeider, S.: Parameterized complexity classes beyond para-NP. J. Comput. Syst. Sci. **87**, 16–57 (2017)
18. Kalmár, L.: Über die erfüllbarkeit derjenigen zählausdrücke, welche in der normalform zwei benachbarte allzeichen enthalten. Math. Ann. **108**(1), 466–484 (1933)
19. Lewis, H.R.: Complexity results for classes of quantificational formulas. J. Comput. Syst. Sci. **21**(3), 317–353 (1980)
20. Löb, M.: Decidability of the monadic predicate calculus with unary function symbols. J. Symb. Log. **32**, 563 (1967)
21. Löwenheim, L.: Über möglichkeiten im relativkalkül. Math. Ann. **76**(4), 447–470 (1915)
22. Lück, M., Meier, A., Schindler, I.: Parametrised complexity of satisfiability in temporal logic. ACM Trans. Comput. Log. (TOCL) **18**(1), 1 (2017)
23. Papadimitriou, C.H.: Computational Complexity. Wiley, Hoboken (2003)
24. Pfandler, A., Rümmele, S., Wallner, J.P., Woltran, S.: On the parameterized complexity of belief revision. In: IJCAI, pp. 3149–3155 (2015)
25. Rabin, M.O.: Decidability of second-order theories and automata on infinite trees. Trans. Am. Math. Soc. **141**, 1–35 (1969)
26. Ramachandran, D., Amir, E.: Compact propositional encodings of first-order theories. In: AAAI, pp. 340–345 (2005)
27. Schütte, K.: Untersuchungen zum entscheidungsproblem der mathematischen logik. Math. Ann. **109**(1), 572–603 (1934)
28. Shelah, S.: Decidability of a portion of the predicate calculus. Isr. J. Math. **28**(1–2), 32–44 (1977)
29. Szeider, S.: On fixed-parameter tractable parameterizations of SAT. In: Giunchiglia, E., Tacchella, A. (eds.) SAT 2003. LNCS, vol. 2919, pp. 188–202. Springer, Heidelberg (2004). https://doi.org/10.1007/978-3-540-24605-3_15

Unification Modulo Builtins

Ştefan Ciobâcă[✉], Andrei Arusoaie, and Dorel Lucanu

Alexandru Ioan Cuza University, Iaşi, Romania
{stefan.ciobaca,arusoaie.andrei,dlucanu}@info.uaic.ro

Abstract. Combining rewriting modulo an equational theory and SMT solving introduces new challenges in the area of term rewriting. One such challenge is unification of terms in the presence of equations and of uninterpreted and interpreted function symbols. The interpreted function symbols are part of a builtin model which can be reasoned about using an SMT solver. In this article, we formalize this problem, that we call unification modulo builtins. We show that under reasonable assumptions, complete sets of unifiers for unification modulo builtins problems can be effectively computed by reduction to usual E-unification problems and by relying on an oracle for SMT solving.

Keywords: Unification · Constrained term rewriting systems
Satisfiability modulo theories

1 Introduction

A recent line of work [1,3,5–7,11,16,17,20] aims at combining rewriting modulo E with SMT solving. The goal is to enable the modelling and analysis of systems beyond what is possible by rewriting modulo E or by SMT solving alone. The unifying idea among the approaches above is that rewriting is constrained by some logical formula and that it also involves elements like integers, reals, bitvectors, arrays, etc. that are handled by the SMT solver. As an example, consider the following constrained rewrite system computing the Collatz sequence:

$$
\begin{array}{lll}
\mathsf{n} \mapsto N & \Rightarrow \mathsf{cnt} \mapsto 0, \mathsf{n} \mapsto N & \text{if } \top, \\
\mathsf{n} \mapsto 2 \times N + 1, \mathsf{cnt} \mapsto C & \Rightarrow \mathsf{cnt} \mapsto C + 1, \mathsf{n} \mapsto 6 \times N + 4 & \text{if } N > 0, \\
\mathsf{n} \mapsto 2 \times N, \mathsf{cnt} \mapsto C & \Rightarrow \mathsf{cnt} \mapsto C + 1, \mathsf{n} \mapsto N & \text{if } N > 0, \\
\mathsf{n} \mapsto 1, \mathsf{cnt} \mapsto C & \Rightarrow \mathsf{result} \mapsto C & \text{if } \top.
\end{array}
$$

The rewrite system above reduces a state of the form $\mathsf{n} \mapsto N$, where N is a natural number, to a final state of the form $\mathsf{result} \mapsto C$, where C is the number of steps taken by N to reach 1 in the Collatz transformation. If the Collatz conjecture is false, then $\mathsf{n} \mapsto N$ does not necessarily terminate.

The symbols appearing in the rewrite system are the following: 1. $\_ \mapsto \_$ (the underscore denotes the place of an argument) is a two-argument free symbol, and $\mathsf{n}, \mathsf{cnt}, \mathsf{result}$ are free constants; 2. $\_, \_$ is an ACI symbol (making, e.g., $\mathsf{cnt} \mapsto$

© Springer-Verlag GmbH Germany, part of Springer Nature 2018
L. S. Moss et al. (Eds.): WoLLIC 2018, LNCS 10944, pp. 179–195, 2018.
https://doi.org/10.1007/978-3-662-57669-4_10

$0, \mathsf{n} \mapsto N$ the same as $\mathsf{n} \mapsto N, \mathsf{cnt} \mapsto 0$); 3. N, C are integer variables and the symbols $\times, +, >, 0, 1, 2, \ldots$ have their usual mathematical interpretation.

Unification modulo builtins is motivated by the computation of the possible successors of a term with variables in a constrained rewrite system such as the one above. Assuming that we are in a symbolic state of the form $\mathsf{cnt} \mapsto C' + N', \mathsf{n} \mapsto N' + 3$, which of the four constrained rewrite rules above could be applied to this state? To answer this question, we should first solve the following equations:

$$
\begin{aligned}
\mathsf{n} \mapsto N &= \mathsf{cnt} \mapsto C' + N', \mathsf{n} \mapsto N' + 3, \\
\mathsf{n} \mapsto 2 \times N + 1, \mathsf{cnt} \mapsto C &= \mathsf{cnt} \mapsto C' + N', \mathsf{n} \mapsto N' + 3, \\
\mathsf{n} \mapsto 2 \times N, \mathsf{cnt} \mapsto C &= \mathsf{cnt} \mapsto C' + N', \mathsf{n} \mapsto N' + 3, \\
\mathsf{n} \mapsto 1, \mathsf{cnt} \mapsto C &= \mathsf{cnt} \mapsto C' + N', \mathsf{n} \mapsto N' + 3.
\end{aligned}
$$

Solving such an equation is *E-unification modulo builtins*. Whereas usual (*E-*)unification is solving an equation of the form $t_1 = t_2$ in the algebra of terms (or in the quotient algebra of terms in the case of *E*-unification), *E*-unification modulo builtins is solving an equation of the form $t_1 = t_2$ in an algebra combining three types of symbols: 1. free symbols (such as $\_ \mapsto \_$), 2. symbols satisfying an equational theory E (such as $\_, \_$, satisfying ACI) and 3. builtin symbols such as integers, bitvectors, arrays or others (handled by an SMT solver).

Unlike regular syntactic unification problems (or *E*-unification problems), where the solution to a unification problem is a unifier (or complete set of unifiers), the solution to an *E*-unification modulo builtins problem is a logical constraint. We provide an algorithm that reduces *E*-unification modulo builtins to usual *E*-unification. As an example, consider the third equation above:

$$
\mathsf{n} \mapsto 2 \times N, \mathsf{cnt} \mapsto C = \mathsf{cnt} \mapsto C' + N', \mathsf{n} \mapsto N' + 3.
$$

As the symbol $\_, \_$ is commutative, the equation reduces to solving the builtin constraint $2 \times N = N' + 3 \wedge C = C' + N'$ (for example, by relying on the SMT solver). We show that *any* *E*-unification modulo builtins problem reduces to a set of logical constraints involving only builtin symbols, plus some substitutions in the range of which there are only non-builtin symbols. Our approach works for *any set of builtins*, not just integers.

Contributions. 1. We formalize the problem of *E-unification modulo builtins*, which appears naturally in the context of combining rewriting and SMT solving; 2. We define the notions of *E-unifier modulo builtins*, which generalizes the usual notion of *E*-unifier, and of *complete set of E-unifiers*; 3. We propose an algorithm for the problem of *E*-unification modulo builtins, which works by reduction to regular *E*-unification problems, given an oracle for SMT solving; 4. The algorithm not only decides *E*-unification modulo builtins, but it can also construct a complete set of *E*-unifiers modulo builtins; 5. We prove that the algorithm is correct and we also implement the algorithm as a Maude prototype.

Related Work. Our algorithm relies on abstractions of terms, various forms of which have been known for a long time and used, for example, in algorithms for combining decision procedures for theories in [15]. The abstractions of terms used here were defined and used for the first time in [2] for automatically obtaining a symbolic execution framework for a given program language definition. In parallel, abstractions were used in [17] for rewriting modulo SMT, where a builtin equational theory is used instead of the data sub-signature and its model. Both approaches use an SMT solver to check satisfiability of the constraints. In [1], Aguirre et al. introduce narrowing for proving reachability in rewriting logic enriched with SMT capabilities. In [20], Skeirik et al. extend the idea from [14] and show how Reachability Logic (defined in [8]) can be generalized to rewrite theories with SMT solving and how safety properties can be encoded as reachability properties. In [17], Rocha et al. were the first to combine rewriting and SMT solving in order to model and analyze an open system; they solve a particular case of the problem of E-unification modulo builtins that can be reduced to matching. In [5], Bae and Rocha introduce guarded terms, which generalize constrained terms. Logically constrained term rewriting systems (LCTRSs), which combine term rewriting and SMT constraints are introduced in [11,13]. LCTRSs generalize previous formalisms like TRSs enriched with numbers and Presburger constraints (e.g., as in [10]) by allowing arbitrary theories that can be handled by SMT solvers. Early ideas on adding constraints to deduction rules in general and unification in particular date back to the 1990s, with articles such as [9,12]. Mixed terms, defined in [9], are similar to our terms that mix free symbols with builtins, except that in [9] builtins are treated as constants. Constrained unification, introduced in [9] as a particular type of constrained deduction, can be seen as a sound and complete (but not necessarily terminating) proof system for a special case of E-unification modulo builtins, where the equational theory E is empty. Additionally, in both [9] and [12], it is assumed that constraints always have a solved form, assumption that is critical in order to present the deduction rules. As we do not make this assumption, and only rely on off-the-shelf SMT solvers, our approach is more general from this point of view. Our work can be seen in the abstract as a combination of E-unification and SMT solving, similar to combinations of unification algorithms for (disjoint) theories (e.g., as in [4,19]).

Organization. In Sect. 2 we formalize constrained terms, which occur naturally in rewriting modulo builtins. Section 3 introduces the concept of E-unification modulo builtins and the notion of (complete set of) E-unifiers. In Sect. 4 we present an algorithm for solving E-unification modulo builtins by reduction to usual E-unification. Section 5 concludes the paper and provides possible directions for future work.

2   Constrained Terms

We first formalize *builtins*, which represent the parts of the model handled by an SMT solver. This section extends our formalism introduced in [6] by the presence of an equational theory E and by the introduction of *defined operations*.

Definition 1 (Builtin Signature). *A builtin signature* $\Sigma^b \triangleq (S^b, F^b)$ *is any many-sorted signature that includes the following distinguished objects: 1. a sort Bool, together with two constants $\top$ and $\bot$ of sort Bool; 2. the propositional operation symbols* $\neg : Bool \rightarrow Bool, \wedge, \vee, \rightarrow\ : Bool \times Bool \rightarrow Bool$ *and 3. an equality predicate symbol* $=\ :s \times s \rightarrow Bool$ *for each sort* $s \in S^b$.

Example 1. We consider the builtin signature $\Sigma^{INT} = (S^{INT}, F^{INT})$, where $S^{INT} = \{Bool, Int, Arr, Id\}$ and F^{INT} includes, in addition to the required symbols (boolean connectives $\neg, \wedge, \vee$, boolean constants $\top$ and $\bot$ and the equality predicates for *Bool* and *Int*), the following function symbols:

$$\begin{aligned}
\mathsf{cnt}, \mathsf{result}, \mathsf{n}, \ldots : &\ \rightarrow Id & + : Int \times Int \rightarrow Int; \\
\times : Int \times Int \rightarrow Int; & & mod : Int \times Int \rightarrow Int; \\
\leq : Int \times Int \rightarrow Bool; & & get : Arr \times Int \rightarrow Int; \\
put : Arr \times Int \times Int \rightarrow Arr; & & 0, 1, 2, \ldots : \rightarrow Int.
\end{aligned}$$

When working over this builtin signature, we take the liberty to write terms using infix notation for function symbols, so that $x + y$ is a term of sort *Int* and $x + y \leq z$ is a term of sort *Bool* whenever x, y and z are of sort *Int*. We also use infix notation for the boolean operations: $\neg b, a \wedge b, a \vee b, a \rightarrow b$.

Definition 2 (Builtin Model). *A builtin model M^b is a model of a builtin signature* Σ^b, *where the interpretation of the distinguished objects of the builtin signature is fixed as follows:* $M^b_{Bool} = \{\top, \bot\}, M^b_\top = \top,\ M^b_\bot = \bot, M^b_=(a,b) = \top$ *iff* $a = b,\ M^b_\neg(\top) = \bot, M^b_\neg(\bot) = \top,\ M^b_\wedge(\top, b) = M^b_\wedge(b, \top) = b,\ M^b_\wedge(\bot, b) = M^b_\wedge(b, \bot) = \bot$, *and so on.*

Example 2. Continuing Example 1, we consider the builtin model M^{INT} that interprets *Bool* as required in Definition 2, the sort *Int* as the set of integers: $M^{INT}_{Int} = \mathbb{Z}$, the sort *Id* as the set of identifiers (strings) and the sort *Arr* as the set of arrays, where both the indices and the values are integers. The builtin model M^{INT} also interprets $+, \times, mod$ and $\leq$ as expected: integer addition, integer multiplication, remainder (defined arbitrarily when the divisor is 0) and respectively the less-than-or-equal relation on integers. The symbol *get* is interpreted as the selection of an array element from a given index, and *put* is interpreted as the update of an array on a given index with a new given element. First-order logical constraints over this model can be solved by an SMT solver implementing the theories of integers, booleans and arrays.

Next, we introduce a formalization of terms extended with builtins.

Definition 3 (Signature Modulo a Builtin Model). *A* signature modulo a builtin model *is a tuple* $\Sigma \triangleq (S, \leq, F, M^{\mathsf{b}})$ *consisting of 1. an order-sorted signature* $(S, \leq, F)$, *and 2. a builtin* Σ^{b}-*model* M^{b}, *where* $\Sigma^{\mathsf{b}} \triangleq (S^{\mathsf{b}}, F^{\mathsf{b}})$ *is a builtin subsignature of* $(S, \leq, F)$, *and 3. the set* $F \setminus F^{\mathsf{b}}$ *is partitioned into two subsignatures:* constructors F^{c} *and defined operations* F^{d} *such that* $F^{\mathsf{c}}_{w,s} = \emptyset$ *for each* $s \in S^{\mathsf{b}}$.

We further assume that the only builtin constant symbols in Σ *are the elements of the builtin model, i.e.,* $F^{\mathsf{b}}_{\varepsilon,s} = M^{\mathsf{b}}_s$. Σ^{b} *is called the* builtin subsignature *of* Σ *and* $\Sigma^{\mathsf{c}} = (S, \leq, F^{\mathsf{c}} \cup \bigcup_{s \in S^{\mathsf{b}}} F^{\mathsf{b}}_{\varepsilon,s})$ *the* constructor subsignature *of* Σ.

Terms over only one of the builtin subsignature and respectively constructor subsignature are *pure*. Terms mixing both constructors and builtins have the builtins as *alien* terms. Due to the restrictions on builtins, we cannot arbitrary nest of builtins and constructors, as is the case when combining arbitrary disjoint signatures (in our case, constructors cannot occur under builtins).

Example 3. We consider the signature $\Sigma = (S, \leq, F, M^{INT})$, where the set of sorts $S = \{Id, Int, Bool, Arr, Val, State\}$ consists of the four builtin sorts *Int, Bool, Id* and *Arr*, together with the additional sorts *State* and *Val*, where the subsorting relation $\leq = \{Int \leq Val, Arr \leq Val\} \subseteq S \times S$, and where the set of function symbols F includes, in addition to the builtin symbols in F^{INT}, the following function symbols:

$$\mathsf{emp} : \; \to State, \qquad \_\mapsto\_ : Id \times Val \to State,$$
$$\_,\_ : State \times State \to State, \quad keyOcc : Id \times State \to Int.$$

The set F^{c} of constructor symbols consists of the symbols defined by the first three declarations above and the set of defined symbols is $F^{\mathsf{d}} = \{keyOcc\}$.

Note that a ground term t defined on the constructor subsignature has the property that any of its builtin subterms is an element of $\Sigma^{\mathsf{b}}_{\varepsilon,s} = M^{\mathsf{b}}_s$. In the rest of this section, $\Sigma = (S, \leq, F, M^{\mathsf{b}})$ denotes a signature modulo a builtin model.

Definition 4 (Model M^{Σ} Generated by a Signature Modulo a Builtin Model). *Let* $\Sigma \triangleq (S, \leq, F, M^{\mathsf{b}})$ *be a signature modulo a builtin model. The model* M^{b} *is extended to a* $(S, \leq, F)$-*model* M^{Σ}, *defined as follows: 1.* $M^{\Sigma}_s = T_{\Sigma^{\mathsf{c}},s}$ *for each* $s \in S \setminus S^{\mathsf{b}}$, *i.e.* M^{Σ}_s *includes the constructor terms; 2.* $M^{\Sigma}_f = M^{\mathsf{b}}_f$ *for each* $f \in F^{\mathsf{b}}$; *3.* M^{Σ}_f *is the term constructor* $M^{\Sigma}_f(t_1, \ldots, t_n) = f(t_1, \ldots, t_n)$ *for each* $f \in F^{\mathsf{c}}$; *4.* M^{Σ}_f *is a function* $M^{\Sigma}_f : M^{\Sigma}_{s_1} \times \cdots \times M^{\Sigma}_{s_n} \to M^{\Sigma}_s$ *for each* $f \in F^{\mathsf{d}}_{s_1 \ldots s_n, s}$.

Note that elements of the carrier sets of M^{Σ} are ground terms of the appropriate sort over the signature Σ (as mentioned earlier, builtin elements are constants in Σ). We make the standard assumption that $M_s \neq \emptyset$ for any $s \in S$. Since the defined function symbols can be interpreted in various ways, it follows that M^{Σ} is not uniquely defined, but its carrier sets are uniquely defined by the builtin model and the constructors.

Example 4. Continuing Example 3, we consider the model M^{Σ} obtained by extending M^{INT}. We have that $M^{\Sigma}_{Val} = M^{\Sigma}_{Int} \cup M^{\Sigma}_{Arr}$, M_{State} is the set of finite sets of the form $\{x_1 \mapsto v_1, \dots, x_n \mapsto v_n\}$ with $n > 0$, $x_i \in M^{\Sigma}_{Id}$ and $v_i \in M^{\Sigma}_{Val}$ for $i = 1, \dots, n$, $M^{\Sigma}_{\_\mapsto\_}(X, Y)$ is the singleton set $\{X \mapsto Y\}$, and $M^{\Sigma}_{\_;\_}(S_1, S_2)$ is the union of the sets S_1 and S_2.

The defined function M^{Σ}_{keyOcc} is recursively defined as follows:

$$M^{\Sigma}_{keyOcc}(X, \mathrm{emp}) = 0, \qquad M^{\Sigma}_{keyOcc}(X, (Y \mapsto V)) = \delta_{X,Y},$$
$$M^{\Sigma}_{keyOcc}(X, (S_1, S_2)) = M^{\Sigma}_{+}(M^{\Sigma}_{keyOcc}(X, S_1), M^{\Sigma}_{keyOcc}(X, S_2)),$$

where δ is the Kronecker delta. We consider a set E of identities such as associativity, commutativity or idempotence for certain function symbols in Σ. We let $\cong$ to be the equivalence relation induced by E on M^{Σ}. We make the assumption that $\cong$ is a congruence on M^{Σ}, i.e., that it is compatible with the functions, including the defined ones. We define $M^{\Sigma}_{\cong} \triangleq M^{\Sigma}/\cong$ to be the quotient algebra induced by the congruence $\cong$ on M^{Σ}.

Example 5. Continuing Example 4, we consider the model M^{Σ}. Let E consist of the ACI identities for the operation $\_,\_$. We have that the equivalence relation $\cong$ induced by E on M^{Σ} is a congruence.

Definition 5 (Constraint Formulas). *The set* $\mathrm{CF}(\Sigma, \mathcal{X})$ *of constraint formulas over variables* $\mathcal{X}$ *is inductively defined as follows:*

$$\phi ::= b \mid t_1 = t_2 \mid \exists x.\phi' \mid \neg\phi' \mid \phi_1 \wedge \phi_2,$$

where b *ranges over* $T_{\Sigma,Bool}(\mathcal{X})$, t_i *over* $T_{\Sigma,s_i}(\mathcal{X})$.

That is, the constraints are the usual formulas in first-order logic with equality. The only non-standard feature, which does not restrict generality, is that we use terms of sort *Bool* as atomic formulas. The role of predicates is played by functions returning *Bool*. As usual, we may also use the following formulas defined as sugar syntax: 1. $t_1 \neq t_2$ for $\neg(t_1 = t_2)$, 2. $\forall x.\phi$ for $\neg\exists x.\neg\phi$, 3. $\forall X.\phi$ for $\forall x_1. \dots .\forall x_n.\phi$, where $X = \{x_1, \dots, x_n\}$, 4. $\phi_1 \vee \phi_2$ for $\neg(\neg\phi_1 \wedge \neg\phi_2)$, 5. $\phi_1 \rightarrow \phi_2$ for $\neg\phi_1 \vee \phi_2$. We denote by $var(\phi)$ the set of variables freely occurring in ϕ.

Example 6. The following formulas are in $\mathrm{CF}(\Sigma, \mathcal{X})$:

$$\phi_1 \triangleq \forall I. 0 \leq I \wedge I < N \rightarrow get(A, I) \geq 0,$$
$$\phi_2 \triangleq \forall X. keyOcc(X, S) \leq 1,$$
$$\phi_3 \triangleq \exists U. (U > 1 \wedge U < N \wedge mod(N, U) = 0).$$

We have $var(\phi_1) = \{N, A\}$, $var(\phi_2) = \{S\}$ and $var(\phi_3) = \{N\}$, assuming that $\{I, U, V, N\} \subseteq \mathcal{X}_{Int}$, $A \in \mathcal{X}_{Arr}$, and $X \in \mathcal{X}_{Id}$.

Definition 6 (Semantics of Constraint Formulas). *The satisfaction relation* $\vDash$ *is inductively defined over the model* $M^{\Sigma}_{\cong}$, *valuations* $\alpha : \mathcal{X} \rightarrow M^{\Sigma}_{\cong}$, *and formulas* $\phi \in \mathrm{CF}(\Sigma, \mathcal{X})$, *as follows:*

1. $M_{\cong}^{\Sigma}, \alpha \vDash b$ *iff* $\alpha(b) = \top$, *where* $b \in T_{\Sigma, Bool}(\mathcal{X})$;
2. $M_{\cong}^{\Sigma}, \alpha \vDash t_1 = t_2$ *iff* $\alpha(t_1) = \alpha(t_2)$;
3. $M_{\cong}^{\Sigma}, \alpha \vDash \exists x.\phi$ *iff* $\exists a \in M_s$ *(where* $x \in \mathcal{X}_s$*) such that* $M_{\cong}^{\Sigma}, \alpha[x \mapsto a] \vDash \phi$;
4. $M_{\cong}^{\Sigma}, \alpha \vDash \neg\phi$ *iff* $M_{\cong}^{\Sigma}, \alpha \nvDash \phi$;
5. $M_{\cong}^{\Sigma}, \alpha \vDash \phi_1 \wedge \phi_2$ *iff* $M_{\cong}^{\Sigma}, \alpha \vDash \phi_1$ *and* $M_{\cong}^{\Sigma}, \alpha \vDash \phi_2$,

where $\alpha[x \mapsto a]$ *denotes the valuation* α' *defined by* $\alpha'(y) = \alpha(y)$, *for all* $y \neq x$, *and* $\alpha'(x) = a$.

Example 7. Continuing the previous example, the formula ϕ_3 is satisfied by the model $M_{\cong}^{\Sigma}$ defined in Example 5 and any valuation α such that $\alpha(N)$ is a composite number.

Definition 7 (Builtin Constraint Formulas). *The set* $\mathrm{CF}^b(\Sigma, \mathcal{X})$ *of* builtin constraint formulas *over the variables* $\mathcal{X}$ *is the subset of* $\mathrm{CF}(\Sigma, \mathcal{X})$ *defined inductively as follows:*

$$\phi ::= b \mid t_1 = t_2 \mid \exists x.\phi' \mid \neg\phi' \mid \phi_1 \wedge \phi_2,$$

where b *ranges over* $T_{\Sigma^b, Bool}(\mathcal{X}), t_i$ *over* $T_{\Sigma, s}(\mathcal{X})$ *such that* s *is a builtin sort, and* x *ranges over all variables of builtin sort.*

Note that in builtin constraint formulas, no symbol that is not builtin is allowed. The constraint formulas ϕ_1, ϕ_3 in the previous example are builtin.

Definition 8 (Constrained Terms). *A* constrained term φ *of sort* $s \in S$ *is a pair* $\langle t \mid \phi \rangle$ *with* $t \in T_{\Sigma, s}(\mathcal{X})$ *and* $\phi \in \mathrm{CF}(\Sigma, \mathcal{X})$.

Let $\mathrm{CT}(\Sigma, \mathcal{X})$ denote the set of constrained terms defined over the signature Σ and the variables $\mathcal{X}$.

Example 8. In the context of the previous examples, we have the following constrained terms in $\mathrm{CT}(\Sigma, \mathcal{X})$: 1. arrays with N nonnegative values: $\langle A \mid \phi_1 \rangle$ 2. legal states, where an Id is bound to at most one value: $\langle S \mid \phi_2 \rangle$ 3. states where the Id n is bound to a composite integer: $\langle \mathsf{n} \mapsto N, S \mid \phi_3 \rangle$.

Definition 9 (Valuation Semantics of Constraints). *The* valuation semantics *of a constraint* ϕ *is the set of valuations* $\lfloor \phi \rfloor \triangleq \{\alpha : \mathcal{X} \to M_{\cong}^{\Sigma} \mid M_{\cong}^{\Sigma}, \alpha \vDash \phi\}$.

Definition 10 (Semantics of Constrained Terms). *The* semantics *of a constrained term* $\langle t \mid \phi \rangle$ *is defined by*

$$[\![\langle t \mid \phi \rangle]\!] \triangleq \{\alpha(t) \mid \alpha \in \lfloor \phi \rfloor\}.$$

Note that the semantics of constrained terms cannot distinguish between constrained terms with different sets of free variables.

3 E-Unification Modulo Builtins

We discuss as an example the four E-unification modulo builtin problems in the Introduction:

1. $n \mapsto N = \text{cnt} \mapsto C' + N', n \mapsto N' + 3$
   Even if the $\_,\_$ symbol is ACI, there are no values of $N, N','$ that make the lhs equal to the rhs (even if $C' + N' = N' + 3$, the atoms $\text{cnt} \mapsto C' + N'$ and respectively $n \mapsto N' + 3$ cannot be identified by the idempotence axiom because n and cnt are two different *Identifiers*). Therefore, the solution to this E-unification problem is $\bot$ (false), i.e., there is no solution.
2. $n \mapsto 2 \times N + 1, \text{cnt} \mapsto C = \text{cnt} \mapsto C' + N', n \mapsto N' + 3$
   There are values for N, C, N', C' that make the terms above equal. In particular, any values that satisfy $C' + N' = C \wedge N' + 3 = 2 \times N + 1$ make the terms equal. Therefore, the constraint $C' + N' = C \wedge N' + 3 = 2 \times N + 1$ is the solution of the E-unification problem.
3. $n \mapsto 2 \times N, \text{cnt} \mapsto C = \text{cnt} \mapsto C' + N', n \mapsto N' + 3$
   Similar to the case above, the solution is $C' + N' = C \wedge N' + 3 = 2 \times N$.
4. $n \mapsto 1, \text{cnt} \mapsto C = \text{cnt} \mapsto C' + N', n \mapsto N' + 3$
   The constraint $N' + 3 = 1 \wedge C' + N' = C$ makes the two terms equal and is the solution to the E-unification problem.

It is helpful to split the solution to a E-unification modulo builtins problem into two parts: a substitution and a logical constraint:

Definition 11 (E-Unifiers Modulo Builtins). *An E-unifier modulo builtins (E-umb) of two terms t_1, t_2 is a pair $u = (\sigma, \phi)$, where σ is a substitution and ϕ is a builtin constraint, such that*

$$M_{\cong}^{\Sigma} \models \phi \rightarrow \sigma(t_1) = \sigma(t_2).$$

Note that we require ϕ to be a *builtin* constraint (see Definition 7), not just a constraint. In particular, ϕ is not allowed to contain any non-builtin symbols. This requirement allows to handle ϕ by using an SMT solver. If ϕ is unsatisfiable, then (σ, ϕ) is vacuously an E-unifier of any two terms.

Definition 12 (Complete Set of E-Unifiers Modulo Builtins). *A set C of pairs of substitutions and builtin logical constraints is called a* complete set of E *-unifiers of t_1 and t_2 if:*

1. *each pair $(\sigma, \phi) \in C$ is an E-umb of t_1 and t_2: $M_{\cong}^{\Sigma} \models \phi \rightarrow \sigma(t_1) = \sigma(t_2)$;*
2. *for any partial valuation $\alpha : var(t_1, t_2) \rightarrow M_{\cong}^{\Sigma}$ such that $\alpha(t_1) = \alpha(t_2)$, there is an E-unifier $(\sigma, \phi) \in C$ and a valuation α^r such that $M_{\cong}^{\Sigma}, \alpha^r \models \phi$ and $\alpha = (\alpha^r \circ \sigma)|_{var(t_1, t_2)}.$*

Example 9. Consider the E-unification modulo builtins problem $n \mapsto 1, \text{cnt} \mapsto C = \text{cnt} \mapsto C', n \mapsto N'$. A complete set of E-unifiers modulo builtins is the singleton set $\{(id, C = C' \wedge 1 = N')\}$, where id denotes the identity substitution.

Example 10. Consider the E-unification modulo builtins problem

$$(n \mapsto 1, cnt \mapsto C) = (I \mapsto C', J \mapsto N').$$

A complete set of E-unifiers modulo builtins is

$$\{(id, I = cnt \wedge J = n \wedge C = C' \wedge 1 = N'),$$
$$(id, I = n \wedge J = cnt \wedge 1 = C' \wedge C = N')\}.$$

Example 11. Consider the E-unification modulo builtins problem

$$n \mapsto 1, cnt \mapsto 42 = M, n \mapsto 1.$$

A complete set of E-unifiers modulo builtins is the singleton set

$$\{(\sigma, I = cnt \wedge N = 42\},$$

where $dom(\sigma) = \{M\}$ and $\sigma(M) = I \mapsto N$.

4   An Algorithm for E-Unification Modulo Builtins

We propose an algorithm that computes, given two terms t_1 and t_2 without any defined operation symbols, a finite and complete set of E-unifiers modulo builtins of t_1 and t_2, assuming that a finitary E-unification algorithm exists. The algorithm critically relies on the notion of abstraction. In order to formally introduce abstractions, we first require a technical helper definition:

Definition 13 (Substitutions as Formulas). *Each substitution* $\sigma : \mathcal{X} \to T_\Sigma(\mathcal{X})$ *defines a constraint formula* $\phi^\sigma = \bigwedge_{x \in dom(\sigma)} x = \sigma(x)$.

Intuitively, an abstraction of a term t is a pair (s, σ) such that $t = \sigma(s)$, where all subterms of builtin sorts have been "moved" from s into the substitution σ, and where the domain of σ consists of fresh variables. Formally:

Definition 14 (Abstractions). *An* abstraction *of a term* $t \in T_{\Sigma \setminus \Sigma^d}(\mathcal{X})$ *w.r.t.* Y, *where* $var(t) \subseteq Y$, *is a constrained term* $\langle t^\circ \,|\, \phi^{\sigma^\circ} \rangle$, *where the pair* (t°, σ°) *is inductively defined as follows:* – *if* $t \in T_{\Sigma^b}(\mathcal{X})$ *(i.e., t is a builtin subterm), then* t° *is a fresh variable w.r.t.* Y *and* $\sigma^\circ(t^\circ) = t$; – *if* $t = f(t_1, \ldots, t_n)$ *and* $f \in \Sigma \setminus \Sigma^b$, *then* $t^\circ = f(t_1^\circ, \ldots, t_n^\circ)$ *and* $\sigma^\circ = \sigma_1^\circ \uplus \cdots \uplus \sigma_n^\circ$, *where* $\langle t_i^\circ \,|\, \phi^{\sigma_i^\circ} \rangle$ *is the abstraction of* t_i *w.r.t.* $Y \cup \bigcup_{j<i} var(\langle t_j^\circ \,|\, \phi^{\sigma_j^\circ} \rangle)$ *(i.e. each argument* t_i *has its own fresh variables).*

In Definition 14, we assume for simplicity that any occurrence of a builtin term is replaced by a fresh variable. Therefore, two different occurrences of the same builtin term are abstracted by two different variables. However, this is not critical to the soundness of our approach: all results in the paper hold, even if the same abstracting variable is used for several occurrences of the same builtin term.

Algorithm 1. Algorithm for E-Unification Modulo Builtins

1: **function** UNIFICATION(t_1, t_2)
2:     ▷ returns: a complete set of E-unifiers modulo builtins of t_1 and t_2
3:     compute $\langle s_1 \mid \phi^{\sigma_1}\rangle$, an abstraction of t_1
4:     compute $\langle s_2 \mid \phi^{\sigma_2}\rangle$, an abstraction of t_2
5:     compute $\{\tau_1, \ldots, \tau_n\}$, a complete set of E-unifiers of s_1 and s_2
6:     **for** $i \in \{1, \ldots, n\}$ **do**
7:         $\tau_i' \leftarrow \tau_i|_{\mathcal{X}\backslash\mathcal{X}^b}$
8:         $\phi_i' \leftarrow \phi^{\sigma_1} \wedge \phi^{\sigma_2} \wedge \bigwedge_{x\in dom(\tau_i)\cap\mathcal{X}^b} \tau_i(x) = x$
9:     **return** $\{(\tau_1', \phi_1'), \ldots, (\tau_n', \phi_n')\}$

Algorithm 1 computes a complete set of E-unifiers. Note that, if $x \in \mathcal{X}^b$ is a builtin variable, then $\tau_i(x)$ can only be a builtin variable (by the construction of the abstraction). Due to our assumptions on the builtin sorts, any builtin variable can only be unified with another builtin variable. Therefore builtin variables are treated by the E-unification algorithm used on Line 5 of Algorithm 1 as real variables, and not as free constants. As an optimization of the algorithm, any pair (τ_i', ϕ_i') can be dropped (without losing completeness) if ϕ_i' is unsatisfiable.

Algorithm 1 produces a complete set of E-unifiers modulo builtins. By Definition 11, any satisfiable instance $\alpha(\phi_i')$ of a constraint ϕ_i' induces the concrete E-umb $\alpha(\tau_i')$ of t_1 and t_2. Therefore, two terms are E-unifiable modulo builtins iff at least one of the constraints ϕ_i' in the result of Algorithm 1 is satisfiable.

Example 12. Consider the E-unification modulo builtins problem

$$(\mathsf{n} \mapsto 1, \mathsf{cnt} \mapsto 42) = (Z, \mathsf{n} \mapsto 1).$$

Let $t_1 = \mathsf{n} \mapsto 1, \mathsf{cnt} \mapsto 42$ and $t_2 = Z, \mathsf{n} \mapsto 1$. Let (s_1, σ_1) be an abstraction of t_1 defined as follows: $s_1 = I \mapsto N, J \mapsto M$ and $\sigma_1(I) = \mathsf{n}, \sigma_1(N) = 1, \sigma_1(J) = \mathsf{cnt}, \sigma_1(M) = 42$. Let (s_2, σ_2) be an abstraction of t_2 defined as follows: $s_2 = Z, K \mapsto L$ and $\sigma_2(K) = \mathsf{n}$ and $\sigma_2(L) = 1$. A complete set of ACI-unifiers of s_1 and s_2 is the set $\{\tau_1, \tau_2\}$, where:

1. $dom(\tau_1) = \{Z, K, L\}$ and $\tau_1(Z) = I \mapsto N$, $\tau_1(K) = J, \tau_1(L) = M$ (the first mapping in s_1 is identified to the first mapping in s_2 and the second mapping in s_1 to the second mapping in s_2);
2. $dom(\tau_2) = \{Z, K, L\}$ and $\tau_2(Z) = J \mapsto M$, $\tau_2(K) = I, \tau_2(L) = N$ (the first mapping in s_1 is identified to the second mapping in s_2 and the second mapping in s_1 to the first mapping in s_2);

The case where all four mappings are identified is subsumed by any of the two cases. For this example, the algorithm computes (τ_i', ϕ_i') as follows:

1. $dom(\tau_1') = \{Z\}, \tau_1'(Z) = I \mapsto N$ and ϕ_1' is
$$\underbrace{I = \mathsf{n} \wedge N = 1 \wedge J = \mathsf{cnt} \wedge M = 42}_{\phi^{\sigma_1}} \wedge \underbrace{K = \mathsf{n} \wedge L = 1}_{\phi^{\sigma_2}} \wedge \underbrace{K = J \wedge L = M}_{\text{builtins from } \tau_1};$$

2. $dom(\tau_2') = \{Z\}, \tau_2'(Z) = J \mapsto M$ and ϕ_2' is
$$\underbrace{I = \mathsf{n} \wedge N = 1 \wedge J = \mathsf{cnt} \wedge M = 42}_{\phi^{\sigma_1}} \wedge \underbrace{K = \mathsf{n} \wedge L = 1}_{\phi^{\sigma_2}} \wedge \underbrace{K = I \wedge L = N}_{\text{builtins from } \tau_2};$$

The algorithm returns the complete set $\{(\tau_1', \phi_1'), (\tau_2', \phi_2')\}$ of E-unifiers modulo builtins of the terms $t_1 = \mathsf{n} \mapsto 1, \mathsf{cnt} \mapsto 42$ and $t_2 = Z, \mathsf{n} \mapsto 1$. As ϕ_1' is unsatisfiable, the first unifier can be pruned away and therefore $\{(\tau_2', \phi_2')\}$ is also a complete set of unifiers of $t_1 = \mathsf{n} \mapsto 1, \mathsf{cnt} \mapsto 42$ and $t_2 = Z, \mathsf{n} \mapsto 1$.

The next result shows that Algorithm 1 is correct:

Theorem 1. *The set $\{(\tau_1', \phi_1'), \ldots, (\tau_n', \phi_n')\}$ computed by Algorithm 1 is a complete set of E-unifiers modulo builtins of t_1 and t_2.*

Proof (Sketch). Let t_1 and t_2 be two terms without defined function symbols.
Let $\langle s_1 \mid \phi^{\sigma_1} \rangle$ be an abstraction of t_1.
Let $\langle s_2 \mid \phi^{\sigma_2} \rangle$ be an abstraction of t_2.
Let $\{\tau_1, \ldots, \tau_n\}$ be a complete set of E-unifiers of s_1 and s_2.
Let $\tau_i' = \tau_i|_{\mathcal{X} \setminus \mathcal{X}^b}$. Let $\phi_i' = \phi^{\sigma_1} \wedge \phi^{\sigma_2} \wedge \bigwedge_{x \in dom(\tau_i) \cap \mathcal{X}^b} \tau_i(x) = x$.
We show that the set $\{(\tau_1', \phi_1'), \ldots, (\tau_n', \phi_n')\}$ is a complete set of unifiers modulo builtins of t_1 and t_2. Firstly, we have to show soundness: that each pair (τ_1', ϕ_1') is indeed a unifier of t_1 and t_2 (easy).

Secondly, we show completeness. Let $\alpha : var(t_1, t_2) \to M_{\cong}^{\Sigma}$ be a partial valuation such that $\alpha(t_1) = \alpha(t_2)$. We will work with the analogous approach, with $\alpha : var(t_1, t_2) \to M^{\Sigma}$ such that $\alpha(t_1) \cong \alpha(t_2)$. Note that, by our definition of the model generated by a signature modulo builtins, $M^{\Sigma} = \mathcal{T}_{\Sigma^c}(\emptyset)$, and therefore valuations such as α can also be seen as substitutions.
Let $\alpha' : var(s_1, s_2) \to \mathcal{T}_{\Sigma^c}(\emptyset)$ be defined as follows:

$$\alpha'(x) = \begin{cases} \alpha(x) & \text{if } x \in var(t_1, t_2) \setminus (dom(\sigma_1) \cup dom(\sigma_2)); \\ \alpha(\sigma_1(x)) & \text{if } x \in dom(\sigma_1); \\ \alpha(\sigma_2(x)) & \text{if } x \in dom(\sigma_2). \end{cases}$$

We have $\alpha'(s_1) \cong \alpha'(s_2)$. By construction, we also have: 1. $\alpha'|_{var(s_1)} = \alpha \circ \sigma_1$; 2. $\alpha'|_{var(s_2)} = \alpha \circ \sigma_2$. As α' is a substitution unifying s_1 and s_2 modulo E (recall that $\cong$ is generated by E), it follows that there exists a unifier $\tau_i \in \{\tau_1, \ldots, \tau_n\}$ of s_1 and s_2 and a substitution α^c such that: 1. $\tau_i(s_1) \cong \tau_i(s_2)$; 2. $\alpha' = (\alpha^c \circ \tau_i)|_{var(t_1, t_2)}$. Note that, as $\alpha'|_{var(s_j)} = \alpha \circ \sigma_j$ $(1 \le j \le 2)$, we also have that $(\alpha^c \circ \tau_i)|_{var(s_1)} = \alpha \circ \sigma_1$ and $(\alpha^c \circ \tau_i)|_{var(s_2)} = \alpha \circ \sigma_2$.
Let $\tau_i' = \tau_i|_{\mathcal{X} \setminus \mathcal{X}^b}$. Let $\alpha^r : var(t_1, t_2, \phi) \to \mathcal{T}_{\Sigma^c}(\emptyset)$ be defined as follows:

$$\alpha^r(x) = \begin{cases} \alpha'(x) \text{ if } x \in var(s_1, s_2) \\ \alpha(x) \text{ if } x \in var(t_1, t_2) \setminus var(s_1, s_2) \\ \alpha^c(y) \text{ otherwise, where } y \text{ is such that } \tau_i(x) = y \end{cases}$$

The substitution α^r defined above satisfies all conditions in Definition 12:

1. $\alpha = (\alpha^r \circ \tau_i')|_{var(t_1,t_2)}$:

   Let $x \in var(t_1, t_2)$. We distinguish two cases: – if $x \in X^b$, then we have $\alpha^r(\tau_i'(x)) = \alpha^r(x) = \alpha(x)$; – if $x \notin X^b$, then $\alpha^r(\tau_i'(x)) = \alpha^r(\tau_i(x)) = \alpha^c(\tau_i(x)) = \alpha'(x) = \alpha(x)$;

2. $M_{\cong}^{\Sigma}, \alpha^r \vDash \phi_i'$:

   (a) $M_{\cong}^{\Sigma}, \alpha^r \vDash \phi^{\sigma_1}$, since: – if $x \notin dom(\sigma_1)$, then $\sigma_1(x) = x$ and therefore trivially $\alpha^r(\sigma_1(x)) = \alpha^r(x)$; – if $x \in dom(\sigma_1)$, then $\alpha^r(\sigma_1(x)) = \alpha(\sigma_1(x)) = \alpha'(x) = \alpha^r(x)$;

   (b) $M_{\cong}^{\Sigma}, \alpha^r \vDash \phi^{\sigma_2}$, analogously;

   (c) $M_{\cong}^{\Sigma}, \alpha^r \vDash \tau_i(x) = x$ for all $x \in dom(\tau_i) \cap X^b$ trivially, by the third branch in the definition of α^r.

We have shown that for any α that unifies modulo builtins t_1 and t_2, there is an E-umb (τ_i', ϕ_i') and a substitution α^r with the properties required in Definition 12, and therefore this proves the completeness of the algorithm.

Complexity. Algorithm 1, which reduces E-unification modulo builtins to E-unification, has a linear-time overhead and therefore the running time is dominated by the algorithm for E-unification and, optionally, by some calls to the SMT solver. Lines 3 and 4 are linear in the size of the input. The running time of Line 5 is determined by the E-unification algorithm. The postprocessing step in lines 6–8 is linear in the size n of the *output* of the E-unification algorithm. Additionally, if we want to check the satisfiability of the constraints ϕ_i' on Line 9, there will be n calls to the SMT solver. In summary, the running time of the algorithm is dominated by one call to the E-unification algorithm, and, optionally, n calls to the SMT solver, where n is the number of unifiers returned by the E-unification algorithm.

5   Conclusion and Future Work

We introduced the problem of E-*unification modulo builtins*. While regular (E)-unification is about solving an equation in the algebra of terms (or the algebra of terms modulo E), E-unification modulo builtins is solving an equation in an algebra combining terms (modulo E) with builtin elements such as booleans, integers, arrays, or any other elements that can be handled by an SMT solver.

Our main contribution is to formalize the problem of E-unification modulo builtins and to provide an algorithm for it that is based on the notion of abstraction of a term. The algorithm reduces E-unification modulo builtins to regular E-unification. This allows to lift all existing E-unification algorithms to E-unification modulo builtins.

Unlike regular E-unification algorithms, which produce (sets of) substitution(s) as their output, algorithms for E-unification modulo builtins produce sets of pairs of substitutions and logical constraints. The equation being solved holds when instantiated by one of the substitutions, in the cases where the attached logical constraint holds. We show how to produce logical constraints that only contain builtins, which means that they can be fully handled by an SMT solver.

We implement our approach as a Maude prototype[1] at the meta-level. Given two terms, the prototype computes a complete set of E-unifiers modulo builtins as described in Algorithm 1. As an optimization, it filters out the E-unifiers modulo builtins that have an unsatisfiable constraint, by using the integrated SMT solver. We describe the prototype in Appendix A. Our result answers in part an open question in [18], namely of finding elements that match two matching logic patterns, which is essential in developing a matching-logic prover.

The main question that needs to be answered in future work is what kind of new applications are enabled by the combination of rewrite and SMT solving. On the theoretical side, all known results in rewriting (confluence, etc.) also need to be developed in the new framework. Another open question is how to perform E-unification modulo builtins in the presence of defined operations.

Acknowledgments. We thank the anonymous reviewers for their valuable suggestions. This work was supported by a grant of the Romanian National Authority for Scientific Research and Innovation, CNCS/CCCDI - UEFISICDI, project number PN-III-P2-2.1-BG-2016-0394, within PNCDI III.

A   Maude Prototype

Since the algorithm needs to manipulate terms (for instance, to compute the abstractions) we use the metalevel capabilities of Maude to implement the following functionalities:

- For a given meta-representation of a term, `getAbstraction` returns the abstraction of a term w.r.t. the set of builtins sorts, which need to be provided explicitly. When computing abstractions, fresh variables are generated;
- The E-unifiers of the abstractions are computed by `unifyAbstractions`;
- The unsatisfiable formulas are filtered out by `filterUnsatUMBs`;
- The E-unification modulo builtins algorithm that we propose is implemented by `completeSetOfUMBs`, which gets as input the meta-representations of two terms and returns the unifiers modulo builtins in two steps: it first generates the substitution-formula pairs with `completeSetOfUMBsUnfiltered`, and then it eliminates the unsatisfiable solutions using the aforementioned `filterUnsatUMBs`.

We show how to use our prototype to find the E-unifiers modulo builtins of the E-unification modulo builtins problem $n \mapsto 1, cnt \mapsto 42 = Z, n \mapsto 1$, introduced in Example 12. First, Maude finds a complete set of ACI-unifiers for the abstractions of the two terms. Then, from these unifiers we generate the pairs $\{(\tau_1', \phi_1'), (\tau_2', \phi_2')\}$, as shown in Example 12):

```
reduce in UNIFICATION-MODULO-BUILTINS :
completeSetOfUMBsUnfiltered(upTerm(n |-> 1,count |-> 42),
```

---

[1] The prototype is available at: https://github.com/andreiarusoaie/unification-modulo-builtins.

```
                         upTerm(Z,n |-> 1), 'STATE) .
rewrites: 2023 in 5ms cpu (6ms real) (338464 rewrites/second)
result UnificationResults:
[
'Z --> %1 |-> %2 |
abs0 #== n /\ (abs1 #== 1 /\ (abs2 #== count /\ abs3 #== 42)) /\
(abs4 #== n /\ abs5 #== 1) /\ (abs0 #== %1 /\ (abs1 #== %2 /\
  (abs2 #== %3 /\ (abs3 #== %4 /\ (abs4 #== %3 /\ abs5 #== %4)))))
],,
[
'Z --> \%1 |-> \%2 |
abs0 #== n /\ (abs1 #== 1 /\ (abs2 #== count /\ abs3 #== 42)) /\
(abs4 #== n /\ abs5 #== 1) /\ (abs0 #== %3 /\ (abs1 #== %4 /\
  (abs2 #== %1 /\ (abs3 #== %2 /\ (abs4 #== %3 /\ abs5 #== %4)))))
]
```

The variables abs0, abs1, ..., are generated during the abstraction process, while %1, %2, ... are generated by the Maude's variant unifier.

Finally, completeSetOfUMBs – the main function in our prototype – filters out the first unifier, which has an unsatisfiable constraint. Because the interaction between Maude and the SMT solver only supports integers and booleans, we have encoded identifiers (of sort Id) into integers before sending the formula to the SMT solver. The solution is:

```
reduce in UNIFICATION-MODULO-BUILTINS :
completeSetOfUMBs(upTerm(n |-> 1,count |-> 42),
                  upTerm(Z,n |-> 1), 'STATE) .
rewrites: 3883 in 17ms cpu (18ms real) (227355 rewrites/second)
result UnificationResults:
[
'Z --> %1 |-> %2 |
abs0 #== n /\ (abs1 #== 1 /\ (abs2 #== count /\ abs3 #== 42)) /\
(abs4 #== n /\ abs5 #== 1.Integer) /\ (abs0 #== %3 /\ (abs1 #== %4
/\(abs2 #== %1 /\ (abs3 #== %2 /\ (abs4 #== %3 /\ abs5 #== %4)))))
]
```

The result is essentially that the variable Z must be cnt $\mapsto$ 42, since %1 = abs2:Id = count and %2 = abs3:Id = 42.

We now show how our prototype solves the four E-unification modulo builtins problems discussed in the Introduction:

1. $\mathsf{n} \mapsto N = \mathsf{cnt} \mapsto C' + N', \mathsf{n} \mapsto N' + 3$
2. $\mathsf{n} \mapsto 2 \times N + 1, \mathsf{cnt} \mapsto C = \mathsf{cnt} \mapsto C' + N', \mathsf{n} \mapsto N' + 3$
3. $\mathsf{n} \mapsto 2 \times N, \mathsf{cnt} \mapsto C = \mathsf{cnt} \mapsto C' + N', \mathsf{n} \mapsto N' + 3$
4. $\mathsf{n} \mapsto 1, \mathsf{cnt} \mapsto C = \mathsf{cnt} \mapsto C' + N', \mathsf{n} \mapsto N' + 3$

The set of unifiers modulo builtins is computed by our Maude prototype for each case as shown below:

```
reduce in UNIFICATION-MODULO-BUILTINS :
completeSetOfUMBs(upTerm(n |-> N),
              upTerm(n |-> N' #+ 3,count |-> N' #+ C'), 'STATE) .
rewrites: 571 in 3ms cpu (3ms real) (148119 rewrites/second)
result UnificationResults: noUMBResults
=========================================
reduce in UNIFICATION-MODULO-BUILTINS :
completeSetOfUMBs(upTerm(n |-> 2 #* N #+ 1,count |-> C),
              upTerm(n |-> N' #+ 3,count |-> N' #+ C'), 'STATE) .
rewrites: 4574 in 15ms cpu (15ms real) (302413 rewrites/second)
result UnificationResults:
[
identity |
abs0 #== n /\ (abs1 #== 2 #* N #+ 1 /\ abs2 #== count) /\
(abs3 #== n /\ (abs4 #== N' #+ 3 /\ (abs5 #== count /\
abs6 #== N' #+ C'))) /\ (C #== %4 /\ (abs0 #== %1 /\
(abs1 #== %2 /\ (abs2 #== %3 /\ (abs3 #== %1 /\
(abs4 #== %2 /\ (abs5 #== %3 /\ abs6 #== %4)))))))
]
=========================================
reduce in UNIFICATION-MODULO-BUILTINS :
completeSetOfUMBs(upTerm(n |-> 2 #* N, count |-> C),
              upTerm(n |-> N' #+ 3,count |-> N' #+ C'), 'STATE) .
rewrites: 4528 in 15ms cpu (16ms real) (298385 rewrites/second)
result UnificationResults:
[
identity |
abs0 #== n /\ (abs1 #== 2 #* N /\ abs2 #== count) /\
(abs3 #== n /\ (abs4 #== N' #+ 3 /\ (abs5 #== count /\
abs6 #== N' #+ C'))) /\ (C #== %4 /\ (abs0 #== %1 /\
(abs1 #== %2 /\ (abs2 #== %3 /\ (abs3 #== %1 /\
(abs4 #== %2 /\ (abs5 #== %3 /\ abs6 #== %4)))))))
]
=========================================
reduce in UNIFICATION-MODULO-BUILTINS :
completeSetOfUMBs(upTerm(n |-> 1,count |-> C),
              upTerm(n |-> N' #+ 3,count |-> N' #+ C'), 'STATE) .
rewrites: 4625 in 15ms cpu (16ms real) (297141 rewrites/second)
result UnificationResults:
[
identity |
abs0 #== n /\ (abs1 #== 1 /\ abs2 #== count) /\
(abs3 #== n /\ (abs4 #== N' #+ 3 /\ (abs5 #== count /\
```

```
abs6 #== N' #+ C'))) /\ (C #== %4 /\ (abs0 #== %1 /\
(abs1 #== %2 /\ (abs2 #== %3 /\ (abs3 #==%1 /\
(abs4 #== %2 /\ (abs5 #== %3 /\ abs6 #== %4))))))
]
```

For the first E-umb problem, the tool returns noUMBResults, which means that it does not have any solution. For the other examples, the prototype finds the unifiers, as expected.

References

1. Aguirre, L., Martí-Oliet, N., Palomino, M., Pita, I.: Conditional narrowing modulo SMT and axioms. In: PPDP, pp. 17–28 (2017)
2. Arusoaie, A., Lucanu, D., Rusu, V.: A generic framework for symbolic execution. In: Erwig, M., Paige, R.F., Van Wyk, E. (eds.) SLE 2013. LNCS, vol. 8225, pp. 281–301. Springer, Cham (2013). https://doi.org/10.1007/978-3-319-02654-1_16
3. Arusoaie, A., Lucanu, D., Rusu, V.: Symbolic execution based on language transformation. Comput. Lang. Syst. Struct. **44**, 48–71 (2015)
4. Baader, F., Schulz, K.U.: Unification in the union of disjoint equational theories: combining decision procedures. JSC **21**(2), 211–243 (1996)
5. Bae, K., Rocha, C.: Guarded terms for rewriting modulo SMT. In: Proença, J., Lumpe, M. (eds.) FACS 2017. LNCS, vol. 10487, pp. 78–97. Springer, Cham (2017). https://doi.org/10.1007/978-3-319-68034-7_5
6. Ciobâcă, Ş., Lucanu, D.: A coinductive approach to proving reachability properties in logically constrained term rewriting systems. IJCAR (2018, to appear)
7. Ciobâcă, Ş., Lucanu, D.: RMT: proving reachability properties in constrained term rewriting systems modulo theories. Technical report TR 16–01, Alexandru Ioan Cuza University, Faculty of Computer Science (2016)
8. Ştefănescu, A., Ciobâcă, Ş., Mereuta, R., Moore, B.M., Şerbănută, T.F., Roşu, G.: All-path reachability logic. In: Dowek, G. (ed.) RTA 2014. LNCS, vol. 8560, pp. 425–440. Springer, Cham (2014). https://doi.org/10.1007/978-3-319-08918-8_29
9. Darlington, J., Guo, Y.: Constrained equational deduction. In: Kaplan, S., Okada, M. (eds.) CTRS 1990. LNCS, vol. 516, pp. 424–435. Springer, Heidelberg (1991). https://doi.org/10.1007/3-540-54317-1_111
10. Falke, S., Kapur, D.: Dependency pairs for rewriting with built-in numbers and semantic data structures. In: Voronkov, A. (ed.) RTA 2008. LNCS, vol. 5117, pp. 94–109. Springer, Heidelberg (2008). https://doi.org/10.1007/978-3-540-70590-1_7
11. Fuhs, C., Kop, C., Nishida, N.: Verifying procedural programs via constrained rewriting induction. ACM TOCL **18**(2), 14:1–14:50 (2017)
12. Kirchner, C., Kirchner, H., Rusinowitch, M.: Deduction with symbolic constraints. Research report RR-1358, INRIA (1990). Projet EURECA
13. Kop, C., Nishida, N.: Term rewriting with logical constraints. In: Fontaine, P., Ringeissen, C., Schmidt, R.A. (eds.) FroCoS 2013. LNCS (LNAI), vol. 8152, pp. 343–358. Springer, Heidelberg (2013). https://doi.org/10.1007/978-3-642-40885-4_24
14. Lucanu, D., Rusu, V., Arusoaie, A., Nowak, D.: Verifying reachability-logic properties on rewriting-logic specifications. In: Martí-Oliet, N., Ölveczky, P.C., Talcott, C. (eds.) Logic, Rewriting, and Concurrency. LNCS, vol. 9200, pp. 451–474. Springer, Cham (2015). https://doi.org/10.1007/978-3-319-23165-5_21

15. Nelson, G., Oppen, D.C.: Simplification by cooperating decision procedures. ACM TOPLAS **1**(2), 245–257 (1979)
16. Nigam, V., Talcott, C., Aires Urquiza, A.: Towards the automated verification of cyber-physical security protocols: bounding the number of timed intruders. In: Askoxylakis, I., Ioannidis, S., Katsikas, S., Meadows, C. (eds.) ESORICS 2016. LNCS, vol. 9879, pp. 450–470. Springer, Cham (2016). https://doi.org/10.1007/978-3-319-45741-3_23
17. Rocha, C., Meseguer, J., Muñoz, C.: Rewriting modulo SMT and open system analysis. JLAMP **86**(1), 269–297 (2017)
18. Roşu, G.: Matching logic. LMCS **13**(4), 1–61 (2017)
19. Schmidt-Schauss, M.: Unification in a combination of arbitrary disjoint equational theories. JSC **8**(1), 51–99 (1989)
20. Skeirik, S., Ştefănescu, A., Meseguer, J.: A constructor-based reachability logic for rewrite theories. CoRR, abs/1709.05045 (2017)

Formalization of the Undecidability of the Halting Problem for a Functional Language

Thiago Mendonça Ferreira Ramos[1]([envelope]), César Muñoz[2]([envelope]),
Mauricio Ayala-Rincón[1]([envelope]), Mariano Moscato[3]([envelope]), Aaron Dutle[2],
and Anthony Narkawicz[2]

[1] University of Brasília, Brasília, Brazil
tramos@aluno.unb.br, ayala@unb.br
[2] NASA Langley Research Center, Hampton, VA, USA
{cesar.a.munoz,aaron.m.dutle,anthony.narkawicz}@nasa.gov
[3] National Institute of Aerospace, Hampton, VA, USA
mariano.moscato@nianet.org

Abstract. This paper presents a formalization of the proof of the undecidability of the halting problem for a functional programming language. The computational model consists of a simple first-order functional language called PVS0 whose operational semantics is specified in the Prototype Verification System (PVS). The formalization is part of a termination analysis library in PVS that includes the specification and equivalence proofs of several notions of termination. The proof of the undecidability of the halting problem required classical constructions such as mappings between naturals and PVS0 programs and inputs. These constructs are used to disprove the existence of a PVS0 program that decides termination of other programs, which gives rise to a contradiction.

1 Introduction

In computer science, program termination is the quintessential example of a property that is undecidable, a fact that is well-known as the *undecidability of the halting problem* [12]. This undecidability implies that it is not possible to build a compiler that would verify whether a program terminates for any given input. Despite this undecidability, it is possible to construct algorithms that partially decide termination, i.e., they correctly answer whether an input program "terminates or not", but may also answer "do not know". Termination analysis of programs is an active area of research. Indeed, substantial progress in this area is regularly presented in meetings such as the International Workshop on Termination and the Annual International Termination Competition.

To formally verify correctness of termination analysis algorithms, it is often necessary to specify and prove equivalence among multiple notions of termination. Given a formal model of computation, one natural notion of termination is specified as *for all inputs there exists an output provided under the operational*

© Springer-Verlag GmbH Germany, part of Springer Nature 2018
L. S. Moss et al. (Eds.): WoLLIC 2018, LNCS 10944, pp. 196–209, 2018.
https://doi.org/10.1007/978-3-662-57669-4_11

semantics of the model. Another notion of termination could be specified considering whether or not the depth of the expansion tree of computation steps for all inputs is finite. These two notions rely on the semantics of the computational model. A more syntactic approach, attributed to Turing [13], is to verify that the actual arguments decrease in any repeating control structure, e.g., recursion, unbounded loop, fix-point, etc., of the program according to some well-founded relation. This notion is used in the majority of proof assistants, where the user must provide the well-founded relation.

The main contribution of this work is the formalization in the Prototype Verification System (PVS) [10] of the theorem of undecidability of the halting problem for a model of computation given by a functional language called PVS0. The formal development includes the definition of PVS0, its operational semantics, and the specification and proof of several concepts used in termination analysis of PVS0 programs. For the undecidability proof of the halting problem, only the semantic notions of termination are used. Turing termination for the language PVS0 is also discussed to show how semantic and syntactic termination criteria are related. The formalization is available as part of the NASA PVS Library under the directory PVS0.[1] All lemmas and theorems presented in this paper were formalized and verified in PVS.

2   Semantic Termination

The PVS0 language is a simple functional language whose expressions are described by the following grammar.

$$expr ::= \mathtt{cnst} \mid \mathtt{vr} \mid op1(expr) \mid op2(expr, expr) \mid rec(expr) \mid \mathtt{ite}(expr, expr, expr)$$

The grammar above is specified in PVS through the abstract data type:

```
PVS0Expr[T:TYPE+] : DATATYPE
BEGIN
  cnst(get_val:T) : cnst?
  vr : vr?
  op1(get_op:nat,get_arg:PVS0Expr) : op1?
  op2(get_op:nat,get_arg1,get_arg2:PVS0Expr) : op2?
  rec(get_arg:PVS0Expr) : rec?
  ite(get_cond,get_if,get_else:PVS0Expr) : ite?
END PVS0Expr
```

A PVS specification of an abstract data type includes the constructors, e.g., ite, the accessors, e.g., get_cond, and the recognizers, e.g., ite?. In this data type, T is a parametric nonempty type that represents the type of values that serve as inputs and outputs of PVS0 programs. Furthermore, cnst is the constructor of constant values of type T, vr is the unique variable constructor, op1 and op2 are constructors of unary and binary operators, respectively, rec is the constructor

---

[1] https://github.com/nasa/pvslib.

of recursion, and `ite` is the constructor of conditional "if-then-else" expressions. The first parameter of the constructors `op1` and `op2` is an index representing built-in unary and binary operators, respectively.

The uninterpreted type `T` and uninterpreted unary and binary operators enable the encoding of arbitrary first-order PVS functions as programs in PVS0. Indeed, the operational semantics of `PVS0Expr` is given in terms of a non-empty set Val, which interprets the type `T`. The type $PVS0[Val]$ of PVS0 programs with values in Val consists of all 4-tuples of the form $(O_1, O_2, \bot, expr)$, such that

- O_1 is a list of PVS functions of type $Val \rightarrow Val$, where $O_1(i)$, i.e., the i-th element of the list O_1, interprets the index i in the constructor `op1`,
- O_2 is a list of PVS functions of type $Val \times Val \rightarrow Val$, where $O_2(i)$, i.e., the i-th element of the list O_2, interprets the index i in the constructor `op2`,
- $\bot$ is a constant of type Val representing the Boolean value false in the conditional construction `ite`, and
- $expr$ is a $PVS0Expr[Val]$, which is the syntactic representation of the program itself.

Henceforth, $|O_1|$ and $|O_2|$ represent the length of the lists O_1 and O_2, respectively. The choice of lists of functions for interpreting unary and binary operators helps in the enumeration of PVS0 programs, which is necessary in the undecidability proof.

Given a program $(O_1, O_2, \bot, e_f)$ of type $PVS0[Val]$, the semantic evaluation of an expression e of type $PVS0Expr[Val]$ is given by the curried inductive relation ε of type $PVS0[Val] \rightarrow (PVS0Expr[Val] \times Val \times Val) \rightarrow$ `bool` defined as follows.

Intuitively, the relation $\varepsilon(O_1, O_2, \bot, e_f)(e, v_i, v_o)$ defined below holds when given a program $(O_1, O_2, \bot, e_f)$ the evaluation of the expression e on the input value v_i is the value v_o.

$$
\begin{aligned}
\varepsilon(O_1, O_2, \bot, e_f)(e, v_i, v_o) := \text{ CASES } & e \text{ OF} \\
\text{cnst}(v) \; &: v_o = v; \\
\text{vr} \; &: v_o = v_i; \\
\text{op1}(j, e_1) \; &: j < |O_1| \wedge \exists v' \in Val : \\
& \varepsilon(O_1, O_2, \bot, e_f)(e_1, v_i, v') \wedge v_o = O_1(j)(v'); \\
\text{op2}(j, e_1, e_2) \; &: j < |O_2| \wedge \exists v', v'' \in Val : \\
& \varepsilon(O_1, O_2, \bot, e_f)(e_1, v_i, v') \wedge \\
& \varepsilon(O_1, O_2, \bot, e_f)(e_2, v_i, v'') \wedge \\
& v_o = O_2(j)(v', v''); \\
\text{rec}(e_1) \; &: \exists v' \in Val : \varepsilon(O_1, O_2, \bot, e_f)(e_1, v_i, v') \wedge \\
& \varepsilon(O_1, O_2, \bot, e_f)(e_f, v', v_o) \\
\text{ite}(e_1, e_2, e_3) \; &: \exists v' : \varepsilon(O_1, O_2, \bot, e_f)(e_1, v_i, v') \wedge \\
& \text{IF } v' \neq \bot \text{ THEN } \varepsilon(O_1, O_2, \bot, e_f)(e_2, v_i, v_o) \\
& \text{ELSE } \varepsilon(O_1, O_2, \bot, e_f)(e_3, v_i, v_o).
\end{aligned}
$$

In the definition of ε, the parameters e_f and e are needed since the evaluation of a program $(O_1, O_2, \bot, e_f)$ leads to evaluation of sub expressions e of e_f and, when a recursive call is evaluated, the whole expression e_f should be considered again (see the recursive case $\text{rec}(e_1)$ above).

For example, consider below a PVS0 program that computes the Ackermann function.

Example 1. Let *Val* be the set $\mathbb{N} \times \mathbb{N}$ of pairs of natural numbers, $\top = (1,0)$, $\bot = (0,0)$, and a be the PVS0 program $(O_1, O_2, \bot, e_a)$, where

$$
\begin{aligned}
O_1(0)(m,n) \qquad &:= \text{IF } m = 0 \text{ THEN } \top \text{ ELSE } \bot, \\
O_1(1)(m,n) \qquad &:= \text{IF } n = 0 \text{ THEN } \top \text{ ELSE } \bot, \\
O_1(2)(m,n) \qquad &:= (n+1, 0), \\
O_1(3)(m,n) \qquad &:= \text{IF } m > 0 \text{ THEN } (m-1, 1) \text{ ELSE } \bot, \\
O_1(4)(m,n) \qquad &:= \text{IF } n > 0 \text{ THEN } (m, n-1) \text{ ELSE } \bot, \\
O_2(0)((m,n),(i,j)) &:= \text{IF } m > 0 \text{ THEN } (m-1, i) \text{ ELSE } \bot, \\
e_a \qquad &:= \text{ite}(\text{op1}(0, \text{vr}), \text{op1}(2, \text{vr}), \\
&\quad\; \text{ite}(\text{op1}(1, \text{vr}), \text{rec}(\text{op1}(3, \text{vr})), \\
&\quad\; \text{rec}(\text{op2}(0, \text{vr}, \text{rec}(\text{op1}(4, \text{vr})))))).
\end{aligned}
$$

It is proved in PVS that a computes the Ackermann function, i.e., for any $n, m, k \in \mathbb{N}$, $ackermann(m,n) = k$ if and only if $\varepsilon(a)(e_a, (n,m), (k,i))$, for some i, where $ackermann$ is the recursive function defined in PVS as

$$
\begin{aligned}
ackermann(m,n) := \;&\text{IF } m = 0 \text{ THEN } n + 1 \\
&\text{ELSIF } n = 0 \text{ THEN } ackermann(m-1, 1) \\
&\text{ELSE } ackermann(m-1, ackermann(m, n-1)).
\end{aligned}
$$

In the definition of a, the type *Val* encodes the two inputs of the Ackermann function, but also the output of the function, which is given by the first entry of the second pair.

The proof of one of the implications in the statement of Example 1 proceeds by induction using a lexicographic order on (m,n). The other implication is proved using the induction schema generated for the inductive relation ε. Although it is not logically deep, this proof is tedious and long. However, it is mechanizable assuming that the PVS function and the PVS0 program share the same syntactical structure. As part of the work presented in this paper, a PVS strategy that automatically discharges equivalences between PVS functions and PVS0 programs was developed. This strategy is convenient since one of the objectives of this work is to reason about computational aspects of PVS functions through their embeddings in PVS0.

Example 1 also illustrates the use of built-in operators in PVS0. Despite its simplicity, this language is not minimal from a fundamental point of view. Indeed, since the type $\top$ is generic, any PVS function can be used as a building block in the construction of a PVS0 program. This feature is justified since all PVS functions are total. Therefore, they can be considered as atomic built-in operators. However, in contrast to proof assistants based on constructive logic, PVS allows for the definition of non-computable functions. The consequences of these features will be clear in the undecidability proof of the halting problem.

The following lemma states that the semantic evaluation relation ε is *deterministic*.

Lemma 1. *Let pvso be a program of type* PVSO[*Val*]. *For any expression e of type* PVSOExpr[*Val*] *and all values* $v_i, v_o', v_o'' \in Val$,

$$\varepsilon(pvso)(e, v_i, v_o') \text{ and } \varepsilon(pvso)(e, v_i, v_o'') \text{ implies } v_o' = v_o''.$$

The proof of this lemma uses the induction schema generated for the inductive relation ε.

The relation ε is functional but not total, i.e., there are programs *pvso* and values v_i, for which there is no value v_o that satisfies $\varepsilon(pvso)(pvso_e, v_i, v_o)$, where $pvso_e$ is the program expression in *pvso*. This suggests the following definition of the *semantic termination predicate*.

$$T_\varepsilon(pvso, v_i) := \exists\, v_o \in Val : \varepsilon(pvso)(pvso_e, v_i, v_o).$$

This predicate states that for a given program *pvso* and input v_i, the evaluation of the program's expression $pvso_e$ on the value v_i terminates with the output value v_o. The program *pvso* is *total with respect to* ε if it satisfies the following predicate.[2]

$$T_\varepsilon(pvso) := \forall\, v \in Val : T_\varepsilon(pvso, v).$$

Semantic termination can also be specified by a function χ of type PVSO[*Val*] $\rightarrow$ (PVSOExpr[*Val*] $\times$ *Val* $\times$ $\mathbb{N}$ $\rightarrow$ *Val* $\cup$ $\{\Diamond\}$) defined as follows.

$$
\begin{aligned}
\chi(O_1, O_2, \bot, e_f)(e, v_i, n) := \ &\text{IF } n = 0 \text{ THEN } \Diamond \text{ ELSE } \text{ CASES } e \text{ OF} \\
\text{cnst}(v) \ :\ & v; \\
\text{vr} \ :\ & v_i; \\
\text{op1}(j, e_1) \ :\ & \text{IF } j < |O_1| \text{ THEN} \\
& \quad \text{LET } v' = \chi(O_1, O_2, \bot, e_f)(e_1, v_i, n) \text{ IN} \\
& \quad \text{IF } v' = \Diamond \text{ THEN } \Diamond \text{ ELSE } O_1(j)(v') \\
& \text{ELSE } \Diamond; \\
\text{op2}(j, e_1, e_2) \ :\ & \text{IF } j < |O_2| \text{THEN} \\
& \quad \text{LET } v' = \chi(O_1, O_2, \bot, e_f)(e_1, v_i, n), \\
& \qquad\quad\ v'' = \chi(O_1, O_2, \bot, e_f)(e_2, v_i, n) \text{ IN} \\
& \quad \text{IF } v' = \Diamond \vee v'' = \Diamond \text{ THEN } \Diamond \text{ ELSE } O_2(j)(v', v'') \\
& \text{ELSE } \Diamond; \\
\text{rec}(e_1) \ :\ & \text{LET } v' = \chi(O_1, O_2, \bot, e_f)(e_1, v_i, n) \text{ IN} \\
& \text{IF } v' = \Diamond \text{ THEN } \Diamond \text{ ELSE } \chi(O_1, O_2, \bot, e_f)(e_f, v', n - 1); \\
\text{ite}(e_1, e_2, e_3) \ :\ & \text{LET } v' = \chi(O_1, O_2, \bot, e_f)(e_1, v_i, n) \text{ IN} \\
& \text{IF } v' = \Diamond \text{ THEN } \Diamond \\
& \text{ELSIF } v' \neq \bot \text{ THEN } \chi(O_1, O_2, \bot, e_f)(e_2, v_i, n) \\
& \text{ELSE } \chi(O_1, O_2, \bot, e_f)(e_3, v_i, n).
\end{aligned}
$$

---

[2] Polymorphism in PVS allow for the use of the same function or predicate name with different types.

In the definition of χ, n is the maximum number of nested recursive calls that are allowed in the evaluation of the recursive program for a given input. If this limit is reached during an evaluation, the function χ returns the symbol $\Diamond$, which represents a "none" value. This function can be used to define an alternative predicate for semantic termination as follows.

$$T_\chi(pvso, v) := \exists n \in \mathbb{N} : \chi(pvso)(pvso_e, v, n) \neq \Diamond.$$

The program $pvso$ is *total with respect to* χ if it satisfies the following predicate.

$$T_\chi(pvso) := \forall v \in Val : T_\chi(pvso, v).$$

The following theorem states that T_ε and T_χ captures the same notion of termination.

Theorem 1. *Let pvso be a PVSO program of type* PVSO[Val]. *The following conditions hold:*

1. *For any $v_i \in Val$ and e of type* PVSOExpr[Val], *$\varepsilon(pvso)(e, v_i, v_o)$ if and only if $v_o = \chi(pvso)(e, v_i, n)$, for some n, where $v_o \neq \Diamond$.*
2. *For any $v \in Val$, $T_\varepsilon(pvso, v)$ if and only if $T_\chi(pvso, v)$.*
3. *$T_\varepsilon(pvso)$ if and only if $T_\chi(pvso)$.*

In Theorem 1, Statement 2 and Statement 3 are consequences of Statement 1. Assuming $T_\chi(pvso, v)$, the proof of the Statement 1 requires the construction of the number $\mu(pvso, v) \in \mathbb{N}^+$ as follows.

$$\mu(pvso, v) := \min(\{n : \mathbb{N} \mid \chi(pvso)(pvso_e, v, n) \neq \Diamond\}).$$

This number satisfies the following property.

Lemma 2. *Let pvso be a program of type* PVSO[Val] *and $v \in Val$ such that $T_\chi(pvso, v)$. For any $n \geq \mu(pvso, v)$, $\chi(pvso)(pvso_e, v, n) \neq \Diamond$.*

A PVSO program $pvso$ that satisfies $T_\chi(pvso)$ (or equivalently, $T_\varepsilon(pvso)$) is said to be *terminating*. The following lemma shows that there are non-terminating PVSO programs.

Lemma 3. *Let $\Delta = (O_1, O_2, \bot, \texttt{rec}(\texttt{vr}))$. For any $v \in Val$, $\neg T_\chi(\Delta, v)$.*

The proof proceeds by showing that if $\chi(\Delta)(\Delta_e, v, n) \neq \Diamond$ for some v, where $n = \mu(\Delta, v)$, then it is also the case that $\chi(\Delta)(\Delta_e, v, n - 1) \neq \Diamond$. By definition of μ, this is a contradiction.

Proving that a PVSO program terminates using the semantic predicates $T_\varepsilon(pvso)$ and $T_\chi(pvso)$ is not very convenient. Indeed, these notions rely on actual computations of the program on a potentially infinite set of inputs. A syntactic criterion here called *Turing termination* relies on the existence of a well-founded relation $<$ over a type A and measure function $\mathcal{M}$ of type $Val \to A$ on the parameters of the recursive function such that $\mathcal{M}$ strictly decreases on every recursive call. This notion is adopted as the meta-theoretical definition of termination in several proof assistants. In PVS, this notion is implemented through the generation of the so called termination TCCs (Type Correctness Conditions), where the measure function and the well-founded relation are provided by the user. This notion is formalized by defining, in PVS, an algorithm that generates termination TCCs for PVS0 programs. It has been proved that this Turing termination is equivalent to $T_\varepsilon(pvso)$ and, therefore, to $T_\chi(pvso)$. The formal infrastructure that is needed for defining Turing termination of PVS0 programs is out of the scope of the present work.

For a PVS0 program $(O_1, O_2, \bot, e_f)$ and two values v_i and v_r, the definition of $v_i \to v_r$ is given by:

$$v_i \to v_r := \varepsilon(O_1, O_2, \bot, e_f)(e_f, v_i, v_r) \wedge \mathcal{M}(v_r) < \mathcal{M}(v_i)$$

The definition above relates an input value v_i of a PVS0 program and the argument v_r of a recursive call such that $\mathcal{M}(v_r) < \mathcal{M}(v_i)$. A key construction that is needed in the equivalence proof between the syntactic and the semantic termination criteria is the definition of the number $\Omega(v)$, where $v \in Val$, as follows.

$$\Omega(v) := \min(\{n : \mathbb{N}^+ \mid \forall v' \in V : \neg(v \to^n v')\}).$$

Intuitively, $\Omega(v)$ is the length of the longest path downwards starting from v. The following lemma states a relation between μ and Ω.

Lemma 4.   *Let pvso be a PVS0 program that satisfies Turing termination for a well-founded relation $<$ over A and a measure function $\mathcal{M}$. For any value $v \in Val$, $\mu(pvso, v) \leq \Omega(v)$.*

3   Partial Recursive Functions

By design, the PVS0 language can directly encode any PVS function f of type $T \to T$, where T is an arbitrary PVS type. This feature enables the use of PVS functions as built-in operators in PVS0 program. The following lemma states that any PVS function can be embedded in a terminating PVS0 program.

Lemma 5. *Let f be a PVS function of type $T \to T$. The program $mk\_pvs0(f) = (O_1, O_2, \bot, e_f)$ of type PVS0, where $e_f = \mathtt{op1}(0, \mathtt{vr})$ and $O_1(0)(t) = f(t)$, satisfies the following properties:*

- *$mk\_pvs0(f)$ is terminating, i.e., $T_\varepsilon(mk\_pvs0(f))$.*
- *For any $t \in T$, $\varepsilon(mk\_pvs0(f))(e_f, t, f(t))$.*

The converse of Lemma 5 is not true in general as illustrated by the following theorem.

Lemma 6. *There is no PVS function f of type $T \rightarrow T$ such that for any $t \in T$, $\varepsilon(\Delta)(\mathtt{rec}(\mathtt{vr}), t, f(t))$, where Δ is the function defined in Lemma 3.*

However, any PVS0 program, even non-terminating ones, can be encoded as a PVS function of type $T \rightarrow T \cup \{\Diamond\}$, as stated in the following lemma.

Lemma 7. *Let pvso be a, possibly non-terminating, program of type PVS0 and $pvso_e$ the PVS0 expression of pvso. The PVS function*

$$f(t) := \mathtt{IF}\ T_\chi(pvso, t)\ \mathtt{THEN}\ \chi(pvso)(pvso_e, t, \mu(pvso, t))\ \mathtt{ELSE}\ \Diamond$$

satisfies the following property for any $t \in T$.

$$T_\varepsilon(pvso, t)\ \text{if and only if}\ \varepsilon(pvso)(pvso_e, t, f(t)).$$

The proofs of the previous lemmas are straightforward applications of the definitions of T_ε, ε, T_χ, and χ.

A consequence of Lemmas 5 and 7 is that is possible to define an oracle of type PVS0[Val], where Val = PVS0, that decides if a program of type PVS0 is terminating or not. The existence of that oracle is stated in the following theorem.

Theorem 2. *The program $Oracle = (O_1, O_2, \bot, e_f)$ of type PVS0[Val], where Val = PVS0, $e_f = mk\_pvs0(\mathtt{LAMBDA}(pvso : \mathtt{PVS0}) : \mathtt{IF}\ T_\varepsilon(pvso)\ \mathtt{THEN}\ \top\ \mathtt{ELSE}\ \bot)$, and $\top \neq \bot$, has the following properties.*

- *Oracle is a terminating PVS0[Val] program, i.e., $T_\varepsilon(Oracle)$.*
- *Oracle decides termination of any PVS0 program, i.e., for any pvso of type PVS0,*

$$\chi(Oracle)(Oracle_e, pvso, \mu(Oracle, pvso)) = \top\ \text{if and only if}\ T_\varepsilon(pvso).$$

This counterintuitive result is possible because PVS, in contrast to proof assistants based on constructive logic, allows for the definition of total functions that are non-computable, e.g., $\mathtt{LAMBDA}(pvso : \mathtt{PVS0}) : \mathtt{IF}\ T_\varepsilon(pvso)\ \mathtt{THEN}\ \top$ $\mathtt{ELSE}\ \bot$. These non-computable functions can be used in the construction of terminating PVS0 programs through the built-in operators. Therefore, in order to formalize the notion of partial recursive functions in PVS0, it is necessary to restrict the way in which programs are built.

First, the basic type T is set to $\mathbb{N}$, i.e., $Val = \mathbb{N}$, where the number 0 represents the value false, i.e., $\bot = 0$. Any value different from 0 represents a true value, in particular $\top = 1$. Second, the built-in operators used in the construction of programs are restricted by a hierarchy of levels: the operators in the first level can only be defined using projections ($\Pi_1(x, y : \mathbb{N}) := x$ and $\Pi_2(x, y : \mathbb{N}) := y$), successor ($succ(x : \mathbb{N}) : x + 1$), and greater or equal than ($ge(x, y : \mathbb{N}) : \mathtt{IF}\ x \geq y\ \mathtt{THEN}\ \top\ \mathtt{ELSE}\ \bot$) functions. Operators in higher levels can only be constructed using programs from the previous level. This idea is formalized by the predicate $pvs0\_level : \mathbb{N} \to \text{PVSO}[\mathbb{N}] \to \text{bool}$ as shown below.

$$pvs0\_level(n)(O_1, O_2, \bot, e_f) :=$$
$$\text{IF } n = 0 \text{ THEN } O_1 = \langle succ \rangle \wedge O_2 = \langle \Pi_1, \Pi_2, ge \rangle$$
$$\text{ELSE } (\ \exists p' \in \text{PVSO}[\mathbb{N}] : pvs0\_level(n-1)(p') \wedge$$
$$\text{LET } (O_1{'}, O_2{'}, \bot{'}, e_f{'}) = p', l_1' = |O_1{'}| \text{ IN}$$
$$|O_1| = l_1' + 1 \wedge$$
$$(\ \forall i \in \mathbb{N} : i < l_1' \Rightarrow O_1(i) = O_1{'}(i)\ ) \wedge$$
$$(\ \forall v \in \mathbb{N} : \varepsilon(p')(e_f{'}, v, O_1(l_1')(v))\ )\ ) \wedge$$
$$(\ \exists p' \in \text{PVSO}[\mathbb{N}] : pvs0\_level(n-1)(p') \wedge$$
$$\text{LET } (O_1{'}, O_2{'}, \bot{'}, e_f{'}) = p', l_2' = |O_2{'}| \text{ IN}$$
$$|O_2| = l_2' + 1 \wedge$$
$$(\ \forall i \in \mathbb{N} : i < l_2' \Rightarrow O_2(i) = O_2{'}(i)\ ) \wedge$$
$$(\ \forall v_1, v_2 \in \mathbb{N} : \varepsilon(p')(e_f{'}, \kappa_2(v_1, v_2), O_2(l_2')(v_1, v_2))\ )\ )\ ),$$

where the function κ_2 is an encoding of pairs of natural numbers onto natural numbers defined as $\kappa_2(m, n) := (m + n + 1) \times (m + n)/2 + n$.

The type `PartialRecursive` is defined to be a subtype of $\text{PVSO}[\mathbb{N}]$ containing all the programs $pvso$ such that there is a natural n for which $pvs0\_level(n)(pvso)$ holds. Additionally, `Computable` is a subtype of `PartialRecursive` containing those elements that are also terminating according to the aforementioned definitions.

The following theorem states that `PartialRecursive`, and thus `Computable`, are enumerable types.

Theorem 3. *There exists a PVS function of type* $\mathbb{N} \to$ `PartialRecursive` *that is surjective.*

As a corollary of this theorem, the inverse of this surjective function, denoted as κ_P is an injective function of type `PartialRecursive` $\to \mathbb{N}$.

The proof of Theorem 3 is technically involved. The proof proceeds by showing that each level is enumerable. Therefore, the function κ_P exists since the countable union of countable sets is also countable. A similar work is presented in [4] by Foster an Smolka. They encode a lambda term into another lambda term such that it represents a natural number using the Scott numbers codification and use that encoding to formalize in Coq the Rice's Theorem.

The function κ_P is used in the proof of the undecidability of the halting problem for PVS0 to encode `PartialRecursive` programs as inputs of a `Computable` program. This function is a key element in the construction of the self-reference argument used in the diagonalization approach of the undecidability proof.

4   Undecidability of the Halting Problem

The classical formalization of the undecidability of the halting problem starts by assuming the existence of an oracle capable of deciding whether a program halts for any input. A Gödelization function transforms the tuple of program and input into a single input to the oracle. After that, using the oracle, another program is created such that if the encoded program halts it enters into an infinite loop. Otherwise, it produces an answer and halts. Passing this program as an input to itself results in the expected contradiction.

Here, the undecidability of the halting problem was formalized using the notions of termination for PVS0 defined in the previous section and the Cantor's *diagonalization* technique.

Theorem 4 (Undecidability of the Halting Problem for PVS0). *There is no program oracle $= (O_1, O_2, \bot, e_o)$ of type* Computable *such that for all pvso $= (O_1', O_2', \bot, e_f)$ of type* PartialRecursive *and for all $n \in \mathbb{N}$,*

$$T_\varepsilon(pvso, n) \text{ if and only if } \neg\varepsilon(oracle)(e_o, \kappa_2(\kappa_P(pvso), n), \bot).$$

Proof. The proof proceeds by assuming the existence of an oracle to derive a contradiction. Suppose there exists a program $oracle = (O_1, O_2, \bot, e_o)$ of type Computable such as the one presented in the statement of the theorem. Then, a PVS0[$\mathbb{N}$] program $pvso = (O_1', O_2', \bot, e_f)$ can be defined, where $O_1'(k) = O_1(k)$, for $k < |O_1|$, $O_2'(k) = O_2(k)$, for $k < |O_2|$, and

- $O_1'(|O_1|)(i) = choose(\{a : \mathbb{N} \mid \varepsilon(oracle)(e_o, i, a)\})$,
- $O_2'(|O_2|)(i, j) = choose(\{a : \mathbb{N} \mid \varepsilon(oracle)(e_o, \kappa_2(i, j), a)\})$, and
- $e_f = \texttt{ite}(\texttt{op2}(|O_2|, \texttt{vr}, \texttt{vr}), rec(\texttt{vr}), \texttt{vr})$.

The PVS function *choose* returns an arbitrary element from a non-empty set. The sets used in the definitions of O_1' and O_2' are non-empty since *oracle* is Computable and, therefore, terminating. The program *pvso* is built in such a way that it belongs to the next level from the level of *oracle*.

Let n be the natural number $\kappa_P(pvso)$. The rest of the proof proceeds by case analysis.

- **Case 1**: $\varepsilon(oracle)(e_o, \kappa_2(n, n), \bot)$. This case holds if and only if $\neg T_\varepsilon(pvso, n)$. Expanding T_ε one obtains

$$\neg\exists(v : \mathbb{N}) : \exists(v_o : \mathbb{N}) : \varepsilon(pvso)(\texttt{op2}(|O_2|, \texttt{vr}, \texttt{vr}), n, v_o) \land \tag{1}$$
$$\text{IF } v_o \neq \bot$$
$$\text{THEN } \varepsilon(pvso)(rec(\texttt{vr}), n, v)$$
$$\text{ELSE } \varepsilon(pvso)(\texttt{vr}, n, v).$$

Expanding ε in $\varepsilon(pvso)(\texttt{op2}(|O_2|, \texttt{vr}, \texttt{vr}), n, v_o)$ yields

$$choose(\{a : \mathbb{N} \mid \varepsilon(oracle)(e_o, \kappa_2(n, n), a)\}) = v_o.$$

Since $\varepsilon(oracle)(e_o, \kappa_2(n, n), \bot)$ holds, $\bot = v_o$. Therefore, Formula (1) is equivalent to

$$\neg\exists(v : \mathbb{N}) : \varepsilon(pvso)(\texttt{vr}, n, v). \tag{2}$$

The predicate $\varepsilon(pvso)(\texttt{vr}, n, v)$ holds if and only if $n = v$. Hence, Formula (2) states that $\neg\exists(v : \mathbb{N}) : n = v$, where n is a natural number. This is a contradiction.

- **Case 2**: $\neg\varepsilon(oracle)(e_o, \kappa_2(n,n), \bot)$. This case holds if and only if $T_\varepsilon(pvso, n)$. From Theorem 1, $T_\chi(pvso, n)$. If the proof starts directly from $T_\varepsilon(pvso, n)$, after expanding and simplifying it, $T_\varepsilon(pvso, n)$ is obtained once again, which implies that there is not such an n, giving a contradiction. However, since PVS does not accept the definition of a function that enters into such an infinite loop, the solution is to apply the equivalence Theorem 1. Expanding the definition of T_χ yields

$$\exists m \in \mathbb{N} : \chi(pvso)(\mathtt{ite}(\mathtt{op2}(|O_2|, \mathtt{vr}, \mathtt{vr}), rec(\mathtt{vr}), \mathtt{vr}), n, m) \neq \Diamond.$$

If there exists such m, it can be chosen as the minimal natural that makes the above proposition hold. Expanding the definition of χ yields

$$\left(\begin{array}{l} \mathtt{IF}\ \chi(pvso)(\mathtt{op2}(|O_2|, \mathtt{vr}, \mathtt{vr}), n, m) \neq \Diamond\ \mathtt{THEN} \\ \quad \mathtt{IF}\ \chi(pvso)(\mathtt{op2}(|O_2|, \mathtt{vr}, \mathtt{vr}), n, m) \neq \bot\ \mathtt{THEN} \\ \quad \chi(pvso)(rec(\mathtt{vr}), n, m) \\ \quad \mathtt{ELSE}\ \chi(pvso)(\mathtt{vr}, n, m) \\ \mathtt{ELSE}\ \Diamond \end{array} \right) \neq \Diamond. \qquad (3)$$

If the condition of the first if-then-else were false, then Formula (3) reduces to $\Diamond \neq \Diamond$, which is a contradiction. Therefore, this condition must be true. After expanding and simplifying χ, $\chi(pvso)(\mathtt{op2}(|O_2|, \mathtt{vr}, \mathtt{vr}), n, m)$ reduces to

$$choose(\{a : \mathbb{N} \mid \varepsilon(oracle)(e_o, \kappa_2(n,n), a)\}).$$

Let $v = choose(\{a : \mathbb{N} \mid \varepsilon(oracle)(e_o, \kappa_2(n,n), a)\})$. If $v = \bot$, then

$$\varepsilon(oracle)(e_o, \kappa_2(n,n), \bot).$$

This is a contradiction since $n = \kappa_P(pvso)$.

Thus, $\chi(pvso)(\mathtt{op2}(|O_2|, \mathtt{vr}, \mathtt{vr}), n, m) \neq \bot$. Then, Formula (3) can be simplified to

$$\chi(pvso)(rec(\mathtt{vr}), n, m) \neq \Diamond.$$

Finally, expanding χ results in $\chi(pvso)(e_f, n, m-1) \neq \Diamond$. This contradicts the minimality of m, completing the proof. $\square$

5   Related Work

Formalization of models of computation is not a novelty. In recent work, Foster and Smolka [4] formalized, in Coq, Rice's Theorem, which states that non-trivial semantic properties of procedures are undecidable. This theorem is a variant of Post's Theorem, which states that a class of procedures is decidable if it is recognizable, co-recognizable, and logically decidable, and the fact that the class of total procedures is not recognizable. This work was done for a model of a functional language presented as a weak call-by-value lambda-calculus in which β-reduction can be applied only if the redex is not below an abstraction and if the argument is an abstraction. Results are adaptable for call-by-value functional languages and Turing-complete models formalized in Coq. Larchey-Wendling [6] gave a formalization that "the Coq type $nat^k \rightarrow nat$ contains every recursive function of arity k which can be proved total in Coq." There, nat is the Coq type for Peano naturals. This is a class of functions between the primitive functions $\mathbb{N}^k \rightarrow \mathbb{N}$ and the partial recursive functions $\mathbb{N}^k \rightharpoonup \mathbb{N}$. Proving that the former class of functions is a set of terms of Coq type $\mathbb{N}^k \rightarrow \mathbb{N}$ is simple since Coq includes all required recursive schemes (basic functions plus composition and recursion). The paper advances into characterizing the type $\mathbb{N}^k \rightharpoonup \mathbb{N}$, which is not straightforward and is related to the class of terminating functions.

Highly relevant related papers include [9] in which Norrish presented a formalization in HOL4 of several results of computability theory using as models recursive functions and the λ-calculus. The mechanizations include proofs of the equivalence of the computational power of both models, undecidability of the halting problem, existence of universal machines, and Rice's Theorem in the λ-calculus. In addition, in [14] Xu et al. presented a formalization of computability theorems in Isabelle/HOL using as models Turing machines, abacus machines (a kind of register machines) and recursive functions. Formalized results include also undecidability of the halting problem and existence of universal Turing machines.

In contrast to those approaches, this paper deals with a computational model that is specified as a concrete functional programing language, namely PVS0. For this language, all the elements required in a classic-style proof of undecidability of termination, such as *Gödelization* of programs and inputs, are developed. Having a concrete programming language such as PVS0 enables the formalization and comparison of different termination analysis techniques for this language. In fact, the current formalization is part of a larger library that relates different termination criteria such as semantic termination, Turing termination [13], dependency pairs [1,2,15], and techniques based on the size change principle [5,7,11] such as calling context graphs [8] and matrix-weighted graphs [3].

6   Conclusion and Future Work

This paper presents the formalization of the undecidability of the halting problem for a simple functional language called PVS0. This formalization required the definition of several notions of termination, which are all proven to be equivalent. Since PVS0 generally allows for the encoding of non-computable functions through the use built-in operators, the undecidability proof is done on a restriction of PVS0 programs. First, the input type is restricted to natural numbers. Then, PartialRecursive and Computable programs are constructed using a layered approach where programs at level $n + 1$ can only depend on programs at level n or below. The existence of a surjective mapping from natural numbers to PVS0 programs constructed using this layered approach is formally verified. This mapping enables the definition of a Gödelization function for the type of programs PartialRecursive, which is crucial in the undecidability proof of the halting problem.

PVS0 is used to study termination and totality properties of PVS functions with the objective of automating the generation of measure functions in PVS recursive definitions. As part of this study, termination analysis techniques for PVS0 programs have been formalized and verified. Furthermore, strategies that automatically prove the equivalence between PVS recursive functions and PVS0 programs and that discharge termination TCCs of PVS recursive functions through their PVS0 counterpart have been developed. Future work includes extending the syntax of the PVS0 minimal language to support constructs such as let-in expressions and extending the proposed framework to support higher-order functions.

Finally, it is conjectured that the PVS0 language is Turing-complete. The proof of this property, which is work in progress, is not conceptually difficult but is technically involved. For simplicity, the PVS0 language only allows for the definition of one recursive function. Composition could be encoded using the lists of built-in operators. However, by definition, built-in operators are terminating. Therefore, to enable composition of (possibly) non-terminating programs, it is necessary to encode it in PVS0 itself using, among other things, the Gödelization of programs. This encoding increases the complexity of the proof of this conjecture.

References

1. Arts, T.: Termination by absence of infinite chains of dependency pairs. In: Kirchner, H. (ed.) CAAP 1996. LNCS, vol. 1059, pp. 196–210. Springer, Heidelberg (1996). https://doi.org/10.1007/3-540-61064-2_38
2. Arts, T., Giesl, J.: Termination of term rewriting using dependency pairs. Theor. Comput. Sci. **236**(1–2), 133–178 (2000)
3. Avelar, A.B.: Formalização da automação da terminação através de grafos com matrizes de medida. Ph.D. thesis, Universidade de Brasília, Departamento de Matemática, Brasília, Distrito Federal, Brasil (2015). In Portuguese
4. Forster, Y., Smolka, G.: Weak call-by-value lambda calculus as a model of computation in Coq. In: Ayala-Rincón, M., Muñoz, C.A. (eds.) ITP 2017. LNCS, vol. 10499, pp. 189–206. Springer, Cham (2017). https://doi.org/10.1007/978-3-319-66107-0_13
5. Krauss, A., Sternagel, C., Thiemann, R., Fuhs, C., Giesl, J.: Termination of Isabelle functions via termination of rewriting. In: van Eekelen, M., Geuvers, H., Schmaltz, J., Wiedijk, F. (eds.) ITP 2011. LNCS, vol. 6898, pp. 152–167. Springer, Heidelberg (2011). https://doi.org/10.1007/978-3-642-22863-6_13
6. Larchey-Wendling, D.: Typing total recursive functions in Coq. In: Ayala-Rincón, M., Muñoz, C.A. (eds.) ITP 2017. LNCS, vol. 10499, pp. 371–388. Springer, Cham (2017). https://doi.org/10.1007/978-3-319-66107-0_24
7. Lee, C.S., Jones, N.D., Ben-Amram, A.M.: The size-change principle for program termination. In: Conference Record of POPL 2001: The 28th ACM SIGPLAN-SIGACT Symposium on Principles of Programming Languages, pp. 81–92 (2001)
8. Manolios, P., Vroon, D.: Termination analysis with calling context graphs. In: Ball, T., Jones, R.B. (eds.) CAV 2006. LNCS, vol. 4144, pp. 401–414. Springer, Heidelberg (2006). https://doi.org/10.1007/11817963_36

9. Norrish, M.: Mechanised computability theory. In: van Eekelen, M., Geuvers, H., Schmaltz, J., Wiedijk, F. (eds.) ITP 2011. LNCS, vol. 6898, pp. 297–311. Springer, Heidelberg (2011). https://doi.org/10.1007/978-3-642-22863-6_22

10. Owre, S., Rushby, J.M., Shankar, N.: PVS: a prototype verification system. In: Kapur, D. (ed.) CADE 1992. LNCS, vol. 607, pp. 748–752. Springer, Heidelberg (1992). https://doi.org/10.1007/3-540-55602-8_217

11. Thiemann, R., Giesl, J.: Size-change termination for term rewriting. In: Nieuwenhuis, R. (ed.) RTA 2003. LNCS, vol. 2706, pp. 264–278. Springer, Heidelberg (2003). https://doi.org/10.1007/3-540-44881-0_19

12. Turing, A.M.: On computable numbers, with an application to the Entscheidungsproblem. Proc. London Math. Soc. **42**(1), 230–265 (1937)

13. Turing, A.M.: Checking a large routine. In: Campbell-Kelly, M. (ed.) The Early British Computer Conferences, pp. 70–72. MIT Press, Cambridge (1989)

14. Xu, J., Zhang, X., Urban, C.: Mechanising Turing machines and computability theory in Isabelle/HOL. In: Blazy, S., Paulin-Mohring, C., Pichardie, D. (eds.) ITP 2013. LNCS, vol. 7998, pp. 147–162. Springer, Heidelberg (2013). https://doi.org/10.1007/978-3-642-39634-2_13

15. Yamada, A., Sternagel, C., Thiemann, R., Kusakari, K.: AC dependency pairs revisited. In: Proceedings 25th EACSL Annual Conference on Computer Science Logic, CSL 2016, LIPIcs, vol. 62, pp. 8:1–8:16. Schloss Dagstuhl - Leibniz-Zentrum fuer Informatik (2016)

Handling Verb Phrase Anaphora
with Dependent Types and Events

Daniyar Itegulov[1,2]([✉]) and Ekaterina Lebedeva[1]

[1] Australian National University, Canberra, Australia
{daniyar.itegulov,ekaterina.lebedeva}@anu.edu.au
[2] ITMO University, St. Petersburg, Russia

Abstract. This paper studies how dependent typed events can be used to treat verb phrase anaphora. We introduce a framework that extends Dependent Type Semantics (DTS) with a new atomic type for neo-Davidsonian events and an extended @-operator that can return new events that share properties of events referenced by verb phrase anaphora.

The proposed framework, along with illustrative examples of its use, are presented after a brief overview of the necessary background and of the major challenges posed by verb phrase anaphora.

1 Introduction

Davidson [3] observed that some verbs can imply the existence of an "action". For example, the sentence "John eats." represents an action of eating. This action can be anaphorically referred from a following sentence (e.g.: "The food is yummy."). Therefore, it is desirable for framework of natural language semantics to encompass the notion of action and a mechanism for action reference. Davidson proposed to equip interpretations of verbs with an additional argument for events. Thus, the sentence "John eats." is interpreted according to Davidson as $\exists e.eats(e, \mathbf{j})$, instead of $eats(\mathbf{j})$.

Parsons [9] and Taylor [14] argued that the approach of event semantics captures the notion of adverbs better than approaches based on higher-order predicates, such as [15], and is easier to work with. For example, adverbial modifiers usually affect only the event and not the entity, as the following example illustrates:

(1)
    a. John buttered the toast slowly, deliberately, in the bathroom, with a knife, at midnight.
    b. $\exists e.butter(e, \mathbf{j}, \mathbf{t}) \wedge slowly(e) \wedge deliberately(e) \wedge in(e, \mathbf{b}) \wedge \exists k.with(e, k) \wedge at(e, \mathbf{m})$

D. Itegulov—This work is supported by ANU Ph.D. (International) Scholarship and HDR Fee Remission Merit Scholarship provided by the Australian National University.

E. Lebedeva—We thank Bruno Woltzenlogel Paleo and Florrie Verity for numerous helpful discussions and the anonymous reviewers for the constructive feedback.

© Springer-Verlag GmbH Germany, part of Springer Nature 2018
L. S. Moss et al. (Eds.): WoLLIC 2018, LNCS 10944, pp. 210–222, 2018.
https://doi.org/10.1007/978-3-662-57669-4_12

Sentence (1a) contains adverbs that modify the event of buttering the toast. The corresponding interpretation in events semantics is shown in (1b).

Additionally to adverbial modifiers, Parsons [9] described two more reasons for introducing events as a new atomic type: perception verbs and reference to events.

Parsons [9], furthermore, proposed a framework based on Davidson's event theory, called neo-Davidsonian event semantics, that extends it as follows:

- event participants are introduced via thematic roles
- state verbs, in addition to action verbs, are handled with an abstract variable
- two concepts, event holding and event culmination, are added
- events are decomposed into subevents.

The differences between Davidsonian and neo-Davidsonian approaches can be seen by comparing interpretations (1b) and (2) of Sentence (1a).

(2) $\exists e.butter(e) \wedge agent(e, \mathbf{j}) \wedge patient(e, \mathbf{t}) \wedge slowly(e) \wedge deliberately(e) \wedge in(e, \mathbf{b}) \wedge$
$\exists k.with(e, k) \wedge at(e, \mathbf{m})$

This paper proposes a framework for solving verb phrase anaphora (also known as verb phrase ellipsis) based on the neo-Davidsonian event semantics and on dependent types; and to adapt the existing techniques of handling the propositional anaphora to Dependent Type Semantics (DTS) framework. Dependent types are already used to express pronominal anaphora in [1].

In Sect. 2, we briefly recall Dependent Type Semantics, which is a theoretical foundation for our framework. In Sect. 3, we discuss major challenges of interpreting verb phrase anaphora. The main contribution of this paper is presented in Sect. 4, which describes an extension of the Dependent Type Semantics, and in Sect. 5, which discusses an application of subtyping in the proposed framework.

2   Recalling Dependent Type Semantics

Dynamic type semantics (DTS) proposed by Bekki in [1] is a framework of discourse semantics based on dependent type theory (Martin-Löf [7]). DTS follows the constructive proof-theoretic approach to semantics established by Sundholm [13], who introduced Sundholmian semantics, and by Ranta [12], who introduced Type Theoretical Grammar.

Definition 1 (Dependent function). *For any* $(s_1, s_2) \in \{(type, type), (type, kind), (kind, type), (kind, kind)\}, s \in \{type, kind\}:$

$$\frac{\begin{array}{c} x:A \\ \vdots \\ A:s_1 \qquad B:s_2 \end{array}}{(x:A) \to B:s_2} \qquad \frac{\begin{array}{c} x:A \\ \vdots \\ A:s \qquad M:B \end{array}}{\lambda x.M:(x:A) \to B} \qquad \frac{M:(x:A) \to B \qquad N:A}{MN:B[N/x]}$$

Definition 2 (Dependent pair). *For any* $(s_1, s_2) \in \{(type, type), (type,$
kind), (kind, kind)\}:

$$\frac{A : s_1 \qquad B : s_2}{\begin{bmatrix} x : A \\ B \end{bmatrix} : s_2} \qquad \frac{\begin{matrix} x : A \\ \vdots \\ M : A \qquad N : B[M/x] \end{matrix}}{(M, N) : \begin{bmatrix} x : A \\ B \end{bmatrix}} \qquad \frac{M : \begin{bmatrix} x : A \\ B \end{bmatrix}}{\pi_1 M : A} \qquad \frac{M : \begin{bmatrix} x : A \\ B \end{bmatrix}}{\pi_2 M : B[\pi_1 M / x]}$$

DTS employs two kinds of dependent types (in addition to simply-typed lambda calculus): *dependent pair type* or Σ-type (notation $(x : A) \to B(x)$) and *dependent function type* or Π-type (notation $(x : A) \times B(x)$). A dependent pair is a generalization of an ordinary pair. By Curry-Howard correspondence between types and propositions, the type $(x : A) \times B(x)$ corresponds to an existentially quantified formula $\exists x^A.B$ and to an ordinary conjunction $A \wedge B$ when $x \notin fv(B)$.[1] A dependent function is a generalization of an ordinary function; the type $(x : A) \to B(x)$ corresponds to $\forall x^A.B$ and to $A \to B$ when $x \notin fv(B)$. Formal definitions are given through inference rules in Definitions 1 and 2.

A comparison between the traditional notation for dependent types and the notation used in DTS can be seen in Fig. 1.

	Π-type	Σ-type
Initial notation	$(\Pi x : A)B(x)$	$(\Sigma x : A)B(x)$
DTS notation	$(x : A) \to B(x)$	$(x : A) \times B(x), \begin{bmatrix} x : A \\ B(x) \end{bmatrix}$
Initial notation, $x \notin fv(B)$	$A \to B$	$A \wedge B$
DTS notation, $x \notin fv(B)$	$A \to B$	$\begin{bmatrix} A \\ B \end{bmatrix}$

Fig. 1. Notation in DTS

The main atomic type in DTS is **entity**, which represents all entities in a discourse. With the employment of dependent type constructors, the entity type can be combined with additional properties. For example, $(\Sigma u : (\Sigma x : \textbf{entity}) \times man(x)) \times enter(\pi_1(u))$ is a valid DTS interpretation of "A man entered". Therefore, in contrast to traditional approaches to semantical interpretation where entities are not distinguished by their types, each entity has its own type in DTS.

In the traditional Montague model-theoretic semantics [8], a proposition denotes a truth value (often defined as an o-type). However, DTS does not follow this convention and instead the meaning of a sentence is represented by

---

[1] $fv(x)$ denotes all free variables in x.

a type. Types in DTS are defined by the inference rules, as shown in Definitions 1 and 2. The rules specify how a dependent type (as a proposition) can be proved under a given context. Thus, the meaning of a sentence in proof-theoretic semantics lies in its *verification condition* similar to the philosophy of language by Dummett [4,5] and Prawitz [10].

To handle anaphora resolution, DTS distinguishes two kinds of propositions: static and dynamic. A static proposition P is called true if it is inhabited, i.e. there exists a term of type P. A dynamic proposition is a function mapping context-proof (a static proposition that is an interpretation of the previous discourse) to a static proposition.

In order to represent anaphoric references and presupposition triggers, Bekki introduced a @-operator. The operator takes the left context of dynamic propositions in which it occurs. For example, Sentence (3a) can be interpreted as (3b) in DTS. The @-operators in (3b) take a context as an argument and try to find a female (due to the interpretation of the word "herself") entity in the context passed to them.

@-operators have a single introduction rule @F (Fig. 2):

$$\frac{A : \textbf{type} \qquad A \; true}{(@_i : A) : A}$$

Fig. 2. @F introduction rule

Different @-operators can take contexts of different types. Therefore, they can be of different types and are distinguished with a numerical subscript. The full type of the @$_i$-operator can look like this: $@_i : \gamma_i \to \textbf{entity}$.

(3)
   a. She loves herself.
   b. $\lambda c.loves(\pi_1(@_1 c : \begin{bmatrix} x : \textbf{entity} \\ female(x) \end{bmatrix}), \pi_1(@_2 c : \begin{bmatrix} x : \textbf{entity} \\ female(x) \end{bmatrix}))$

The *felicity condition* states that in order to be felicitous, an instance of a syntactic category S (i.e. a sentence) has to be of sort $\gamma \to \textbf{type}$ for some new variable γ. To check the felicity condition of a sentence, a type checking algorithm must be evoked. Consider that the interpretation of the sentence "Mary lives in London" is passed to (3b) as the left context. The result is shown in (4).

(4)
$$\begin{bmatrix} v : \begin{bmatrix} u : \begin{bmatrix} \textbf{m} : \textbf{entity} \\ female(\textbf{m}) \end{bmatrix} \\ lives\_in(\pi_1(u), \textbf{l}) \end{bmatrix} \\ \lambda c.loves(\pi_1(@_1 c : \begin{bmatrix} x : \textbf{entity} \\ female(x) \end{bmatrix}), \pi_1(@_2 c : \begin{bmatrix} x : \textbf{entity} \\ female(x) \end{bmatrix})) \end{bmatrix}$$

However, before type-checking this sentence, @-operators should be built using the @F introduction rule. The rule requires the existence of a proof term inhabiting the type of the @-operator. In case of Sentence (4), both @$_1$ and @$_2$

operators are inhabited by the proof term $\lambda c.\pi_1(c)$. In particular, Mary (the first element of the dependent pair in the context) is a valid entity having the property of being female for both pronouns "she" and "herself". Hence, the anaphora resolution process involves proof search and can be done using theorem provers.

3 Verb Phrase Anaphora

Verb phrase anaphora [11] are anaphora with an intentional omission of part of a full-fledged verb phrase when the ellipsed part can be implicitly derived from the context. For example, verb phrase anaphora can be observed in (5a) and (5b):

(5)
   a. John left before Mary did.
   b. John left. Mary did too.

In (5a), the word "did" refers to an action John did before Mary. In (5b), the "did too" clause refers to an action which John and Mary both did. These sentences can be interpreted in event semantics as the following logical expressions:

(6)
   a. $\exists e.agent(e,\mathbf{j}) \wedge left(e) \wedge \exists e'.agent(e',\mathbf{m}) \wedge left(e') \wedge before(e,e')$
   b. $\exists e.agent(e,\mathbf{j}) \wedge left(e) \wedge \neg\exists e'.agent(e',\mathbf{m}) \wedge left(e')$

Furthermore, an anaphoric verb phrase can "inherit" some properties from its referent. Consider Example (7a) where "did too" not only refers to the event of eating performed by John, but also to properties such as "quietly" and "last night". Expression (7b) is the interpretation of this sentence in event semantics.

(7)
   a. John quietly ate the cake last night. Mary did too.
   b. $\exists e.(agent(e,\mathbf{j}) \wedge patient(e,\mathbf{c}) \wedge ate(e) \wedge quietly(e) \wedge at(e,\mathbf{ln})) \wedge$
      $\exists e'.(agent(e',\mathbf{m}) \wedge patient(e',\mathbf{c}) \wedge ate(e') \wedge quietly(e') \wedge at(e',\mathbf{ln}))$

A verb phrase anaphor may have an additional property that can ease the choice of a correct anaphoric referent-event from the context. This phenomenon is exemplified in (8a), where it is explicit that "too" refers to an action connected with eating.

(8) a. John ate pasta and did not feel well. Mary ate too, but nothing happened to her.

An ambiguity between strict and sloppy identity readings of verb phrase anaphora described by Prüst [11] is another intriguing phenomenon. Example (9) illustrates this:

(9) a. John likes his hat. Fred does too.
     b. $\exists x.hat(x) \wedge owner(x,\mathbf{j}) \wedge \exists e.like(e) \wedge agent(e,\mathbf{j}) \wedge patient(e,x) \wedge$
      $\exists e'.like(e') \wedge agent(e',\mathbf{f}) \wedge patient(e',x)$

c. $\exists x.hat(x) \wedge owner(x, \mathbf{j}) \wedge \exists e.like(e) \wedge agent(e, \mathbf{j}) \wedge patient(e, x) \wedge$
$\exists y.hat(y) \wedge owner(y, \mathbf{f}) \wedge \exists e'.like(e') \wedge agent(e', \mathbf{f}) \wedge patient(e', y)$

The anaphoric clause in the second sentence of (9a) can be interpreted as "Fred likes John's hat" (the sloppy identity interpretation (9b)) or as "Fred likes Fred's hat" (the strict interpretation (9c)). A desirable framework should be able to provide both interpretations.

4   Events with Dependent Types

To tackle phenomena discussed in Sect. 3, we propose to extend DTS with a new atomic type **event** for interpreting events. Then, given its left context c, DTS's @-operator can be employed for retrieving a variable of type **event** analogously to its original use for retrieving a referent of type **entity**.

As was shown by Parsons [9], event semantics can be employed to represent propositional anaphora. An example of propositional anaphora is shown in (10a), where "this" refers to the whole proposition expressed in the first sentence. Formula (10b) is an interpretation of (10a).

(10)
   a. John loved Mary. But Mary did not believe this.
   b. $\exists e.agent(e, \mathbf{j}) \wedge patient(e, \mathbf{m}) \wedge loved(e) \wedge \exists e'.believed(e') \wedge agent(e', \mathbf{m}) \wedge$
     $patient(e', e)$

Dependent typed events allow us to handle more complex types of propositional anaphora. Similar to entities, events can have various properties provided by their description. Assume the following three sentences appear in the same discourse, possibly remotely from each other, but with preservation of the order:

(11)
   a. Canberra was hit by a flood on Sunday.
   b. The fair was held in London.
   c. What happened in Canberra is surprising.

Here the anaphoric clause in Sentence (11c) refers to an event discussed earlier. There are however (at least) two potential events for the reference: one given by (11a) and another given by (11b). Since the anaphoric clause in (11c) specifies that it refers to an event happened in Canberra, the anaphor disambiguates to the event in (11a).

The interpretation of verb phrase anaphora is more challenging, however, than the interpretation of propositional anaphora: an anaphoric clause in a verb phrase usually talks about a new event that inherits properties of another event. For example, "John left. Bob did too." conveys two events: one is about John leaving and the second one is about Bob leaving. In cases of pronominal and propositional anaphora, however, there is just a reference to an entity or an event in the context. For example in "John walks. He is slow.", pronoun "he" in the second sentence just refers to the entity "John" from the first sentence.

To handle verb phrase anaphora correctly, it is not enough to just fetch a referenced variable from the left context; instead a new variable of type **event** should be introduced. This new variable *copies* properties from the referred event. Furthermore, the agent of the referred event should be changed to the current agent in the new event. This can be seen in interpretation (7b) of (7a), where the agent John is replaced with Mary.

Although $@_i$-operator has type $\gamma_i \rightarrow$ **entity** in DTS for handling pronominal anaphora, according to DTS syntax for raw terms the operator can be of any type. We therefore suggest a new type of @-operator that guarantees that the returned event has a proper agent, necessary for interpreting verb phrase anaphora:

(12) $@_i : (c : \gamma_i) \rightarrow (x : \textbf{entity}) \rightarrow \begin{bmatrix} e : \textbf{event} \\ agent(e, x) \end{bmatrix}$

Formula (13a) is an interpretation of discourse (5b). The $@_1$-operator in (13b) is applied to its left context c (of type γ_0) and an entity, and returns a *new* event of type **event** with the same properties (apart from the agent property) of the referenced event. Crucially, the event returned by $@_1$-operator in (13b) is not an event that was in the context previously. It is a new event with the same properties (e.g. location, time) as a referenced event from the context, but with a replaced agent.

(13)

a. $\lambda c. \begin{bmatrix} \begin{bmatrix} e : \textbf{event} \\ left(e) \\ agent(e, \mathbf{j}) \end{bmatrix} \end{bmatrix}$

b. $\lambda c.(@_1 c : (x : \textbf{entity}) \rightarrow \begin{bmatrix} e : \textbf{event} \\ agent(e, x) \end{bmatrix})(\mathbf{m})$

Note that the entity accepted by the @-operator defined in (12) is the agent in the new event. For instance, in Example (7a), the interpretation of "did too" using (12) would have the agent "John" of the referenced event replaced by "Mary", but the patient (i.e. the cake) would remain. On the other hand, there exist cases of verb phrase anaphora where the patient in the referenced event should be replaced. This usually depends on the voice (active or passive) of an anaphoric clause, as can be seen from examples in (14).

(14)

a. Mary is loved by John. So is Ann.

b. John loves Mary. So does Bob.

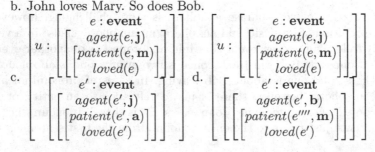

The first sentences in (14a) and (14b) have the same semantics and hence the interpretations given to them in (14c) and (14d) coincide. However, despite the fact that both second sentences are written in the same voice as their first sentences, the second sentences are interpreted differently. Naturally, the second sentence in (14a) means "Ann is loved by John", while the second sentence in (14b) means "Bob loves Mary". Note that they have replaced different participants of the first sentences: in (14a) Mary (patient) was replaced by Ann and in (14b) John (agent) was replaced by Bob.

Furthermore, the interpretation of a sentence may require both the agent and the patient to be replaced, as for example in the sloppy reading of (9c). These possible cases of anaphora resolution can be tackled with the judgements (16) assuming they occur in a global context $\mathcal{K}$.

Another important notion in DTS is the felicity condition. The anaphora resolution for $@_i$ operator is launched by type checking of the following judgement: $\mathcal{K}, \gamma_i : type \vdash @_i : \gamma_i \rightarrow type$. It means that the semantical interpretation of a sentence must be of the sort **type** assuming that the left context is of type γ_i. A requirement of a success of the launching the type checker is called *felicity condition*.

In order to preserve the original DTS invariants, we should show how the felicity condition is being fulfilled in the extended DTS. An example of a felicity-judgement generated by verb phrase anaphora is shown in example (15). It is different from the felicity condition from original DTS notion since the new @-operator has a new type as shown in (12).

(15) $\mathcal{K}, \gamma_i : type \vdash @_i : \gamma_i \rightarrow (x : \textbf{entity}) \rightarrow \begin{bmatrix} e : \textbf{event} \\ agent(e, x) \end{bmatrix}$

Assume that the global context $\mathcal{K}$ contains the judgements from (16). Then one should be able to type check judgements generated by verb phrase anaphora.

(16)

    a. $replaceA : (p : \textbf{entity} \rightarrow (e : \textbf{event}) \rightarrow \textbf{type}) \rightarrow$
        $(original : \textbf{entity}) \rightarrow (new : \textbf{entity}) \rightarrow$
        $(u : \begin{bmatrix} e' : \textbf{event} \\ p\ original\ e' \end{bmatrix}) \rightarrow (v : \begin{bmatrix} e'' : \textbf{event} \\ p\ new\ e'' \end{bmatrix})$

    b. $replaceP : (p : \textbf{entity} \rightarrow (e : \textbf{event}) \rightarrow \textbf{type}) \rightarrow$
        $(original : \textbf{entity}) \rightarrow (new : \textbf{entity}) \rightarrow$
        $(u : \begin{bmatrix} e' : \textbf{event} \\ p\ original\ e' \end{bmatrix}) \rightarrow (v : \begin{bmatrix} e'' : \textbf{event} \\ p\ new\ e'' \end{bmatrix})$

    c. $replaceAP : (p : \textbf{entity} \rightarrow \textbf{entity} \rightarrow (e : \textbf{event}) \rightarrow \textbf{type}) \rightarrow$
        $(oagent : \textbf{entity}) \rightarrow (nagent : \textbf{entity}) \rightarrow$
        $(opatient : \textbf{entity}) \rightarrow (npatient : \textbf{entity}) \rightarrow$
        $(u : \begin{bmatrix} e' : \textbf{event} \\ p\ oagent\ opatient\ e' \end{bmatrix}) \rightarrow (v : \begin{bmatrix} e'' : \textbf{event} \\ p\ nagent\ npatient\ e'' \end{bmatrix})$

    d. $j : \textbf{entity}$

Functions *replaceA*, *replaceP*, *replaceAP* construct a new event v from an existing event u. To express the inheritance of properties and the change of the agent in *replaceA* (or patient in *replaceP*), properties are expressed as a function that accepts two arguments: an agent-entity (or patient-entity in *replaceP*) and an event; and returns a logical expression describing the event using the entity. Function *replaceAP* accounts for cases where both an agent and a patient are replaced.

We can now construct term $@_1$ of type (12), to fulfill the felicity condition of form (15), as shown in (17):

(17) $\mathcal{K}, \gamma_0 : type \vdash @_1 : \gamma_0 \rightarrow (x : \mathbf{entity}) \rightarrow \begin{bmatrix} e : \mathbf{event} \\ agent(e, x) \end{bmatrix} =$

$$\lambda c. \lambda x. replaceA \ (\lambda y. \lambda e. \begin{bmatrix} left(e) \\ agent(e, y) \end{bmatrix}) \ \mathbf{j} \ x \ \pi_1 \pi_2(c)$$

A substitution of $@_1$ in (13b) with its term defined in (17) leads to the following semantical interpretation:

(18) $\begin{bmatrix} \begin{bmatrix} e'' : \mathbf{event} \\ left(e'') \\ agent(e'', \mathbf{m}) \end{bmatrix} \end{bmatrix}$

Since anaphora in DTS are resolved using the type checking procedure, verb phrase anaphora, just like pronominal anaphora, can be resolved in various ways. A type checking algorithm can find different terms which conform to the specified (by felicity condition) type. For example, in order to handle the ambiguity between strict and sloppy identity readings, which were discussed in Example (9), our framework can provide both possible interpretations for Sentence (9a). Term (19) shows a generic interpretation of (9a) in the proposed framework.

(19) $\lambda c.$ $\begin{bmatrix} u : \begin{bmatrix} v : \begin{bmatrix} x : \mathbf{entity} \\ hat(x) \\ owner(x, \mathbf{j}) \end{bmatrix} \\ \begin{bmatrix} e : \mathbf{event} \\ like(e) \\ agent(e, \mathbf{j}) \\ patient(e, \pi_1(v)) \end{bmatrix} \end{bmatrix} \\ @_1(c, u)\mathbf{f} : \begin{bmatrix} e' : \mathbf{event} \\ agent(e', \mathbf{f}) \end{bmatrix} \end{bmatrix}$

In (19), "Fred does too." is interpreted as the term $@_1(c, u)\mathbf{f}$, where u stands for the interpretation of the preceding sentence "John likes his hat.". Recall from Example (9) that the latter sentence has an ambiguous meaning. (20) defines two alternative terms for $@_0$, one for each of the possible meanings. Note that the type of these terms for $@_0$ conforms with the felicity condition.

(20)

a. $\mathcal{K} \vdash @_1 : \gamma_0 \rightarrow (x : \textbf{entity}) \rightarrow \begin{bmatrix} e' : \textbf{event} \\ agent(e', x) \end{bmatrix} =$

$$\lambda c.\lambda f. replaceA \ (\lambda y.\lambda e. \begin{bmatrix} like(e) \\ \begin{bmatrix} agent(e, y) \\ patient(e, x) \end{bmatrix} \end{bmatrix}) \ \textbf{j} \ f \ \pi_1 \pi_2 \pi_2(c)$$

b. $\mathcal{K} \vdash @_1 : \gamma_0 \rightarrow (x : \textbf{entity}) \rightarrow \begin{bmatrix} e' : \textbf{event} \\ agent(e', x) \end{bmatrix} =$

$$\textbf{let} \ p = \lambda y.\lambda z.\lambda e. \begin{bmatrix} like(e) \\ \begin{bmatrix} agent(e, y) \\ patient(e, z) \end{bmatrix} \end{bmatrix}$$

$$\textbf{in} \ \lambda c.\lambda f. \begin{bmatrix} u : \begin{bmatrix} y : \textbf{entity} \\ hat(y) \wedge owner(y, f) \end{bmatrix} \\ replaceAP \ p \ \textbf{j} \ f \ \pi_1 \pi_1 \pi_2(c) \ \pi_1(u) \ \pi_1 \pi_2 \pi_2(c) \end{bmatrix}$$

Both terms are valid substitutions for $@_1$-operator in (19) and they represent strict and sloppy anaphora readings respectively. In (20b) let-in structure is used only as a syntactical sugar for readability and is not actually a part of DTS term syntax.

In line with the original approach of Bekki [1], the verb phrase anaphora resolution for $@$-operator involves proof search and can be done using a theorem prover.

5   Subtyping

The equation in (17) (i.e. an anaphora resolution solution: a proof of existence of a term with the required type under the global context $\mathcal{K}$) is not sound: the type of the right side of the equation is

$$\gamma_0 \rightarrow (x : \textbf{entity}) \rightarrow \begin{bmatrix} e'' : \textbf{entity} \\ \begin{bmatrix} left(e'') \\ agent(e'', x) \end{bmatrix} \end{bmatrix}$$

while the type required by the left side is

$$\gamma_0 \rightarrow (x : \textbf{entity}) \rightarrow \begin{bmatrix} e : \textbf{entity} \\ agent(e, x) \end{bmatrix}$$

The former type is more specific than the latter type because it has the additional property "left".

This is not a problem, as events have a natural subtyping relationship between them. As described by Luo and Soloviev in [6], an event whose agent is a and patient is p, is an event with agent a. Despite a different theory underneath, the techniques described there can be reused for subtyping events in DTS. This leads to the following subtyping relations in event semantics:

(21)

a. $Evt_{AP}(a,p) <: Evt_A(a) <: Event \longleftrightarrow$

$$\begin{bmatrix} e : \textbf{event} \\ \begin{bmatrix} agent(e,a) \\ patient(e,p) \end{bmatrix} \end{bmatrix} <: \begin{bmatrix} e : \textbf{event} \\ agent(e,a) \end{bmatrix} <: \begin{bmatrix} e : \textbf{event} \\ () \end{bmatrix}$$

b. $Evt_{AP}(a,p) <: Evt_P(p) <: Event \longleftrightarrow$

$$\begin{bmatrix} e : \textbf{event} \\ \begin{bmatrix} agent(e,a) \\ patient(e,p) \end{bmatrix} \end{bmatrix} <: \begin{bmatrix} e : \textbf{event} \\ patient(e,p) \end{bmatrix} <: \begin{bmatrix} e : \textbf{event} \\ () \end{bmatrix}$$

Subtyping relations of events can also depend on other properties (e.g. a loud event performed by John is also an event performed by John). We employ Luo and Soloviev's notation to define a new type $Event_{NA}(n,a)$, which is the type of events with agent a and nature n. Nature is a main predicate for each event in neo-Davidsonian semantics (e.g. "left(e)" for an event of leaving, "ate(e)" for an event of eating).

The following transformation shows how the dependent event types in DTS notation from (17) can be converted into dependent event types in the notation of Luo and Soloviev:

(22)

a. $$\begin{bmatrix} e'' : \textbf{event} \\ \begin{bmatrix} left(e'') \\ agent(e'',x) \end{bmatrix} \end{bmatrix} \longleftrightarrow e'' : Event_{DA}(left,x)$$

b. $$\begin{bmatrix} e : \textbf{entity} \\ agent(e,x) \end{bmatrix} \longleftrightarrow e : Event_A(x)$$

A subtyping relationship between these types can be constructed (assuming the appropriate subtyping rules have been added along with type $Event_{DA}(d,a)$).

$$\frac{left : Description \qquad x : Agent}{Event_{DA}(left,x) <: Event_A(x)}$$

The discussed subtyping relationship allows us to obtain (17).

6   Comparison with Previous Approaches

The approach presented here shares the goal of interpreting elliptical constructions, verb phrase ellipses in particular, with the work of Dalrymple et al. [2]. However, there are crucial conceptual differences.

The method of Dalrymple, Shieber and Pereira relies on the parallel structure of sentences involved in verb phrase anaphora (e.g.: "John loves golf" and "Bob does too" are parallel in the sense that the second sentence can be used as a verb phrase anaphoric clause referring to the first sentence). They rely on a black-box mechanism to determine the parallel structures.

Here we rely on a theorem prover to find a proof term for @-operators. In this way we leverage the advances in the extensively researched field of automated theorem proving.

Our approach, furthermore, avoids some problems described in [2]. For example, it supports semantic parallelism, i.e. parallelism between a "logical subject" and an anaphoric clause. Consider Sentence (23), which is a slightly simplified version of sentence (59) from [2].

(23) The material can be presented in an accessible fashion, and often I do.

We interpret the subject in the first part of the sentence as an existentially quantified variable of type **entity**. This variable can be accessed from the second sentence. In other words, our approach introduces a logical subject that can be used for interpreting an anaphoric clause that follows.

7   Conclusion

This paper introduces dependent event types for resolving verb phrase anaphora with DTS as the underlying framework. To tackle verb phrase anaphora, we extend DTS's @-operator, which was originally introduced for handling pronominal anaphora. The paper also addresses strict and sloppy readings of verb phrase anaphora and shows that each of them can be achieved solely by manipulating the interpretation of the @-operator. The previous approaches to handling the propositional anaphora were also adapted to DTS framework.

Techniques described in this paper could be applied to handle other cases of anaphora, such as adjectival anaphora, modal and "do so" anaphora. Another interesting topic would be to study specific behaviours of various thematic roles, such as experiencer, theme and source.

References

1. Bekki, D.: Representing anaphora with dependent types. In: Asher, N., Soloviev, S. (eds.) LACL 2014. LNCS, vol. 8535, pp. 14–29. Springer, Heidelberg (2014). https://doi.org/10.1007/978-3-662-43742-1_2
2. Dalrymple, M., Shieber, S.M., Pereira, F.C.N.: Ellipsis and higher-order unification. Linguist. Philos. **14**(4), 399–452 (1991)
3. Davidson, D.: The logical form of action sentences. In: Rescher, N. (ed.) The Logic of Decision and Action. University of Pittsburgh Press, Pittsburgh (1967)
4. Dummett, M.: What is a theory of meaning? (II). In: Evans, G., McDowell, J. (eds.) Truth and Meaning: Essays in Semantics. Clarendon Press, Oxford (1976)
5. Dummett, M.A.E.: What is a theory of meaning? In: Guttenplan, S. (ed.) Mind and Language. Oxford University Press, Oxford (1975)
6. Luo, Z., Soloviev, S.: Dependent event types. In: Kennedy, J., de Queiroz, R.J.G.B. (eds.) WoLLIC 2017. LNCS, vol. 10388, pp. 216–228. Springer, Heidelberg (2017). https://doi.org/10.1007/978-3-662-55386-2_15
7. Martin-Löf, P., Sambin, G.: Intuitionistic Type Theory. Studies in Proof Theory. Bibliopolis, Berkeley (1984)

8. Montague, R.: Formal Philosophy; Selected Papers of Richard Montague. Yale University Press, New Haven (1974)
9. Parsons, T.: Events in the Semantics of English: A Study in Subatomic Semantics. MIT Press, Cambridge (1990)
10. Prawitz, D.: Intuitionistic logic: a philosophical challenge. In: Von Wright, G.H. (ed.) Logic and Philosophy/Logique et Philosophie, vol. 5, pp. 1–10. Springer, Dordrecht (1980). https://doi.org/10.1007/978-94-009-8820-0_1
11. Prüst, H., Scha, R., van den Berg, M.: Discourse grammar and verb phrase anaphora. Linguist. Philos. **17**(3), 261–327 (1994)
12. Ranta, A.: Type-Theoretical Grammar. Oxford University Press, Oxford (1994)
13. Sundholm, G.: Proof theory and meaning. In: Gabbay, D., Guenthner, F. (eds.) Handbook of Philosophical Logic: Volume III: Alternatives in Classical Logic. SYLI, vol. 166, pp. 471–506. Springer, Dordrecht (1986). https://doi.org/10.1007/978-94-009-5203-4_8
14. Taylor, B.: Modes of occurence, verbs, adverbs and events. Revue Philosophique de la France Et de l'Etranger **176**(3), 406–407 (1986)
15. Verkuyl, H.J.: On the Compositional Nature of the Aspects. D. Reidel Publishing Company, Dordrecht (1972)

Parameterized Complexity for Uniform Operators on Multidimensional Analytic Functions and ODE Solving

Akitoshi Kawamura[1], Florian Steinberg[2], and Holger Thies[3(✉)]

[1] Kyushu University, Fukuoka, Japan
[2] Inria, Rocquencourt, France
[3] University of Tokyo, Tokyo, Japan
info@holgerthies.com

Abstract. Real complexity theory is a resource-bounded refinement of computable analysis and provides a realistic notion of running time of computations over real numbers, sequences, and functions by relying on Turing machines to handle approximations of arbitrary but guaranteed absolute error. Classical results in real complexity show that important numerical operators can map polynomial time computable functions to functions that are hard for some higher complexity class like NP or #P. Restricted to analytic functions, however, those operators map polynomial time computable functions again to polynomial time computable functions. Recent work by Kawamura, Müller, Rösnick and Ziegler discusses how to extend this to uniform algorithms on one-dimensional analytic functions over simple compact domains using second-order and parameterized complexity. In this paper, we extend some of their results to the case of multidimensional analytic functions. We further use this to show that the operator mapping an analytic ordinary differential equations to its solution is computable in parameterized polynomial time. Finally, we discuss how the theory can be used as a basis for verified exact numerical computation with analytic functions and provide a prototypical implementation in the iRRAM C++ framework for exact real arithmetic.

1 Introduction

Computable analysis gives a formal model for reliable computations involving real numbers and other continuous structures. Its origins reach back to Alan Turing and computability theory itself [22]. Later it was extended by complexity considerations [14,15], also known as real complexity theory. The main idea is that real numbers are encoded as functions that give approximations up to any finite precision. Computing a real function f means to approximate the result $f(x)$ up to any desired precision while having acces to arbitrary exact approximations to x. Over compact sets, complexity can be measured as the resources needed to produce approximations to any return value in dependence on their accuracy.

© Springer-Verlag GmbH Germany, part of Springer Nature 2018
L. S. Moss et al. (Eds.): WoLLIC 2018, LNCS 10944, pp. 223–236, 2018.
https://doi.org/10.1007/978-3-662-57669-4_13

Typical problems over real functions involve computing operators such as maximization, integration or derivatives. However, classical results in real complexity theory imply that computing some of the most common operators can already be computationally hard. For example, parametric maximization relates to the P vs. NP problem [15] and integration to the stronger FP vs. #P problem in the sense that the complexity classes are equal if those operators map polynomial time computable functions to polynomial time computable functions [6]. The statement remains true even when restricted to smooth functions, implying that finding efficient algorithms even for basic operators is most likely impossible.

A possible solution is to look at more restrictive classes of functions. Indeed, it is known that the situation improves drastically for analytic functions: Many important operations are known to preserve polynomial time computability in this case [21]. However, these results are typically stated in a non-uniform way, that is, they are of the form "If f is a polynomial time computable function then the function $G(f)$ the operator G returns is also a polynomial time computable function". While for hardness results a non-uniform formulation is particularly strong, for the opposite goal to show that a problem is feasible such a result is not satisfying as the algorithm for G (and therefore also its time complexity) depends in some unspecified ways on the function f.

A *uniform* algorithm on the other hand transforms a description of f to a description of $G(f)$ and therefore requires a full specification of what information about f is needed to compute $G(f)$. The notion of computing with such descriptions and the underlying complexity can be made formal in the framework of representations. We introduce some basic concepts in Sect. 2.

In recent work, Kawamura et al. [12] discuss how to compute uniformly with one-dimensional (complex) analytic functions and analyze the complexity of some important operators in terms of natural discrete parameters of the function. For many applications, however, being able to manipulate also multi-dimensional functions is required. We therefore extend some of their notions to the multidimensional case and show that similar complexity bounds still hold. We follow their approach to first analyze computations with single power series (Sect. 3.1) and then extend to functions analytic on a simple compact subset of the reals (Sect. 3.2).

The main motivation for the multidimensional extension is the problem of solving initial value problems (IVPs) for ordinary differential equations of the form $\dot{y}(t) = f(y(t))$, $y(0) = y_0$ for $f : \mathbb{R}^d \to \mathbb{R}^d$ and $y_0 \in \mathbb{R}^d$ for some $d \geq 2$. It has been proved that, unless P = PSPACE, the solution y may not be polynomial time computable even if the right-hand side function f is polynomial time computable and Lipschitz continuous [10,13]. If, on the other hand, the right-hand side function is a polynomial time computable analytic function, it is well known that y also is a polynomial time computable real function (see e.g. [17]). However, as the actual algorithm can differ for each right-hand side function f and initial value y_0 this statement is very different from saying that there is an efficient ODE solving algorithm on analytic functions. In Sect. 4 we define a uniform ODE solver on top of our definition, analyze its complexity and show that it runs in parameterized polynomial time in terms of parameters defined in Sect. 3.

Computable analysis aims to be a realistic model in the sense that the theory can be a basis for actual exact computations with real numbers (sometimes called exact real arithmetic [1,7]). The ideas in this paper can be used as a basis for data-types for analytic functions and ODE solving in exact real arithmetic. To show that such an implementation is indeed feasible, we provide a prototypical implementation based on iRRAM [20], a C++-library for exact real arithmetic. We discuss some possible optimizations for practical purposes and briefly describe the implementation in Sect. 5.

2   Computability and Complexity in Analysis

Computable analysis combines methods from theoretical computer science, real analysis and numerical analysis to provide a framework for computational problems over real numbers and more general continuous structures. We present some basic notions needed for our purpose here and refer the reader to the extensive literature (e.g. [3,11,14,23]) for deeper understanding.

We fix the finite alphabet $\Sigma = \{0,1\}$ and denote by Σ^* the set of finite strings over Σ. We assume the reader to be familiar with the general notion of computability for functions $\Sigma^* \to \Sigma^*$, e.g., by Turing machines. As discrete structures like natural numbers, rational numbers or graphs can be encoded by finite strings, computability over such structures can be defined as computability on their encodings. Objects from uncountable spaces (such as the real numbers), on the other hand, cannot be encoded in such a way. Instead, we encode them by functions that provide partial information to the object when asked. The idea of such an encoding is formalized by the notion of a represented space. We denote by $\mathcal{B} = (\Sigma^*)^{\Sigma^*}$ the *Baire space* of all string functions $\Sigma^* \to \Sigma^*$.

Definition 1. *A represented space is a pair (X, ξ_X) of a set X and a partial surjective function $\xi_X : \mathcal{B} \to X$. An element $\varphi \in \mathcal{B}$ such that $\xi_X(\varphi) = x$ is called a ξ_X-name of $x \in X$. An element of a represented space is called $(\xi_X\text{-})$computable if it has a computable (ξ_X-)name.*

Let $(X, \xi_X), (Y, \xi_Y)$ be represented spaces and $f : X \to Y$ a function. A function $F : \mathcal{B} \to \mathcal{B}$ mapping names of elements $x \in X$ to names of $f(x)$, i.e., such that $\xi_Y(F(\varphi)) = f(\xi_X(\varphi))$ for all $\varphi \in \text{dom}(\xi_X)$, is called a *realizer* for f (Fig. 1). Computability on Baire space is typically defined by oracle machines. A function $F : \mathcal{B} \to \mathcal{B}$ is computable if there is an oracle machine M s.t. for all $\varphi \in \text{dom}(F)$ and $q \in \Sigma^*$ the machine with input q and oracle φ outputs $F(\varphi)(q)$ (Fig. 2). We now define some standard representations that we need later. For $n \in \mathbb{N}$ we denote by $bin(n) \in \Sigma^*$ the binary encoding of the integer n and by 0^n and 1^n the string consisting of n repetitions of 0 and 1, respectively. We use the following representation for real numbers: A $\xi_{\mathbb{R}}$-name for a real number x is a function $\varphi : \Sigma^* \to \Sigma^*$ such that $\varphi(0^n) = bin(a_n)$ is a binary encoding of some integer a_n with $\left| x - \frac{a_n}{2^n} \right| \leq 2^{-n}$ for all $n \in \mathbb{N}$.

We also have to work with sequences of real numbers. For this we choose the following representation $\xi_{\mathbb{R}^\omega}$: A name of a sequence $(a_i)_{i \in \mathbb{N}} \in \mathbb{R}^\omega$ of reals is

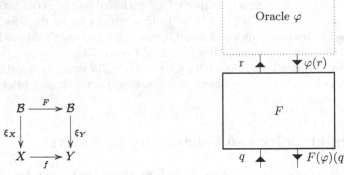

$$\begin{array}{ccc} \mathcal{B} & \xrightarrow{\ F\ } & \mathcal{B} \\ \xi_X \big\downarrow & & \big\downarrow \xi_Y \\ X & \xrightarrow[\ f\ ]{} & Y \end{array}$$

Fig. 1. Computing a function f between represented spaces (X, ξ_X) and (Y, ξ_Y). The function $F : \mathcal{B} \to \mathcal{B}$ is called a realizer for the function $f : X \to Y$.

Fig. 2. Computing a function $F : \mathcal{B} \to \mathcal{B}$ with an oracle Turing machine.

a function $\psi : \Sigma^* \to \Sigma^*$ such that for each i, $n \in \mathbb{N}$ it holds that $\psi(0^i 10^n) = bin(b_{i,n})$ is the binary encoding of some integer $b_{i,n}$ with $\left| a_i - \frac{b_{i,n}}{2^n} \right| \le 2^{-n}$.

Let $\langle \cdot, \cdot \rangle$ denote a bijective, polynomial time computable pairing function for finite strings that has polynomial time computable projections. For represented spaces (X, ξ_X) and (Y, ξ_Y) we can combine their representations to get a representation $\xi_{X \times Y}$ for their product $X \times Y$ by using the pairing function on the names [23, Sect. 3.3]. We thus use the standard representations $\xi_{\mathbb{R}^d} := (\xi_{\mathbb{R}})^d$ for $\mathbb{R}^d$, $\xi_{\mathbb{C}^d} := (\xi_{\mathbb{R}})^{2d}$ for $\mathbb{C}^d$ and $\xi_{\mathbb{C}^\omega}^d := (\xi_{\mathbb{R}^\omega})^{2d}$ for complex multi-sequences.

2.1 Complexity Theory

As elements of represented spaces are computed by ordinary Turing machines, the usual definitions for time-complexity from discrete complexity theory can be applied. For functions, on the other hand, defining complexity becomes more subtle. While it is still possible to define complexity independently from the oracle for real functions restricted to a fixed compact domain, for operators there is no reasonable way to do this and the oracle has to be accounted for. Recently, Kawamura and Cook provided the tools needed to reason about uniform complexity of operators by introducing a framework for doing complexity theory for functions between represented spaces and an appropriate representation for the space of continuous functions [11]. To obtain a reasonable complexity theory it is necessary to restrict to *length-monotone* string functions, i.e., functions $\varphi \in \mathcal{B}$ such that for all $a, b \in \Sigma^*$ it holds that $|\varphi(a)| \le |\varphi(b)|$ whenever $|a| \le |b|$. The *length* of a length-monotone $\varphi \in \mathcal{B}$ is the function $|\varphi| : \mathbb{N} \to \mathbb{N}$ defined by $|\varphi|(|a|) = |\varphi(a)|$ for all $a \in \Sigma^*$. A representation where all names are length-monotone is called *second-order representation*. Most representations and in particular those introduced above can be turned into second-order representations by padding names. In the rest of the paper we assume all representations to be second-order representations.

To bound the running time of an oracle machine both in the size of the input and the size of the oracle, we need the following definition.

Definition 2. *The class of* second-order polynomials *is the class of functions* $\mathbb{N}^{\mathbb{N}} \times \mathbb{N} \to \mathbb{N}$ *defined inductively by*

- *the functions* $(L, n) \mapsto c$ *for any constant* $c \in \mathbb{N}$ *and the function* $(L, n) \mapsto n$ *are second-order polynomials,*
- *for second-order polynomials* P *and* Q, *the functions* $(L, n) \mapsto P(L, n) + Q(L, n)$ *and* $(L, n) \mapsto P(L, n) \cdot Q(L, n)$ *are second-order polynomials, and*
- *for a second-order polynomial* P, *the function* $(L, n) \mapsto L(P(L, n))$ *is a second-order polynomial.*

We can now define when an oracle machine runs in polynomial time.

Definition 3. *An oracle machine* $M^?$ *runs in* polynomial time *if there is a second-order polynomial* P *such that for any length-monotone* $\varphi \in \mathcal{B}$ *and any* $q \in \Sigma^*$, $M^\varphi(q)$ *halts after at most* $P(|\varphi|, |q|)$ *steps.*

A function $F : X \to Y$ between represented spaces is called polynomial time computable if there is a polynomial time computable realizer.

For our purpose we do not need the full framework of second order complexity. We only consider representations that encode a second-order argument together with some discrete information. For this purpose definitions similar to the following are used in [12]. For any $\psi \in \mathcal{B}$ and string $w \in \Sigma^*$ let $\langle w, \psi \rangle \in \mathcal{B}$ denote the function defined by $\langle w, \psi \rangle(q) = \langle w, \psi(q) \rangle$ for all $q \in \Sigma^*$.

Definition 4. *For each* $x \in X$ *let* $L(x) \subseteq \Sigma^*$ *be some non-empty set of finite strings. For a second-order representation* $\xi : \mathcal{B} \to X$, *a* parameterized representation with parameters from L *is defined as follows:* $\varphi \in \mathcal{B}$ *is a name for* $x \in X$ *if* $\varphi = \langle w, \psi \rangle$ *for a* ξ-name ψ *for* x *and some* $w \in L(x)$. *We say a parameterized representation has* polynomial length *if there is a (first-order) polynomial* $p : \mathbb{N} \to \mathbb{N}$ *such that for all names* $\langle w, \psi \rangle \in \mathcal{B}$ *and strings* $q \in \Sigma^*$ *we have* $|\psi(q)| \leq p(|q| + |w|)$.

A parameterized representation enriches a name of $x \in X$ with some discrete information from $L(x)$. For operators between spaces with polynomial-length parameterized representations, it suffices to bound the running time in a (first-order) polynomial in terms of the input length and the length of the parameter to show polynomial time computability.

3   Uniform Computations on Analytic Functions

An analytic function is a function which is locally given by a power series. For our application we are interested in multivariate complex functions $f : \mathbb{C}^d \to \mathbb{C}$. We use multi-index notation to denote d-tuples $\beta = (\beta_1, \ldots, \beta_d) \in \mathbb{N}^d$. For multi-indices $\alpha, \beta \in \mathbb{N}^d$, tuples of complex numbers $z \in \mathbb{C}^d$ and function $f : \mathbb{C}^d \to \mathbb{C}$ we further use the conventions $\alpha + \beta = (\alpha_1 + \beta_1, \ldots, \alpha_d + \beta_d)$, $z^\alpha = z_1^{\alpha_1} z_2^{\alpha_2} \cdots z_d^{\alpha_d}$, $\alpha! = \alpha_1! \alpha_2! \cdots \alpha_d!$, $|\alpha| = \alpha_1 + \alpha_2 + \cdots + \alpha_d$ and $D^\alpha f = \frac{\partial^{|\alpha|} f}{\partial x_1^{\alpha_1} \cdots \partial x_d^{\alpha_d}}$.

Definition 5. *Let $D \subseteq \mathbb{C}^d$ be a subset. A function $f : D \to \mathbb{C}$ is called **analytic**, if for any $t \in D$ there is a complex multi-sequence $(a_\beta)_{\beta \in \mathbb{N}^d} \subseteq \mathbb{C}$ and a neighborhood U of t such that for all $x \in U \cap D$ we have*

$$f(x) = \sum_{\beta \in \mathbb{N}^d} a_\beta (x - t)^\beta.$$

We call $(a_\beta)_{\beta \in \mathbb{N}^d}$ the power series and the coefficients a_β the Taylor coefficients of f around t and denote the class of analytic functions on a set D by $\mathcal{C}^\omega(D)$.

An important fact about analytic functions that we will use several times is Cauchy's integral formula.

Theorem 1. *If $f : D \to \mathbb{C}$ is analytic and $U(\zeta) \subseteq D$ is some polydisc $U(\zeta) = \prod_{i=0}^d B_{r_i}(\zeta_i)$ around $\zeta = (\zeta_1, \ldots, \zeta_d) \in D$ with radius $R = (r_1, \ldots, r_d)$, then*

$$f(z) = \frac{1}{(2\pi i)^d} \int_{\partial U_1} \cdots \int_{\partial U_d} \frac{f(\xi_1, \ldots, \xi_d)}{(\xi_1 - z_1) \ldots (\xi_d - z_d)} d\xi_1 \cdots d\xi_n.$$

In particular, if $|f| \le M$ for all $z \in U(z_0)$, then $|D^\beta f(\zeta)| \le \beta! \frac{M}{R^\beta}$ for all $\beta \in \mathbb{N}^d$.

The reason why we are interested in analytic functions stems from real complexity theory: An analytic function $f : K \to \mathbb{R}$ for $K \subseteq \mathbb{R}$ compact and connected (containing more than one point) is polynomial time computable if and only if the sequence of Taylor coefficients around some point $x_0 \in K \cap \mathbb{Q}$ is polynomial time computable [16,18]. It follows that for a polynomial time computable analytic function f, for example, the derivative function f' is again polynomial time computable. Similarly, the result of many other operators can be shown to be a polynomial time computable function, while in the case of general polynomial time real functions hardness results are known that make efficient algorithms unlikely to exist.

3.1   Representing Power Series

For simplicity we first consider the one-dimensional case and discuss which information has to be encoded. Let $\mathcal{C}^\omega(\overline{B_r(z_0)})$ denote the space of complex analytic functions on a neighborhood of the closed disc $\overline{B_r(z_0)} \subseteq \mathbb{C}$ with radius $r > 0$ around $z_0 \in \mathbb{C}$. For $f \in \mathcal{C}^\omega(\overline{B_r(z_0)})$ let (a_m) denote the series expansion at the point z_0. Consider the operator mapping a power series $(a_m)_{m \in \mathbb{N}}$ and a complex number z to $\sum_{m=0}^\infty a_m (z - z_0)^m$. Any algorithm computing this sum can only read a finite number of coefficients. However, having no further information, it is not possible to determine the number of coefficients necessary to get a good enough approximation of the sum in a computable way [18]. Thus, not even evaluation of a function is uniformly computable if the only information the name provides is the power series. Therefore, additional information similar to the following is often considered in real complexity theory [19]. Let $q, B \in \mathbb{R}$ be such that $q > r$ and $|a_m| \leq \frac{B}{q^m}$ for all $m \in \mathbb{N}$. Then for all $z \in \overline{B_r(0)}$ it holds that

$$\left| \sum_{m=M}^{\infty} a_m z^m \right| \leq \frac{B}{1 - \frac{|z|}{q}} \left(\frac{|z|}{q} \right)^M \tag{1}$$

Thus B, and q can be used to make a tail estimate on how many coefficients are needed to make the error arbitrarily small. While we could encode those reals directly in our representation, using a parameterization by integers allows a simpler characterization of the complexity.

Let us now give a multidimensional extension of the representations for analytic functions in [12]. Note that we always assume the dimension d to be (a small) constant and it is thus not part of our complexity analysis. The time needed by most algorithms presented below is exponential in the dimension. As they manipulate multidimensional power series, this seems inevitable already for combinatorial reasons.

For uniform computations we now fix the domain. Let $\mathcal{C}^\omega(\overline{B_1(0)}^d)$ denote the space of analytic functions on the closed unit polydisc $\overline{B_1(0)}^d$ around 0. We define a parameterized representation according to Definition 4 as follows:

Definition 6. *A name for an* $f \in \mathcal{C}^\omega(\overline{B_1(0)}^d)$ *is a* $\xi_{\mathcal{C}^\omega}^d$*-name for the* d*-dimensional power series* $(a_\beta)_{\beta \in \mathbb{N}^d}$ *of* f *around* 0 *together with a string* $1^k 0 bin(A)$ *for integers* $A, k \in \mathbb{N}$ *such that* $|a_\beta| \leq A 2^{-\frac{|\beta|}{k}}$ *for all* $\beta \in \mathbb{N}^d$.

A $\mathcal{C}^\omega(\overline{B_1(0)}^d)$-name thus enriches the standard representation for multidimensional complex sequences by integer constants A encoded in binary and k encoded in unary. Note that the length of the parameter is given by $k + \log(A)$. As the absolute value of each coefficient is bounded by A, it is easy to see that this parameterized representation has polynomial length and thus second-order polynomial time computability corresponds to the existence of a realizer with time bounded in a first-order polynomial in terms of the input size, k and $\log A$.

Using the bound (1) one can approximate a one dimensional analytic function by a partial sum and by that define a polynomial time realizer for the operation EVAL :$\subseteq \mathcal{C}^\omega(\overline{B_1(0)}) \times \overline{B_1(0)} \to \mathbb{C}$, EVAL$(f, z) = f(z)$ mapping a one-dimensional function and a point in the domain to the function value at that point. Instead of directly extending this to the multidimensional case we first introduce some operators that reduce the dimension.

Theorem 2. *Let* $d \geq 1$. *The following operators are polynomial time computable:*

1. *The operator* $\pi_1^d \colon \mathcal{C}^\omega(\overline{B_1(0)}^{d+1}) \times \mathbb{N} \to \mathcal{C}^\omega(\overline{B_1(0)}^d)$ *that maps a* $(d + 1)$*-dimensional analytic function with power series* $(a_{i_1,\ldots,i_{d+1}})$ *and an index* j *(encoded in unary) to the* d*-dimensional function with power series* $(b_{i_1,\ldots,i_d})$ *where the first index is fixed to* j, *i.e.,* $b_{i_1,\ldots,i_d} = a_{j,i_1,\ldots,i_d}$.
2. *The operator* $\pi_\bullet^d \colon \mathcal{C}^\omega(\overline{B_1(0)}^{d+1}) \times \mathbb{N}^d \to \mathcal{C}^\omega(\overline{B_1(0)})$ *that fixes all but the first index of the power series.*

3. *The operator* $\sigma^d\colon C^\omega\left(\overline{B_1(0)}^{d+1}\right) \times \overline{B_1(0)} \to C^\omega\left(\overline{B_1(0)}^d\right)$ *whose values are given by* $\sigma(f,z)(z_1,\ldots,z_d) = f(z,z_1,\ldots,z_d)$.

Proof. For π_1^d and $\pi_\bullet^d$ it is obvious how to get the coefficients of the power-series. Let $A, k \in \mathbb{N}$ be the parameters encoded in a name for $f \in C^\omega\left(\overline{B_1(0)}^{d+1}\right)$. It easy to see that $k' = k$ and $A' = A$ can be chosen both for $\pi_1^d(f)$ and $\pi_\bullet^d(f)$.

For σ^d note that the coefficient with index $(i_1,\ldots,i_d)$ of the power series of $\sigma(f,z)$ is given by $b_{i_1,\ldots,i_d} = \sum_{j=0}^\infty a_{j,i_1,\ldots,i_d} z^j = \text{EVAL}(\Pi_\bullet(f,i_1,\ldots,i_d),z)$. Thus it can be computed by polynomial time operators defined earlier. Choosing $k' = k$ and $A' = 2Ak$ fulfills the necessary bounds.

Using the substitution operator iteratively gives an algorithm for evaluating d-dimensional analytic functions.

Extending the above framework by further operations is easy: We just have to specify how to compute the power series and the integer constants for the resulting function. It is therefore straight-forward to generalize the results in [12] to show that e.g. addition, multiplication or computing derivatives is polynomial time computable and we omit the details.

3.2   Representing Analytic Functions

We can use the result in the previous section to compute with functions complex analytic on a (possibly very small) ball around some $z_0 \in \mathbb{C}^d$: Assume f is analytic on the ball $B_r(z_0)$ with $r < 1$. Then $g(z) := f(z_0 + \frac{rz}{2})$ is analytic on $\overline{B_2(0)}^d$. A simple transformation thus reduces to the above case and allows, e.g., efficient evaluation on $z \in \mathbb{C}$ with $|z| \leq \frac{r}{2}$.

However, usually we are interested in slightly more complicated domains, such as simple compact subsets of $\mathbb{R}^d$. We therefore next consider the case of functions analytic on the domain $[0,1]^d \subseteq \mathbb{R}^d$. Let $C^\omega([0,1]^d)$ be the set of functions complex analytic on some neighborhood of $[0,1]^d$. We define two different representations for this set, one directly allowing point-wise evaluation of the function and one allowing manipulating power series. They are multidimensional generalizations of those found in Sect. 3.2 of [12].

The first representation just gives a cover of the hypercube $[0,1]^d$ with power series with a uniform bound on the radius of convergence. For $l \in \mathbb{N}$ let $\overline{R_L} := \{x + iy : y \in [-\frac{1}{L}, \frac{1}{L}] \text{ and } x \in [-\frac{1}{L}, 1 + \frac{1}{L}]\}$. For any $f \in C^\omega([0,1]^d)$ there is an $l \in \mathbb{N}$ such that $f \in C^\omega(\overline{R_l}^d)$. We define a (polynomial-length) parameterized representation for $C^\omega([0,1]^d)$ as follows:

Definition 7. *A* series name *of an* $f \in C^\omega([0,1]^d)$ *is given by a* $\xi_{C^\omega}^{2d}$*-name for a sequence* $(a_{m,\beta})_{m,\beta \in \mathbb{N}^d}$ *and a string* $1^k 0 bin(A)$ *for integers* $k, A \in \mathbb{N}$ *such that for all* $m_i \in \{0,\ldots,2k\}$ *and* $\beta \in \mathbb{N}^d$ *it holds* $|a_{m_1,\ldots,m_d,\beta}| \leq Ak^{|\beta|}$ *and* $D^\beta f(\frac{m_1}{2k},\ldots,\frac{m_d}{2k}) = a_{m_1,\ldots,m_d,\beta}\beta!$.

A series name thus encodes a covering of the domain $[0,1]^d$ with $(2k+1)^d$ power series with radius of convergence at least $\frac{1}{k}$. The series are overlapping so that for any $x \in [0,1]^d$ we can easily select one such that x has at most distance $\frac{1}{2k}$ to the boundary and get a $\mathcal{C}^\omega(\overline{B_1(0)}^d)$-name for the (scaled) series.

Instead of encoding the power series it is often more practical to encode function evaluation directly:

Definition 8. *A function name for $f \in C^\omega([0,1]^d)$ is defined by the following:*

1. *A function $\phi \in \mathcal{B}$ that approximates $f(q)$ with arbitrary precision on dyadic rationals $q \in [0,1]^d$. Formally, whenever $w \in \Sigma^*$ is the binary encoding of d natural number $n_1, \ldots, n_d \in \mathbb{N}$ then $\left| \frac{\phi(w)}{2^{|w|+1}} - f\left(\frac{n_1}{2^{|w|+1}}, \ldots, \frac{n_d}{2^{|w|+1}}\right) \right| \leq 2^{-|w|}$.*
2. *A string $1^l 0 bin(B)$ for integers $B, l \in \mathbb{N}$ such that $f \in C^\omega(\overline{R_l})$ and B is an upper bound for $|f|$ on $\overline{R_l}$.*

As the parameters bound the length of the integer part of the approximation function, the above defines a polynomial-length parameterized representation where the second-order part is given by the approximation function.

The following Lemma can be derived from Cauchy's integral formula.

Lemma 1. *If $B, l \in \mathbb{N}$ are constants as in Definition 8 for some $f \in C^\omega([0,1]^d)$ then for all $\beta \in \mathbb{N}^d$ and $x \in [0,1]^d$ it holds $|f^{(\beta)}(x)| \leq \beta! B l^\beta$. Further, f is Lipschitz-continuous on $\overline{R_{2l}}$ with Lipschitz constant $L = 2\sqrt{d} B l$.*

We show that both representations are polynomial time equivalent, i.e., given a name of one representation it is possible to compute a name of the other in polynomial time. Given a series name and some $z \in \overline{R_{4k}}$ we can choose a power series with index $(m_1, \ldots, m_d)$ that converges for z and it holds

$$\left| \sum_{\beta \in \mathbb{N}^d} a_{m_1, \ldots, m_d, \beta} \left(z_1 - \frac{m_1}{2k}\right)^{\beta_1} \cdots \left(z_d - \frac{m_d}{2k}\right)^{\beta_d} \right| \leq \sum_{\beta \in \mathbb{N}^d} Ak^\beta (2k)^{-\beta} = 2^d A.$$

Thus a function name with constants $B = 2^d A$ and $l = 4k$ can be computed from a series name. For the other way round note that the power series around any point in $[0,1]^d$ can be computed in polynomial time from the information in a function name (see e.g. [19]). Further by Lemma 1 we can choose $k = l$ and $A = B$ as constants.

To show that an operator is polynomial time computable we can therefore use either of the representations or even a combination of both.

Theorem 3. *The following operators on multidimensional analytic functions are polynomial time computable:*

1. *Evaluation $EVAL \colon C^\omega([0,1]^d) \times [0,1] \to \mathbb{R}$, $(f,x) \mapsto f(x)$,*
2. *Addition and Multiplication $+, \times \colon C^\omega([0,1]^d) \times C^\omega([0,1]^d) \to C^\omega([0,1]^d)$,*
3. *Partial derivatives $D \colon C^\omega([0,1]^d) \times \mathbb{N}^d \to C^\omega([0,1]^d)$, $(f, \alpha) \mapsto D^\alpha f$ where $\alpha_1, \ldots, \alpha_d$ are given in unary.*

Proof. We only show the third part as for the other operators a multidimensional generalization of results in [12] is straightforward. Assume we are given a series

name. Let $(a_{m,\beta})_{\beta \in \mathbb{N}^d}$ be the power series with index $m \in \mathbb{N}^d$. Then $b_{m,\beta} := \frac{(\alpha+\beta)!}{\beta!} a_{m,\alpha+\beta}$ gives the Taylor coefficient of $D^\alpha f$ around $z_m = \frac{m}{2k}$. Since for all $a, b \in \mathbb{N}$ it is $a^b \leq (2b)^b (\frac{4}{3})^a$ it is $\frac{(\alpha+\beta)!}{\beta!} \leq (\alpha+\beta)^\alpha \leq (2\alpha)^\alpha (\frac{4}{3})^{|\alpha+\beta|}$. Thus $|b_{m,\beta}| \leq (3|\alpha| k)^{|\alpha|} (\frac{4}{3}k)^{|\beta|}$ holds. This can be used to approximate $D^\alpha f(q)$ for all $q \in [0,1]^d$ with $\|q - z_m\|_\infty \leq \frac{1}{2k}$ with precision 2^{-n} in time polynomial in $n+k+\log A$ yielding the approximation function for the function name. Similarly, for all $z \in \overline{R_{4k}}$ it is $|D^\alpha f(z)| \leq 3^d (3|\alpha|)^{|\alpha|} A$. Thus $B' = 3^d (3|\alpha|)^{|\alpha|} A$ and $l' = 4k$ are constants for a function name for the derivative.

4   Ordinary Differential Equations

In this section we show how to use the above representations to define a solver for initial value problems of the form

$$\dot{y} = F(y), \qquad y(0) = y_0 \tag{2}$$

for analytic $F : \mathbb{R}^d \to \mathbb{R}^d$ and $y_0 \in \mathbb{R}^d$. Non-autonomous ODEs where the the right-hand side function F explicitly depends on the time t can be expressed in this form by increasing the dimension by 1 (this is one reason why we are mostly interested in multidimensional functions).

We first show how to compute a solution corresponding to a single power series, i.e., $F = (F_1, \ldots, F_d)$ with $F_i \in \mathcal{C}^\omega(\overline{B_1(0)})^d$ and $y_0 \in \overline{B_0(1)}^d$. In this case the solution is of the form $y = (y_1, \ldots, y_d)$ with $y_i : \mathbb{R} \to \mathbb{R}$ analytic but not necessarily defined on all of $\overline{B_0(1)}$. We first show that it is possible to compute *some* local solution defined on a possibly very small radius in polynomial time. To fit in our framework we scale the solution to get a name according to Definition 6. That is, for $i = 1, \ldots, d$ consider the operator $\mathtt{LSolve}_i : \mathcal{C}^\omega(\overline{B_1(0)})^d \times \overline{B_1(0)}^d \to \mathcal{C}^\omega(\overline{B_1(0)}) \times \mathbb{R}$ that maps a function $F \in \mathcal{C}^\omega(\overline{B_1(0)})^d$ and initial value $y_0 \in \overline{B_1(0)}^d$ to a function $\overline{y_i}$ and a real number $r \in \mathbb{R}$ such that the function y_i defined by $y_i(z) = \overline{y_i}(\frac{2z}{r})$ is a solution to the IVP (2).

Theorem 4. *$\mathtt{LSolve}_i$ is polynomial time computable for $i = 1, \ldots, d$.*

Proof. Let us first show how to compute the Taylor coefficients for each y_i. We can use the power series method [5]: For $i = 1, \ldots, d$ we inductively define the functions $f_{i,m} : \mathbb{C}^d \to \mathbb{C}$ for $k \in \mathbb{N}$ by letting $f_{i,0}(z) = z_i$ and

$$f_{i,m+1}(z) = \frac{1}{m+1} \left(\sum_{j=1}^d \frac{\partial f_{i,m}}{\partial y_j} F_j(z) \right). \tag{3}$$

We can use $f_{i,m}$ to express the mth derivative of y_i as $y_i^{(m)}(t) = m! f_{i,m}(y(t))$ holds. Thus, for the mth Taylor coefficient $a_{i,m}$ of y_i around 0 it holds $a_{i,m} = f_{i,m}(y_0)$. Each of the functions given by (3) is analytic and can be computed using polynomial time computable operators defined in the previous section. Therefore the operator mapping f, y_0, i and m to $a_{i,m}$ is polynomial time computable.

Let A and k be the constants from the name. From the Picard-Lindelöf theorem it follows that the solution is valid on a radius of at least $r = \frac{\sqrt[k]{2}-1}{(d+1)A}$. Thus scaling the solution function accordingly yields a polynomial time computable operator.

The function given by the LSolve operator can be used to evaluate the solution efficiently (i.e. approximate up to precision 2^{-n} in time polynomial in $n + k + \log A$) on $z \in \mathbb{C}$ with $|z| \le \frac{r}{2}$. Note that the radius r depends on the constants of the name and is usually very small. We therefore also call this the local solution.

Let us now show how to extend the local solution. The idea is to iteratively use the local solution operator to compute new initial values and thereby stepwise increase the time. Of course this only works as long as the solution takes values in the domain of F. Thus, let us now assume that $F \in C^\omega([0,1]^d)$ and for $y_0 \in [0,1]$ the solution $y(t)$ exists for $t \in [0,1]$ and only takes values in $[0,1]^d$.

Theorem 5. *Let $y \in C^\omega([0,1])$ be the solution to the IVP (2) for $F \in C^\omega([0,1])$ and $y_0 \in [0,1]$. Given a series name of F with parameters A and k and $\xi_{\mathbb{R}^d}$-names of $y_0, t \in [0,1]$ the solution $y(t)$ can be approximated up to precision 2^{-n} in time $poly(n + A + k)$ for each $n \in \mathbb{N}$.*

Proof. Given some $y_0 \in [0,1]^d$ and a series name with constants A and k for F we can select a power series centered around a point with distance at most $\frac{1}{2k}$ from y_0. Combining this with the above solution operator can be used to approximate a local solution $y(t)$ for $t \le \frac{1}{2(d+1)kA}$ up to error 2^{-m} for any $m \in \mathbb{N}$ in time polynomial in $m + k + \log(A)$. Let us fix some $m \in \mathbb{N}$ and let z_0 be a 2^{-m} approximation of y_0 and z_{i+1} a 2^{-m} approximation of $\mathrm{Solve}_i(F, z_i)$ evaluated at $t := (\frac{1}{2(d+1)kA})$. Thus after at most $2(d+1)kA$ steps we reach any time $t \in [0,1]$.

It remains to show that it suffices to choose m polynomial in $n + k + A$. Let $t_i := \frac{i}{2(d+1)kA}$. Using the Lipschitz-bound from Lemma 1 and the well known Grönwall Lemma it can be shown that if z_i differs by at most ε from the correct solution $y(t_i)$ then z_{i+1} differs by at most 2ε from $y(t_{i+1})$ (see e.g. [2]). As we need at most $2(d+1)Ak$ steps, the total error is bounded by $2^{2(d+1)Ak+1-m}$. Thus choosing $m > n + 2(d+1)Ak + 1$ suffices to guarantee precision 2^{-n}.

The above can be easily extended to an operator mapping F and y_0 to the function $y \in C^\omega([0,1])$. Note, that the algorithm already needs linearly in A many steps and is therefore not polynomial time computable by the strict definition that requires the complexity in A to be logarithmic. In general, however, we can not expect to get a better algorithm as the solution to the ODE can already take values with magnitude exponentially large in A.

5   Practical Considerations

The ideas of Sect. 3 can be easily translated to an implementation in an object oriented programming language. We provide a very simple, prototypical implementation of our ideas based on the iRRAM C++ library[1]. iRRAM already has a class REAL for exact computations with real numbers and many standard operations that we could built on. Our library extends iRRAM by a class for real analytic functions of arbitrary dimension and operators for, e.g., evaluation, addition, subtraction, multiplication, partial derivatives and composition of analytic functions. It also contains an ODE solver using the ideas of the step-wise solver described in Sect. 4. All operations are performed in exact real arithmetic, making it possible to evaluate functions or compute series coefficients with any desired precision and guaranteed correctness.

Currently, our implementation only covers functions given by single power series, similarly to the definitions in Sect. 3.1. However, we made a few adjustments to make our implementation more useable for practical applications. First, we do not fix domain in advance, i.e., we consider power series of arbitrary radius $r > 0$ of type REAL around some point $x_0 \in \mathbb{R}^d$ and let r be part of the description. Further instead of a single integer we encode the maximum of the function on each ball of radius $r' < r$ as a function $M :$ REAL $\to$ REAL. While this does not allow a simple parameterized complexity characterization it removes the restriction to the unit ball which seems artificial for practical purposes and also allows to consider total functions.

Further, in the previous sections our goal was to make the representations as simple as possible to guarantee polynomial complexity bounds. However, for a practical implementation it can be helpful to encode more than this minimal information. For many special functions, for example, many more efficient evaluation algorithms than evaluating the power series exist. Therefore, our implementation also allows to provide an alternative evaluation algorithm. This makes it possible to have very efficient implementations of standard functions while still allowing to define general functions only by specifying how to compute the minimal information.

As heuristics as the above provide information that is redundant from a complexity theoretic viewpoint and do not always lead to better results, it is hard to quantify their usefulness. Instead of a formal analysis an experimental approach choosing an appropriate set of benchmark functions seems more feasible.

6   Conclusion

We have shown how to extend some basic representations for analytic functions on simple domains to the multidimensional case and applied that to define a uniform solution operator for ordinary differential equations that computes the solution of an initial value problem in (parameterized) polynomial time. We further discussed how the ideas can be refined for an actual implementation for a library for computations on multidimensional analytic functions and gave a prototypical implementation of such a library in iRRAM. As the main purpose of

---

[1] The complete source code for our implementation including some test functions is publicly available on GitHub: https://www.github.com/holgerthies/iRRAM-analytic.

the considered results was showing polynomial time computability, it is obvious that most algorithms can be improved in terms of efficiency. An important detail we omitted in our analysis is how to actually perform the operations on the power series coefficients as for polynomial time computability the most simple and well known classical algorithms already suffice. There are of course much more efficient algorithms on power series (see e.g. [4,8,9]) and for a practical implementation this can make a crucial difference.

In future work we plan to analyze the complexity of our algorithms in more detail in terms of bit-complexity and improve them by using more sophisticated approaches. Further we plan to systematically evaluate the running time of our algorithms for a set of benchmark functions and compare them to similar approaches in numerics and interval analysis.

Acknowledgements. This work was supported by JSPS KAKENHI Grant Numbers JP18H03203 and JP18J10407 and by the Japan Society for the Promotion of Science (JSPS), Core-to-Core Program (A. Advanced Research Networks).

References

1. Boehm, H.J., Cartwright, R., Riggle, M., O'Donnell, M.J.: Exact real arithmetic: a case study in higher order programming. In: Proceedings of the 1986 ACM Conference on LISP and Functional Programming, pp. 162–173. ACM (1986)
2. Bournez, O., Graça, D.S., Pouly, A.: On the complexity of solving initial value problems. In: ISSAC 2012-Proceedings of the 37th International Symposium on Symbolic and Algebraic Computation, pp. 115–121. ACM, New York (2012). https://doi.org/10.1145/2442829.2442849
3. Brattka, V., Hertling, P., Weihrauch, K.: A tutorial on computable analysis. In: Cooper, S.B., Löwe, B., Sorbi, A. (eds.) New Computational Paradigms: Changing Conceptions of What is Computable, pp. 425–491. Springer, New York (2008). https://doi.org/10.1007/978-0-387-68546-5_18
4. Brent, R.P., Kung, H.T.: Fast algorithms for manipulating formal power series. J. ACM (JACM) **25**(4), 581–595 (1978)
5. Chang, Y., Corliss, G.: ATOMFT: solving ODEs and DAEs using Taylor series. Comput. Math. Appl. **28**(10–12), 209–233 (1994)
6. Friedman, H.: The computational complexity of maximization and integration. Adv. Math. **53**(1), 80–98 (1984)
7. Geuvers, H., Niqui, M., Spitters, B., Wiedijk, F.: Constructive analysis, types and exact real numbers. Math. Struct. Comput. Sci. **17**(1), 3–36 (2007)
8. van der Hoeven, J.: Relax, but don't be too lazy. J. Symb. Comput. **34**(6), 479–542 (2002)
9. van der Hoeven, J.: On effective analytic continuation. Math. Comput. Sci. **1**, 111–175 (2007)
10. Kawamura, A.: Lipschitz continuous ordinary differential equations are polynomial-space complete. Comput. Complex. **19**(2), 305–332 (2010)
11. Kawamura, A., Cook, S.: Complexity theory for operators in analysis. ACM Trans. Comput. Theory **4**(2), 5:1–5:24 (2012)
12. Kawamura, A., Müller, N., Rösnick, C., Ziegler, M.: Computational benefit of smoothness: parameterized bit-complexity of numerical operators on analytic functions and Gevrey's hierarchy. J. Complex. **31**(5), 689–714 (2015)

13. Ko, K.I.: On the computational complexity of ordinary differential equations. Inf. Control **58**(1–3), 157–194 (1983)
14. Ko, K.I.: Complexity theory of real functions: Progress in Theoretical Computer Science. Birkhäuser Boston Inc., Boston (1991)
15. Ko, K.I., Friedman, H.: Computational complexity of real functions. Theor. Comput. Sci. **20**(3), 323–352 (1982)
16. Ko, K.I., Friedman, H.: Computing power series in polynomial time. Adv. Appl. Math. **9**(1), 40–50 (1988)
17. Moiske, B., Müller, N.: Solving initial value problems in polynomial time. In: Proceedings of the 22th JAIIO-PANEL, vol. 93, pp. 283–293 (1993)
18. Müller, N.T.: Uniform computational complexity of Taylor series. In: Ottmann, T. (ed.) ICALP 1987. LNCS, vol. 267, pp. 435–444. Springer, Heidelberg (1987). https://doi.org/10.1007/3-540-18088-5_37
19. Müller, N.T.: Constructive aspects of analytic functions. In: Proceedings of Workshop on Computability and Complexity in Analysis, InformatikBerichte, vol. 190, pp. 105–114. FernUniversität Hagen (1995)
20. Müller, N.T.: The iRRAM: exact arithmetic in C++. In: Blanck, J., Brattka, V., Hertling, P. (eds.) CCA 2000. LNCS, vol. 2064, pp. 222–252. Springer, Heidelberg (2001). https://doi.org/10.1007/3-540-45335-0_14
21. Pour-El, M.B., Richards, J.I.: Computability in analysis and physics: Perspectives in Mathematical Logic. Springer-Verlag, Berlin (1989)
22. Turing, A.: On computable numbers, with an application to the Entscheidungsproblem. Proc. Lond. Math. Soc. **2**(42), 230–265 (1936). https://doi.org/10.1112/plms/s2-42.1.230
23. Weihrauch, K.: Computable Analysis. Springer, Heidelberg (2000). https://doi.org/10.1007/978-3-642-56999-9

Advanced Kripke Frame for Quantum Logic

Tomoaki Kawano(✉)

Tokyo Institute of Technology, 2-12-1 Ookayama,
Meguro-ku, Tokyo 152-8550, Japan
kawano.t.af@m.titech.ac.jp

Abstract. Quantum logic has been studied with orthomodular lattices. The semantics required for quantum logic can also be provided by *OM-models*, whose nature is almost equivalent to the notion of orthomodular lattices. However, the development of OM-models is in its infancy, and important notions of orthomodular lattices such as OM laws, atomicity and covering laws cannot yet be fully described. Thus, in this paper, we develop OM-models in an attempt to solve these problems.

Keywords: Quantum logic · Orthomodular lattice · Kripke frame
Dynamic quantum logic

1 Introduction

Quantum logic describes the propositional spaces of quantum physics, which are related to Hilbert spaces (see [2] for details). The major semantics required for quantum logic are covered by *orthomodular lattices*, as all closed subspaces of a Hilbert space constitute such lattices. *Orthomodular(OM)-models* and *OM-frames* also provide suitable semantics for quantum logic, and their nature is almost equivalent to the notion of orthomodular lattices [2,6]. OM-models are constructed by *Ortho(O)-models* + the orthomodular law.

Orthomodular lattices have been studied in terms of the detailed conditions required for Hilbert spaces. The *orthomodular law* is required to ensure the consistency of the dimension of Hilbert spaces. The *atomicity* expresses the need for each closed subspace of a Hilbert space to be expressed by a space enclosed by one-dimensional closed subspaces. The *covering law* expresses that there is no closed subspace between the n-dimensional subspace S and the $n+1$-dimensional subspace S' such that $S \subseteq S'$.

These conditions should be defined as conditions on the relations in frames. However, O-frames are too simple to deal with above conditions. They have to be stated directly without using any conditions on the relations in frames. For example, in [6], the orthomodular law is directly expressed by using subsets of elements in frames. This sort of problem has been discussed in [3,5,9].

In this paper, an advanced version of OM-frames is introduced. That is, we add the new relations to OM-frames to make it easier to deal with various

© Springer-Verlag GmbH Germany, part of Springer Nature 2018
L. S. Moss et al. (Eds.): WoLLIC 2018, LNCS 10944, pp. 237–249, 2018.
https://doi.org/10.1007/978-3-662-57669-4_14

conditions. These new frames are referred to as *dynamic orthomodular frames* (DOM-frames) because they are related to the notion of *dynamic quantum logic* [1]. The main result of this paper is to express all of the above conditions as conditions on the relations in frames (Theorems 2 and 3). These results may give new prospects to quantum logic.

In the context of dynamic quantum logic, models known as *Quantum Transition Systems* (**QTS**) are used [1]. Almost all complex dynamical concepts of projections of Hilbert spaces are expressly included in **QTS**. In contrast, as DOM-frames are essentially equivalent to OM-frames, the dynamical concepts of projection are *implicitly* definable, as we will see in the theorems presented in this paper.

For future study, a new sequent calculus for quantum logic, **GOMS**, is introduced and its relationship with DOM-frames is discussed. There are several types of sequent calculus for quantum logic; however, in general, they do not include the notion of implication, primarily because there are multiple candidates for quantum implication, but no unified opinion as to which is most relevant. For example, the sequent calculus **GOM,nis1** only has rules for negation and conjunction, and the sequent calculus **GMQL** [7,8] only has rules for negation, conjunction and disjunction. Although there are many such candidates, the *Sasaki arrow* $A \to_S B \equiv \neg A \vee (A \wedge B)$ [4] has attracted the most attention, as the meaning and behavior of this implication is useful for both quantum physics and quantum logic. Intuitively, this formula can be translated as "After state is projected to closed subspace A, B is true." This arrow is adopted here because DOM-frames are somewhat compatible with it. **GOMS** is constructed from **GOM** and the rules for the Sasaki arrow.

2   Basics

This section states some basic notions of quantum logic. A language containing an infinite set of propositional variables, a propositional constant $\bot$, a unary connective $\neg$, and a binary connective $\wedge$ is used. Formulas are constructed using the traditional method. The formulas $A \wedge B$ and $B \wedge A$ are not distinguished. The abbreviation $A \vee B$ represents $\neg(\neg A \wedge \neg B)$. The symbols $p, q, r, \ldots$ describe propositional variables, whereas $A, B, C, \ldots$ denote composite formulas. The symbols $\Gamma, \Delta, \Sigma, \ldots$ denote the sets of formulas. A *sequent* is defined by $\Gamma \Rightarrow \Delta$, where Γ and Δ are finite sets of formulas.

An *ortho lattice* $(L, \leq, \sqcap, \sqcup, ', \mathbf{1}, \mathbf{0})$ is a lattice that satisfies the following conditions.

$\forall a \in L, \exists! a',$
$a'' = a,$
If $a \leq b$, then $b' \leq a',$
$a \sqcap a' = \mathbf{0},$
$a \sqcup a' = \mathbf{1}.$

For a given ortho lattice $(L, \leq, \sqcap, \sqcup, ', \mathbf{1}, \mathbf{0})$, if all $a, b \in X$ satisfy the following *orthomodular law*, then $(L, \leq, \sqcap, \sqcup, ', \mathbf{1}, \mathbf{0})$ is referred to as an *orthomodular lattice*. These three conditions are equivalent.

$a \sqcap (a' \sqcup (a \sqcap b)) \leq b.$
$a \sqcap (a' \sqcup (a \sqcap b)) = a \sqcap b.$
If $a \leq b$, then $b \leq a \sqcup (a' \sqcap b)$.

The most definitive feature of ortho lattices and orthomodular lattices is that the *distributive law* (shown below) does not hold.

$a \sqcap (b \sqcup c) = (a \sqcap b) \sqcup (a \sqcap c)$
$a \sqcup (b \sqcap c) = (a \sqcup b) \sqcap (a \sqcup c)$

A *valuation* of a formula A in an ortho lattice (or orthomodular lattice) $(L, \leq, \sqcap, \sqcup, ', \mathbf{1}, \mathbf{0})$ is a function v from formulas to L that satisfies the following conditions.

$v(A \wedge B) = v(A) \sqcap v(B)$
$v(\neg A) = v(A)'$

Formula A is *valid* in an ortho lattice (or orthomodular lattice) $(L, \leq, \sqcap, \sqcup, ', \mathbf{1}, \mathbf{0})$ if $V(A) = \mathbf{1}$ for all v.

An *O-frame* is defined by a pair $(X, \perp)$, where X is a non-empty set and $\perp$ is a binary relation on X that is irreflexive and symmetric. Traditionally, the symbol $\perp$ is used both as a relation and as a formula in the contexts of quantum logic. We write $x \perp Y$ if, for all $y \in Y$, $x \perp y$. Given $Y \subseteq X$, the set $Y^{\perp}$ is defined by $\{x \in X | x \perp Y\}$. We say Y is $\perp$-*closed* if $Y^{\perp\perp} = Y$. We write $x \not\perp y$ if $x \perp y$ does not hold. Because $\perp$ is an irreflexive and symmetric relation, $\not\perp$ can be regarded as a reflexive and symmetric relation on X.

$Y \sqcup Z$ is an abbreviation for $(Y^{\perp} \cap Z^{\perp})^{\perp}$. Disjunction $\sqcup$ of an infinite set $Y_i (i \in I)$ is defined by $\bigsqcup_{i \in I} Y_i = (\bigcap_{i \in I} Y_i^{\perp})^{\perp}$.

Lemma 1. (Fundamental natures of $\perp$- closed sets). *For any set $Y \subseteq X$, $Y^{\perp}$ is $\perp$-closed. For any set $Y \subseteq X$, $Y \subseteq Y^{\perp\perp}$. If all $Y_i \subseteq X (i \in I)$ are $\perp$-closed, then $\bigcap_{i \in I} Y_i$ is $\perp$-closed. $\bigcup_{i \in I} Y_i \subseteq \bigsqcup_{i \in I} Y_i$. $\bigsqcup_{i \in I} Y_i$ is the minimum $\perp$-closed set that include all Y_i.*

Proof. We omit the details here. See [2]. □

An *OM-frame* is an O-frame $(X, \perp)$ that satisfies a condition called the *orthomodular law (for frames)*:

For all $\perp$-closed subsets Y, Z of X, $Y \cap (Y^{\perp} \sqcup (Y \cap Z)) \subseteq Z$.

A generally known property is that, although orthomodular lattices and OM-frames do not satisfy the distributive law, part of the distributive law $Y \sqcup (Z \cap W) \subseteq (Y \sqcup Z) \cap (Y \sqcup W)$ and $(Y \cap Z) \sqcup (Y \cap W) \subseteq Y \cap (Z \sqcup W)$ and its generalized form $Y \sqcup (\bigcap_i Z_i) \subseteq \bigcap_i (Y \sqcup Z_i)$ and $\bigsqcup_i (Y \cap Z_i) \subseteq Y \cap (\bigsqcup_i Z_i)$ are always satisfied [2].

An *O-model* is a triple $(X, \perp, V)$, where $(X, \perp)$ is an O-frame and V is a function assigning each propositional variable p to an $\perp$-closed subset of X. The set $\|A\|$ is defined by induction on the composition of A.

$$\|p\| = V(p)$$
$$\|A \wedge B\| = \|A\| \cap \|B\|$$
$$\|\neg A\| = \|A\|^{\perp}$$
$$\|\perp\| = \emptyset$$

Formula A is *true* at $x \in X$ if $x \in \|A\|$, and we write $x \models A$. Formula A is *false* at $x \in X$ if A is not true at x. Formula A is valid in an O-model $(X, \perp, V)$ if $\|A\| = X$. An *OM-model* is an O-model $(X, \perp, V)$ in which $(X, \perp)$ is an OM-frame.

The definition of an OM-frame and an OM-model in this paper are slightly different from those in [2,6]. We discuss this difference in Sect. 4, and we can confirm that there is no essential difference between these definitions.

A formula A is valid in all orthomodular lattices if and only if (iff) A is valid in all OM-models [2].

We say a relation R' in a frame satisfies *partial functionality* if R' has the following nature.

For all $x, y, z \in X$, if $x(R')y$ and $x(R')z$, then $y = z$.

3   DOM-Frame

In this section, a *DOM-frame* and a *DOM-model* are introduced as advanced models for quantum logic.

A *DO-frame* $(X, \perp, R)$ is derived from an O-frame $(X', \perp')$ as follows:

$X = X'$ and $\perp = \perp'$.
R is the set of all $R_{Y?}$. Y is any $\perp$-closed subset of X.
$R_{Y?}$ is the binary relation on X such that, for any $x \in X$ and $y \in X$, $(x, y) \in R_{Y?}$ iff $Y \cap (Y^{\perp} \sqcup Z)$ includes y for any $\perp$-closed subset Z of X that includes x.

A DO-frame $(X, \perp, R)$ is said to be a DOM-frame if $(X', \perp')$ is an OM-frame. We denote $x(Y?)y$ for $(x, y) \in R_{Y?}$.

A DOM-frame is essentially identical to an OM-frame because $X = X'$, $\perp = \perp'$, and R is defined from X and $\perp$.

A *DO-model* $(X, \perp, R, V)$ is defined using the same definition of V as in an O-model. A DO-model $(X, \perp, R, V)$ is defined as a *DOM-model* if $(X', \perp')$ is an OM-frame.

The set $Y \cap (Y^\perp \sqcup Z)$ (or formula $A \wedge (\neg A \vee B)$) is called the *Sasaki projection*. In a Hilbert space, this formula is true after a state in which B is true is projected to a subspace in which A is true. Therefore, intuitively, R expresses the change in state given by the projection. Note that this explanation is just a intuition because $R_{Y?}$ does not necessarily satisfy partial functionality. For example, the relation $\{c\}^\perp?$ in the DOM-frame which is derived from the OM-frame $(\{a, b, c, d, e\}, \{a \perp b, c \perp d, d \perp e, e \perp c\})$ does not satisfy partial functionality because $a(\{c\}^\perp?)d$ and $a(\{c\}^\perp?)e$.

A DOM-model is essentially identical to an OM-model because $X = X'$, $\perp = \perp'$, and the definition of V is identical to that in an OM-model. Therefore, the same definition of truth and validity can be used for a formula. Hence, the next Theorem holds.

Theorem 1. *A formula A is valid in all DOM-models iff A is valid in all OM-models.*

In any DOM-frame, for any $\perp$-closed subset Y, if $x \in Y$, then $x(Y?)x$ because for any $\perp$-closed subset Z that includes x, $Z \subseteq Z \sqcup Y^\perp$ and $Y \subseteq Y$. We say a $\perp$-closed set Y is *self-adjoint* in a DO-frame $(X, \perp, R)$ if for every $x, y, z \in X$, if $x(Y?)y$ and $y \not\perp z$, there exists some $w \in X$ such that $z(Y?)w$ and $w \not\perp x$. This idea can be used like a frame condition for the orthomodular law. That is, a DO-frame $(X, \perp, R)$ is actually a DOM-frame iff all $\perp$-closed sets Y are self-adjoint in $(X, \perp, R)$. To prove this theorem, the following lemmas are required.

Lemma 2. *In any OM-frame, for any $\perp$-closed subset Y, Z, W_i, if $Y \cap (\bigcap_i (W_i \sqcup Y^\perp)) \subseteq Z$, then $\bigcap_i W_i \subseteq Y^\perp \sqcup (Y \cap Z)$.*

Proof. Suppose $Y \cap (\bigcap_i (W_i \sqcup Y^\perp)) \subseteq Z$. From $Y \cap (\bigcap_i (W_i \sqcup Y^\perp)) \subseteq Y$, $Y \cap (\bigcap_i (W_i \sqcup Y^\perp)) \subseteq Y \cap Z$. Therefore, $Y^\perp \sqcup (Y \cap (\bigcap_i (W_i \sqcup Y^\perp))) \subseteq Y^\perp \sqcup (Y \cap Z)$. From the (part of) distributive law and orthomodular law, $\bigcap_i W_i = (\bigsqcup_i W_i^\perp)^\perp \subseteq (Y \cap (Y^\perp \sqcup (Y \cap (\bigsqcup_i W_i^\perp))))^\perp \subseteq (Y \cap (Y^\perp \sqcup (\bigsqcup_i (Y \cap W_i^\perp))))^\perp = Y^\perp \sqcup (Y \cap (\bigcap_i (W_i \sqcup Y^\perp)))$. Therefore, $\bigcap_i W_i \subseteq Y^\perp \sqcup (Y \cap Z)$. □

Lemma 3. *In any DOM-frame $(X, \perp, R)$, if $x \notin Y$, then there exists some $y \in X$ such that $x(Y^\perp?)y$.*

Proof. Suppose $x \in Y^\perp$. Then, x is chosen as y, because $x(Y^\perp?)x$ if $x \in Y^\perp$.

Suppose $x \notin Y^\perp$. The expression $\{Z_0, Z_1, ...\}$ denotes the set of all $\perp$-closed subsets that include x. For the sake of contradiction, suppose that there are no $y \in X$ such that $y \in Y^\perp \cap (Y \sqcup Z)$ for all $Z \in \{Z_0, Z_1, ...\}$. That is, $\bigcap_i (Y^\perp \cap (Y \sqcup Z_i))$ is the empty set. From Lemma 2, $Z_0 \cap Z_1 \cap ... \subseteq Y$ because $Y^{\perp\perp} \sqcup (Y^\perp \cap \phi) = Y$. This contradicts $x \notin Y$. □

We now introduce new sets $[Y?]Z$ which are similar to those in *dynamic logic*.

$$[Y?]Z = \{x \in X | \text{ for all } y \in X, \text{ if } x(Y?)y, \text{ then } y \in Z. \}$$

The theorem that is similar to the following theorem is already shown in the context of dynamic quantum logic [1]. However, since the premise is different, we have to prove it here as well.

Lemma 4 (Fundamental nature of the Sasaki arrow). *In any DO-frame* $(X, \perp, R)$ *and for any* $\perp$*-closed subset* Y *that is self-adjoint,* $[Y?]Z = Y^{\perp} \sqcup (Y \cap Z)$ *for any* $\perp$*-closed subset* Z.

Proof. Suppose $x \in X$ and $x \notin Y^{\perp} \sqcup (Y \cap Z)$. Then, there exists some $y \in X$ such that $x \not\perp y$ and $y \in Y \cap (Y \cap Z)^{\perp}$. From $y(Y?)y$ and the self-adjoint property, there exists some $z \in X$ such that $x(Y?)z$ and $z \not\perp y$. As $y \in (Y \cap Z)^{\perp}$, $z \notin (Y \cap Z)$. Therefore, $z \notin Z$ and $z \in Y$ because $x(Y?)z$. Therefore, $x \notin [Y?]Z$.

Suppose $x \in X$ and $x \notin [Y?]Z$. Then, there exists some $y \in X$ such that $x(Y?)y$ and $y \notin Z$. Therefore, there exists some $z \in X$ such that $y \not\perp z$ and $z \in Z^{\perp}$. From the self-adjoint property, there exists $w \in X$ such that $z(Y?)w$ and $w \not\perp x$. From $z \in Z^{\perp}$ and $z(Y?)w$, $w \in Y \cap (Y^{\perp} \sqcup Z^{\perp})$. Therefore, $x \notin (Y \cap (Y^{\perp} \sqcup Z^{\perp}))^{\perp} = Y^{\perp} \sqcup (Y \cap Z)$ because $w \not\perp x$. □

Theorem 2. *A DO-frame* $(X, \perp, R)$ *is a DOM-frame iff all* $\perp$*-closed subsets of* X *are self-adjoint.*

Proof. ($\Rightarrow$) Suppose $(X, \perp, R)$ is a DOM-frame and suppose x, y and z satisfy $x, y, z \in X$, $x(Y?)y$, and $y \not\perp z$. From $y \in Y$, $z \notin Y^{\perp}$. From Lemma 3, there exists some $w \in X$ that satisfies $z(Y^{\perp\perp}?)w(= z(Y?)w)$. For the sake of contradiction, suppose there is no w such that $w \not\perp x$. Now, $\{Z_0, Z_1, ...\}$ can be denoted as the set of all $\perp$-closed subsets of X that include z. Then, $x \in (Y \cap (Y^{\perp} \sqcup Z_0) \cap (Y^{\perp} \sqcup Z_1)...)^{\perp}$ because x is orthogonal to all w that are included in $(Y \cap (Y^{\perp} \sqcup Z_0) \cap (Y^{\perp} \sqcup Z_1)...)$. From the (part of) distributive law, $(Y \cap (Y^{\perp} \sqcup Z_0) \cap (Y^{\perp} \sqcup Z_1)...)^{\perp} \subseteq (Y \cap (Y^{\perp} \sqcup (Z_0 \cap Z_1 \cap ...)))^{\perp} = Y^{\perp} \sqcup (Y \cap (Z_0 \cap Z_1 \cap ...)^{\perp})$. From $x(Y?)y$ and the orthomodular law, $y \in Y \cap (Y^{\perp} \sqcup (Y^{\perp} \sqcup (Y \cap (Z_0 \cap Z_1 \cap ...)^{\perp}))) = Y \cap (Y^{\perp} \sqcup (Y \cap (Z_0 \cap Z_1 \cap ...)^{\perp})) \subseteq (Z_0 \cap Z_1 \cap ...)^{\perp}$, which contradicts $y \not\perp z$.

($\Leftarrow$) Suppose all $\perp$-closed subsets are self-adjoint in a DO-frame $(X, \perp, R)$. Then, from Lemma 4, $Y \cap Y^{\perp} \sqcup (Y \cap Z) = Y \cap [Y?]Z$. Therefore, if $x \in Y \cap Y^{\perp} \sqcup (Y \cap Z)$, then $x \in Z$ for any $x \in X$ because $x(Y?)x$ for any Y if $x \in Y$. □

In the context of dynamic quantum logic, relations for projections are introduced from the beginning, and self-adjointness is introduced as a condition on the frames for **QTS** [1]. In the case of a DOM-frame, self-adjointness is the result of a theorem, not a condition. OM-frames do not include the notion of projections, so their, self-adjointness cannot be easily confirmed. From Theorems 1 and 2, it can be concluded that a DOM-frame is a natural expansion of an OM-frame.

Now we define the atomicity and covering law. In an orthomodular lattice $(L, \leq, \sqcap, \sqcup, ', \mathbf{1}, \mathbf{0})$, we define the relation $a \lessdot b$ as follows.

$a \lessdot b \Leftrightarrow a < b$ and if $a \leq c \leq b$, then $c = a$ or $c = b$.

An *atomic element* of $(L, \leq, \sqcap, \sqcup, ', \mathbf{1}, \mathbf{0})$ is defined as a such that $0 \lessdot a$. We say an orthomodular lattice satisfies the atomicity if the lattice satisfies following condition.

For all $a \in L$, $a = \bigsqcup \{b | b \leq a$ and b is an atomic element$\}$.

We say an orthomodular lattice satisfies the covering law if the lattice satisfies following condition.

For all $a \in L$ and all atomic elements $b \in L$, if $b \not\leq a$, then $a \lessdot a \sqcup b$.

Atomicity and covering law is defined in almost the same way in OM-frames. We define $Y \prec Z$ for $\perp$-closed subsets Y, Z as follows.

$Y \prec Z \Leftrightarrow Y \subset Z$ and if $Y \subseteq W \subseteq Z$ and W is $\perp$-closed, then $W = Y$ or $W = Z$.

A $\perp$-closed subset Y is defined as an *atomic set* if $\emptyset \prec Y$. We say that an OM-frame has atomicity if the frame satisfies the following condition.

For all $\perp$-closed subset $Y \subseteq X$, $Y = \bigsqcup \{Z | Z \subseteq Y$ and Z is an atomic set.$\}$

We say that an OM-model satisfies the covering law if the frame satisfies the following condition.

For all $\perp$-closed subsets Y and all atomic sets Z, if $Z \not\subseteq Y$, then $Y \prec Y \sqcup Z$.

We now introduce the concept of *strong atomicity* because it is appropriate for DOM-frames. This change does not affect the essence of our argument. We say that an OM-model has strong atomicity if all singletons $\{x\}(x \in X)$ are $\perp$-closed. This notion is justified by the next lemma.

Lemma 5. *If an OM-frame $(X, \perp, R)$ has strong atomicity, then $(X, \perp, R)$ has atomicity.*

Proof. $\bigsqcup_{y \in Y} \{y\} = (\bigcap_{y \in Y} \{y\}^{\perp})^{\perp} = (\{z | \text{for all } y \in Y, y \perp z\})^{\perp} = (Y^{\perp})^{\perp} = Y.$ □

The definition of $R_{Y?}$ for a DO-frame can be replaced by the following if the model has strong atomicity, because if $Z \subseteq Z'$, then $Y \cap (Y^{\perp} \sqcup Z) \subseteq Y \cap (Y^{\perp} \sqcup Z')$.

$R_{Y?}$ is the binary relation on X such that, for any $x \in X$ and $y \in X$, $(x, y) \in R_{Y?}$ iff $Y \cap (Y^{\perp} \sqcup \{x\})$ includes y.

Strong atomicity and the covering law can be represented by simple condition on a DOM-frame.

Lemma 6. *In any DOM-frame* $(X, \perp, R)$, *if* $x(Y?)y$, *then* $x \notin Y^{\perp}$.

Proof. If $x(Y?)y$ and $x \in Y^{\perp}$, then $y \in Y \cap (Y^{\perp} \sqcup Y^{\perp}) = \emptyset$. This is a contradiction. $\qquad\square$

Theorem 3. *An OM-frame* $(X, \perp)$ *satisfies both strong atomicity and the covering law iff all relations in* R *for the DOM-frame derived from* $(X, \perp)$ *satisfy partial functionality.*

Proof. ($\Rightarrow$) For the sake of contradiction, suppose that the DOM-frame $(X, \perp, R)$ satisfies both strong atomicity and the covering law, but there exist some $x \in X$ and $W \subseteq X$ such that $2 \leq |\{y | x(W?)y\}|$. From the strong atomicity, the definition of $W?$, and Lemma 5, $\{y | x(W?)y\} = W \cap (W^{\perp} \sqcup \{x\})$. From Lemma 6 and $y(W?)y$, $x, y \notin W^{\perp}$. Then, from the covering law, $W^{\perp} \prec W^{\perp} \sqcup \{x\}$ and, for all $y \in W \cap (W^{\perp} \sqcup \{x\})$, $W^{\perp} \prec W^{\perp} \sqcup \{y\}$. From $\{y\} \subseteq W^{\perp} \sqcup \{x\}$ and $W^{\perp} \subseteq W^{\perp} \sqcup \{x\}$, $W^{\perp} \sqcup \{y\} \subseteq W^{\perp} \sqcup \{x\}$. Therefore, from $W^{\perp} \prec W^{\perp} \sqcup \{x\}$, $W^{\perp} = W^{\perp} \sqcup \{y\}$ or $W^{\perp} \sqcup \{y\} = W^{\perp} \sqcup \{x\}$. Suppose that $W^{\perp} = W^{\perp} \sqcup \{y\}$. Then, $W \cap W^{\perp} \neq \emptyset$ because $\{y\} \subseteq W \cap W^{\perp}$. This is a contradiction. Suppose that $W^{\perp} \sqcup \{y\} = W^{\perp} \sqcup \{x\}$. Then, $W \cap (W^{\perp} \sqcup \{x\}) = W \cap (W^{\perp} \sqcup \{y\})$. From the orthomodular law and $\{y\} \subseteq W$, $\{y\}^{\perp} \subseteq W^{\perp} \sqcup (W \cap \{y\}^{\perp}) = (W \cap (W^{\perp} \sqcup \{y\}))^{\perp}$. Therefore, $W \cap (W^{\perp} \sqcup \{x\}) = W \cap (W^{\perp} \sqcup \{y\}) \subseteq \{y\}$. This result contradicts $2 \leq |W \cap (W^{\perp} \sqcup \{x\})|$.

($\Leftarrow$) Suppose all relations in R for the DOM-frame $(X, \perp, R)$ satisfies partial functionality. From the definition of R and $X \cap (X^{\perp} \sqcup Y) = Y$, $x(X?)y$ iff, for all Y such that $x \in Y$, $y \in Y$. Therefore, from $x(X?)x$ and the partial functionality, if $x \neq y$, there exists some $\perp$-closed subset Z such that $x \in Z$ and $y \notin Z$. The conjunction of all such subsets is $\{x\}$. Therefore, from Lemma 1, $\{x\}$ is $\perp$-closed. Therefore, $(X, \perp, R)$ has strong atomicity.

For the covering law, for the sake of contradiction, suppose W, U are $\perp$-closed, $W \subset U \subset W \sqcup \{x\}$, and $x \notin W$. From the nature of the least upper bound $\sqcup$, $W \sqcup \{x\} = U \sqcup \{x\}$. From $U^{\perp} \subseteq W^{\perp}$, $U^{\perp} \cap (U \sqcup \{x\}) = U^{\perp} \cap (W \sqcup \{x\}) \subseteq W^{\perp} \cap (W \sqcup \{x\})$. From the partial functionality and strong atomicity, $W^{\perp} \cap (W \sqcup \{x\})$ is a singleton. Therefore, $U^{\perp} \cap (U \sqcup \{x\}) = \emptyset$ or $U^{\perp} \cap (U \sqcup \{x\}) = W^{\perp} \cap (W \sqcup \{x\})$. Suppose that $U^{\perp} \cap (U \sqcup \{x\}) = \emptyset$. Then, from the orthomodular law, $U^{\perp} \cap \{x\}^{\perp} = U^{\perp} \cap ((U^{\perp} \cap (U \sqcup \{x\}))^{\perp}) = U^{\perp} \cap \emptyset^{\perp} = U^{\perp}$. Therefore, $U = U \sqcup \{x\}$. This result contradicts $U \neq W \sqcup \{x\}$. Suppose that $U^{\perp} \cap (U \sqcup \{x\}) = W^{\perp} \cap (W \sqcup \{x\})$. Then, $W^{\perp} \cap (W \sqcup \{x\}) \subseteq U^{\perp}$. Therefore, from the orthomodular law and $(W \sqcup \{x\})^{\perp} \subseteq W^{\perp}$, $W^{\perp} \subseteq (W \sqcup \{x\})^{\perp} \sqcup (W^{\perp} \cap (W \sqcup \{x\})) \subseteq (W \sqcup \{x\})^{\perp} \sqcup U^{\perp} = U^{\perp}$. This result contradicts $W \subset U$. $\qquad\square$

Actually, the OM-frame $(\{a, b, c, d, e\}, \{a \perp b, c \perp d, d \perp e, e \perp c\})$ does not satisfy the covering law because $\{c\} \subset \{d\}^{\perp} \subset \{a\} \sqcup \{c\}$.

In the context of dynamic quantum logic, atomicity, the covering law, and partial functionality are separately defined. The relations of projections are expressly defined in **QTS**. Therefore, we have to define each condition individually. In the case of DOM-frames, these conditions are mutually related.

4   Related Sequent Calculus and Future Work

In this section, a sequent calculus including the Sasaki arrow is constructed and the relation with DOM-models is discussed. Some issues are still under discussion.

We say that a sequent $\Gamma \Rightarrow \Delta$ is *true* at $x \in X$ of an OM-model if, whenever A is true at x for all $A \in \Gamma$, B is true at x for some $B \in \Delta$. If $\Gamma \Rightarrow \Delta$ is not true at x, then it is *false* at x. If $\Gamma \Rightarrow \Delta$ is true at all x in an OM-model, then $\Gamma \Rightarrow \Delta$ is *valid* in the OM-model.

Before introducing the sequent calculus, we argue a subtle differences between the definitions of OM-frames (models) in this paper and those in [2,6]. A *TOM-frame* (Traditional OM-frame), which is called OM-frame in [2,6], is defined as a triple $(X, \perp, \Psi)$ where $(X, \perp)$ is an O-frame, and Ψ is a nonempty collection of $\perp$-closed subsets of X that satisfies following conditions:

Ψ is closed under set intersection and the operation $^\perp$.
For all $Y, Z \in \Psi$, $Y \cap (Y^\perp \sqcup (Y \cap Z)) \subseteq Z$.

A *TOM-model* (Traditional OM-model), which is called OM-model in [2,6] is defined as a 4-tuple $(X, \perp, \Psi, V)$ where $(X, \perp, \Psi)$ is a TOM-frame and V is a function assigning each propositional variable p to an element of Ψ. In a TOM-model, only Ψ is required to be an orthomodular. We adopted an OM-frame in this paper because it is appropriate to deal with new relations $R_{Y?}$.

OM-models and TOM-models are essentially equivalent in the sense of the next lemma.

Lemma 7. *A formula A is valid in all OM-models iff A is valid in all TOM-models.*

Proof. ($\Rightarrow$) Suppose A is false at some $x \in X$ in a TOM-model $(X, \perp, \Psi, V)$. We make an OM-model $(X_2, \perp_2, V_2)$ from $(X, \perp, \Psi, V)$ which preserve the truth of formulas. First, we construct a lattice $(L, \leq, \sqcap, \sqcup, ', \mathbf{1}, \mathbf{0})$ and v from $(X, \perp, \Psi, V)$ as follows:

$$L = \{Y \subseteq X | Y = \|B\| \text{ for some formulas } B\}, Y \leq Z \Leftrightarrow Y \subseteq Z, Y \sqcap Z \Leftrightarrow$$
$$Y \cap Z, Y \sqcup Z \Leftrightarrow Y \sqcup Z, Y' \Leftrightarrow Y^\perp, \mathbf{1} = \mathbf{X}, \mathbf{0} = \emptyset, v(p) = V(p).$$

It is easy to confirm that this lattice is an orthomodular lattice and $v(A) \neq \mathbf{1}$. Next, we construct an OM-model $(X_2, \perp_2, V_2)$ from $(L, \leq, \sqcap, \sqcup, ', \mathbf{1}, \mathbf{0})$ as follows:

$$X_2 = L - \mathbf{0}, \ x \perp_2 y \Leftrightarrow x \leq y', \ V_2(p) = \{x \in X_2 | x \leq v(p)\}.$$

By using similar notions in [2], we can check that $(X_2, \perp_2, V_2)$ is an O-model, $\{Y \subseteq X_2 | Y = \|B\| \text{ for some formulas } B\}$ (we denote $\mathcal{F}$ for this set) is identical to above orthomodular lattice, and $\|A\| \neq X$. Furthermore, in this model, not only $V_2(p)$ but also all $\|A\|$ are expressed by $\{x \in X_2 | x \leq v(A)\}$. We can prove that this model is actually an OM-model by proving that there are no $\perp$-closed sets other than those in $\mathcal{F}$ in this model.

For the sake of contradiction, suppose that Y is a $\perp$-closed set of X_2 and $Y \notin \mathcal{F}$. From the nature of this model described above, each $y \in Y$ correspond to an element of L, and $\{x \in X_2 | x \leq y\}$ (we denote $U(y)$ for this set) is identical to some $\|B\|$. From $y \in U(y)$, $Y \subseteq \bigcup_{y \in Y} U(y)$ is satisfied.

Suppose $Y \subset \bigcup_{y \in Y} U(y)$. Then, there exist $y \in Y$ and $z \in U(y)$ such that $U(y) \not\subseteq Y$ and $z \notin Y$. From Lemma 1, $U(y) \cap Y$ must be a $\perp$-closed set. However, $U(y) \cap Y$ is not $\perp$-closed because from the definition of the relation $\perp_2$ and $y \perp_2 (U(y) \cap Y)^\perp$, $z \perp_2 (U(y) \cap Y)^\perp$. This is a contradiction.

Suppose $Y = \bigcup_{y \in Y} U(y)$. Then, from Lemma 1, $Y \subseteq \bigsqcup_{y \in Y} U(y)$. If $Y = \bigsqcup_{y \in Y} U(y)$, then $Y \in \mathcal{F}$. This is a contradiction. If $Y \subset \bigsqcup_{y \in Y} U(y)$, since Y is $\perp$-closed, this is also a contradiction because of the last statement of Lemma 1.

($\Leftarrow$) It is obvious because an OM-model is essentially a TOM-model if we define Ψ as a set of all $\perp$-closed sets of the OM-model. $\square$

GOMS

Axiom:
$$A \Rightarrow A \quad \perp \Rightarrow$$
Rules:
$$\frac{\Gamma \Rightarrow \Delta, A \quad A, \Pi \Rightarrow \Sigma}{\Gamma, \Pi \Rightarrow \Delta, \Sigma} \text{ (cut)}$$

$$\frac{\Gamma \Rightarrow \Delta}{\Pi, \Gamma \Rightarrow \Delta, \Sigma} \text{ (weakening)}$$

$$\frac{A, \Gamma \Rightarrow \Delta}{A \wedge B, \Gamma \Rightarrow \Delta} (\wedge L) \quad \frac{B, \Gamma \Rightarrow \Delta}{A \wedge B, \Gamma \Rightarrow \Delta} (\wedge L) \quad \frac{\Gamma \Rightarrow \Delta, A \quad \Gamma \Rightarrow \Delta, B}{\Gamma \Rightarrow \Delta, A \wedge B} (\wedge R)$$

$$\frac{\Gamma \Rightarrow \Delta, A}{\neg A, \Gamma \Rightarrow \Delta} (\neg L) \quad \frac{A \Rightarrow \Delta}{\neg \Delta \Rightarrow \neg A} (\neg R)$$

$$\frac{A, \Gamma \Rightarrow \Delta}{\neg\neg A, \Gamma \Rightarrow \Delta} (\neg\neg L) \quad \frac{\Gamma \Rightarrow \Delta, A}{\Gamma \Rightarrow \Delta, \neg\neg A} (\neg\neg R)$$

$$\frac{\Gamma \Rightarrow \Delta, A \quad B, \Gamma \Rightarrow \Delta}{\Gamma, A \rightarrow_s B \Rightarrow \Delta} (\rightarrow L) \quad \frac{\Gamma_{P(A)}, A \Rightarrow B}{\Gamma \Rightarrow A \rightarrow_s B} (\rightarrow R)$$

$$\frac{\neg B \Rightarrow \neg A \quad \neg A, B \Rightarrow}{\neg A \Rightarrow \neg B} \text{ (OM)}$$

where $\Gamma_{P(A)} = \{(\neg A \vee C) \wedge A \mid C \in \Gamma\}$.

This sequent calculus is the expansion of **GOM** in [6].

Theorem 4 (Soundness theorem for GOMS). *If $\Gamma \Rightarrow \Delta$ is provable in* **GOMS**, *then $\Gamma \Rightarrow \Delta$ is valid in all OM-models.*

Proof. We proceed by induction on the construction of a proof of the sequent $\Gamma \Rightarrow \Delta$. In the cases of the rules for $\neg$ and $\wedge$, see [6]. For rule $(\rightarrow R)$, the proof is obvious from Lemma 2. In the case of rule $(\rightarrow L)$, the proof uses the fact that the orthomodular law implies that the Sasaki arrow satisfies modus ponens. $\square$

This calculus satisfies an important theorem called the disjunction theorem. Note that this theorem is different from the *disjunction property* in intuitionistic logic. The disjunction property concerns the provability of formulas, whereas the disjunction theorem is concerned with the provability of sequents.

Theorem 5 (Disjunction theorem for GOMS). *If* $\Gamma \Rightarrow \Delta$ *is provable in* **GOMS**, *there exists some* $A \in \Delta$ *such that* $\Gamma \Rightarrow A$ *is provable in* **GOMS**.

Proof. We proceed by induction on the construction of a proof of the sequent $\Gamma \Rightarrow \Delta$. This theorem is already established in the case of **GOM** [6]. Therefore, we deal with the cases of the rules for the implication. For ($\rightarrow$R), the proof is easy because the right side of the sequent in this rule includes at most one formula. For ($\rightarrow$L), suppose $\Gamma \Rightarrow C$ is provable. Then, $\Gamma, A \rightarrow_S B \Rightarrow C$ is provable by weakening. Next, suppose $\Gamma \Rightarrow A$ is provable and $B, \Gamma \Rightarrow C$ is provable. Then, $\Gamma, A \rightarrow_S B \Rightarrow C$ is provable by ($\rightarrow$L).   □

Lemma 8. $\neg A \vee (A \wedge B) \Rightarrow A \rightarrow_S B$ *and* $A \rightarrow_S B \Rightarrow \neg A \vee (A \wedge B)$ *are provable in* **GOMS**.

Proof. In the proofs below, some deductions that are obvious from the soundness and completeness theorem for **GOM** have been omitted.

$$\frac{\dfrac{\vdots \qquad\qquad\qquad \vdots}{(\neg A \vee (\neg A \vee (A \wedge B))) \wedge A \Rightarrow (\neg A \vee (A \wedge B)) \wedge A \quad (\neg A \vee (A \wedge B)) \wedge A \Rightarrow B}}{\dfrac{(\neg A \vee (\neg A \vee (A \wedge B))) \wedge A, A \Rightarrow B}{\neg A \vee (A \wedge B) \Rightarrow A \rightarrow_S B}}$$

$$\frac{\dfrac{\dfrac{\dfrac{\dfrac{A \Rightarrow A \qquad\qquad \dfrac{\dfrac{\dfrac{\dfrac{\dfrac{A \Rightarrow A \quad B \Rightarrow B}{B, A \Rightarrow A \quad B, A \Rightarrow B}}{B, A \Rightarrow A \wedge B}}{B, A, \neg(A \wedge B) \rightarrow}}{B, A, A \wedge \neg(A \wedge B) \Rightarrow}}{A \wedge \neg(A \wedge B) \Rightarrow A \qquad B, A \wedge \neg(A \wedge B) \Rightarrow}}{A \rightarrow_S B, A \wedge \neg(A \wedge B) \Rightarrow}}{\dfrac{\neg A \vee (A \wedge B) \Rightarrow A \rightarrow_S B \qquad\qquad \neg\neg(A \rightarrow_S B), A \wedge \neg(A \wedge B) \Rightarrow}{}}}{\dfrac{\neg(A \wedge \neg(A \wedge B)) \Rightarrow \neg\neg(A \rightarrow_S B)}{\neg\neg(A \rightarrow_S B) \Rightarrow \neg A \vee (A \wedge B)}} \text{(OM)}$$

$$\vdots$$

$$\frac{A \rightarrow_S B \Rightarrow \neg\neg(A \rightarrow_S B) \quad \neg\neg(A \rightarrow_S B) \Rightarrow \neg A \vee (A \wedge B)}{A \rightarrow_S B \Rightarrow \neg A \vee (A \wedge B)} \text{(cut)}$$

□

The completeness theorem for **GOMS** is proved using Lemmas 7, 8, and the fact that **GOM** is complete with respect to TOM-models. Note that Lemma 7 can be easily expanded for validity of sequent.

Theorem 6 (Completeness theorem for GOMS). *If* $\Gamma \Rightarrow \Delta$ *is valid in all OM-models, then* $\Gamma \Rightarrow \Delta$ *is provable in* **GOMS**.

Proof. The sequent $\Gamma' \Rightarrow \Delta'$ is constructed from $\Gamma \Rightarrow \Delta$ by changing all appearances of the Sasaki arrow $A \rightarrow_S B$ to $\neg A \vee (A \wedge B)$ for every A and B. As $\Gamma \Rightarrow \Delta$ is valid in all OM-models, $\Gamma' \Rightarrow \Delta'$ is provable in **GOM** because of Lemma 7 and the completeness theorem for **GOM** in [6]. Suppose T is a proof of $\Gamma' \Rightarrow \Delta'$ in **GOM**. In **GOMS**, the formula $\neg A \vee (A \wedge B)$ can always be changed to $A \rightarrow_S B$ in T because of the cut rule and Lemma 8. Therefore, the proof of $\Gamma \Rightarrow \Delta$ in **GOMS** can be constructed using T. $\square$

The following analysis cannot be used in the proof of the completeness theorem, but it will be useful for understanding the rule ($\rightarrow$R) and DOM-models. In general, a completeness theorem is proved using a canonical model $(X_c, \perp_c, V_c)$. Such a model is constructed using a class of sets of formulas or a set of sequents. In the case of **GOM**, Nishimura used the set of unprovable sequents [6]. In this model, A is true at $\Gamma \Rightarrow \Delta \in X_c$ iff $A \in \Gamma$. From this analysis, we can see the features of the rule ($\rightarrow$R), as shown below. In a canonical model, if $(\Gamma \Rightarrow A \rightarrow_S B, \Delta) \in X_C$, then $\Gamma \Rightarrow A \rightarrow_S B, \Delta$ is not provable in **GOMS**. Thus, $\Gamma_{P(A)}, A \Rightarrow B$ is also not provable. Then, there is a sequent $(\Gamma', \Gamma_{P(A)}, A \Rightarrow B, \Delta') \in X_C$ that is not provable and an expansion of $\Gamma_{P(A)}, A \Rightarrow B$. Therefore, $A \rightarrow_S B$ is not true at $(\Gamma \Rightarrow A \rightarrow_S B, \Delta) \in X_C$, as this element is related to $(\Gamma', \Gamma_{P(A)}, A \Rightarrow B, \Delta') \in X_C$ by $R_{\|A\|?}$ because of Γ and $\Gamma_{P(A)}$. We cannot use these notions in the proof of completeness theorem because $\Gamma_{P(A)}$ are not subformulas of $\Gamma \Rightarrow A \rightarrow_S B$. That is, we cannot confirm the truth of $\Gamma_{P(A)}$ before the argument of $\Gamma \Rightarrow A \rightarrow_S B$. Furthermore, we cannot confirm the orthomodular law for all $\perp$- closed sets of X_c because formulas express only a part of them.

It is conjectured that the set of all valid formulas (or sequents) in all OM-models is equal to the set of all valid formulas (or sequents) in all OM-models that satisfy atomicity and the covering law. That is, it is conjectured that atomicity and the covering law cannot be expressed by formulas. It is difficult to express the relation $\prec$ by formulas of quantum logic. Using the above notion, the rule for the partial functionality of $R_{Y?}$ can be constructed as follows.

$$\frac{\Gamma_{P(A)}, A \Rightarrow \Delta}{\Gamma \Rightarrow \Delta_{S(A)}} \; (\rightarrow_S \text{RP})$$

where $\Delta_{S(A)} = \{A \rightarrow_S B \mid B \in \Delta\}$.

The concept of this rule is identical to the rule for partial functionality in modal logic, which states that multiple modal formulas are allowed on the right side of the sequent.

$$\frac{\Gamma \Rightarrow A}{\Box \Gamma \Rightarrow \Box A} \qquad \rightarrow \qquad \frac{\Gamma \Rightarrow \Delta}{\Box \Gamma \Rightarrow \Box \Delta} \; (\text{PF})$$

However, ($\rightarrow_S$ RP) is admissible in **GOMS** because of Theorem 5, ($\rightarrow$R) and the weakening rule. This result strengthens the assurance of the proposition that atomicity and the covering law cannot be expressed by formulas, sequents or rules in quantum logic. The strict proof of this proposition remains under discussion because of detailed problems. In the context of dynamic quantum logic, it is already known that a similar condition can be expressed by rules or formulas using *classical negation* [1].

Acknowledgements. I would like to thank R. Kashima and the anonymous reviewers for helpful comments on earlier version of this paper.

References

1. Baltag, A., Smets, S.: The dynamic turn in quantum logic. Synthese **186**(3), 753–773 (2012)
2. Chiara, M.L.D., Giuntini, R.: Quantum logics. In: Gabbay, D.M., Guenthner, F. (eds.) Handbook of Philosophical Logic, vol. 6, 2nd edn, pp. 129–228. Springer, Heidelberg (2002). https://doi.org/10.1007/978-94-017-0460-1_2
3. Foulis, D.J., Randall, C.H.: Lexicographic orthogonality. J. Comb. Theory **11**, 157–162 (1971)
4. Hardegree, G.M.: Material implication in orthomodular (and Boolean) lattices. Notre Dame J. Formal Logic **22**(2), 163–182 (1981)
5. Hedlíková, J., Pulmannová, S.: Orthogonality spaces and atomistic orthocomplemented lattices. Czech. Math. J. **41**(1), 8–23 (1991)
6. Nishimura, H.: Sequential method in quantum logic. J. Symbolic Logic **45**(2), 339–352 (1980)
7. Nishimura, H.: Proof theory for minimal quantum logic I. Int. J. Theor. Phys. **33**(1), 103–113 (1994)
8. Nishimura, H.: Proof theory for minimal quantum logic II. Int. J. Theor. Phys. **33**(7), 1427–1443 (1994)
9. Zhong, S.: A formal state-property duality in quantum logic. Stud. Logic. **10**(2), 112–133 (2017)

The Undecidability of Orthogonal and Origami Geometries

J. A. Makowsky[✉]

Department of Computer Science, Technion - Israel Institute of Technology,
Haifa, Israel
janos@cs.technion.ac.il

Abstract. In the late 1950s A. Tarski published an abstract stating that
the set of first order consequences of projective (and also affine) incidence
geometry is undecidable. Although his theorem is cited in many follow
up papers, a detailed proof was not published. To the best of our knowledge, we have not found a detailed complete proof in the literature. In
this paper we analyze what is needed to give a correct proof which is
reconstructible by practitioners of AI and automated theorem proving
and extend the undecidability to many other axiomatizations of geometry. These include the geometry of Hilbert and Eulidiean planes, Wu's
geometry and Origami geometry. We also discuss applications to automated theorem proving.

1 Introduction and Summary of Results

This paper is the result of my efforts in giving a complete proof for graduate students of Logic in Computer Science of the impossibility of building a Geometry
engin. I included this fact in my graduate and advanced undergraduate course on
Automated Theorem proving, which I use to teach regularly for the last 25 years.
To show this impossibility, one only needs techniques from mathematical logic
and classical geometry, which were well known in the 1950s. Nowadays, these
techniques are not anymore part of the general knowledge outside the highly
specialized communities of mathematical logic and geometry. My search in the
literature for a proof, which would be reconstructible for graduate students in
Computer Science and AI, did not lead me to a satisfactory source. Therefore,
I set out to provide for my students all the details needed, and extended the
known results to Wu's orthogonal geometry and to Origami geometry.

Since the beginning of the computer era automated theorem proving in geometry remained a central topic and challenge for artificial intelligence. Already in
the late 1950s, [Gel59, GHL60], H. Gelernter presented a machine implementation of a theorem prover for Euclidean Geometry.

J. A. Makowsky—Partially supported by a grant of Technion Research Authority.
Work done in part while the author was visiting the Simons Institute for the Theory
of Computing in Fall 2016.

L. S. Moss et al. (Eds.): WoLLIC 2018, LNCS 10944, pp. 250–270, 2018.
https://doi.org/10.1007/978-3-662-57669-4_15

The very first idea for mechanizing theorem proving in Euclidean geometry came from the fact that till not long ago high-school students were rather proficient in proving theorems in planimetry using Euclidean style deductions. A modern treatment of Euclidean Geometry was initiated by D. Hilbert at the end of the 19th century [Hil02], and a modern reevaluation of Euclidean Geometry can be found in [Har00].

On the high-school level one distinguishes between *Analytic Geometry* which is the geometry using coordinates ranging over the real numbers, and *Synthetic Geometry* which deals with points and lines with their incidence relation augmented by various other relations such as *equidistance, orthogonality, betweenness, congruence of angles* etc. A geometric statement is, in the most general case, a formula in *second order logic* SOL using these relations. However, it is more likely that for practical purposes full second order logic is rarely used. In fact, all the geometrical theorems proved in [Hil02] are expressible by formulas of of first order logic FOL with very few quantifier alternations, cf. also [ADM09, Mil07]. Instead, one uses statements expressed in a suitable fragment $\mathfrak{F}$ of second order logic, which can be full first order logic FOL (the *Restricted Calculus* in the terminology of [HA50]) or an even more restricted fragment, such as the universal ∀-formulas $\mathfrak{U}$, the existential ∃-formulas $\mathfrak{E}$, or ∀∃-Horn formulas $\mathfrak{H}$ of first order logic.

Many variants of Synthetic Euclidean Geometry are axiomatized in the language of first order logic by a finite set of axioms or axiom schemes $T \subseteq$ FOL if *continuity requirements are discarded*. It follows from the Completeness Theorem of first order logic that the first order consequences of T are recursively (computably) enumerable. If full continuity axioms, which not FOL-expressible, are added even the first order consequences of T are not necessarily recursively enumerable.

A first order statement in the case of Analytic Geometry over the reals is a first order formula ϕ in the language of ordered fields and we ask whether ϕ is true in the ordered field of real numbers. By a celebrated theorem of A. Tarski announced in [Tar31], and proven in [Tar51], this question is mechanically decidable using quantifier elimination. However, the complexity of the decision procedure given by Tarski uses an exponential blowup for each elimination of a quantifier. This has been dramatically improved by Collins in 1975, reprinted in [CJ12], giving a doubly exponential algorithm in the size of the input formula. Further progress was and is slow. For a state of the art discussion, cf. [BPR03, CJ12]. However, it is unlikely that a polynomial time algorithm exists for quantifier elimination for existential formulas over the ordered field of real numbers, because this would imply that in the computational model of Blum-Shub-Smale over the reals $\mathbb{R}$, [BCSS98], we would have $\mathbf{P}_{\mathbb{R}} = \mathbf{NP}_{\mathbb{R}}$, [Poi95, Pru06], which is one of the open Millennium Problems, [CJW06].

So what can a geometry engine for Euclidean Geometry try to achieve?

For a fixed fragment $\mathfrak{F}$ of SOL in the language of Analytic or Synthetic Geometry we look at the following possibilities:

Analytic Tarski Machine ATM($\mathfrak{F}$):

    Input: A first order formula $\phi \in \mathfrak{F}$ in the language of ordered fields.

    Output: true if ϕ is true in the ordered field of real numbers, and **false** otherwise.

Synthetic Tarski Machine STM($\mathfrak{F}$):

    Input: A first order formula $\psi \in \mathfrak{F}$ in a language of synthetic geometry.

    Output I: a translation $\phi = cart(\psi)$ of ψ into the language of analytic geometry.

    Output II: true if ϕ is true in the ordered field of real numbers, and **false** otherwise.

Geometric Theorem Generator GTG($\mathfrak{F}$):

    Input: A recursive set of first order formulas $\mathbf{T} \subseteq$ FOL (not necessarily in $\mathfrak{F}$), in the language of some synthetic geometry.

    Output: A non-terminating sequence $\phi_i :\in \mathbb{N}$ of formulas in $\mathfrak{F}$ the language of the same synthetic geometry which are consequences of $\mathbf{T}$.

Geometric Theorem Checker GTC($\mathfrak{F}$):

    Input: A recursive set of first order formulas $\mathbf{T}$ and another formula $\phi \in \mathfrak{F}$ in the language of some synthetic geometry.

    Output: true if ϕ is a consequence of $\mathbf{T}$, and **false** otherwise.

In the light of the complexity of quantifier elimination over the real numbers, [Poi95, Pru06], designing *computationally feasible* Analytic or Synthetic Tarski Machines for various fragments $\mathfrak{F}$ with the exception of $\mathfrak{U}$ is a challenge both for Automated Theorem Proving (ATP) as well as for Symbolic Computation (SymbComp). Designing Geometric Theorem Generators GTG($\mathfrak{F}$) is possible but seems pointless, because it will always output long sub-sequences of geometric theorems in which we are not interested.

In this paper we will concentrate on the challenge of designing Geometric Theorem Checkers GTC($\mathfrak{F}$). This is possible only for very restricted fragments $\mathfrak{F}$ of FOL, such as the universal formulas $\mathfrak{U}$.

The main purpose of this paper is to bring *negative results* concerning Geometric Theorem Checkers GTC($\mathfrak{F}$) to a *wider audience not trained in the tradition of mathematical logic.*

The negative results are based on a correspondence between sufficiently strong axiomatizations of Synthetic Euclidean Geometries and certain theories of fields consistent with the theory of the ordered field of real numbers.

A model of incidence geometry is an incidence structure which satisfies the axioms I-1, I-2 and I-3 from Sect. 2. An *affine plane* is a model of incidence geometry satisfying the Parallel Axiom (ParAx). An affine plane is *Pappian* is it additionally satisfies the axiom of Pappus (Pappus). In this paper an axiomatization of geometry T is sufficiently strong if all its models are affine planes.

Let $\mathcal{F}$ be a field of characteristic 0. One can construct an Cartesian plane $\Pi(\mathcal{F})$ over $\mathcal{F}$ which satisfies the Pappian axiom, and where all the lines are infinite. This construction is an example of a transduction as defined in Appendix B. On the other side, if Π is an Pappian plane which has no finite lines then one can define inside Π its coordinate field $\mathcal{F}(\Pi)$ which is of characteristic 0.

Proposition 1 (Schur [Sch09] and Artin [Art57])

(i) $\mathcal{F}$ is a field of characteristic 0 iff $\Pi(\mathcal{F})$ is a Pappian plane with no finite lines.

(ii) Π is a Pappian plane with no finite lines iff $\mathcal{F}(\Pi)$ is a field of characteristic 0.

(iii) The fields $\mathcal{F}$ and $\mathcal{F}(\Pi(\mathcal{F}))$ are isomorphic.

(iv) The Pappian planes Π and $\Pi(\mathcal{F}(\Pi))$ are isomorphic as incidence structures.

A theory (set of formulas) $T \subseteq \mathrm{FOL}(\tau)$ is *axiomatizable* if the set of consequences of T is computably enumerable. T is *decidable* if the set of consequences of T is computable. T is *undecidable* if it is not decidable. T is *complete* if for every formula $\phi \in \mathrm{FOL}(\tau)$ without free variables either $T \models \phi$ or $T \models \neg\phi$. We note that if T is axiomatizable and complete, the T is decidable.

On the side of theories of fields we have several undecidability results:

Proposition 2 (Robinson [Rob49])

(i) The theory of fields is undecidable. The same holds for fields of characteristic 0.

(ii) The theory of ordered fields is undecidable.

(iii) The theory of the field of rational numbers $\langle \mathbb{Q}, +, \times \rangle$ is undecidable.

To show that the first order theory of affine geometry is undecidable we would like to use a classical tool from decidability theory.

Proposition 3 (Folklore, based on [TMR53]). *Let I be a first order translation scheme with associated transduction I^* which maps τ-structures into σ-structures. Furthermore, let S be an undecidable first order theory over a relational vocabulary σ and let T be a theory over τ. Assume that I^* maps the models of T **onto** the models of S, and that S is undecidable, then T is also undecidable.*

The onto-condition needed for our purpose appears in [Rab65], but is rarely stated in textbooks. However, it is explicitly stated in [EFT80, Hod93].

Propositions 1 and 2 are not enough to prove that first order theory of affine geometry is undecidable. We have to verify all the conditions of Proposition 3.

In particular, we have to show:

(A) There is a first order translation scheme RF_{field} such that for every Pappian plane Π the structure $RF^*_{field}(\Pi)$ is a field.

(B) There is a first order translation scheme $PP_\in$ such that for every field $\mathcal{F}$ the structure $PP^*_\in(\Pi)$ is an Pappian plane.

(C) For every field $\mathcal{F}$ we have

$$RF^*_{field}(PP^*_\in(\mathcal{F})) \simeq \mathcal{F}.$$

All this is shown in detail [Mak17]. While the existence of $PP_{\in}$ is rather straight-forward, the existence of RF_{field} with the necessary properties (B) and (C) requires the first order definability of the coordinatization of affine planes. If Π is a Hilbert plane or a Euclidean plane, coordinatization can be achieved through segment arithmetic, which can be achieved via a first order translation scheme FF_{field}, which is somehow simpler that RR_{field}.

Only after having established (A) and (C) we can conclude:

Theorem 1

(i) The first order theory of Pappian planes undecidable.

(ii) The first order theory of affine geometry is undecidable.

(ii) follows from (i) because Pappian planes are obtained from affine planes by adding a finite number of axioms in the language of incidence geometry.

The ingredients for proving Theorem 1 were all implicitly available when Proposition 2 was published. I would also assume that Theorem 1 was known in Berkeley, but no detailed proof was written down. A. Tarski presented the result for projective planes at the 11th Meeting of the Association of Symbolic Logic already in 1949, [Zor49]. An incomplete sketch of a proof Theorem 1 was published in 1961 by Rautenberg [Rau61]. His more detailed proof of the projective case in [Rau62] uses Proposition 2 and Lemma 2 from Appendix B, but fails to note that something like Theorem 6 is needed to complete the argument. It also seems that Szmielew planned to include a proof of Theorem 1 in her unfinished and posthumously published [Szm83]. The only complete proof of Theorem 1 I could find in the literature appears in [BGKV07]. However, the arguments contain some fixable errors[1]. One of the purposes of this paper is to give a conceptually clear account of what is needed to prove Theorem 1.

To repeat this argument for other axiomatizations of extensions of affine geometry we need the following theorem of M. Ziegler:

Theorem 2 (Ziegler [Zie82, Bee]). Let T be a finite subset of the theory of the reals $\langle \mathbb{R}, +, \times \rangle$ and let $T^* = T \cup \{\mathbf{n} \neq 0, n \in \mathbb{N}\}$, where $\mathbf{n}$ is shorthand for $\underbrace{1 + \ldots + 1}_{n}$. Then both T and T^* are undecidable.

The same holds if T is a finite subset of the theory of the complex numbers $\langle \mathbb{C}, +, \times \rangle$.

We paraphrase this theorem, following [SV14], by saying that the theory of real closed (algebraically closed) fields of characteristic 0 is finitely hereditarily undecidable.

Theorem 2 was conjectured[2] by Tarski, but only proved in 1982 by Ziegler. Ziegler's Theorem remained virtually unnoticed, having been published in German in the Festschrift in honor of Ernst Specker's 60th birthday, published as

---

[1] In [BGKV07, Sect. 7] the undecidability of affine and projective spaces is stated (Corollary 7.38). It also discusses Proposition 1 in [BGKV07, Sect. 6], but fails to mention that Proposition 1 is needed to prove Theorem 1 (Theorem 1.37 in [BGKV07, Sect. 7]). It also attributes Proposition 2 erroneously to A. Tarski.

[2] In [HR74] a proof was announced, which later was found containing in irreparable mistake, cf. [Zie82].

a special issue of *L'Enseignement Mathématique*. In [SST83] the significance of the results of [Zie82] is recognized. However, the book is written in German and is usually quoted for its presentation of Tarskian geometry. The discussion of Theorem 2 is buried there in the second part of the book dealing with meta-mathematical questions of geometry. This part of the book is difficult to absorb, both because of its pedantic style and its length. In short, the only reference to Theorem 2 within the framework of ATP and SC is [Bee13]. A very short and casual mention of Theorem 2 can also be found in [BGKV07].

Main Contribution of this Paper

The present paper gives a survey on the status of decidability of various axiomatizations of Euclidean Geometry. This includes the seemingly new cases of Wu's metric geometry and the Origami geometry which are all undecidable, see Theorems 7, 9 and 11. None of these theorems are technically new. They all could have been proven with the tools used in the proof of Theorem 1 together with Ziegler's Theorem 2. However, Theorem 7 is stated and proved only in [SST83], and Theorems 9 and 11 could not have been stated before the corresponding geometries were axiomatized. For Wu's orthogonal geometry this would be 1984 respectively 1994, when the first translation from Chinese appeared [Wu94], or 1986 [Wu86]. For Origami geometry this would be at the earliest in 1989, [Jus89], but rather in 2000 with [Alp00].

The purpose of this paper is to discuss undecidability results in geometry addressing practitioners in Automated Theorem Proving, Articial Intelligence, and Symbolic Computation. Although many variants of these results were stated and understood already in the early 1950s, I could not find references with detailed proofs which could be easily understood and reconstructed by graduate students of Logic in Computer Science. On the other hand the techniques described in this paper are well known in the mathematical logic community. Theorem 1 and some of its variations are given as an exercise in [Hod93, Exercise 10 of Sect. 5.4]. Although the Theorems 9 and 11 are strictly speaking new, their proofs use the same techniques, together with Ziegler's Theorem 2 from 1982.

We hope that our presentation of this material is sufficiently concise and transparent in showing the limitations of automatizing theorem proving in affine geometry. We restrict our discussion here to theories of affine Euclidean geometries. However, the methods can be extended to projective and hyperbolic geometries.

2 Plane Geometry

Models of plane geometry are called *planes*. These models differ in their basic relations. The universe is always two-sorted, consisting of Points and Lines and the most basic relation is *incidence* $\in$ with $p \in \ell$ to be interpreted as a point p is coincident with a line ℓ. Other relations are

Equidissstant: $Eq(p_1, p_2, p_1', p_2')$ to be interpreted as two pairs of points p_1, p_2 and p_1', p_2' define congruent segments.
Orthogonality: $Or(\ell_1, \ell_2)$ to be interpreted as two lines are orthogonal (perpendicular).
Equiangular: $An(p_1, p_2, p_3, p_1', p_2', p_3')$ to be interpreted as two triples of points define congruent angles.
Betweenness: $Be(p_1, p_2, p_3)$ to be interpreted as three distinct points are on the same line and p_2 is between p_1 and p_3.
P-equidistant: $Peq(\ell_1, p, \ell_2)$ to be interpreted as the point p has the same distance from two lines ℓ_1 and ℓ_2.
L-equidistant: $Leq(p_1, \ell, p_2)$ to be interpreted as the two points p_1 and p_2 have the same distance from the line ℓ.
Symmetric Line: $SymLine(p_1, \ell, p_2)$ to be interpreted as the two points p_1 and p_2 are symmetric with respect to the line ℓ.

We define now the following vocabularies:

$\tau_\in$: The vocabulary of incidence geometry, which uses incidence alone, possibly extended with a few symbols for specific constants.
$\tau_{hilbert}$: The vocabulary of Hilbertian style geometry: Incidence, Betweenness, Equidistance and Equiangularity [Har00].
τ_{wu}: The vocabulary of Wu's Orthogonal geometry: Incidence, Equidistance, Orthogonality [Wu94, Chap. 2].
$\tau_{origami}$: The vocabulary used to describe Origami constructions: Incidence, Symmetric Line, L-equidistant, Orthogonality, [GIT+07].
$\tau_{o-origami}$: An alternative version for describing Origami constructions. Incidence, Equidistance, Orthogonality, hence $\tau_{o-origami} = \tau_{wu}$.

We note that all these vocabularies contain the symbol $\in$ for the incidence relation.

We shall look at various axiomatization of plane geometry. The axioms are given in the Appendix A. They are all first order definable in their respective vocabularies.

The axioms which can be formulated using only the incidence relation are the incidence axioms (I-1, I-2, I-3), the parallel axiom (ParAx) and Pappus' axiom (Pappus), and the Desargues axioms (D-1, D-2). The axiom of infinity (InfLines) states that there are no finite lines. It actually is an infinite set of axioms.

Affine plane: Let $\tau_\in \subseteq \tau$ be a vocabulary of geometry. A τ-structure Π is an *(infinite) affine plane* if it satisfies (I-1, I-2, I-3 and the parallel axiom (ParAx) and (InfLines). We denote the set of these axioms by T_{affine}.
Pappian plane: Π is a *Pappian plane* if additionally it satisfies the Axiom of Pappus (Pappus). We denote the set of these axioms by T_{pappus}.

The axioms using the notion of equidistance and betweenness are (B-1, ..., B-4) and (C-1, ..., C-6), and axiom (E).

Hilbert plane: Let τ with $\tau_{hilbert} \subseteq \tau$ be a vocabulary of geometry. A τ-structure Π is an *(infinite) Hilbert plane* if it satisfies (I-1, I-2, I-3), (B-1, B-2, B-3, B-4) and (C-1, C-2, C-3, C-4, C-5, C-6) (and (InfLines)).
We denote the set of these axioms by $T_{hilbert}$

P-Hilbert plane: Π is a *P-Hilbert plane* if it additionally satisfies (ParAx).
We denote the set of these axioms by $T_{p-hilbert}$

Euclidean plane: Π is a *Euclidean plane* if it is a P-Hilbert plane which also satisfies Axiom E.
We denote the set of these axioms by T_{euclid}

The congruence axioms using orthogonality instead of equidistance are (O-1, O-2, O-3, O-4 and O-5). There are two more axioms: (AxSymAx) and (AxTrans).

Orthogonal Wu plane: Let τ with $\tau_{Wu} \subseteq \tau$ be a vocabulary of geometry. A τ-structure Π is an *orthogonal Wu plane* if it satisfies (I-1, I-2, I-3), (O-1, O-2, O-3, O-4, O-5), the axiom of infinity (InfLines), (ParAx), and the two axioms of Desargues (D-1) and (D-2).
We denote the set of these axioms by T_{o-wu} and note that every orthogonal Wu plane is an affine plane.

Metric Wu plane: Π is a *metric Wu plane* if it satisfies additionally the axiom of symmetric axis (AxSymAx) or, equivalently, the axiom of transposition (AxTrans).
We denote the set of these axioms by T_{m-wu}

The origami axioms are (H-1, H-2, H-3, H-4, H-5, H-6 and H-7).
These axioms do not define a geometry, they merely state closure under certain folding operations. To apply these operations, we always assume that we are in an affine plane.

Affine Origami plane: Let τ with $\tau_{origami} \subseteq \tau$ be a vocabulary of geometry. A τ-structure Π is an *affine Origami plane* if it satisfies (I-1, I-2, I-3), the axiom of infinity (InfLines), (ParAx) and the Huzita-Hatori axioms (H-1) - (H-7).
We denote the set of these axioms by $T_{a-origami}$

3   Background on Fields

Let τ_{field} be the purely relational vocabulary consisting of a ternary relation $Add(x, y, z)$ for addition with $Add(x, y, z)$ holds if $x + y = z$, a ternary relation $Mult(x, y, z)$ for multiplication with $Mult(x, y, z)$ holds if $x \cdot y = z$, and two constants for the neutral elements 0 and 1. A field $\mathcal{F} = \langle A, Add_A, Mult_A, 0_A, 1_A \rangle$ is a τ_{field}-structure satisfying the usual field axioms, which we write for convenience in the usual notation with $+$ and $\cdot$. Let τ_{ofield} be the purely relational vocabulary $\tau_{field} \cup \{\leq\}$ where $\leq$ is a binary relation symbol. An ordered field $\mathcal{F} = \langle A, Add_A, Mult_A, 0_A, 1_A, \leq_A \rangle$ is a τ_{ofield}-structure satisfying the usual axioms of ordered fields.

We sometimes also look at (ordered) fields as structures over a vocabulary containing function symbols. Let $\tau_{f-field}$ be the vocabularies with binary functions for addition and multiplication, unary functions for negatives $-x$ and inverses $\frac{1}{x}$, and $\tau_{f-ofields} = \tau_{f-field} \cup \{\leq\}$.

The difference between the relational and functional version lies in the notion of substructure. In the functional version substructures of (ordered fields are (ordered) fields. Formulas in the functional version can be translated into formulas in the relational version but this requires the use of existential quantifiers.

Let $B(x_1, \ldots, x_m, \bar{y})$ be a quantifier free formula with free variables $x_1, \ldots, x_m, \bar{y}$. A formula ϕ with free variables $\bar{y}$ is *universal* if it is of the form

$$\phi = \forall x_1, \ldots, \forall x_m B(x_1, \ldots, x_m, \bar{y}).$$

A formula ψ is *existential* if it is of the form

$$\psi = \exists x_1, \ldots, \exists x_m B(x_1, \ldots, x_m, \bar{y})$$

Note that when translating a quantifier-free formula in $\text{FOL}_{f-field}$ into an equivalent formula in FOL_{field}, the result is not quantifier-free but in general an existential formula. Translating an universal formula results in an $\forall\exists$-formula.

Let $\mathcal{F}$ be a field. $\mathcal{F}$ is of *characteristic p* if for every prime p we have $\underbrace{1 + \ldots + 1}_{p} = 0$, and $\mathcal{F}$ is of *characteristic 0* if for all $n \in \mathbb{N}$ we have that $\underbrace{1 + \ldots + 1}_{n} \neq 0$. $\mathcal{F}$ is *Pythagorean* if every sum of two squares is a square. $\mathcal{F}$ is a *Vieta field* if every polynomial with coefficients in $\mathcal{F}$ of degree at most 3 has a root in $\mathcal{O}$. $\mathcal{F}$ is *formally real* if 0 cannot be written as a sum of nonzero squares. $\mathcal{F}$ is *algebraically closed* if every non-constant polynomial with coefficients in $\mathcal{F}$ has a root in $\mathcal{F}$. We denote by ACF_0 the first order sentences describing an algebraically closed field of characteristic 0.

An ordered field $\mathcal{O}$ is a field $\mathcal{F}$ with an additional binary relation $\leq$ which is compatible with the arithmetic relations of $\mathcal{F}$. An ordered fields is always of characteristic 0.

Let $\mathcal{O}$ be an ordered field. $\mathcal{O}$ is *Euclidean* if every positive element has a square root. It is Pythagorean (Vieta, formally real) if it is an ordered field and as a field is Pythagorean (Vieta, formally real). $\mathcal{O}$ is *real closed* if $\mathcal{O}$ is formally real, every positive element in $\mathcal{O}$ has a square root. and every polynomial of odd degree with coefficients in $\mathcal{O}$ has a root in $\mathcal{O}$. We denote by RCF the first order sentences of ordered fields describing a real closed field.

4   From Fields to Planes

Given a field, one can define inside the field a plane using first order transductions. This goes back to Descartes. But to prove undecidability results, one has the verify more properties of these transductions.

Without going into details (which are given in [Mak17] in full) we state here the relevant facts.

Depending on the vocabularies we have three transductions given by quantifier-free translation schemes: $PP_\in$, PP_{wu} and $PP_{hilbert}$.

Theorem 3 (Correctness of $PP_\in$, PP_{wu} and $PP_{hilbert}$).

(i) [Har00, 14.1] *If $\mathcal{F}$ is a field, then $PP_\in^*(\mathcal{F})$ satisfies the incidence axioms (I_1)–(I_3), the Parallel Axiom and the Pappus Axiom.*

(ii) [Har00, 14.4] *If $\mathcal{F}$ is a field of characteristic 0, then $PP_\in^*(\mathcal{F})$ satisfies additionally the Axioms (InfLines), i.e., is a Pappian plane without finite lines.*

(iii) [Wu94] *If $\mathcal{F}$ is a Pythagorean field of characteristic 0, then $PP_{Wu}^*(\mathcal{F})$ satisfies (I-1)–(I-3), (O-1)–(O-5), the Parallel Axiom, the Axioms (InfLines), the Axiom of Desargues and the Axiom of Symmetric Axis, which are axioms of a metric Wu plane.*

(iv) [Alp00] *If $\mathcal{F}$ is a Vieta field, then $PP_{Wu}^*(\mathcal{F})$ satisfies the Huzita-Hatori axioms (H-1)–(H-7).*

(v) [Har00, 17.3] *If $\mathcal{O}$ is an ordered Pythagorean field, then $PP_{hilbert}^*(\mathcal{O})$ satisfies (I-1)–(I-3), (B-1)–(B-4) (C-1)–(C-6) and the Parallel Axiom, which are axioms of a Hilbert Plane which satisfies the parallel axiom.*

(vi) [Har00, 17.3] *If $\mathcal{O}$ is an ordered Euclidean field, then $PP_{hilbert}^*(\mathcal{F})$ is a Hilbert Plane which satisfies the parallel axiom and Axiom E.*

5   From Planes to Fields

Given an affine plane, one can define inside the plane a field using first order transductions. In the presence of equidistance this can be done using segment arithmetic. But if we have only an incidence relation, one has to use planar ternary rings, introduced by von Staudt [vS47] a student of Gauss, before Hilbert's [Hil02]. The first modern treatment of coordinatization for affine and projective planes was given by Hall [Hal43].

Definition 1. *Let τ a vocabulary for geometry, $T \subseteq \mathrm{FOL}(\tau)$ a set of axioms of geometry, T_f be a set of axioms for fields in τ_{fields} or $\tau_{ofields}$. We say that the models of T have a first order coordinatization in fields satisfying T_f if there exists a first order translation scheme CC_{field} such that*

(a) *for every Π which satisfies T the structure $CC_{field}^*(\Pi)$ $(CC_{o-field}^*(\Pi))$ is a field which satisfies T;*

(b) *for every field $\mathcal{F}$ which satisfies T_f, the τ-structure $PP_\tau(\mathcal{F})$ satisfies T;*

(c) *For every field $\mathcal{F}$ which satisfies T_f we have*

$$CC_{field}(PP_\tau(\mathcal{F})) \simeq \mathcal{F};$$

(d) *For every τ-structure Π which satisfies T we have*

$$PP_\tau(CC_{field}(\Pi)) \simeq \Pi.$$

There are first order translation schemes FF_{field} and FF_{ofield} defining coordinates in certain planes using segment arithmetic:

Theorem 4 (Correctness of FF_{field} and FF_{ofield}). *Let Π be a Hilbert Plane which satisfies the Parallel Axiom.*

(i) *$FF^*_{field}(\Pi)$ is a field of characteristic 0 which can be uniquely ordered to be an ordered field $\mathcal{F}_\Pi$.*

(ii) *Let $\mathcal{F}_\Pi = FF^*_{field}(\Pi)$ be the ordered field of segment arithmetic in Π. Then $\mathcal{F}$ is Pythagorean and $PP^*_{hilbert}(\mathcal{F})$ is isomorphic to Π.*

(iii) *An ordered field $\mathcal{O}$ is Pythagorean iff $PP^*_{hilbert}(\mathcal{O})$ is a Pappian Hilbert Plane which satisfies the Parallel Axiom.*

(iv) *Π is a Euclidean plane iff $FF^*_{field}(\Pi)$ is a Euclidean field.*

(v) *$\mathcal{F}$ is a Euclidean field iff $PP^*_{hilbert}(\mathcal{F})$ is a Euclidean plane.*

If we have only an incidence relation, we can use instead translation schemes $RR_{ptr} = (lines, Ter, add_T, mult_T)$ and $RF_{field} = (add_T, mult_T)$.

Theorem 5 (Correctness of RR_{ptr} and RF_{field}). *Let Π be plane satisfying I-1, I-2, and I-3 with distinguished lines ℓ, m, d and points $O = (0,0)$ and $I = (1,0)$.*

(i) *$RR^*_{ptr}(\Pi)$ is a planar ternary ring.*

(ii) *Π is a Pappian plane (without finite lines) iff $RF^*_{field}(\Pi)$ is a field (of characteristic 0).*

A detailed proof may be found in [Blu80, Szm83].

6   Undecidable Geometries

6.1   Incidence Geometries

First we look $\tau_\in$-structures, i.e., at models of the incidence relation alone. To proved undecidability, the correctness of the translation scheme RR_{field}, Theorem 5, is not enough. We still have to show that RR^*_{field} is onto as a transduction from Pappus planes to fields.

Theorem 6 ([Szm83, Sect. 4.5])

(i) *If Π is a Pappus plane there is a field $\mathcal{F}_\Pi$ such that $PP^*_\in(\mathcal{F}_\Pi)$ is isomorphic to Π.*

(ii) *If additionally Π satisfies (Inf), $\mathcal{F}_\Pi$ is a field of characteristic 0.*

In fact, $\mathcal{F}_\Pi$ can be chosen to be $RR^*_{field}(\Pi)$.

Corollary 1. *RR^*_{field} is onto as a transduction from Pappus planes to fields.*

We now can use Proposition 2(i) to prove Theorem 1, which states that the theory T_{pappus} of Pappus planes is undecidable.

6.2   Hilbert Planes and Euclidean Planes

Similarly, the correctness of the translation scheme FF^*_{field}, Theorem 4, is not enough. We need one more lemma[3]:

Lemma 1. *Let $\mathcal{F}$ be a Pythagorean field and $\Pi_{\mathcal{F}} = PP^*_{hilbert}(\mathcal{F})$. Then $FF^*_{field}(\Pi_{\mathcal{F}})$ is isomorphic to $\mathcal{F}$.*

Corollary 2

(i) FF^*_{field} *is onto as a transduction from Hilbert planes to ordered Pythagorean fields.*

(ii) FF^*_{field} *is onto as a transduction from Euclidean planes to ordered Euclidean fields.*

Using Ziegler's Theorem 2, Lemma 2, from Appendix B, Theorem 4 and Corollary 2 we conclude:

Theorem 7

(i) *The consequence problem for the Hilbert plane which satisfies the Parallel Axiom is undecidable.*

(ii) *The consequence problem for the Hilbert plane is undecidable.*

(iii) *The consequence problem for the Euclidean Plane is undecidable.*

6.3   Wu's Geometry

Recall that in Wu's orthogonal geometry we have as basic relation incidence $A \in l$ and Orthogonality $Or(l_1, l_2)$, but neither betweenness nor equidistance.

There are two systems of orthogonal geometry, Wu's orthogonal geometry T_{o-wu} and Wu's metric geometry T_{m-wu}.

Theorem 8. *Let $\mathcal{F}$ be a Pythagorean field of characteristic 0. Then $PP_{wu}(\mathcal{F})$ is a metric Wu plane.*

Conversely, let Π be a metric Wu plane then $RF^*_{field}(\Pi)$ is a Pythagorean field of characteristic 0. In fact $RF^*_{field}(PP^*(\Pi)) \simeq \Pi$.

Corollary 3. RR^*_{field} *maps metric Wu planes onto Pythagorean fields.*

Theorem 9

(i) *The consequence problem for (Wu-metric) is undecidable.*

(ii) *The consequence problem for (Wu-orthogonal) is undecidable.*

Proof

(i): Use Ziegler's Theorem 2 in the version for the complex numbers, Lemma 2, from Appendix B, Theorem 5 and Corollary 3.

(ii): We observe that (Wu-metric) is obtained from (Wu-orthogonal) by adding one more axiom. This gives that if (Wu-orthogonal) were decidable, so would (Wu-metric) be decidable, which contradicts (i). □

---

[3] In [BGKV07] it is overlooked that Lemma 1 from Appendix B is needed in order to apply Lemma 2. In [SST83] it is used properly but not explicitly stated.

6.4 Origami Geometry

In Origami Geometry we have also points and lines, the incidence relation $A \in l$, the orthogonality relation $Or(l_1, l_2)$, a relation $SymP(A, l, B)$ and a relation $d(A, l_1, l_2)$.

The intended interpretation of $SymP(A, l, B)$ states that A and B are symmetric with respect to l, i.e., l is perpendicular to the line AB and intersects AB at a point C such that $Eq(A, C, B, C)$.

The intended interpretation of $d(A, l_1, l_2)$ states that the point A has the same perpendicular distance from l_1 and l_2.

Clearly, $SymP(A, l, B)$ and $d(A, l_1, l_2)$ are definable in a Hilbert plane using $\in, Eq, Or$.

An *Origami Plane* is a Pappian plane which satisfies additionally the axioms (H-1) to (H-7).

Theorem 10 ([Alp00])

(i) If Π is an Origami plane, then $RR^*_{field}(\Pi)$ is a Vieta field.

(ii) Conversely, for every Vieta field $\mathcal{F}$ the structure $PP^*_{origami}(\mathcal{F})$ is an Origami field.

(iii) RR^*_{field} maps Origami planes onto Vieta fields.

Using Ziegler's Theorem 2, Lemma 2 from Appendix B, and 3 we can now apply Theorem 10 to conclude:

Theorem 11

(i) The consequence problem for Origami Planes is undecidable.

(ii) The consequence problem for the Huzita axioms (H-1)–(H-7) is undecidable.

A   Axioms for Affine Geometries

In this appendix we collect some of Hilbert's axioms of geometry which we used, and which are all true when one considers the analytic geometry of the plane with real coordinates.

A.1   Incidence Geometries

Axioms Using Only the Incidence Relation

(I-1): For any two distinct points A, B there is a unique line l with $A \in l$ and $B \in l$.

(I-2): Every line contains at least two distinct points.

(I-3): There exists three distinct points A, B, C such that no line l contains all of them.

They can be formulated in FOL using the incidence relation only.

Parallel Axiom. We define: $Par(l_1, l_2)$ or $l_1 \parallel l_2$ if l_1 and l_2 have no point in common.

(ParAx): For each point A and each line l there is at most one line l' with $l \parallel l'$ and $A \in l'$.

$Par(l_1, l_2)$ can be formulated in FOL using the incidence relation only, hence also the Parallel Axiom.

Pappus' Axiom

(Pappus): Given two lines l, l' and points $A, B, C \in l$ and $A', B', C' \in l'$ such that $AC' \parallel A'C$ and $BC' \parallel B'C$. Then also $AB' \parallel A'B$.

Axioms of Desargues and of Infinity

(InfLines): Given distinct A, B, C and l with $A \in l, B, C \notin l$ we define $A_1 = Par(AB, C) \times l$, and inductively, $A_{n+1} = Par(A_n B, C) \times l$. Then all the A_i are distinct.
  Note that this axiom is stronger than just saying there infinitely many points. It says that there are no lines which have only finitely many points.
(De-1): If AA', BB', CC' intersect in one point or are all parallel, and $AB \parallel A'B'$ and $AC \parallel A'C'$ then $BC \parallel B'C'$.
(De-2): If $AB \parallel A'B'$, $AC \parallel A'C'$ and $BC \parallel B'C'$ then AA', BB', CC' are all parallel.

The axiom (InfLies) is not first order definable but consists of an infinite set of first order formulas with infinitely many new constant symbols for the points A_i, and the incidence relation. The two Desargues axioms are first order definable using the incidence relation only.

Affine plane: Let $\tau_{\in} \subseteq \tau$ be a vocabulary of geometry. A τ-structure Π is an *(infinite) affine plane* if it satisfies (I-1, I-2, I-3 and the parallel axiom (ParAx) and (InfLines). We denote the set of these axioms by T_{affine}.
Pappian plane: Π is a *Pappian plane* if additionally it satisfies the Axiom of Pappus (Pappus). We denote the set of these axioms by T_{pappus}.

In the literature the definition of affine planes vary. Sometimes the parallel axiom is included, and sometimes not. We always include the parallel axiom, unless indicated explicitly otherwise.

A.2   Hilbert Style Geometries

Axioms of Betweenness

(B-1): If $Be(A, B, C)$ then there is l with $A, B, C \in l$.
(B-2): For every A, B there is C with $Be(A, B, C)$.

(B-3): For each distinct $A, B, C \in l$ exactly one point of the points A, B, C is between the two others.

(B-4): (Pasch) Assume the points A, B, C and l in general position, i.e. the three points are not on one line, none of the points is on l. Let D be the point at which l and the line AB intersect. If $Be(A, D, B)$ there is $D' \in l$ with $Be(A, D', C)$ or $Be(B, D', C)$.

The axioms of betweenness are all first order expressible in the language with incidence relation and the betweenness relation.

Congruence Axioms: Equidistance. We write for $Eq(A, B, C, D)$ the usual $AB \cong CD$.

(C-0): $AB \cong AB \cong BA$.
(C-1): Given A, B, C, C', l with $C, C' \in l$ there is a unique $D \in l$ with $AB \cong CD$ and $B(C, C', D)$ or $B(C, D, C')$.
(C-2): If $AB \cong CD$ and $AB \cong EF$ then $CD \cong EF$.
(C-3): (Addition) Given A, B, C, D, E, F with $Be(A, B, C)$ and $Be(D, E, F)$, if $AB \cong DE$ and $BC \cong EF$, then $AC \cong DF$.

Note that (C-1) and (C-3) use the betweenness relation Be. Hence they are first order definable using the incidence, betweenness and equidistance relation.

Congruence Axioms: Equiangularity. We denote by $\boldsymbol{AB}$ the directed ray from A to B, and by $\angle(ABC)$ the angle between $\boldsymbol{AB}$ and $\boldsymbol{BC}$. For the congruence of angles $An(A, B, C, A', B', C')$ we write $\angle(ABC) \cong \angle(A'B'C')$.

(C-4): Given rays $\boldsymbol{AB}$, $\boldsymbol{AC}$ and $\boldsymbol{DE}$ there is a unique ray $\boldsymbol{DF}$ with $\angle(BAC) \cong \angle(EDF)$.
(C-5): Congruence of angles is an equivalence relation.
(C-6): (Side-Angle-Side) Given two triangles ABC and $A'B'C'$ with $AB \cong A'B'$, $AC \cong A'C'$ and $\angle BAC \cong \angle B'A'C'$ then $BC \cong B'C'$, $\angle ABC \cong \angle A'B'C'$ and $\angle ACB \cong \angle A'C'B'$.

Axiom E. Let A be a point and BC be a line segment. A circle $\Gamma(A, BC)$ is the set of all points U such that $E(A, U, B, C)$. A point D is inside the circle $\Gamma(A, BC)$ if there is U with $E(A, U, B, C)$ and $Be(A, D, U)$. A point D is outside the circle $\Gamma(A, BC)$ if there is U with $E(A, U, B, C)$ and $Be(A, U, D)$.

(AxE): Given two circles Γ, Δ such that Γ contains at least one point inside, and one point outside Δ, then $\Gamma \cap \Delta \neq \emptyset$.

Hilbert plane: Let τ with $\tau_{hilbert} \subseteq \tau$ be a vocabulary of geometry. A τ-structure Π is an *(infinite) Hilbert plane* if it satisfies (I-1, I-2, I-3), (B-1, B-2, B-3, B-4) and (C-1, C-2, C-3, C-4, C-5, C-6).
We denote the set of these axioms by $T_{hilbert}$
P-Hilbert plane: Π is a *P-Hilbert plane* if it additionally satisfies (ParAx).
We denote the set of these axioms by $T_{p-hilbert}$
Euclidean plane: Π is a *Euclidean plane* if it is a P-Hilbert plane which also satisfies Axiom E.
We denote the set of these axioms by T_{euclid}

A.3   Axioms of Orthogonal Geometry

Congruence Axioms: Orthogonality. We denote by $l_1 \perp l_2$ the orthogonality of two lines $Or(l_1, l_2)$. We call a line l *isotropic* if $l \perp l$. Note that our definitions do not exclude this.

(O-1): $l_1 \perp l_2$ iff $l_2 \perp l_1$.
(O-2): Given O and l_1, there exists exactly one line l_2 with $l_1 \perp l_2$ and $O \in l_2$.
(O-3): $l_1 \perp l_2$ and $l_1 \perp l_3$ then $l_2 \parallel l_3$.
(O-4): For every O there is an l with $O \in l$ and $l \not\perp l$.
(O-5): The three heights of a triangle intersect in one point.

Axiom of Symmetric Axis and Transposition

(AxSymAx): Any two intersecting non-isotropic lines have a symmetric axis.
(AxTrans): Let l, l' be two non-isotropic lines with $A, O, B \in l$, $AO \cong OB$ and $O' \in l'$ there are exactly two points $A', B' \in l'$ such that $AB \cong A'B' \cong B'A'$ and $A'O' \cong O'B'$.

The two axioms are equivalent in geometries satisfying the Incidence, Parallel, Desargues and Orthogonality axioms together with the axiom of infinity.

Orthogonal Wu plane: Let τ with $\tau_{Wu} \subseteq \tau$ be a vocabulary of geometry. A τ-structure Π is an *orthogonal Wu plane* if it satisfies (I-1, I-2, I-3), (O-1, O-2, O-3, O-4, O-5), the axiom of infinity (InfLines), (ParAx), and the two axioms of Desargues (D-1) and (D-2).
  We denote the set of these axioms by T_{o-wu}
Metric Wu plane: Π is a *metric Wu plane* if it satisfies additionally the axiom of symmetric axis (AxSymAx) or, equivalently, the axiom of transposition (AxTrans).
  We denote the set of these axioms by T_{m-wu}

The axiomatization of orthogonal is due to Wu [Wu86, Wu94, WG07], see also [Pam07].

A.4   The Origami Axioms

A line which is obtained by folding the paper is called a *fold*. The first six axioms are known as Huzita's axioms. Axiom (H-7) was discovered by K. Hatori. Jacques Justin and Robert J. Lang also found axiom (H-7), [Wik]. The axioms (H-1)-(H-7) only express closure under folding operations, and do not define a geometry. To make it into an axiomatization of geometry we have to that these operations are performed on an affine plane.
  We follow here [GIT+07]. The original axioms and their expression as first order formulas in the vocabulary $\tau_{origami}$ are as follows:

(H-1): Given two points P_1 and P_2, there is a unique fold (line) that passes through both of them.

$$\forall P_1, P_2 \exists^{=1} l (P_1 \in l \wedge P_2 \in l)$$

(H-2): Given two points P_1 and P_2, there is a unique fold (line) that places P_1 onto P_2.

$$\forall P_1, P_2 \exists^{=1} l \, SymLine(P_1, l, P_2)$$

(H-3): Given two lines l_1 and l_2, there is a fold (line) that places l_1 onto l_2.

$$\forall l_1, l_2 \exists k \forall P \, (P \in k \rightarrow Peq(l_1, P, l_2))$$

(H-4): Given a point P and a line l_1, there is a unique fold (line) perpendicular to l_1 that passes through point P.

$$\forall P, l \exists^{=1} k \forall P (P \in k \wedge Or(l, k))$$

(H-5): Given two points P_1 and P_2 and a line l_1, there is a fold (line) that places P_1 onto l_1 and passes through P_2.

$$\forall P_1, P_2 l_1 \exists l_2 \forall P (P_2 \in l_2 \wedge \exists P_2 (SymLine(P_1, l_2, P_2) \wedge P_2 \in l_1))$$

(H-6): Given two points P_1 and P_2 and two lines l_1 and l_2, there is a fold (line) that places P_1 onto l_1 and P_2 onto l_2.

$$\forall P_1, P_2 l_1, l_2 \exists l_3 \, ((\exists Q_1 SymLine(P_1, l_3, Q_1) \wedge Q_1 \in l_1) \wedge$$
$$(\exists Q_2 SymLine(P_2, l_3, Q_2) \wedge Q_2 \in l_2))$$

(H-7): Given one point P and two lines l_1 and l_2, there is a fold (line) that places P onto l_1 and is perpendicular to l_2.

$$\forall P, l_2, l_2 \exists l_3 \, (Or(l_2, l_3) \wedge (\exists Q SymLine(P, l_3, Q) \wedge Q \in l_1))$$

Affine Origami plane: Let τ with $\tau_{origami} \subseteq \tau$ be a vocabulary of geometry. A τ-structure Π is an *affine Origami plane* if it satisfies (I-1, I-2, I-3), the axiom of infinity (InfLines), (ParAx) and the Huzita-Hatori axioms (H-1) - (H-7).
We denote the set of these axioms by $T_{a-origami}$

Proposition 4. *The relations SymLine and Peq are first order definable using Eq and Or with existential formulas over $\tau_{f-field}$: Hence the axioms (H-1)–(H-7) are first order definable in* $\mathrm{FOL}(\tau_{wu})$.

Proof

(i) $SymLine(P_1, \ell, P_2)$ iff there is a point $Q \in \ell$ such that $Or((P_1, Q), \ell)$, $Or((P_2, Q), \ell)$ and $Eq(P_1, Q, P_2, Q)$.
(ii) $Peq(\ell_1, P, \ell_2)$ iff there exist points Q_1, Q_2 such that $Or((P, Q_1), \ell_1)$, $Or((P, Q_2), \ell_2)$, $Eq(P, Q_1)$ and $Eq(P, Q_2)$. □

B   Translation Schemes

We first introduce the formalism of *translation schemes, transductions and translation*. In [TMR53] this was first used, but not spelled out in detail. The details appear first in [Rab65]r. Our approach follows [Mak04, Sect. 2]. To keep it notationally simple we explain on an example. Let τ be a vocabulary consisting of one binary relation symbol R, σ be a vocabulary consisting of one ternary relation symbol S. We want to interpret a σ structure on k-tuples of elements of a τ-structure.

A $\tau - \sigma$-*translation scheme* $\Phi = (\phi, \phi_S)$ consists of a formula $\phi(\bar{x})$ with k free variables and a formula ϕ_S with $3k$ free variables. Φ is quantifier-free if all its translation formulas are quantifier-free.

Let $\mathcal{A} = \langle A, R^A \rangle$ be a τ-structure. We define a σ-structure $\Phi^*(\mathcal{A}) = \langle B, S^B \rangle$ as follows: The universe is given by

$$B = \{\bar{a} \in A^k : \mathcal{A} \models \phi(\bar{a})\}$$

and the relation is given by

$$S^B = \{\bar{b} \in A^{k \times 3} : \mathcal{A} \models \phi_S(\bar{b})\}.$$

Φ^* is called a *transduction*.

Let θ be a σ-formula. We define a τ-formula $\Phi^\sharp(\theta)$ inductively by substituting occurrences of $S(\bar{b}$ by their definition via ϕ_S where the free variables are suitable named. $\Phi^\sharp$ is called a *translation*.

The fundamental property of translation schemes, transductions and translation is the following:

Proposition 5 (Fundamental Property of Translation Schemes).   *Let Φ be a $\tau - \sigma$-translation scheme, and θ be a σ-formulas.*

$$\mathcal{A} \models \Phi^\sharp(\theta) \text{ iff } \Phi^*(\mathcal{A}) \models \theta$$

If θ has free variables, the assignment have to be chosen accordingly. Furthermore, if Φ is quantifier-free, and θ is a universal formula, $\Phi^\sharp(\theta)$ is also universal.

In order to use translation schemes to prove decidability and undecidability of theories we need two lemmas.

Lemma 2. *Let Φ be a $\tau - \sigma$-translation scheme.*

(i) *Let $\mathcal{A}$ be a τ-structure. If the complete first order theory T_0 of $\mathcal{A}$ is decidable, so is the complete first order theory T_1 of $\Phi^*(\mathcal{A})$.*

(ii) *There is a τ-structure $\mathcal{A}$ such that the complete first order theory T_1 of $\Phi^*(\mathcal{A})$ is decidable, but the complete first order theory T_0 of $\mathcal{A}$ is undecidable.*

(iii) *If however, $\Phi^\sharp$ is onto, i.e., for every $\phi \in \text{FOL}(\tau)$ there is a formula $\theta \in \text{FOL}(\sigma)$ with $\Phi^\sharp(\theta)$ logically equivalent to ϕ, then the converse of (i) also holds.*

(iv) Let $T \subseteq \mathrm{FOL}(\tau)$ be a decidable theory and $T' \subseteq \mathrm{FOL}(\sigma)$ and Φ^* be such that $\Phi^*|_{Mod(T)} : Mod(T) \rightarrow Mod(T')$ be onto. Then T' is decidable.

Remark 1. The condition that $\Phi^\sharp$, resp. Φ^* have to be onto is often overlooked in the literature[4].

We shall need one more observation:

Lemma 3. *Let $T \subseteq \mathrm{FOL}(\tau)$ and $\phi \in \mathrm{FOL}(\tau)$. Assume T is decidable. Then $T \cup \{\phi\}$ is also decidable.*

References

[ADM09] Avigad, J., Dean, E., Mumma, J.: A formal system for Euclid's elements. Rev. Symb. Log. **2**(4), 700–768 (2009)

[Alp00] Alperin, R.C.: A mathematical theory of origami constructions and numbers. New York J. Math. **6**(119), 133 (2000)

[Art57] Artin, E.: Geometric Algebra. Interscience Tracts in Pure and Applied Mathematics, vol. 3. Interscience Publishers, Geneva (1957)

[BCSS98] Blum, L., Cucker, F., Shub, M., Smale, S.: Complexity and Real Computation. Springer, New York (1998). https://doi.org/10.1007/978-1-4612-0701-6

[Bee] Beeson, M.: Some undecidable field theories. Translation of [Zie82]. www.michaelbeeson.com/research/papers/Ziegler.pdf

[Bee13] Beeson, M.: Proof and computation in geometry. In: Ida, T., Fleuriot, J. (eds.) ADG 2012. LNCS (LNAI), vol. 7993, pp. 1–30. Springer, Heidelberg (2013). https://doi.org/10.1007/978-3-642-40672-0_1

[BGKV07] Balbiani, P., Goranko, V., Kellerman, R., Vakarelov, D.: Logical theories for fragments of elementary geometry. In: Aiello, M., Pratt-Hartmann, I., Van Benthem, J. (eds.) Handbook of Spatial Logics, pp. 343–428. Springer, Dordrecht (2007). https://doi.org/10.1007/978-1-4020-5587-4_7

[Blu80] Blumenthal, L.M.: A Modern View of Geometry. Courier Corporation, North Chelmsford (1980)

[BPR03] Basu, S., Pollack, R., Roy, M.-F.: Algorithms in Real Algebraic Geometry. Algorithms and Computation in Mathematics, vol. 10. Springer, Heidelberg (2003). https://doi.org/10.1007/978-3-662-05355-3

[CJ12] Caviness, B.F., Johnson, J.R.: Quantifier Elimination and Cylindrical Algebraic Decomposition. Springer, Heidelberg (2012)

[CJW06] Carlson, J.A., Jaffe, A., Wiles, A.: The Millennium Prize Problems. American Mathematical Society, Providence (2006)

[EFT80] Ebbinghaus, H.-D., Flum, J., Thomas, W.: Mathematical Logic. Undergraduate Texts in Mathematics. Springer, Heidelberg (1980)

[Gel59] Gelernter, H.: Realization of a geometry theorem proving machine. In: IFIP Congress, pp. 273–281 (1959)

---

[4] Theorems 1.36 and 1.37 as stated in [BGKV07] are only true when one notices that their Theorems 1.20 and 1.21 imply that the particular transductions used in Theorems 1.36 and 1.37 are indeed onto. However, this is not stated there, although it follows from the cited results from geometry.

[GHL60] Gelernter, H., Hansen, J.R., Loveland, D.W.: Empirical explorations of the geometry theorem machine. Papers Presented at the May 3–5, 1960, Western Joint IRE-AIEE-ACM Computer Conference, pp. 143–149. ACM (1960)

[GIT+07] Ghourabi, F., Ida, T., Takahashi, H., Marin, M., Kasem, A.: Logical and algebraic view of Huzita's origami axioms with applications to computational origami. In: Proceedings of the 2007 ACM Symposium on Applied Computing, pp. 767–772. ACM (2007)

[HA50] Hilbert, D., Ackermann, W.: Principles of Mathematical Logic. Chelsea Publishing Company, White River Junction (1950)

[Hal43] Hall, M.: Projective planes. Trans. Am. Math. Soc. **54**(2), 229–277 (1943)

[Har00] Hartshorne, R.: Geometry: Euclid and Beyond. Springer, Heidelberg (2000). https://doi.org/10.1007/978-0-387-22676-7

[Hil02] Hilbert, D.: The Foundations of Geometry. Open Court Publishing Company, Chicago (1902)

[Hod93] Hodges, W.: Model Theory. Encyclopedia of Mathematics and its Applications, vol. 42. Cambridge University Press, Cambridge (1993)

[HR74] Hauschild, K., Rautenberg, W.: Rekursive unentscheidbarkeit der theorie der pythagoräischen körper. Fundamenta Mathematicae, **82**(3), 191–197 (1974). The name of the coauthor W. Rautenberg was ommitted for political reasons, but appears on the page headers of the paper

[Jus89] Justin, J.: Résolution par le pliage de équation du troisieme degré et applications géométriques. In: Proceedings of the First International Meeting of Origami Science and Technology, Ferrara, Italy, pp. 251–261 (1989)

[Mak04] Makowsky, J.A.: Algorithmic uses of the Feferman-Vaught theorem. Ann. Pure Appl. Log. **126**(1–3), 159–213 (2004)

[Mak17] Makowsky, J.A.: Can one design a geometry engine? On the (un)decidability of affine Euclidean geometries (2017). https://arxiv.org/abs/1712.07474

[Mil07] Miller, N.: Euclid and His Twentieth Century Rivals: Diagrams in the Logic of Euclidean Geometry. CSLI Publications Stanford, Stanford (2007)

[Pam07] Pambuccian, V.: Orthogonality as a single primitive notion for metric planes. Contrib. Algebra Geom. **49**, 399–409 (2007)

[Poi95] Poizat, B.: Les Petits Cailloux: Une approche modèle-théorique de l'algorithmie. Aléas, Paris (1995)

[Pru06] Prunescu, M.: Fast quantifier elimination means P = NP. In: Beckmann, A., Berger, U., Löwe, B., Tucker, J.V. (eds.) CiE 2006. LNCS, vol. 3988, pp. 459–470. Springer, Heidelberg (2006). https://doi.org/10.1007/11780342_47

[Rab65] Rabin, M.O.: A simple method for undecidability proofs and some applications. In: Bar-Hillel, Y. (ed.) Logic, Methodology and Philosophy of Science, pp. 58–68. North-Holland Publishing Company, Amsterdam (1965)

[Rau61] Rautenberg, W.: Unentscheidbarkeit der Euklidischen Inzidenzgeometrie. Math. Log. Q. **7**(1–5), 12–15 (1961)

[Rau62] Rautenberg, W.: Über metatheoretische Eigenschaften einiger geometrischer Theorien. Math. Log. Q. **8**(1–5), 5–41 (1962)

[Rob49] Robinson, J.: Definability and decision problems in arithmetic. J. Symb. Log. **14**(2), 98–114 (1949)

[Sch09] Schur, F.: Grundlagen der Geometrie. BG Teubner (1909)

[SST83] Schwabhäuser, W., Szmielew, W., Tarski, A.: Metamathematische Methoden in der Geometrie. Springer (1983)

[SV14] Shlapentokh, A., Videla, C.: Definability and decidability in infinite algebraic extensions. Ann. Pure Appl. Log. **165**(7), 1243–1262 (2014)

[Szm83] Szmielew, W.: From Affine to Euclidean Geometry, an Axiomatic Approach. Polish Scientific Publishers, Warszawa and D. Reidel Publishing Company, Dordrecht (1983)

[Tar31] Tarski, A.: Sur les ensembles définissables de nombre réels. Fundamenta Mathematicae **17**, 210–239 (1931)

[Tar51] Tarski, A.: A Decision Method for Elementary Algebra and Geometry. University of California Press, Berkeley (1951)

[TMR53] Tarski, A., Mostowski, A., Robinson, R.M.: Undecidable Theories. Studies in Logic and the Foundations of Mathematics. North Holland, Amsterdam (1953)

[vS47] von Staudt, K.G.C.: Geometrie der lage. Bauer und Raspe (1847)

[WG07] Wu, W., Gao, X.: Mathematics mechanization and applications after thirty years. Front. Comput. Sci. China **1**(1), 1–8 (2007)

[Wik] Wikipedia. Huzita-Hatori axioms. Wikipedia entry: https://en.wikipedia.org/wiki/Huzita-Hatori_axioms

[Wu86] Wu, W.-T.: Basic principles of mechanical theorem proving in elementary geometries. J. Autom. Reason. **2**(3), 221–252 (1986)

[Wu94] Wu, W.-T.: Mechanical Theorem Proving in Geometries. Springer, Heidelberg (1994). https://doi.org/10.1007/978-3-7091-6639-0. (Original in Chinese 1984)

[Zie82] Ziegler, M.: Einige unentscheidbare Körpertheorien. In: Strassen, V., Engeler, E., Läuchli, H. (eds.) Logic and Algorithmic, An International Symposium Held in Honour of E. Specker, pp. 381–392. L'enseignement mathématique (1982)

[Zor49] Zorn, M.: Eleventh meeting of the association for symbolic logic. J. Symb. Log. **14**(1), 73–80 (1949)

Algebraic Semantics for Nelson's Logic $\mathcal{S}$

Thiago Nascimento[1(✉)], Umberto Rivieccio[2], João Marcos[2],
and Matthew Spinks[3]

[1] Programa de Pós-graduação em Sistemas e Computação, UFRN, Natal, Brazil
thiagnascsilva@gmail.com
[2] Departamento de Informática e Matemática Aplicada, UFRN, Natal, Brazil
{urivieccio,jmarcos}@dimap.ufrn.br
[3] Università degli Studi di Cagliari, Cagliari, Italy
mspinksau@yahoo.com.au

Abstract. Besides the better-known Nelson's Logic and Paraconsistent
Nelson's Logic, in "Negation and separation of concepts in constructive
systems" (1959), David Nelson introduced a logic called $\mathcal{S}$ with the aim
of analyzing the constructive content of provable negation statements in
mathematics. Motivated by results from Kleene, in "On the Interpretation of Intuitionistic Number Theory" (1945), Nelson investigated a
more symmetric recursive definition of truth, according to which a formula could be either primitively verified or refuted. The logic $\mathcal{S}$ was
defined by means of a calculus lacking the contraction rule and having
infinitely many schematic rules, and no semantics was provided. This
system received little attention from researchers; it even remained unnoticed that on its original presentation it was inconsistent. Fortunately,
the inconsistency was caused by typos and by a rule whose hypothesis and conclusion were swapped. We investigate a corrected version of
the logic $\mathcal{S}$, and focus on its propositional fragment, showing that it is
algebraizable in the sense of Blok and Pigozzi (in fact, implicative) with
respect to a certain class of involutive residuated lattices. We thus introduce the first (algebraic) semantics for $\mathcal{S}$ as well as a finite Hilbert-style
calculus equivalent to Nelson's presentation; we also compare $\mathcal{S}$ with the
other two above-mentioned logics of the Nelson family. Our approach is
along the same lines of (and partly relies on) previous algebraic work on
Nelson's logics due to M. Busaniche, R. Cignoli, S. Odintsov, M. Spinks
and R. Veroff.

Keywords: Nelson's logics · Involutive residuated lattices
Algebraic semantics · Algebraic logic

1 Introduction

To study the notion of constructible falsity, David Nelson introduced a number
of systems of non-classical logic that combine an intuitionistic approach to truth
with a dual-intuitionistic treatment of falsity. Nelson's logics ($\mathcal{S}, \mathcal{N}3$, and $\mathcal{N}4$)

© Springer-Verlag GmbH Germany, part of Springer Nature 2018
L. S. Moss et al. (Eds.): WoLLIC 2018, LNCS 10944, pp. 271–288, 2018.
https://doi.org/10.1007/978-3-662-57669-4_16

accept some notable theorems of classical logic, such as $\sim\sim\varphi \Leftrightarrow \varphi$, while rejecting others, such as $(\varphi \Rightarrow (\varphi \Rightarrow \psi)) \Rightarrow (\varphi \Rightarrow \psi)$ and $(\varphi \wedge \sim\varphi) \Rightarrow \psi$. Nelson introduced these logics with the aim of studying constructive proofs in Number Theory. To such an end, he gave a definition of truth [14, Definition 1] (analogous to Kleene's [12, p. 112]) according to which either a formula or its negation should be realized by some natural number.

Nelson's logic $\mathcal{N}3$ was introduced in [14] and $\mathcal{N}4$, a paraconsistent version of $\mathcal{N}3$, was introduced in [1]. $\mathcal{N}3$ is in fact an axiomatic extension of $\mathcal{N}4$ by the axiom[1] $\sim\varphi \to (\varphi \to \psi)$. The logic $\mathcal{N}3$ is by now well studied, both via a proof-theoretic approach and through algebraic methods; in particular, Odintsov [16] proved that $\mathcal{N}4$ (thus also $\mathcal{N}3$) is algebraizable à la Blok-Pigozzi [3].

In [15] Nelson also introduced the logic $\mathcal{S}$, aimed at the study of realizability. As suggested by Humberstone [11, Chap. 8.2, p. 1239–1240], the introduction of $\mathcal{S}$ can perhaps also be viewed as an attempt to remedy what some logicians consider an undesirable feature of $\mathcal{N}3$ (and $\mathcal{N}4$), namely the fact that there are formulas φ, ψ in $\mathcal{N}3$ (and $\mathcal{N}4$) that are mutually interderivable but such that their negations $\sim\varphi$, $\sim\psi$ fail to be interderivable; and also a formula such as $(\varphi \to \psi) \wedge (\psi \to \varphi)$ is a theorem of $\mathcal{N}3$ (and $\mathcal{N}4$) but $(\sim\varphi \to \sim\psi) \wedge (\sim\psi \to \sim\varphi)$ is not. It is useful to recall that these two phenomena are in general disassociated; the latter stems from the failure of the contraposition law for the so-called *weak implication* connective $\to$ of $\mathcal{N}3$ (and $\mathcal{N}4$), while the former entails that $\mathcal{N}3$ and $\mathcal{N}4$ are non-congruential (or, as other authors say, non-self-extensional) logics: that is, the logical interderivability relation fails to be a congruence of the formula algebra. Now, while $\mathcal{S}$ is also a non-congruential logic, its implication connective (here denoted $\Rightarrow$) does satisfy the contraposition law: in fact in $\mathcal{S}$ one has that $(\varphi \Rightarrow \psi) \wedge (\psi \Rightarrow \varphi)$ is a theorem if and only if $(\sim\varphi \Rightarrow \sim\psi) \wedge (\sim\psi \Rightarrow \sim\varphi)$ is a theorem. In other words $\mathcal{S}$, although non-congruential if we look at its interderivability relation, enjoys at least $\Leftrightarrow$-*congruentiality* in Humberstone's terminology (relative to the bi-implication $\Leftrightarrow$ defined, in the usual way, as follows: $\varphi \Leftrightarrow \psi := (\sim\varphi \Rightarrow \sim\psi) \wedge (\sim\psi \Rightarrow \sim\varphi))$[2].

Nelson's original presentation of $\mathcal{S}$ has infinitely many schematic rules and no algebraic semantics; [15] also leaves unclear whether $\mathcal{N}3$ is comparable with $\mathcal{S}$ (and if so, which of the two is stronger). Unlike its relatives $\mathcal{N}3$ and $\mathcal{N}4$, the logic $\mathcal{S}$ received little attention after [15] and basic questions about it were left open, for example: Is $\mathcal{S}$ algebraizable? Can $\mathcal{S}$ be finitely axiomatized? What are the exact relations between $\mathcal{S}$ and $\mathcal{N}3$, and between $\mathcal{S}$ and $\mathcal{N}4$? In the present paper we will use the modern techniques of algebraic logic to answer these questions.

Our study will follow the same lines of previous papers by Busaniche, Cignoli, Odintsov, Spinks and Veroff on (algebraic models of) $\mathcal{N}3$ and $\mathcal{N}4$ (see,

---

[1] The presence of two implications, the *strong* one ($\Rightarrow$) mentioned earlier and the *weak* one ($\to$), is a distinctive feature of Nelson's logics; more on this below.

[2] Actually, we now know that $\mathcal{N}3$ (and $\mathcal{N}4$) are also $\Leftrightarrow$-congruential for a suitable choice of implication $\Rightarrow$ (called *strong implication*) that can be defined using the weak one $\to$; but Nelson may well not have been aware of this while writing [15].

e.g., [5,6,16,19,20]), which in turn rely on classic work by H. Rasiowa on the algebraization of non-classical logics. These investigations have shown that the algebraic approach to Nelson's logics may be particularly insightful, as it allows to view them as either conservative expansions of the negation-free fragment of intuitionistic logic by the addition of a new unary logical connective of *strong negation* ($\sim$) or as axiomatic extensions of well-known substructural/relevance logics. The first perspective allows us to establish a particularly useful link between algebraic models of $\mathcal{N}3/\mathcal{N}4$ and models of intuitionistic logic (via the so-called *twist-structure* construction—see especially [16]), while the second affords the possibility of exploiting general results and techniques that have been introduced in the study of residuated structures; this is the approach of [19,20] as well as [13], and that we shall also take in the present paper (see especially Subsect. 3.2).

The paper is organized as follows. In Sect. 2 we present the propositional fragment of the logic $\mathcal{S}$ and highlight some of its theorems, which will later be used to establish its algebraizability. In Sect. 3 we prove that $\mathcal{S}$ is algebraizable and present its equivalent algebraic semantics. In Sect. 4 we provide another calculus for $\mathcal{S}$, one that has a finite number of schematic axioms and only one schematic rule (*modus ponens*). We point out that having only one rule makes it easy to prove the Deduction Metatheorem in the standard way using induction over derivations. In Sect. 5 we prove that $\mathcal{N}3$ is a proper axiomatic extension of $\mathcal{S}$, and that $\mathcal{S}$ and $\mathcal{N}4$ are not extensions of each other. Proofs of some of the main new results are to be found in an Appendix to this paper.

2   Nelson's Logic $\mathcal{S}$

In this section we recall Nelson's original presentation of the propositional fragment of $\mathcal{S}$ [15] and we highlight some theorems of $\mathcal{S}$ that will be used further on to establish its algebraizability.

As is now usual, here we take a sentential logic $\mathcal{L}$ to be a structure containing a substitution-invariant consequence relation $\vdash_{\mathcal{L}}$ defined over an algebra of formulas **Fm** freely generated by a denumerable set of propositional variables $\{p, q, r, \ldots\}$ over a given language Σ. We will henceforth refer to algebras using boldface strings (such as **Fm** and **A**), and use the corresponding italicized version of these same strings (such as Fm and A) to refer to their corresponding carriers. Fixing a given logic, we will use φ, ψ and γ, possibly decorated with subscripts, to refer to arbitrary formulas of it.

Definition 1. *Nelson's logic* $\mathcal{S} = \langle \mathbf{Fm}, \vdash_{\mathcal{S}} \rangle$ *is the sentential logic in the language* $\langle \wedge, \vee, \Rightarrow, \sim, \bot \rangle$ *of type* $\langle 2, 2, 2, 1, 0 \rangle$ *defined by the Hilbert-style calculus with the schematic axioms and rules listed below. As usual,* $\varphi \Leftrightarrow \psi$ *will be used to abbreviate* $(\varphi \Rightarrow \psi) \wedge (\psi \Rightarrow \varphi)$.

Axioms

(A1) $\varphi \Rightarrow \varphi$

(A2) $\bot \Rightarrow \varphi$

(A3) $\sim\varphi \Rightarrow (\varphi \Rightarrow \bot)$

(A4) $\sim\bot$

(A5) $(\varphi \Rightarrow \psi) \Leftrightarrow (\sim\psi \Rightarrow \sim\varphi)$

Rules

$$\dfrac{\Gamma \Rightarrow (\varphi \Rightarrow (\psi \Rightarrow \gamma))}{\Gamma \Rightarrow (\psi \Rightarrow (\varphi \Rightarrow \gamma))} \ (\text{P}) \qquad \dfrac{\varphi \Rightarrow (\varphi \Rightarrow (\varphi \Rightarrow \gamma))}{\varphi \Rightarrow (\varphi \Rightarrow \gamma)} \ (\text{C}) \qquad \dfrac{\Gamma \Rightarrow \varphi \quad \varphi \Rightarrow \gamma}{\Gamma \Rightarrow \gamma} \ (\text{E})$$

$$\dfrac{\Gamma \Rightarrow \varphi \quad \psi \Rightarrow \gamma}{\Gamma \Rightarrow ((\varphi \Rightarrow \psi) \Rightarrow \gamma)} \ (\Rightarrow 1) \qquad \dfrac{\gamma}{\varphi \Rightarrow \gamma} \ (\Rightarrow \mathbf{r}) \qquad \dfrac{\varphi \Rightarrow \gamma}{(\varphi \wedge \psi) \Rightarrow \gamma} \ (\wedge 11)$$

$$\dfrac{\psi \Rightarrow \gamma}{(\varphi \wedge \psi) \Rightarrow \gamma} \ (\wedge 12) \qquad \dfrac{\Gamma \Rightarrow \varphi \quad \Gamma \Rightarrow \psi}{\Gamma \Rightarrow (\varphi \wedge \psi)} \ (\wedge \mathbf{r}) \qquad \dfrac{\varphi \Rightarrow \gamma \quad \psi \Rightarrow \gamma}{(\varphi \vee \psi) \Rightarrow \gamma} \ (\vee 11)$$

$$\dfrac{\varphi \Rightarrow^2 \gamma \quad \psi \Rightarrow^2 \gamma}{((\varphi \vee \psi) \Rightarrow^2 \gamma)} \ (\vee 12) \qquad \dfrac{\Gamma \Rightarrow \varphi}{\Gamma \Rightarrow (\varphi \vee \psi)} \ (\vee \mathbf{r}1) \qquad \dfrac{\Gamma \Rightarrow \psi}{\Gamma \Rightarrow (\varphi \vee \psi)} \ (\vee \mathbf{r}2)$$

$$\dfrac{(\varphi \wedge \sim\psi) \Rightarrow \gamma}{\sim(\varphi \Rightarrow \psi) \Rightarrow \gamma} \ (\sim \Rightarrow 1) \qquad \dfrac{\Gamma \Rightarrow^2 (\varphi \wedge \sim\psi)}{\Gamma \Rightarrow^2 \sim(\varphi \Rightarrow \psi)} \ (\sim \Rightarrow \mathbf{r}) \qquad \dfrac{(\sim\varphi \vee \sim\psi) \Rightarrow \gamma}{\sim(\varphi \wedge \psi) \Rightarrow \gamma} \ (\sim \wedge 1)$$

$$\dfrac{\Gamma \Rightarrow (\sim\varphi \vee \sim\psi)}{\Gamma \Rightarrow \sim(\varphi \wedge \psi)} \ (\sim \wedge \mathbf{r}) \qquad \dfrac{(\sim\varphi \wedge \sim\psi) \Rightarrow \gamma}{\sim(\varphi \vee \psi) \Rightarrow \gamma} \ (\sim \vee 1) \qquad \dfrac{\Gamma \Rightarrow (\sim\varphi \wedge \sim\psi)}{\Gamma \Rightarrow \sim(\varphi \vee \psi)} \ (\sim \vee \mathbf{r})$$

$$\dfrac{\varphi \Rightarrow \gamma}{\sim\sim\varphi \Rightarrow \gamma} \ (\sim\sim 1) \qquad \dfrac{\Gamma \Rightarrow \varphi}{\Gamma \Rightarrow \sim\sim\varphi} \ (\sim\sim\mathbf{r})$$

In the above rules, following Nelson's notation, $\Gamma = \{\varphi_1, \varphi_2, \ldots, \varphi_n\}$ is a finite set of formulas and the following abbreviations are employed:

$\Gamma \Rightarrow \varphi := \varphi_1 \Rightarrow (\varphi_2 \Rightarrow (\ldots \Rightarrow (\varphi_n \Rightarrow \varphi)\ldots))$

$\varphi \Rightarrow^2 \psi := \varphi \Rightarrow (\varphi \Rightarrow \psi)$

$\Gamma \Rightarrow^2 \varphi := \varphi_1 \Rightarrow^2 (\varphi_2 \Rightarrow^2 (\ldots \Rightarrow^2 (\varphi_n \Rightarrow^2 \varphi)\ldots))$

Moreover, when $\Gamma = \emptyset$, we take $\Gamma \Rightarrow \varphi := \varphi$.

Notice that we have fixed obvious infelicities in the rules ($\wedge 12$), ($\wedge \mathbf{r}$) and ($\sim \Rightarrow \mathbf{r}$) as they appear in [15, pp. 214–215]. For example, the original rule ($\wedge 12$) in Nelson's paper was:

$$\dfrac{(\varphi \wedge \psi) \Rightarrow \gamma}{\psi \Rightarrow \gamma} \ (\wedge 12)$$

This clearly makes the logic inconsistent. Indeed, taking $\varphi = \gamma$, we have:

$$\dfrac{(\gamma \wedge \psi) \Rightarrow \gamma}{\psi \Rightarrow \gamma} \ (\wedge 12)$$

Now, since $(\gamma \wedge \psi) \Rightarrow \gamma$ is a theorem (see Proposition 1, below), $\psi \Rightarrow \gamma$ is a theorem too. Choosing ψ as an axiom, we would conclude thus that γ is a theorem for any formula γ.

We note in passing that the rule (C), called *weak condensation* by Nelson, replaces (and is indeed a weaker form of) the usual contraction rule:

$$\frac{\varphi \Rightarrow (\varphi \Rightarrow \psi)}{\varphi \Rightarrow \psi}$$

Rule (C) is also known in the literature as *3-2 contraction* [17, p. 389] and corresponds, on algebraic models, to the property of *three-potency* (see Subsect. 3.2).

Also, do note that we obtain *modus ponens*, (MP), by taking $\Gamma = \emptyset$ in rule (E):

$$\frac{\varphi \quad \varphi \Rightarrow \gamma}{\gamma}$$

It is worth noticing that, despite appearances, Nelson's system $\mathcal{S}$ is a Hilbert-style calculus, rather than a sequent system. Its underlying notion of derivation, $\vdash_{\mathcal{S}}$, is the usual one. Henceforth, for any logic $\mathcal{L}$ and any set of formulas $\Gamma \cup \Pi$, we shall write $\Gamma \vdash_{\mathcal{L}} \Pi$ to say that $\Gamma \vdash_{\mathcal{L}} \pi$ for every $\pi \in \Pi$. By $\Gamma \dashv\vdash_{\mathcal{L}} \Pi$ we will abbreviate the double assertion $\Gamma \vdash_{\mathcal{L}} \Pi$ and $\Pi \vdash_{\mathcal{L}} \Gamma$.

One of the crucial steps in proving that a logic is algebraizable (in the sense of Blok and Pigozzi [3, Definition 2.2]) is to prove that it satisfies certain congruence properties. In the present context, this entails checking that $\varphi \Leftrightarrow \psi \vdash_{\mathcal{S}} {\sim}\varphi \Leftrightarrow {\sim}\psi$ and $\{\varphi_1 \Leftrightarrow \psi_1, \varphi_2 \Leftrightarrow \psi_2\} \vdash_{\mathcal{S}} (\varphi_1 \bullet \varphi_2) \Leftrightarrow (\psi_1 \bullet \psi_2)$ for each connective $\bullet \in \{\wedge, \vee, \Rightarrow\}$. The following auxiliary results will be used to prove that much, in the next section.

Proposition 1. *The following formulas are theorems of $\mathcal{S}$:*

1. $(\varphi \wedge \psi) \Rightarrow \varphi$
2. $(\varphi \wedge \psi) \Rightarrow \psi$
3. $\varphi \Rightarrow (\varphi \vee \psi)$
4. $\psi \Rightarrow (\varphi \vee \psi)$
5. $(\varphi \Rightarrow (\psi \Rightarrow \gamma)) \Leftrightarrow (\psi \Rightarrow (\varphi \Rightarrow \gamma))$

Proof. All justifying derivations are straightforward. We detail the first item, as an example:

$$\frac{\dfrac{}{\varphi \Rightarrow \varphi} \text{ (A1)}}{(\varphi \wedge \psi) \Rightarrow \varphi} \text{ (}\wedge\text{11)}$$

Proposition 2. $\{\varphi \Leftrightarrow \psi\} \dashv\vdash_{\mathcal{S}} \{\varphi \Rightarrow \psi, \psi \Rightarrow \varphi\}$.

Proof. Such a logical equivalence is easily justified by Proposition 1(1–2) and by considering the rule $(\wedge\mathbf{r})$ with $\Gamma = \emptyset$.

3 S Is Algebraizable

In this section we prove that S is algebraizable in the sense of Blok and Pigozzi (it is, in fact, implicative [8, Definition 2.3]), and we give two alternative presentations for its equivalent algebraic semantics (to be called 'S-algebras'). The first one is obtained via the algorithm of [3, Theorem 2.17], while the second one is closer to the usual axiomatizations of classes of residuated lattices, which are the algebraic counterparts of many logics in the substructural family. As a particular advantage, the second presentation of S-algebras will allow us to see at a glance that they form an equational class, and will also make it easier to compare them with other known classes of algebras for substructural logics.

Definition 2. *An* implicative logic *is a sentential logic $\mathcal{L}$ whose underlying algebra of formulas in a language Σ has a term $\alpha(p,q)$ in two variables that satisfies the following conditions:*
[IL1] $\vdash_{\mathcal{L}} \alpha(p,p)$
[IL2] $\alpha(p,q), \alpha(q,r) \vdash_{\mathcal{L}} \alpha(p,r)$
[IL3] $p, \alpha(p,q) \vdash_{\mathcal{L}} q$
[IL4] $q \vdash_{\mathcal{L}} \alpha(p,q)$
[IL5] *for each n-ary $\bullet \in \Sigma$,*
  $\bigcup_{i=1}^{n}\{\alpha(p_i,q_i), \alpha(q_i,p_i)\} \vdash_{\mathcal{L}} \alpha(\bullet(p_1,\dots,p_n), \bullet(q_1,\dots,q_n))$

We call any such α an $\mathcal{L}$-implication.

Given an algebra of formulas **Fm** of the language Σ, the associated set $Fm \times Fm$ of *equations* will henceforth be denoted by Eq; we will write $\varphi \approx \psi$ rather than $(\varphi, \psi) \in$ Eq. Let **A** be an algebra with the same similarity type as **Fm**. A homomorphism $\mathcal{V} \colon \mathbf{Fm} \to \mathbf{A}$ is called a *valuation in* **A**. We say that a valuation $\mathcal{V}$ in **A** *satisfies* $\varphi \approx \psi$ *in* **A** when $\mathcal{V}(\varphi) = \mathcal{V}(\psi)$; we say that an algebra **A** *satisfies* $\varphi \approx \psi$ when all valuations in **A** satisfy it.

Definition 3. *A logic $\mathcal{L}$ is* algebraizable *if and only if there are equations $\mathsf{E}(\varphi) \subseteq$ Eq and a transform $\mathsf{Eq} \overset{\rho}{\longmapsto} 2^{Fm}$, denoted by $\Delta(\varphi, \psi) := \rho(\varphi \approx \psi)$, such that $\mathcal{L}$ respects the following conditions:*
[Alg] $\varphi \dashv\vdash_{\mathcal{L}} \Delta(\mathsf{E}(\varphi))$
[Ref] $\vdash_{\mathcal{L}} \Delta(\varphi, \varphi)$
[Sym] $\Delta(\varphi, \psi) \vdash_{\mathcal{L}} \Delta(\psi, \varphi)$
[Trans] $\Delta(\varphi, \psi) \cup \Delta(\psi, \gamma) \vdash_{\mathcal{L}} \Delta(\varphi, \gamma)$
[Cong]   *for each n-ary $\bullet \in \Sigma$,*
  $\bigcup_{i=1}^{n} \Delta(\varphi_i, \psi_i) \vdash_{\mathcal{L}} \Delta(\bullet(\varphi_1, \dots, \varphi_n), \bullet(\psi_1, \dots, \psi_n))$

We call any such $\mathsf{E}(\varphi)$ the set of defining equations *and any such $\Delta(\varphi, \psi)$ the set of* equivalence formulas *of $\mathcal{L}$.*

Clarifying the notation in [Alg], recall that the set $\mathsf{E}(\varphi)$ contains pairs of formulas and we write $\varphi \approx \psi$ simply as syntactic sugar for a pair (φ, ψ) belonging to this set. Now, $\Delta(\varphi, \psi)$ transforms an equation into a set of formulas. Accordingly, we take $\Delta(\mathsf{E}(\varphi))$ as $\bigcup\{\Delta(\varphi_1, \varphi_2) \mid (\varphi_1, \varphi_2) \in \mathsf{E}(\varphi)\}$. Similarly, we shall let $\mathsf{E}(\Delta(\varphi, \psi))$ stand for $\bigcup\{\mathsf{E}(\chi) \mid \chi \in \Delta(\varphi, \psi)\}$.

Definition 4. *Let $\mathcal{L}$ be an implicative logic in the language Σ, having an $\mathcal{L}$-implication α. An $\mathcal{L}$-algebra $\mathbf{A}$ is a Σ-algebra such that $1 \in A$ and:*

[LALG1] *For all $\Gamma \cup \{\varphi\} \subseteq Fm$ and every valuation $\mathcal{V}$ in $\mathbf{A}$,*
    if $\Gamma \vdash_{\mathcal{L}} \varphi$ and $\mathcal{V}(\Gamma) \subseteq \{1\}$, then $\mathcal{V}(\varphi) = 1$.
[LALG2] *For all $a, b \in A$, if $\alpha(a, b) = 1$ and $\alpha(b, a) = 1$, then $a = b$.*

The class of $\mathcal{L}$-algebras is denoted by $Alg^{*}\mathcal{L}$.

Every implicative logic $\mathcal{L}$ is algebraizable with respect to the class $Alg^{*}\mathcal{L}$ [8, Proposition 3.15], and such algebraizability is witnessed by the defining equations $\mathsf{E}(\varphi) := \{\varphi \approx \alpha(\varphi, \varphi)\}$ and the equivalence formulas $\Delta(\varphi, \psi) := \{\alpha(\varphi, \psi), \alpha(\psi, \varphi)\}$. These are in fact the sets of defining equations and of equivalence formulas that we will use in the remainder of the present paper.

We can now prove (the details are to be found in the Appendix) that:

Theorem 1. *The calculus $\vdash_{\mathcal{S}}$ is implicative and thus algebraizable. The $\mathcal{S}$-implication is given by $\Rightarrow$, that is, $\alpha(p, q) := p \Rightarrow q$.*

In the case of $\mathcal{S}$ we have thus that $\mathsf{E}(\varphi) = \{\varphi \approx \varphi \Rightarrow \varphi\}$ and $\Delta(\varphi, \psi) = \{\varphi \Rightarrow \psi, \psi \Rightarrow \varphi\}$.

3.1 $\mathcal{S}$-algebras

By Blok-Pigozzi's algorithm ([3, Theorem 2.17], see also [8, Proposition 3.44]), we know that the equivalent algebraic semantics of $\mathcal{S}$ is the class of algebras given by Definition 5 below. We denote by Ax the set of axioms and denote by Inf R the set the set of inference rules of $\mathcal{S}$, given in Definition 1.

Definition 5. *An $\mathcal{S}$-algebra is a structure $\mathbf{A} = \langle A, \wedge, \vee, \Rightarrow, \sim, 0, 1 \rangle$ of type $\langle 2, 2, 2, 1, 0, 0 \rangle$ that satisfies the following equations and quasiequations:*

1. $\mathsf{E}(\Delta(\varphi, \varphi))$
2. $\mathsf{E}(\Delta(\varphi, \psi))$ *implies* $\varphi \approx \psi$
3. $\mathsf{E}(\varphi)$, *for each* $\varphi \in \mathsf{Ax}$
4. $\bigcup\limits_{i=1}^{n} \mathsf{E}(\gamma_i)$ *implies* $\varphi \approx 1$ *for each* $\gamma_1, \cdots, \gamma_n \vdash_{\mathcal{S}} \varphi \in \mathsf{Inf\ R}$

Regarding the notation in the above definition, $\mathsf{E}(\Delta(\varphi, \varphi))$ stands for the equation $\varphi \Rightarrow \varphi \approx (\varphi \Rightarrow \varphi) \Rightarrow (\varphi \Rightarrow \varphi)$. Item 2 is the quasiequation: $(\varphi \Rightarrow \psi) \approx (\varphi \Rightarrow \psi) \Rightarrow (\varphi \Rightarrow \psi)$ and $(\psi \Rightarrow \varphi) \approx (\psi \Rightarrow \varphi) \Rightarrow (\psi \Rightarrow \varphi)$ implies $\varphi \approx \psi$; $\mathsf{E}(\varphi)$ is the equation $\varphi \approx \varphi \Rightarrow \varphi$ for each axiom φ of $\mathcal{S}$. In fact, these conditions are telling us that for each axiom φ of $\mathcal{S}$ we have the equation $\varphi \approx 1$, and for each rule $\varphi \vdash_{\mathcal{S}} \psi$ of $\mathcal{S}$, in the corresponding algebras we have the quasiequation: if $\varphi \approx 1$, then $\psi \approx 1$.

We shall denote by $\mathsf{E}(\mathsf{An})$ the equation given in Definition 5(3) for the axiom An (for $1 \leq \mathsf{n} \leq 5$), and by $\mathsf{Q}(\mathsf{R})$ the quasiequation given in Definition 5(4) for the rule R of $\mathcal{S}$. From this point on, in this subsection, in order to make the

propositions and their proofs shorter, we shall also use the following abbreviations:

$x * y := \sim(x \Rightarrow \sim y)$

$x^2 := x * x$

$x^n := x * (x^{n-1})$, for $n > 2$

The following result, whose proof may be found in the Appendix, will help us in
checking that the class of $\mathcal{S}$-algebras forms a variety.

Proposition 3. *Let* **A** *be an* $\mathcal{S}$-algebra *and let* $a, b, c \in A$. *Then:*

1. $a \Rightarrow a = 1 = \sim 0$.
2. *The relation* $\leq$ *defined by setting* $a \leq b$ *iff* $a \Rightarrow b = 1$, *is a partial order with maximum* 1 *and minimum* 0.
3. $a \Rightarrow b = \sim b \Rightarrow \sim a$.
4. $a \Rightarrow (b \Rightarrow c) = b \Rightarrow (a \Rightarrow c)$.
5. $\sim\sim a = a$ *and* $a \Rightarrow 0 = \sim a$.
6. $\langle A, *, 1 \rangle$ *is a commutative monoid.*
7. $(a * b) \Rightarrow c = a \Rightarrow (b \Rightarrow c)$.
8. *The pair* $(*, \Rightarrow)$ *is residuated with respect to* $\leq$, *i.e.,* $a * b \leq c$ *iff* $b \leq a \Rightarrow c$.
9. $a^2 \leq a^3$.
10. $\langle A, \wedge, \vee \rangle$ *is a lattice with order* $\leq$.
11. $(a \vee b)^2 \leq a^2 \vee b^2$.

In the next section we introduce an equivalent presentation of $\mathcal{S}$-algebras
which takes precisely the properties of Proposition 3 above as postulates.

3.2   Alternative Presentation of $\mathcal{S}$-algebras

We start here by recalling the following standard definition [9, p. 185]:

Definition 6. *A* commutative integral bounded residuated lattice (CIBRL) *is
an algebra* $\mathbf{A} = \langle A, \wedge, \vee, *, \Rightarrow, 0, 1 \rangle$ *of type* $\langle 2, 2, 2, 2, 0, 0 \rangle$ *such that:*

1. $\langle A, \wedge, \vee, 0, 1 \rangle$ *is a bounded lattice with ordering* $\leq$, *minimum element* 0 *and maximum element* 1.
2. $\langle A, *, 1 \rangle$ *is a commutative monoid.*
3. *The pair* $(*, \Rightarrow)$ *is residuated with respect to* $\leq$, *i.e.,* $a * b \leq c$ *iff* $b \leq a \Rightarrow c$.

In the context of the above definition, the *integrality* condition corresponds to
having 1 not only as a maximum but also as the multiplicative unit of the
operation $*$, that is, $x * 1 = x$. For a CIBRL this condition immediately follows
from Definition 6(1–2).

Setting $\sim x := x \Rightarrow 0$, we say that a residuated lattice is *involutive* [10, p. 186]
when $\sim\sim a = a$ (in such a case, it follows that $a \Rightarrow b = \sim b \Rightarrow \sim a$). We say that
a residuated lattice is *3-potent* when it satisfies the equation $x^2 \leq x^3$. While we
have earlier defined $*$ from $\Rightarrow$, and now $*$ is a primitive operation, now we can
show that every CIBRL satisfies $x * y = \sim(x \Rightarrow \sim y)$ (see [10, Lemma 5.1]).

Definition 7. *An* $\mathcal{S}'$-algebra *is an involutive 3-potent CIBRL.*

The proof of the following result may be found in the Appendix:

Lemma 1. *1. Any CIBRL satisfies the equation $(x \vee y) * z \approx (x * z) \vee (y * z)$.*
2. Any CIBRL satisfies $x^2 \vee y^2 \approx (x^2 \vee y^2)^2$.
3. Any 3-potent CIBRL satisfies $(x \vee y^2)^2 \approx (x \vee y)^2$.
4. Any 3-potent CIBRL satisfies $(x \vee y)^2 \approx x^2 \vee y^2$.

Since involutive residuated lattices form an equational class [9, Theorem 2.7], it is obvious that $\mathcal{S}'$-algebras are also an equational class. From Proposition 3, we immediately conclude the following:

Proposition 4. *Let $\mathbf{A} = \langle A, \wedge, \vee, \Rightarrow, \sim, 0, 1 \rangle$ be an $\mathcal{S}$-algebra. Defining $x * y :=$
$\sim(x \Rightarrow \sim y)$, we have that $\mathbf{A} = \langle A, \wedge, \vee, *, \Rightarrow, 0, 1 \rangle$ is an $\mathcal{S}'$-algebra.*

Conversely, we are going to see that every $\mathcal{S}'$-algebra gives rise to an $\mathcal{S}$-algebra by checking that all (quasi) equations introduced in Definition 5 are satisfied (the proof may be found in the Appendix):

Proposition 5. *Let $\mathbf{A} = \langle A, \wedge, \vee, *, \Rightarrow, 0, 1 \rangle$ be an $\mathcal{S}'$-algebra. Defining $\sim x :=$
$x \Rightarrow 0$, we have that $\mathbf{A} = \langle A, \wedge, \vee, \Rightarrow, \sim, 0, 1 \rangle$ is an $\mathcal{S}$-algebra.*

Thus, the classes of $\mathcal{S}$-algebras and of $\mathcal{S}'$-algebras are term-equivalent.

The presentation given in Definition 7 has several advantages in what concerns the study of the semantics of $\mathcal{S}$. For example, it is now straightforward to check that the three-element MV-algebra [7] is a model of Nelson's logic $\mathcal{S}$. This in turn allows one to prove that the formulas which Nelson claims not to be derivable in $\mathcal{S}$ [15, p. 213] are indeed not valid (see [13]).

4   A Finite Hilbert-Style Calculus for $\mathcal{S}$

In this section we introduce a finite Hilbert-style calculus (which is an extension of the calculus $IPC^* \backslash c$, called *intuitionistic logic without contraction*, of [4]) that is algebraizable with respect to the class of $\mathcal{S}'$-algebras.

We are thus going to have two logics that are both algebraizable with respect to the same variety with the same defining equations and equivalence formulas; from this we will obtain an equivalence between our calculus and Nelson's.

The logic $\mathcal{S}' = \langle \mathbf{Fm}, \vdash_{\mathcal{S}'} \rangle$ is the sentential logic in the language $\langle \wedge, \vee, \Rightarrow, *, \sim, \bot, \top \rangle$ of type $\langle 2, 2, 2, 2, 1, 0, 0 \rangle$ defined by the Hilbert-style calculus with the following schematic axioms and with *modus ponens* as the only rule:

(A1') $(\varphi \Rightarrow \psi) \Rightarrow ((\gamma \Rightarrow \varphi) \Rightarrow (\gamma \Rightarrow \psi))$
(A2') $(\varphi \Rightarrow (\psi \Rightarrow \gamma)) \Rightarrow (\psi \Rightarrow (\varphi \Rightarrow \gamma))$
(A3') $\varphi \Rightarrow (\psi \Rightarrow \varphi)$
(A4') $(\varphi \Rightarrow \gamma) \Rightarrow ((\psi \Rightarrow \gamma) \Rightarrow ((\varphi \vee \psi) \Rightarrow \gamma))$
(A5') $\varphi \Rightarrow (\varphi \vee \psi)$
(A6') $\psi \Rightarrow (\varphi \vee \psi)$
(A7') $(\varphi \wedge \psi) \Rightarrow \varphi$

(A8') $(\varphi \wedge \psi) \Rightarrow \psi$

(A9') $\varphi \Rightarrow (\psi \Rightarrow (\varphi \wedge \psi))$

(A10') $((\gamma \Rightarrow \varphi) \wedge (\gamma \Rightarrow \psi)) \Rightarrow (\gamma \Rightarrow \varphi(\wedge\psi))$

(A11') $\varphi \Rightarrow (\psi \Rightarrow (\varphi * \psi))$

(A12') $(\varphi \Rightarrow (\psi \Rightarrow \gamma)) \Rightarrow ((\varphi * \psi) \Rightarrow \gamma)$

(A13') $\sim\varphi \Rightarrow (\varphi \Rightarrow \psi)$

(A14') $(\varphi \Rightarrow \psi) \Leftrightarrow (\sim\psi \Rightarrow \sim\varphi)$

(A15') $\varphi \Leftrightarrow \sim\sim\varphi$

(A16') $\perp \Rightarrow \varphi$

(A17') $\varphi \Rightarrow \top$

(A18') $\varphi^2 \Rightarrow \varphi^3$

As before, $\varphi \Leftrightarrow \psi$ abbreviates $(\varphi \Rightarrow \psi) \wedge (\psi \Rightarrow \varphi)$, while the connective $*$ is here taken as primitive.

Axioms (A1')–(A13'), (A14'($\Rightarrow$)), (A15'($\Rightarrow$)), (A16') and (A17') of our calculus are the same as those of $IPC^*\backslash c$ as presented in [4, Table 3.2], where it is proven that $IPC^*\backslash c$ is algebraizable. We added the converse implication in axioms (A14') and (A15') to characterize involution and we added the axiom (A18') to characterize 3-potency. As algebraizability is preserved by axiomatic extensions (cf. [8, Proposition 3.31]) we have the following results:

Theorem 2. *The calculus S' is algebraizable (with the same defining equations and equivalence formulas as S) with respect to the class of S'-algebras.*

Proof. We know from [4, Theorem 5.1] that $IPC^*\backslash c$ is algebraizable with respect to the class of commutative integral bounded residuated lattices with the same defining equations and equivalence formulas already considered above. The axioms that were now added imply that the algebraic semantics of our extension is involutive and 3-potent, i.e., it is an S'-algebra.

Corollary 1. *S and S' define the same logic.*

Proof. Let K_S be the class of S-algebras. Thanks to Propositions 4 and 5 we know that K_S is also the class of S'-algebras. The result follows now from [8, Proposition 3.47], that gives us an algorithm to find a Hilbert-style calculus for an algebraizable logic from its quasivariety, defining equations and equivalence formulas. As S-algebras and S'-algebras are the same class of algebras and their defining equations and equivalence formulas are the same, the Hilbert-style calculus given by the algorithm must do the same job as the one we had before.

Working with Nelson's original presentation of S, it can be hard to directly prove some version of the Deduction Metatheorem. Indeed, if we prove it, as usual, by way of an induction over the structure of the derivations, we need to apply the inductive hypothesis over each rule of the system. The advantage of S', in employing such a strategy, is that it has only one inference rule. This allows us to establish:

Theorem 3 (Deduction Metatheorem). *If $\Gamma \cup \{\varphi\} \vdash \psi$, then $\Gamma \vdash \varphi^2 \Rightarrow \psi$.*

Proof. Thanks to [9, Corollary 2.15] we have a version of the Deduction Metatheorem for substructural logics which says that $\Gamma \cup \{\varphi\} \vdash \psi$ iff $\Gamma \vdash \varphi^n \Rightarrow \psi$ for some n. In view of (A'18) it is easy to see that in $\mathcal{S}$ we can always choose $n = 2$.

5   Comparing $\mathcal{S}$ with $\mathcal{N}3$ and $\mathcal{N}4$

As mentioned before, Nelson introduced two other better-known logics, $\mathcal{N}3$ and $\mathcal{N}4$, which are also algebraizable with respect to classes of residuated structures (namely, the so-called $\mathcal{N}3$-lattices and $\mathcal{N}4$-lattices). A question that immediately arises concerns the precise relation between $\mathcal{S}$ and these other logics, or (equivalently) between $\mathcal{S}$-algebras and $\mathcal{N}3$- and $\mathcal{N}4$-lattices. In what follows it is worth taking into account that not all $\mathcal{S}$-algebras are distributive (see [13, Example 5.1]).

5.1 $\mathcal{N}4$

Definition 8. $\mathcal{N}4 = \langle \mathbf{Fm}, \vdash_{\mathcal{N}4} \rangle$ *is the sentential logic in the language* $\langle \wedge, \vee, \rightarrow, \sim \rangle$ *of type* $\langle 2, 2, 2, 1 \rangle$ *defined by the Hilbert-style calculus with the following schematic axioms and modus ponens as the only schematic rule. Below,* $\varphi \leftrightarrow \psi$ *will be used to abbreviate* $(\varphi \rightarrow \psi) \wedge (\psi \rightarrow \varphi)$.

(N1) $\varphi \rightarrow (\psi \rightarrow \varphi)$

(N2) $(\varphi \rightarrow (\psi \rightarrow \gamma)) \rightarrow ((\varphi \rightarrow \psi) \rightarrow (\varphi \rightarrow \gamma))$

(N3) $(\varphi \wedge \psi) \rightarrow \varphi$

(N4) $(\varphi \wedge \psi) \rightarrow \psi$

(N5) $(\varphi \rightarrow \psi) \rightarrow ((\varphi \rightarrow \gamma) \rightarrow (\varphi \rightarrow (\psi \wedge \gamma)))$

(N6) $\varphi \rightarrow (\varphi \vee \psi)$

(N7) $\psi \rightarrow (\varphi \vee \psi)$

(N8) $(\varphi \rightarrow \gamma) \rightarrow ((\psi \rightarrow \gamma) \rightarrow ((\varphi \vee \psi) \rightarrow \gamma))$

(N9) $\sim\sim\varphi \leftrightarrow \varphi$

(N10) $\sim(\varphi \vee \psi) \leftrightarrow (\sim\varphi \wedge \sim\psi)$

(N11) $\sim(\varphi \wedge \psi) \leftrightarrow (\sim\varphi \vee \sim\psi)$

(N12) $\sim(\varphi \rightarrow \psi) \leftrightarrow (\varphi \wedge \sim\psi)$

The implication $\rightarrow$ in $\mathcal{N}4$ is usually called *weak implication*, in contrast to the *strong implication* $\Rightarrow$ that is defined in $\mathcal{N}4$ as follows:

$$\varphi \Rightarrow \psi := (\varphi \rightarrow \psi) \wedge (\sim\psi \rightarrow \sim\varphi).$$

As the notation suggests, it is the strong implication that we shall compare with the implication of $\mathcal{S}$. This appears indeed to be the most meaningful choice, for otherwise, since the weak implications of both $\mathcal{N}4$ and $\mathcal{N}3$ fail to satisfy contraposition (which holds in $\mathcal{S}$), we would have to say that $\mathcal{S}$ is incomparable with both logics.

The logic $\mathcal{N}4$ is algebraizable (though not implicative) with equivalence formulas $\{\varphi \Rightarrow \psi, \psi \Rightarrow \varphi\}$ and defining equation $\varphi \approx \varphi \rightarrow \varphi$ [18, Theorem 2.6].

We notice in passing that the implication in this defining equation could as well be taken to be the strong one, so $\varphi \approx \varphi \Rightarrow \varphi$ would work too; in contrast, $\{\varphi \to \psi, \psi \to \varphi\}$ would not be a set of equivalence formulas, due precisely to the failure of contraposition. The equivalent algebraic semantics of $\mathcal{N}4$ is the class of $\mathcal{N}4$-lattices defined below [16, Definition 8.4.1]:

Definition 9. *An algebra* $\mathbf{A} = \langle A, \vee, \wedge, \to, \sim \rangle$ *is an $\mathcal{N}4$-lattice if it satisfies the following properties:*

1. $\langle A, \vee, \wedge, \sim \rangle$ *is a De Morgan algebra.*
2. *The relation* $\preceq$ *defined, for all* $a, b \in A$, *by* $a \preceq b$ *iff* $(a \to b) \to (a \to b) = (a \to b)$ *is a pre-order on* $\mathbf{A}$.
3. *The relation* $\equiv$ *defined, for all* $a, b \in A$ *as* $a \equiv b$ *iff* $a \preceq b$ *and* $b \preceq a$ *is a congruence relation with respect to* $\wedge, \vee, \to$ *and the quotient algebra* $\mathbf{A}_{\bowtie} := \langle A, \vee, \wedge, \to \rangle / \equiv$ *is an implicative lattice.*
4. *For any* $a, b \in A$, $\sim(a \to b) \equiv a \wedge \sim b$.
5. *For any* $a, b \in A$, $a \leq b$ *iff* $a \preceq b$ *and* $\sim b \preceq \sim a$, *where* $\leq$ *is the lattice order for* $\mathbf{A}$.

A very simple example of an $\mathcal{N}4$-lattice is the four-element algebra $\mathbf{A_4}$ whose lattice reduct is the four-element diamond De Morgan algebra. This algebra has carrier $A_4 = \{0, 1, b, n\}$, the maximum element of the lattice order being 1, the minimum 0, and b and n being incomparable. The negation (Fig. 1) is given by $\sim b := b$, $\sim n := n$, $\sim 1 := 0$ and $\sim 0 := 1$. The weak implication is given, for all $a \in A_4$, by $1 \to a = b \to a := a$ and $0 \to a = n \to a := 1$. One can check that $\mathbf{A_4}$ satisfies all properties of Definition 9 (in particular, the quotient $\mathbf{A_4}/\equiv$ is the two-element Boolean algebra).

Proposition 6. *$\mathcal{N}4$ and $\mathcal{S}$ are incomparable, that is, neither of them extends the other.*

Proof. We show that not every $\mathcal{S}$-algebra is an $\mathcal{N}4$-lattice, and that no $\mathcal{N}4$-lattice is an $\mathcal{S}$-algebra. The first claim follows from the fact that $\mathcal{N}4$-lattices have a distributive lattice reduct, whereas $\mathcal{S}$-algebras need not be distributive. As to the second, it is sufficient to observe that the equation $x \Rightarrow x \approx y \Rightarrow y$ is satisfied in all $\mathcal{S}$-algebras but does not hold in the four-element $\mathcal{N}4$-lattice $\mathbf{A_4}$. There we have $1 \Rightarrow 1 \neq b \Rightarrow b$ because $1 \Rightarrow 1 = (1 \to 1) \wedge (\sim 1 \to \sim 1) = 1 \wedge 1 = 1$ but $b \Rightarrow b = (b \to b) \wedge (\sim b \to \sim b) = b \wedge b = b$. Since both $\mathcal{N}4$ and $\mathcal{S}$ are algebraizable logics, this immediately entails that neither $\mathcal{N}4 \leq \mathcal{S}$ nor $\mathcal{S} \leq \mathcal{N}4$. In logical terms, one can check that the distributivity axiom is valid in $\mathcal{N}4$ but not in $\mathcal{S}$, whereas the formula $(\varphi \Rightarrow \varphi) \Rightarrow (\psi \Rightarrow \psi)$ is valid in $\mathcal{S}$ but not in $\mathcal{N}4$.

$\to$	0	n	b	1
0	1	1	1	1
n	1	1	1	1
b	0	n	b	1
1	0	n	b	1

$\sim$	
0	1
n	n
b	b
1	1

Fig. 1. A_4

5.2 $\mathcal{N}3$

Definition 10. *Nelson's logic* $\mathcal{N}3 = \langle \mathbf{Fm}, \vdash_{\mathcal{N}3} \rangle$ *is the axiomatic extension of* $\mathcal{N}4$ *obtained by adding the following axiom:*

(N13) $\sim\!\varphi \rightarrow (\varphi \rightarrow \psi)$.

Proposition 7. $\mathcal{N}3$ *is a proper extension of* $\mathcal{S}$.

Proof. It is known from [19] that every $\mathcal{N}3$-lattice (the algebraic counterpart of $\mathcal{N}3$) satisfies all properties of our Definition 7, and therefore every $\mathcal{N}3$-lattice is an $\mathcal{S}$-algebra. On the other hand, the logic $\mathcal{N}3$ was defined as an axiomatic extension of $\mathcal{N}4$, therefore it is distributive too, whereas $\mathcal{S}$-algebras need not be distributive (see [13, Example 5.1]).

6 Future Work

We have studied $\mathcal{S}$ in two directions, through a proof-theoretic approach and through algebraic methods. Concerning the proof-theoretic approach, we have introduced a finite Hilbert-style calculus for $\mathcal{S}$. An interesting question that still remains is about other types of calculi. In this sense we would find it attractive to be able to present a sequent calculus for $\mathcal{S}$ enjoying a cut-elimination theorem, so that it could be used to determine, among other things, whether $\mathcal{S}$ is decidable and enjoys the Craig interpolation theorem.

As observed in Theorem 3, if we let $\varphi \rightarrow \psi := \varphi \Rightarrow (\varphi \Rightarrow \psi)$, then the weak implication $\rightarrow$ enjoys a version of the Deduction Metatheorem; this suggests that the connective $\rightarrow$ has a special logical role within $\mathcal{S}$, whereas $\Rightarrow$ is the key operation on the corresponding algebras. It is well known that the logic $\mathcal{N}3$ as well as its algebraic counterpart can be equivalently axiomatized by taking either the weak or the strong implication as primitive, defining $\rightarrow$ from $\Rightarrow$ as shown above. The analogous result for $\mathcal{N}4$ has been harder to prove (see [20]), and the corresponding definition is $\varphi \rightarrow \psi := (\varphi \wedge (((\varphi \wedge (\psi \Rightarrow \psi)) \Rightarrow \psi) \Rightarrow ((\varphi \wedge (\psi \Rightarrow \psi)) \Rightarrow \psi))) \Rightarrow ((\varphi \wedge (\psi \Rightarrow \psi)) \Rightarrow \psi)$.

We can ask a similar question about the logic $\mathcal{S}$ and its algebraic counterpart: namely, given that Nelson's axiomatization as well as ours have $\Rightarrow$ as primitive, is it also possible to axiomatize $\mathcal{S}$(-algebras) by taking the weak implication $\rightarrow$ as primitive? This question is related to certain algebraic properties that $\rightarrow$ enjoys on $\mathcal{S}$-algebras. In fact, we have shown in [13, Theorem 4.5] that, analogously to $\mathcal{N}3$-lattices, $\mathcal{S}$-algebras are a variety of *weak Brouwerian semilattices with filter-preserving operations* [2, Definition 2.1], which means that they possess an intuitionistic-like internal structure, where a *weak relative pseudo-complementation* operation (an intuitionistic-like implication) is given precisely by the weak implication. This suggests that one may in fact hope to be able to view (and axiomatize) $\mathcal{S}$ as a conservative expansion of some intuitionistic-like positive logic by a strong (involutive) negation, as has been the case of $\mathcal{N}3$ and $\mathcal{N}4$.

As hinted above, a more detailed study of $\mathcal{S}$-algebras can be found in the companion paper [13]. Some questions regarding the variety of $\mathcal{S}$-algebras, its extensions, congruences and more relations between $\mathcal{S}$-algebras and other well-known algebras are investigated there. Another question that is still open is which logic (class of algebras) is the infimum of $\mathcal{S}$(-algebras) and $\mathcal{N}4$(-lattices)—it is easy to see that the least logic extending $\mathcal{S}$ and $\mathcal{N}4$ is precisely $\mathcal{N}3$.

Appendix: Proofs of the Some of the Main Results

Theorem 1. The calculus $\vdash_\mathcal{S}$ is implicative, and thus algebraizable.

Proof. In the case of $\mathcal{S}$, the term $\alpha(\varphi, \psi)$ may be chosen to be $\varphi \Rightarrow \psi$. We will make below free use of Proposition 2.

IL1 follows immediately from axiom **(A1)**, while **IL2** follows from rule **(E)**. **IL3** follows from **(MP)** and **IL4** follows from $(\Rightarrow r)$. We are left with proving that $\Rightarrow$ respects **IL5** for each connective $\bullet \in \{\wedge, \vee, \Rightarrow, \sim\}$.

($\sim$) $\{(\varphi \Leftrightarrow \psi), (\psi \Leftrightarrow \varphi)\} \vdash_\mathcal{S} {\sim}\varphi \Leftrightarrow {\sim}\psi$ holds by axiom **(A5)** and the (derived) rule **(MP)**.

($\wedge$) We must prove that $\{(\varphi_1 \Leftrightarrow \psi_1), (\varphi_2 \Leftrightarrow \psi_2)\} \vdash_\mathcal{S} (\varphi_1 \wedge \varphi_2) \Leftrightarrow (\psi_1 \wedge \psi_2)$. From Proposition 1(1–2) we have $\vdash_\mathcal{S} (\varphi_1 \wedge \varphi_2) \Rightarrow \varphi_1$ and $\vdash_\mathcal{S} (\varphi_1 \wedge \varphi_2) \Rightarrow \varphi_2$. Then:

$$\cfrac{\cfrac{\overline{(\varphi_1 \wedge \varphi_2) \Rightarrow \varphi_1}\ \ \varphi_1 \Rightarrow \psi_1}{(\varphi_1 \wedge \varphi_2) \Rightarrow \psi_1}\,(\text{E}) \quad \cfrac{\overline{(\varphi_1 \wedge \varphi_2) \Rightarrow \varphi_2}\ \ \varphi_2 \Rightarrow \psi_2}{(\varphi_1 \wedge \varphi_2) \Rightarrow \psi_2}\,(\text{E})}{(\varphi_1 \wedge \varphi_2) \Rightarrow (\psi_1 \wedge \psi_2)}\,(\wedge \text{r})$$

The remainder of the proof is analogous.

($\vee$) We must prove that $\{(\varphi_1 \Leftrightarrow \psi_1), (\varphi_2 \Leftrightarrow \psi_2)\} \vdash_\mathcal{S} (\varphi_1 \vee \varphi_2) \Leftrightarrow (\psi_1 \vee \psi_2)$. From Proposition 1(3–4), $\psi_1 \Rightarrow (\psi_1 \vee \psi_2)$ and $\psi_2 \Rightarrow (\psi_1 \vee \psi_2)$ are derivable. Then:

$$\cfrac{\cfrac{\varphi_1 \Rightarrow \psi_1\ \ \overline{\psi_1 \Rightarrow (\psi_1 \vee \psi_2)}}{\varphi_1 \Rightarrow (\psi_1 \vee \psi_2)}\,(\text{E}) \quad \cfrac{\varphi_2 \Rightarrow \psi_2\ \ \overline{\psi_2 \Rightarrow (\psi_1 \vee \psi_2)}}{\varphi_2 \Rightarrow (\psi_1 \vee \psi_2)}\,(\text{E})}{(\varphi_1 \vee \varphi_2) \Rightarrow (\psi_1 \vee \psi_2)}\,(\vee \text{l1})$$

The remainder of the proof is analogous.

($\Rightarrow$) We must prove that $\{(\theta \Leftrightarrow \varphi), (\psi \Leftrightarrow \gamma)\} \vdash_\mathcal{S} (\theta \Rightarrow \psi) \Leftrightarrow (\varphi \Rightarrow \gamma)$. This time, we have:

$$\cfrac{\varphi \Rightarrow \theta \quad \psi \Rightarrow \gamma}{\varphi \Rightarrow ((\theta \Rightarrow \psi) \Rightarrow \gamma)}\,(\Rightarrow 1)$$

Taking ψ as $\theta \Rightarrow \psi$ in Proposition 1(5), we have:

$$\cfrac{\varphi \Rightarrow ((\theta \Rightarrow \psi) \Rightarrow \gamma) \quad \overline{(\varphi \Rightarrow ((\theta \Rightarrow \psi) \Rightarrow \gamma)) \Rightarrow ((\theta \Rightarrow \psi) \Rightarrow (\varphi \Rightarrow \gamma))}}{(\theta \Rightarrow \psi) \Rightarrow (\varphi \Rightarrow \gamma)}\,(\text{MP})$$

The remainder of the proof is analogous.

Proposition 3. Let **A** be an $\mathcal{S}$-algebra and let $a, b, c \in A$. Then:

1. $a \Rightarrow a = 1 = {\sim}0$.
2. The relation $\leq$ defined by setting $a \leq b$ iff $a \Rightarrow b = 1$, is a partial order with maximum 1 and minimum 0.
3. $a \Rightarrow b = {\sim}b \Rightarrow {\sim}a$.
4. $a \Rightarrow (b \Rightarrow c) = b \Rightarrow (a \Rightarrow c)$.
5. ${\sim}{\sim}a = a$ and $a \Rightarrow 0 = {\sim}a$.
6. $\langle A, *, 1 \rangle$ is a commutative monoid.
7. $(a * b) \Rightarrow c = a \Rightarrow (b \Rightarrow c)$.
8. The pair $(*, \Rightarrow)$ is residuated with respect to $\leq$, i.e., $a * b \leq c$ iff $b \leq a \Rightarrow c$.
9. $a^2 \leq a^3$.
10. $\langle A, \wedge, \vee \rangle$ is a lattice with order $\leq$.
11. $(a \vee b)^2 \leq a^2 \vee b^2$.

Proof. 1. This follows from the fact that $\mathcal{S}$ is an implicative logic, see [8, Lemma 2.6]. In particular, ${\sim}0 = 0 \Rightarrow 0 = 1$.
2. By E(A2) we have that 0 is the minimum element with respect to the order $\leq$. The rest easily follows from the fact that $\mathcal{S}$ is implicative.
3. This follows from E(A5) and item 2 above.
4. By Q(P) and item 2 above, we have that $d \leq a \Rightarrow (b \Rightarrow c)$ implies $d \leq b \Rightarrow (a \Rightarrow c)$ for all $d \in A$. Then, taking $d = a \Rightarrow (b \Rightarrow c)$, we have $a \Rightarrow (b \Rightarrow c) \leq b \Rightarrow (a \Rightarrow c)$, which easily implies the desired result.
5. The identity ${\sim}{\sim}a = a$ follows from item 2 above together with Q(${\sim}{\sim}$l) and Q(${\sim}{\sim}$r). By item 3 above, $a \Rightarrow 0 = {\sim}0 \Rightarrow {\sim}a = 1 \Rightarrow {\sim}a = {\sim}a$. The last identity holds good because, on the one hand, by Q($\Rightarrow$ 1) we have that $1 \leq 1$ and ${\sim}a \leq {\sim}a$ implies $1 \Rightarrow {\sim}a \leq {\sim}a$. On the other hand, by item 1 we have ${\sim}a \Rightarrow {\sim}a \leq 1$ and so we can apply Q($\Rightarrow$ r) to obtain $1 \Rightarrow ({\sim}a \Rightarrow {\sim}a) = 1$. By item 4, we have $1 \Rightarrow ({\sim}a \Rightarrow {\sim}a) = {\sim}a \Rightarrow (1 \Rightarrow {\sim}a)$, hence we conclude that ${\sim}a \Rightarrow (1 \Rightarrow {\sim}a) = 1$ and so, by item 2, ${\sim}a \leq 1 \Rightarrow {\sim}a$.
6. As to commutativity, using items 3 and 5 above, we have $a * b = {\sim}(a \Rightarrow {\sim}b) = {\sim}({\sim}{\sim}b \Rightarrow {\sim}a) = {\sim}(b \Rightarrow {\sim}a) = b * a$. As to associativity, using 3, 5, Q(${\sim}{\sim}$r) and Q(${\sim}{\sim}$l), we have $(a * b) * c = {\sim}({\sim}(a \Rightarrow {\sim}b) \Rightarrow {\sim}c) = {\sim}({\sim}{\sim}c \Rightarrow {\sim}{\sim}(a \Rightarrow {\sim}b)) = {\sim}(c \Rightarrow (a \Rightarrow {\sim}b)) = {\sim}(a \Rightarrow (c \Rightarrow {\sim}b)) = {\sim}(a \Rightarrow (b \Rightarrow {\sim}c)) = {\sim}(a \Rightarrow {\sim}{\sim}(b \Rightarrow {\sim}c)) = a * (b * c)$. As to 1 being the neutral element, using items 1 and 5 above, we have $a * 1 = a * {\sim}0 = {\sim}(a \Rightarrow {\sim}{\sim}0) = {\sim}(a \Rightarrow 0) = {\sim}{\sim}a = a$.
7. Using items 2, 3, 5 and 6 above, we have $(a * b) \Rightarrow c = {\sim}(a \Rightarrow {\sim}b) \Rightarrow c = {\sim}c \Rightarrow {\sim}{\sim}(a \Rightarrow {\sim}b) = {\sim}c \Rightarrow (a \Rightarrow {\sim}b) = a \Rightarrow ({\sim}c \Rightarrow {\sim}b) = a \Rightarrow ({\sim}{\sim}b \Rightarrow {\sim}{\sim}c) = a \Rightarrow (b \Rightarrow c)$.
8. By item 2 above, we have $a * b \leq c$ iff $(a * b) \Rightarrow c = 1$ iff, by item 7, $a \Rightarrow (b \Rightarrow c) = 1$ iff, by item 6, $b \Rightarrow (a \Rightarrow c) = 1$ iff, by 2 again, $b \leq a \Rightarrow c$.
9. By Q(C) we have that $a^3 \leq c$ implies $a^2 \leq c$ for all $c \in A$. Then, taking $c = a^3$, we have $a^2 \leq a^3$.
10. We check that $a \wedge b$ is the infimum of the set $\{a, b\}$ with respect to $\leq$. First of all, we have $a \wedge b \leq a$ and $a \wedge b \leq b$ by Q($\wedge$11), Q($\wedge$12) and item 2 above. Then, assuming $c \leq a$ and $c \leq b$, we have $c \leq a \wedge b$ by Q($\wedge$r). An

analogous reasoning, using Q($\lor$r1), Q($\lor$r2) and Q($\lor$11) shows that $a \lor b$ is the supremum of $\{a, b\}$.

11. By item 10 we have that $a^2 \leq a^2 \lor b^2$ and $b^2 \leq a^2 \lor b^2$. Hence, by item 8, we have $a \leq a \Rightarrow (a^2 \lor b^2)$ and $b \leq b \Rightarrow (a^2 \lor b^2)$. By item 2 we have then $a \Rightarrow (a \Rightarrow (a^2 \lor b^2)) = b \Rightarrow (b \Rightarrow (a^2 \lor b^2)) = 1$, hence we can use Q($\lor$12) to obtain $(a \lor b) \Rightarrow ((a \lor b) \Rightarrow (a^2 \lor b^2)) = 1$. Then items 2 and 8 give us $(a \lor b)^2 \leq a^2 \lor b^2$, as was to be proved.

Lemma 1. 1. Any CIBRL satisfies the equation $(x \lor y) * z \approx (x * z) \lor (y * z)$.
2. Any 3-potent CIBRL satisfies $x^2 \lor y^2 \approx (x^2 \lor y^2)^2$.
3. Any 3-potent CIBRL satisfies $(x \lor y^2)^2 \approx (x \lor y)^2$.
4. Any 3-potent CIBRL satisfies $(x \lor y)^2 \approx x^2 \lor y^2$.

Proof. 1. See [9, Lemma 2.6].
2. Let a, b be arbitrary elements of a given 3-potent CIBRL. From $a^2 \leq (a^2 \lor b^2)$ and $b^2 \leq (a^2 \lor b^2)$, using monotonicity of $*$, we have $a^4 \leq (a^2 \lor b^2)^2$ and $b^4 \leq (a^2 \lor b^2)^2$. Using 3-potency, the latter inequalities simplify to $a^2 \leq (a^2 \lor b^2)^2$ and $b^2 \leq (a^2 \lor b^2)^2$. Thus, $a^2 \lor b^2 \leq (a^2 \lor b^2)^2$.
3. We have $(a \lor b^2) \leq (a \lor b)$ from monotonicity of $*$ and supremum of $\lor$, therefore $(a \lor b^2)^2 \leq (a \lor b)^2$. For the converse, we have that $(a * b) \leq a$, whence $(a * b) \leq (a \lor b^2)$. Also $a^2 \leq (a \lor b^2)$ and $b^2 \leq (a \lor b^2)$. By supremum of $\lor$, $(a^2 \lor (a*b) \lor b^2) \leq (a \lor b^2)$. But $(a^2 \lor (a*b) \lor b^2) = (a \lor b)^2$ by Lemma 1(1), so $(a \lor b)^2 \leq (a \lor b^2)$. Using the monotonicity of $*$, $(a \lor b)^4 \leq (a \lor b^2)^2$ and from 3-potency we have $(a \lor b)^2 \leq (a \lor b^2)^2$.
4. From Lemma 1(2) we have $(a^2 \lor b^2) = (a^2 \lor b^2)^2$, and from Lemma 1(3) we have $(a^2 \lor b^2)^2 = (a^2 \lor b)^2 = (b^2 \lor a^2)^2 = (b \lor a)^2$.

Proposition 5. Let $\mathbf{A} = \langle A, \land, \lor, *, \Rightarrow, 0, 1 \rangle$ be an $\mathcal{S}'$-algebra. Defining $\sim x := x \Rightarrow 0$, we have that $\mathbf{A} = \langle A, \land, \lor, \Rightarrow, \sim, 0, 1 \rangle$ is an $\mathcal{S}$-algebra.

Proof. Let $\mathbf{A}$ be an $\mathcal{S}'$-algebra. We first consider the equations corresponding to the axioms of $\mathcal{S}$. As $a \leq b$ iff $a \Rightarrow b = 1$, we will write the former rather than the latter.

Equations
The equation E(A1) easily follows from integrality. We have E(A2) from the fact that 0 is the minimum element of $\mathbf{A}$. From the definition of $\sim$ in $\mathcal{S}'$ and from E(A1) we see that E(A3) holds. We know that $1 := \sim 0$, therefore we have E(A4). As $\mathbf{A}$ is involutive, it follows that E(A5) holds. We are still to prove the equation E($\Delta(\varphi, \varphi)$). For that, see that we need to prove the identity $(\varphi \Rightarrow \varphi) \land (\varphi \Rightarrow \varphi) = 1$, and we already know that $\varphi \Rightarrow \varphi = 1$, therefore also $(\varphi \Rightarrow \varphi) \land (\varphi \Rightarrow \varphi) = 1$.

Quasiequations
Q(P) follows from the commutativity of $*$ and from the identity $(a*b) \Rightarrow c = a \Rightarrow (b \Rightarrow c)$. Q(C) follows from 3-potency: since $a^2 \leq a^3$, we have that $a^3 \Rightarrow b = 1$ implies $a^2 \Rightarrow b = 1$.

Q(E) follows from the fact that $\mathbf{A}$ comprises a partial order $\leq$ that is determined by the implication $\Rightarrow$. To prove Q($\Rightarrow$ 1), suppose $a \leq b$ and $c \leq d$. From

$c \leq d$, as $b \Rightarrow c \leq b \Rightarrow c$, using residuation we have that $b * (b \Rightarrow c) \leq c \leq d$, therefore $b * (b \Rightarrow c) \leq d$ and therefore $b \Rightarrow c \leq b \Rightarrow d$. Note that as $a \leq b$, using residuation we have that $a * (b \Rightarrow d) \leq b * (b \Rightarrow d) \leq d$, therefore $b \Rightarrow d \leq a \Rightarrow d$ and then $b \Rightarrow c \leq a \Rightarrow d$. Now, since $b \Rightarrow c \leq a \Rightarrow d$ iff $a * (b \Rightarrow c) \leq d$ iff $a \leq (b \Rightarrow c) \Rightarrow d$, we obtain thus the desired result.

For $Q(\Rightarrow r)$ we need to prove that if $d = 1$, then $b \Rightarrow d = 1$. This follows immediately from integrality.

Quasiequations $Q(\wedge 11)$, $Q(\wedge 12)$, $Q(\wedge r)$, $Q(\vee 11)$, $Q(\vee r1)$ and $Q(\vee r2)$ follow straightforwardly from the fact that $\mathbf{A}$ is partially ordered and the order is determined by the implication.

In order to prove $Q(\vee 12)$, notice that $(b \vee c)^2 \leq b^2 \vee c^2$ by Lemma 1(4). Suppose $b^2 \leq d$ and $c^2 \leq d$, then since $\mathbf{A}$ is a lattice, we have $b^2 \vee c^2 \leq d$ and as $(b \vee c)^2 \leq b^2 \vee c^2$ we conclude that $(b \vee c)^2 \leq d$ and thus $(b \vee c)^2 \Rightarrow d = 1$.

As to $Q(\sim \Rightarrow 1)$, by integrality we have $b*c \leq b$ and $b*c \leq c$. Thus $b*c \leq b \wedge c$. Now, if $b \wedge c \leq d$, then $b * c \leq d$.

In order to prove $Q(\sim \Rightarrow r)$, suppose $d^2 \leq b \wedge c$. Using monotonicity of $*$, we have $d^2 * d^2 \leq (b \wedge c) * (b \wedge c)$, i.e., $d^4 \leq (b \wedge c)^2$. Using 3-potency, we have $d^4 = d^2$, therefore $d^2 \leq (b \wedge c)^2$. Since $(b \wedge c)^2 \leq b * c$, we have $d^2 \leq (b \wedge c)^2 \leq (b * c)$, i.e., $d^2 \leq (b * c)$.

$Q(\sim \wedge 1)$, $Q(\sim \wedge r)$, $Q(\sim \vee 1)$ and $Q(\sim \vee 1)$ follow from the De Morgan's Laws (cf. [9, Lemma 3.17]).

Finally, we have $Q(\sim\sim 1)$ and $Q(\sim\sim r)$ from $\mathbf{A}$ being involutive.

It remains to be proven that the quasiequation $E(\Delta(\varphi, \psi))$ implies $\varphi \approx \psi$, that is, if $((\varphi \Rightarrow \psi) \wedge (\psi \Rightarrow \varphi)) = 1$, then $\varphi = \psi$. As 1 is the maximum of the algebra, we have that $(\varphi \Rightarrow \psi) = 1$ and $(\psi \Rightarrow \varphi) = 1$, therefore $\varphi \leq \psi$ and $\psi \leq \varphi$. As $\leq$ is an order relation, it follows that $\varphi = \psi$.

References

1. Almukdad, A., Nelson, D.: Constructible falsity and inexact predicates. J. Symb. Log. **49**(1), 231–233 (1984)
2. Blok, W.J., Köhler, P., Pigozzi, D.: On the structure of varieties with equationally definable principal congruences II. Algebra Universalis **18**, 334–379 (1984)
3. Blok, W.J., Pigozzi, D.: Algebraizable Logics. Memoirs of the American Mathematical Society, vol. 396. A.M.S., Providence (1989)
4. Bou, F., García-Cerdaña, À., Verdú, V.: On two fragments with negation and without implication of the logic of residuated lattices. Arch. Math. Log. **45**(5), 615–647 (2006)
5. Busaniche, M., Cignoli, R.: Constructive logic with strong negation as a substructural logic. J. Log. Comput. **20**, 761–793 (2010)
6. Cignoli, R.: The class of Kleene algebras satisfying an interpolation property and Nelson algebras. Algebra Universalis **23**(3), 262–292 (1986)
7. Cignoli, R., D'Ottaviano, I.M.L., Mundici, D.: Algebraic Foundations of Many-Valued Reasoning. Trends in Logic—Studia Logica Library, vol. 7. Kluwer Academic Publishers, Dordrecht (2000)
8. Font, J.M.: Abstract Algebraic Logic. An Introductory Textbook. College Publications, London (2016)

9. Galatos, N., Jipsen, P., Kowalski, T., Ono, H.: Residuated Lattices: An Algebraic Glimpse at Substructural Logics. Studies in Logic and the Foundations of Mathematics, vol. 151. Elsevier, Amsterdam (2007)
10. Galatos, N., Raftery, J.G.: Adding involution to residuated structures. Studia Logica **77**(2), 181–207 (2004)
11. Humberstone, L.: The Connectives. MIT Press, Cambridge (2011)
12. Kleene, S.C.: On the interpretation of intuitionistic number theory. J. Symb. Log. **10**, 109–124 (1945)
13. Nascimento, T., Marcos, J., Rivieccio, U., Spinks, M.: Nelson's logic $\mathcal{S}$. https://arxiv.org/abs/1803.10851
14. Nelson, D.: Constructible falsity. J. Symb. Log. **14**, 16–26 (1949)
15. Nelson, D.: Negation and separation of concepts in constructive systems. Studies in Logic and the Foundations of Mathematics, vol. 39, pp. 205–225 (1959)
16. Odintsov, S.P.: Constructive Negations and Paraconsistency. Springer, Dordrecht (2008). https://doi.org/10.1007/978-1-4020-6867-6
17. Restall, G.: How to be *really* contraction free. Studia Logica **52**(3), 381–391 (1993)
18. Rivieccio, U.: Paraconsistent modal logics. Electron. Notes Theor. Comput. Sci. **278**, 173–186 (2011)
19. Spinks, M., Veroff, R.: Constructive logic with strong negation is a substructural logic. II. Studia Logica **89**(3), 401–425 (2008)
20. Spinks, M., Veroff, R.: Paraconsistent constructive logic with strong negation as a contraction-free relevant logic. In: Czelakowski, J. (ed.) Don Pigozzi on Abstract Algebraic Logic, Universal Algebra, and Computer Science. Outstanding Contributions to Logic, vol. 16, pp. 323–379. Springer, Cham (2018). https://doi.org/10.1007/978-3-319-74772-9_13

Beliefs Based on Evidence and Argumentation

Chenwei Shi[✉], Sonja Smets[✉], and Fernando R. Velázquez-Quesada[✉]

Institute for Logic, Language and Computation, Universiteit van Amsterdam,
Amsterdam, The Netherlands
{C.Shi,S.J.L.Smets,F.R.VelazquezQuesada}@uva.nl

Abstract. In this paper, we study doxastic attitudes that emerge on the basis of argumentational reasoning. In order for an agent's beliefs to be called 'rational', they ought to be well-grounded in strong arguments that are constructed by combining her available evidence in a specific way. A study of how these rational and grounded beliefs emerge requires a new logical setting. The language of the logical system in this paper serves this purpose: it is expressive enough to reason about concepts such as *factive combined evidence*, *correctly grounded belief*, and *infallible knowledge*, which are the building blocks on which our notions of *argument* and *grounded belief* can be defined. Building further on previous work, we use a topological semantics to represent the structure of an agent's collection of evidence, and we use input from abstract argumentation theory to single out the relevant sets of evidence to construct the agent's beliefs. Our paper provides a sound and complete axiom system for the presented logical language, which can describe the given models in full detail, and we show how this setting can be used to explore more intricate epistemic notions.

Keywords: Evidence-based beliefs · Topological models
Abstract argumentation theory · Doxastic logic

1 Introduction

Propositional attitudes such as knowledge and belief are extensively studied in a number of areas, ranging from artificial intelligence (in particular in multi-agent systems) and computer science (in the study of distributed systems) to philosophy (in epistemology). In these studies we encounter the need to find a mechanism for distinguishing between different attitudes, ranging from weak and stronger forms of belief to infallible knowledge. In order to answer this question, philosophers have proposed a number of additional ingredients (including justifications, evidence, arguments) as well as principles (e.g. safety or stability) on the basis of which one can conduct this comparison and ultimately decide when an item of belief is strong enough to qualify as a piece of knowledge. The topic of this paper relates directly to this discussion on distinguishing different

© Springer-Verlag GmbH Germany, part of Springer Nature 2018
L. S. Moss et al. (Eds.): WoLLIC 2018, LNCS 10944, pp. 289–306, 2018.
https://doi.org/10.1007/978-3-662-57669-4_17

propositional attitudes. Our starting point is based on making explicit the *reasons* (e.g., evidence, arguments, justifications) on which beliefs are grounded. In this way, our work contributes to the different representations of beliefs in the literature, which includes not only purely qualitative structures (e.g., the *KD45* approach in doxastic logic [1]; the plausibility models of [2,3]), but also quantitative frameworks (e.g., ranking-based plausibility representations [4]; conditional probabilistic spaces [5,6]). Yet, while the mentioned representations are useful for discussing the properties of belief, they cannot capture the support the agent has for such beliefs, i.e. her available *evidence* and the way she combines it to build *arguments* for and against a given proposition, and how such arguments may yield *justifications* to adopt certain beliefs.[1]

In order to explicitly represent the reasons on which beliefs are based, this paper brings together two different formal frameworks. On the one hand we use input from abstract argumentation theory [9] and relate our work to the use of abstract argumentation within modal logic. As such our work relates to [10–12]. On the other hand we contribute to the semantic approaches for representing evidence-based reasoning; this stands in contrast to the syntactic approach of e.g. justification logic [13]. The semantic approach traces back to [14,15], representing evidence as a set of possible worlds (in terms of so-called *evidence models*) and defining beliefs in terms of the maximally consistent ways in which evidence can be combined. The work in [16] follows the latter direction, adding a topological structure to describe how combined evidence is generated, then using topological notions to single out relevant sets of *combined* pieces of evidence. Note that the two mentioned semantic proposals, using the agent's available evidence to define her beliefs, consider all pieces of combined evidence equally important. This simplifies some definitions, as contradictory evidence is kept separated, yet it also means that in the presence of contradicting evidence, the agent will not make a choice. As such the existing work on evidence-based beliefs doesn't capture one important aspect: the *argumentative* stage in which the agent weighs her (possibly contradicting) evidence in order to make sense of it. Indeed, in [9]'s words (p. 323),

> *[...] a statement is believable if it can be argued successfully against attacking arguments. In other words, whether or not a rational agent believes in a statement depends on whether or not [an] argument supporting this statement can be successfully defended against the counterarguments.*

The work in this paper explicitly incorporates argumentational reasoning. We build on the investigation that was first initiated in [17], by equipping our models with an extra argumentative layer over the topological setting of [16]. This allows us to single out (even in the presence of conflict) a meaningful family of combined pieces of evidence (i.e., arguments) on which *grounded beliefs* can be defined. The combination of a topological semantics with abstract argumentation theory gives

---

[1] An exception are the so called *truth maintenance systems* [7,8], which keep track of natural-deduction-style *syntactic* justifications.

raise to a wide spectrum of epistemic notions, including not only known concepts such as *evidence, argument, justified belief* and *infallible knowledge*, but also new ones, such as *(correctly) grounded belief* and *full support belief*.

On the syntactic side, our formal language differs from the logic in [17] as it has the expressive power to reason explicitly about concepts such as *factive combined evidence, correctly grounded belief*, and *infallible knowledge*, which are the building blocks for our notions of *argument* and *grounded belief*. Even more: the main technical result of this paper, a sound and complete axiom system for this new language with respect to the given structures, is useful in two important ways. First, it characterises the basic properties of the language's primitive concepts (e.g., grounded beliefs are mutually consistent) as well as the essential relationship between them (e.g., justified beliefs are grounded beliefs), which allow us to find further connections. Second, the axiomatisation can be used as a tool for exploring more intricate epistemic notions and their relationship with grounded belief and justified belief.

The paper starts with Sect. 2 recalling the frameworks on which this proposal is based; then Sect. 3 introduces a topological argumentation model and a formal language to describe it, together with a sound and complete axiom system. Sect. 4 uses the axiomatisation to explore further epistemic notions, and Sect. 5 summarises the proposal, outlining some directions for further work.

2   Preliminaries

Evidence-Based Belief. Let At be a countable set of atomic propositions.

Definition 1 (Evidence model [14]). *A (uniform) evidence model is a tuple* $M = (W, \mathcal{E}_0, V)$ *where (i)* $W \neq \varnothing$ *is a set of possible worlds; (ii)* $\mathcal{E}_0 \subseteq 2^W - \{\varnothing\}$ *is a family of non-empty subsets of* W *(with* $W \in \mathcal{E}_0$*) called the collection of pieces of basic evidence; (iii)* $V : \mathsf{At} \to 2^W$ *is a valuation function.*

Intuitively, $\mathcal{E}_0$ contains the pieces of evidence the agent has acquired. The requirements over $\mathcal{E}_0$ state that a contradiction cannot be taken as evidence ($\varnothing \notin \mathcal{E}_0$) and that, if knowledge is defined as truth in all possible worlds, the agent knows what is the full range of possibilities ($W \in \mathcal{E}_0$).

Note that some pieces of basic evidence can contradict each other: there may be $P, Q \in \mathcal{E}_0$ with $P \cap Q = \varnothing$. However, this does not mean that the agent accepts contradictions. Collecting evidence is not the end of the story: the agent should be able to *combine* her basic evidence in a meaningful way, and thus the agent's beliefs should ideally not be taken directly from her basic pieces of evidence. The strategy in [14] is that beliefs arise from the maximal consistent ways these basic pieces of evidence can be combined.

Definition 2 (Body of evidence). *Let* $M = (W, \mathcal{E}_0, V)$ *be an evidence model.*

- *A family* $\mathcal{U} \subseteq 2^W$ *has the* finite intersection property *iff the intersection of every finite subset of* $\mathcal{U}$ *is non-empty.*

- A body of evidence *is a subfamily* $\mathcal{F} \subseteq \mathcal{E}_0$ *satisfying the finite intersection property.*
- *A body of evidence is* maximal *iff it cannot be properly extended.*

By combining her available evidence in this way, the agent gets a set

$$\mathcal{MC} := \{\bigcap \mathcal{F} \subseteq W \mid \mathcal{F} \text{ is a maximal body of evidence}\}$$

with all its elements in conflict with each other. It is precisely this set which will define the agent's beliefs, yet there are different ways in which this can be done. The choice in [14] is, in some sense, conservative: the agent will believe only what is *supported* by *all* the elements of $\mathcal{MC}$.

Definition 3 (Evidence-based belief [14]). *Let* $M = (W, \mathcal{E}_0, V)$ *be an evidence model. The agent believes a proposition* $P \subseteq W$ *(notation:* $B^e P$*) if and only if the combination of* every *maximal body of evidence supports* P, *i.e.,*

$$B^e P \quad iff_{def} \quad E \subseteq P \text{ for all } E \in \mathcal{MC}$$

Justified Belief. Even though the agent's beliefs are given by the combination of maximally consistent pieces of evidence, contradictions may occur.

Example 1 ([18]). Consider the evidence model $(\mathbb{N}, \mathcal{E}_0 = \{[n, +\infty) \mid n \in \mathbb{N}\}, \varnothing)$. Note how $\mathcal{E}_0$ itself is a body of evidence and, moreover, is the unique maximal one. But $\bigcap \mathcal{E}_0 = \varnothing$, and thus the agent believes $\varnothing$.

The reason for this is that maximal bodies of evidence $\mathcal{F}$ are only *finitely* consistent. However, in order to determine whether they support a given proposition, the agent uses arbitrary intersections ($\bigcap \mathcal{F}$ in the definition of $\mathcal{MC}$). In order to reconcile this discrepancy, [16] uses a different strategy; it uses the *topology* generated by $\mathcal{E}_0$.[2]

Definition 4 (Topological evidence model [16]). *A topological evidence model* $M = (W, \mathcal{E}_0, \tau_{\mathcal{E}_0}, V)$ *extends an evidence model* $(W, \mathcal{E}_0, V)$ *(Definition 1) with* $\tau_{\mathcal{E}_0}$, *the topology over* W *generated by* $\mathcal{E}_0$.[3]

Arguments. Open sets in τ are unions of *finite* intersections of elements of $\mathcal{E}_0$, and can be seen as the agent's logical manipulation of her basic evidence. Following [16], non-empty open sets in τ are called *arguments* ([19, Subsect. 5.2.2] justifies the use of this term). Note how not every maximal body of evidence defines an argument, as even if all finite intersections of its elements are non-empty, arbitrary intersections might not (Example 1). Thus, the definition of beliefs changes in [16]: instead of asking for all maximal bodies of evidence to support P, it is required that all *arguments* (i.e., finite bodies of evidence) can be strengthened (i.e., combined with further evidence) to yield an argument supporting P.

---

[2] A *topology* over a non-empty domain X is a family $\tau \subseteq 2^X$ containing both X and $\varnothing$, and is closed under both *finite* intersections and *arbitrary* unions. The elements of a topology are called *open sets*. The *topology generated by* a given $\mathcal{Y} \subseteq 2^X$ is the smallest topology $\tau_{\mathcal{Y}}$ over X such that $\mathcal{Y} \subseteq \tau_{\mathcal{Y}}$.

[3] When no confusion arises, $\tau_{\mathcal{E}_0}$ will be denoted simply by τ.

Definition 5 (Justified belief [16]**).** *Let* $M = (W, \mathcal{E}_0, \tau, V)$ *be a topological evidence model. The agent has a* justified belief *of a proposition* $P \subseteq W$ *(notation:* $\mathsf{B}^j P$*) if and only if every argument* T *can be* strengthened *to an argument* T' *that supports* P*, that is,*

$$\mathsf{B}^j P \quad \textit{iff}_{def} \quad \textit{for all } T \in \tau \setminus \{\varnothing\} \textit{ there is } T' \in \tau \setminus \{\varnothing\} \textit{ s.t. } T' \subseteq T \textit{ and } T' \subseteq P$$

Given a topology τ over a set X, an open $T \in \tau$ is *dense* if and only if it has a non-empty intersection with all the other non-empty opens. Then, as stated in [16, Proposition 2].

Proposition 1. *Let* $M = (W, \mathcal{E}_0, \tau, V)$ *be a topological evidence model. Then,* $\mathsf{B}^j P$ *holds in* M *if and only if there is a dense open* $T \in \tau$ *such that* $T \subseteq P$.

Hence from Definition 5 and Proposition 1 it follows that the agent justifiably believes P if and only if P is supported by an argument that is consistent with any other argument. In this setting, every argument is equally important when deciding what to believe. In order for an agent to weigh her arguments differently we bring in argumentational reasoning in the next section.

3 Belief, Evidence, Argumentation, and Their Logic

The proposal of [17] extends the topological evidence model of [16] with a further semantic component coming from the abstract argumentation framework of [9] to elaborate on relations between arguments.

Definition 6 (Topological argumentation model [17]**).** *A* topological argumentation (*TA*) model $M = (W, \mathcal{E}_0, \tau, \hookleftarrow, V)$ *extends a topological evidence model* $(W, \mathcal{E}_0, \tau, V)$ *(Definition 4) with a relation* $\hookleftarrow \subseteq (\tau \times \tau)$*, the* attack relation *on* τ *(where* $T_1 \hookleftarrow T_2$ *reads as "T_2 attacks T_1 "), required to satisfy the following:*

1. *for every* $T_1, T_2 \in \tau$: $T_1 \cap T_2 = \varnothing$ *if and only if* $T_1 \hookleftarrow T_2$ *or* $T_2 \hookleftarrow T_1$;
2. *for every* $T, T_1, T_1' \in \tau$: *if* $T_1 \hookleftarrow T$ *and* $T_1' \subseteq T_1$, *then* $T_1' \hookleftarrow T$;
3. *for every* $T \in \tau \setminus \{\varnothing\}$: $\varnothing \hookleftarrow T$ *and* $T \not\hookleftarrow \varnothing$.

The first condition states (right to left) that attack implies conflict (i.e., empty intersection), but also (left to right) that, while conflict implies attack, the attack does not need to be mutual. The second asks that, if T attacks T_1, then it should also attack any stronger T_1'. The last establishes that, while the empty set is attacked by all non-empty opens, it does not attack any of them.[4]

In a *TA* model, the topology τ represents the arguments the agent has in her mind, and the attack relation $\hookleftarrow$ can be understood as inducing a form of preference over conflicting combined evidence. Together, τ and $\hookleftarrow$ form the basis of the agent's *argumentation framework*. Yet how can the agent use this framework to form her beliefs? Abstract argumentation theory [9] provides useful tools; here are the required notions.

---

[4] In fact, as the first condition implies, it only attacks itself.

Definition 7 (Characteristic (defense) function). *Let* $M = (W, \mathcal{E}_0, \tau, \hookleftarrow$ $, V)$ *be a TA model, with* $A_\tau = (\tau, \hookleftarrow)$ *its argumentation framework. A subset* $T \subseteq \tau$ *is said to* defend $T \in \tau$ *iff any open* T' *attacking* T *(i.e., for all* $T' \in \tau$ *with* $T \hookleftarrow T'$ *) is attacked by some open in* T *(i.e., there is* $T'' \in T$ *with* $T' \hookleftarrow T''$ *). Then, the* characteristic function *of* A_τ, *denoted by* d_τ, *also called the* defense *function, receives a set of opens* $T \subseteq \tau$ *and returns the set of opens it defends:*

$$d_\tau(T) := \{T \in \tau \mid T \text{ is defended by } T\}$$

The characteristic function d_τ is monotonic [9, Lemma 19], so it has a least fixed point $\mathsf{LFP}_\tau \subseteq \tau$ (i.e., LFP_τ is the smallest subset of τ satisfying $\mathsf{LFP}_\tau = d_\tau(\mathsf{LFP}_\tau)$ [20,21]). Since LFP_τ can defend all ($\subseteq$) and only ($\supseteq$) its members against any attack, and it is also conflict-free (i.e., there are no $T, T' \in \mathsf{LFP}_\tau$ such that $T \hookleftarrow T'$), it provides a reasonable definition for the relevant family of open sets in τ over which beliefs will be defined. This set LFP_τ, called *the grounded extension* in abstract argumentation, is never empty in this setting, as W is never attacked (it is in conflict only with the empty set, which does not attack anybody) and therefore it is always in LFP_τ.

Definition 8 (Grounded belief [17]). *Let* $M = (W, \mathcal{E}_0, \tau, \hookleftarrow, V)$ *be a TA model. The agent believes a proposition* $P \subseteq W$ *(notation:* $\mathfrak{B}^g P$*) iff there is an open set in* LFP_τ *supporting* P, *that is*

$$\mathfrak{B}^g P \quad iff_{def} \quad \text{there exists } F \in \mathsf{LFP}_\tau \text{ such that } F \subseteq P.$$

Grounded belief is a fully introspective and mutually consistent notion, closed under conjunction elimination, but not under conjunction introduction [17].

To illustrate grounded belief's failure of conjunction introduction while also showing how a *TA* model can be used to model 'real life' scenarios, we recall Example 3.1 of [17].

Example 2. The zoo in Tom's town bought a new animal and will show it soon to the public. Tom is curious about what species the animal is, so he asks his colleagues. However, he gets different answers. Some tell him that the animal is a penguin ($\{1\}$), some tell him that the animal is a pterosaur ($\{2\}$) and some tell him that the animal is a bat ($\{3\}$). Moreover, two other colleagues, who he really trusts, tell him that the animal can fly ($\{2,3\}$) and the animal is not a mammal ($\{1,2\}$). After receiving all these pieces of information, Tom is very puzzled. Although "the animal can fly" and "the animal is not a mammal" imply that the animal is a neither a penguin nor a bat, it is still hard to imagine that there can be a pterosaur living in the world. Intuitively, in such a situation, Tom comes to believe that the animal can fly and the animal is not a mammal. However, it seems that his evidence is not strong enough to support the claim that the animal is a pterosaur.

Let $\mathsf{At} = \{p, t, b\}$ be a set of atomic propositions (p: "the animal is a penguin"; t: "the animal is a pterosaur"; b: "the animal is a bat"). The following *TA* model M describes Tom's evidence, arguments and doxastic situation.

$$M = (W = \{1,2,3\}, \mathcal{E}_0 = \{\{1\}, \{2\}, \{3\}, \{1,2\}, \{2,3\}\}, \tau = 2^W, \hookleftarrow, V) \quad (1)$$

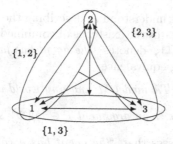

Fig. 1. Grounded beliefs are not closed under conjunction.

with $V = \{(p, \{1\}), (t, \{2\}), (b, \{3\})\}$ and $\leftharpoonup$ given by the union of *(i)*singletons attacking one another, *(ii)* $\{\varnothing \leftharpoonup T \mid T \in \tau\}$ and *(iii)* $\{\{3\} \leftharpoonup \{1,2\}, \{1\} \leftharpoonup \{2,3\}, \{2\} \leftharpoonup \{1,3\}, \{1,3\} \leftharpoonup \{2\}\}$, as shown in Fig. 1.[5] According to the definition, $\mathsf{LFP}_\tau = \{\{1,2\}, \{2,3\}, \{1,2,3\}\}$ (a set that is not closed under intersection); together with the definition of grounded belief, this confirms the intuition that Tom can come to believe that the animal can fly ($\{2,3\} \in \mathsf{LFP}_\tau$) and that the animal is not a mammal ($\{1,2\} \in \mathsf{LFP}_\tau$), but he does not come to believe that the animal is a pterosaur (no subset of $\{2\}$ is in LFP_τ).

3.1   The Logic of Belief, Evidence and Argumentation

In order to reveal the relationship between evidence, arguments, and grounded beliefs, this paper introduces a richer language to describe *TA* models compared to the logic studied in [17]. It relies on the notions of *infallible knowledge, factive combined evidence* and *correctly grounded belief*, from which notions as *argument* and *grounded belief* can be defined. This language can be used not only for providing a more detailed description of the models; it can be also used to explore more intricate epistemic notions and their interrelationship, as in Sect. 4.

Definition 9 (Language $\mathcal{L}_{\Box,\mathrm{K},\mathfrak{T}}$). *The language $\mathcal{L}_{\Box,\mathrm{K},\mathfrak{T}}$ is generated by:*

$$\varphi ::= p \mid \neg\varphi \mid \varphi \wedge \varphi \mid \Box\varphi \mid \mathrm{K}\varphi \mid \mathfrak{T}\varphi$$

with atoms $p \in \mathsf{At}$ (define $\Diamond\varphi := \neg\Box\neg\varphi$, $\widehat{\mathrm{K}}\varphi := \neg\mathrm{K}\neg\varphi$, and $\widehat{\mathfrak{T}}\varphi := \neg\mathfrak{T}\neg\varphi$). For its semantics, given a TA model M, atoms and Boolean operators are interpreted as usual. For cases involving operators, we first define $\mathcal{F}^c$, the set containing the combination of finite bodies of evidence:

$$\mathcal{F}^c := \{\textstyle\bigcap \mathcal{F} \subseteq W \mid \mathcal{F} \text{ is a finite body of evidence}\}.$$

$$M, w \models \Box\varphi \quad \textit{iff}_{def} \quad \textit{there exists } E \in \mathcal{F}^c \textit{ such that } w \in E \textit{ and } E \subseteq [\![\varphi]\!]$$
$$M, w \models \mathrm{K}\varphi \quad \textit{iff}_{def} \quad W \subseteq [\![\varphi]\!]$$
$$M, w \models \mathfrak{T}\varphi \quad \textit{iff}_{def} \quad \textit{there exists } F \in \mathsf{LFP}_\tau \textit{ such that } w \in F \textit{ and } F \subseteq [\![\varphi]\!]$$

---

[5] Attack edges involving the empty set are not drawn.

The modality K can be understood as describing the agent's *infallible knowledge*. The operator $\square$ indicates the existence of combined evidence ($E \in \mathcal{F}^c$) that is *factive* ($w \in E$) [16], so $\square\varphi$ denotes *"the agent has factive combined evidence for φ"*. Note the following equivalence.

Proposition 2. *Given a TA model M and any world w in it,*

$$M, w \models \square\varphi \quad \text{iff} \quad \text{there is an argument } T \in \tau\backslash\{\varnothing\} \text{ such that } w \in T \subseteq [\![\varphi]\!].$$

Hence, $\square\varphi$ also expresses that *"the agent has a correct argument for φ"*. Finally, to understand the interpretation of the operator $\mathfrak{T}$, compare it to the operator for grounded belief $\mathfrak{B}^g$ in Definition 8. The difference is that, while the truth condition of $\mathfrak{T}\varphi$ requires that there is a *correct* argument F in LFP_τ ($w \in F$), the truth condition of $\mathfrak{B}^g\varphi$ does not require correctness. Thus, $\mathfrak{T}\varphi$ is read as *"the agent has correctly grounded belief of φ"*, from which the operator for grounded belief can be defined as

$$\mathfrak{B}^g\,\varphi := \widehat{K}\mathfrak{T}\varphi.$$

Choosing $\mathfrak{T}$ instead of $\mathfrak{B}^g$ as a basic operator in the language is only a matter of technical convenience. There is no difference between the two choices, as

Proposition 3. $\models \mathfrak{T}\varphi \leftrightarrow (\mathfrak{B}^g\,\varphi \wedge \square\varphi)$.

This equivalence also shows that correctly grounded belief $\mathfrak{T}\varphi$ is different from *grounded true belief*, which can be expressed in the language as $\mathfrak{B}^g\,\varphi \wedge \varphi$. While the latter only requires the belief to be true, a correctly grounded belief requires a *correct argument* in LFP supporting the belief.

Axiom System. It has been proved [16, Theorem 4] that the validities of $\mathcal{L}_{\square,K}$ with respect to topological evidence models (Definition 4) are characterised by *(i)* propositional tautologies and Modus Ponens, *(ii)* the S4 axioms and rules for $\square$; *(iii)* the S5 axioms and rules for K, *(iv)* $K\varphi \to \square\varphi$. The challenge here is to find a proper axiom system characterising the validities of $\mathcal{L}_{\square,K,\mathfrak{T}}$ (which extends $\mathcal{L}_{\square,K}$ with $\mathfrak{T}$) over *TA* models (which extend topological evidence models with an attack relation on τ); Table 1 shows our proposal.

Axioms and rules in the upper block of Table 1 are self-explanatory, with exception of the last two, describing the interaction between $\mathfrak{T}$ and K. For the first, recall that $\widehat{K}\mathfrak{T}$ is the operator for grounded belief; then, $\widehat{K}\mathfrak{T}\varphi \to \neg\widehat{K}\mathfrak{T}\neg\varphi$ states that grounded beliefs are mutually consistent. The second, $(\mathfrak{T}\varphi \wedge K\psi) \to \mathfrak{T}(\varphi \wedge K\psi)$, is the 'pullout' axiom[6] for $\mathfrak{T}$, and states that a correctly grounded belief of φ and infallible knowledge of ψ give the agent a correctly grounded belief of the conjunction of φ and her infallible knowledge of ψ. The axiom can be used to derive easier-to-read validities describing the interaction between $\mathfrak{T}$ and K. An example is the following one, a variation of the famous K axiom, indicating that infallible knowledge of an implication and a correctly grounded belief of the antecedent gives the agent a correctly grounded belief of the consequent.

---

[6] The 'pullout' axiom is from [14], where it is used with the operator for evidence $\square$.

Table 1. Axiom system $\mathsf{L}_{\square,\mathrm{K},\mathfrak{T}}$, for $\mathcal{L}_{\square,\mathrm{K},\mathfrak{T}}$ w.r.t. topological argumentation models.

• Propositional Tautologies and Modus Ponens	
• The S5 axioms and rules for K	• The S4 axioms and rules for $\square$
• $\mathfrak{T}\top$	• $\mathfrak{T}\varphi \to \varphi$
• $\mathfrak{T}\varphi \to \mathfrak{T}\mathfrak{T}\varphi$	• From $\varphi \to \psi$ infer $\mathfrak{T}\varphi \to \mathfrak{T}\psi$
• $\widehat{\mathrm{K}}\mathfrak{T}\varphi \to \neg\widehat{\mathrm{K}}\mathfrak{T}\neg\varphi$	• $(\mathfrak{T}\varphi \wedge \mathrm{K}\psi) \to \mathfrak{T}(\varphi \wedge \mathrm{K}\psi)$
• $\mathfrak{T}\varphi \to \square\varphi$	
• $\mathfrak{T}\varphi \to \mathrm{K}(\square\varphi \to \mathfrak{T}\varphi)$	• $\mathrm{K}\lozenge\square\varphi \to \widehat{\mathrm{K}}\mathfrak{T}\varphi$
• $(\widehat{\mathrm{K}}\mathfrak{T}\varphi \wedge \neg\widehat{\mathrm{K}}\mathfrak{T}\psi \wedge \mathrm{K}((\varphi \wedge \psi) \to \square(\varphi \wedge \psi))) \to \widehat{\mathrm{K}}\square(\varphi \wedge \neg\psi)$	

Proposition 4. $\vdash \mathrm{K}(\varphi \to \psi) \to (\mathfrak{T}\varphi \to \mathfrak{T}\psi)$.

Proof
(1) $\vdash (\mathfrak{T}\varphi \wedge \mathrm{K}(\varphi \to \psi)) \to \mathfrak{T}(\varphi \wedge \mathrm{K}(\varphi \to \psi))$ Instance of the 'pullout' axiom
(2) $\vdash (\varphi \wedge \mathrm{K}(\varphi \to \psi)) \to \psi$ Axioms T for K, $\mathfrak{T}$; Modus Ponens
(3) $\vdash \mathfrak{T}(\varphi \wedge \mathrm{K}(\varphi \to \psi)) \to \mathfrak{T}\psi$ (2) and rule for $\mathfrak{T}$
(4) $\vdash \mathrm{K}(\varphi \to \psi) \to (\mathfrak{T}\varphi \to \mathfrak{T}\psi)$ (1), (3) and Modus Ponens

The four axioms in the lower block of Table 1 describe the relationship between different modalities. Axiom $\mathfrak{T}\varphi \to \square\varphi$, tells us that a correctly grounded belief of φ implies that the agent has a correct argument for φ. The axiom, $\mathfrak{T}\varphi \to \mathrm{K}(\square\varphi \to \mathfrak{T}\varphi)$, states that if the agent has a correctly grounded belief of φ then she infallibly knows that a correct argument for φ implies a correctly grounded belief of φ; it describes a form of strong (K) introspection of grounded arguments ($\mathfrak{T}$) in the presence of 'normal' arguments ($\square$).

To understand axiom $\mathrm{K}\lozenge\square\varphi \to \widehat{\mathrm{K}}\mathfrak{T}\varphi$, recall first that $\widehat{\mathrm{K}}\mathfrak{T}$ characterises grounded belief. Then, note that justified belief (Definition 5) is characterised by $\mathrm{K}\lozenge\square$ [16, Proposition 2], that is, $\mathsf{B}^j \varphi := \mathrm{K}\lozenge\square\varphi$. Hence, the axiom indicates that justified belief implies grounded belief. Finally, the fourth axiom, $\big(\widehat{\mathrm{K}}\mathfrak{T}\varphi \wedge \neg\widehat{\mathrm{K}}\mathfrak{T}\psi \wedge \mathrm{K}((\varphi \wedge \psi) \to \square(\varphi \wedge \psi))\big) \to \widehat{\mathrm{K}}\square(\varphi \wedge \neg\psi)$, states that if the agent has a grounded belief of φ but not of ψ, and she at the same time has argument $[\![\varphi \wedge \psi]\!]$,[7] then the agent must have an argument for $\varphi \wedge \neg\psi$.

The soundness of the axiom system is proved by verifying that the axioms are valid and the rules are validity-preserving. Most of the cases are relatively simple; here we focus on the last three axioms, to give the reader a better grasp of the modalities' semantic interpretation.

Proposition 5. $\models \mathfrak{T}\varphi \to \mathrm{K}(\square\varphi \to \mathfrak{T}\varphi)$.

---

[7] Note that "the agent has argument $[\![\varphi]\!]$" is different from "the agent has *an* argument *for* $[\![\varphi]\!]$". The former is expressed by $\mathrm{K}(\varphi \to \square\varphi)$, semantically stating that there is an argument T such that $T = [\![\varphi]\!]$; the latter corresponds to $\widehat{\mathrm{K}}\square\varphi$, semantically stating that there is an argument T such that $T \subseteq [\![\varphi]\!]$.

Proof. Let $M = (W, \mathcal{E}_0, \tau, \leftarrow, V)$ be a *TA* model; take $w \in W$, and suppose $M, w \models \mathfrak{T}\varphi$. Then, there exists $F \in \mathsf{LFP}_\tau$ such that $w \in F$ and $F \subseteq \llbracket\varphi\rrbracket$, that is, there is $F \in \mathsf{LFP}_\tau$ such that $F \subseteq \llbracket\varphi\rrbracket$. Now, take any $u \in W$: if there is $T_u \in \mathcal{F}^c$ such that both $u \in T_u$ and $T_u \subseteq \llbracket\varphi\rrbracket$, then $F \cup T_u$ is not only a factive (at u, as $u \in (F \cup T_u)$) argument supporting φ (clearly, $(F \cup T_u) \subseteq \llbracket\varphi\rrbracket$); it is also in LFP_τ, as $F \in \mathsf{LFP}_\tau$ and [17, Proposition 3.1] indicates that, for all $T, T' \in \tau$, if $T \subseteq T'$ and $T \in \mathsf{LFP}_\tau$ then $T' \in \mathsf{LFP}_\tau$. Therefore, $M, u \models \mathfrak{T}\varphi$.

Proposition 6. $\models \mathrm{K}\Diamond\Box\varphi \to \widehat{\mathrm{K}}\mathfrak{T}\varphi$.

Proof. By Proposition 1, justified belief of φ, $\mathrm{K}\Diamond\Box\varphi$, implies the existence of a dense open T supporting φ. But dense opens intersect with all non-empty opens, so they are not attacked at all; hence, T must be in LFP_τ.

Proposition 7. $\models \big(\widehat{\mathrm{K}}\mathfrak{T}\varphi \wedge \neg\widehat{\mathrm{K}}\mathfrak{T}\psi \wedge \mathrm{K}((\varphi \wedge \psi) \to \Box(\varphi \wedge \psi))\big) \to \widehat{\mathrm{K}}\Box(\varphi \wedge \neg\psi)$.

Proof. Let $M = (W, \mathcal{E}_0, \tau, \leftarrow, V)$ be a *TA* model; take $w \in W$. Suppose

$$M, w \models \widehat{\mathrm{K}}\mathfrak{T}\varphi \wedge \neg\widehat{\mathrm{K}}\mathfrak{T}\psi \wedge \mathrm{K}((\varphi \wedge \psi) \to \Box(\varphi \wedge \psi))$$

From the first conjunct, there is $F \in \mathsf{LFP}_\tau$ such that $F \subseteq \llbracket\varphi\rrbracket$. But, from the second, no $F' \in \mathsf{LFP}_\tau$ is such that $F' \subseteq \llbracket\psi\rrbracket$; in particular, $\llbracket\varphi \wedge \psi\rrbracket \notin \mathsf{LFP}$. However, by the third conjunct, $\llbracket\varphi \wedge \psi\rrbracket \in \tau$. So there must be an argument keeping $\llbracket\varphi \wedge \psi\rrbracket$ out of LFP, that is, there is a non-empty $T \in \tau$ such that $\llbracket\varphi \wedge \psi\rrbracket \leftarrow T$ and T intersects with all arguments in LFP_τ, with the former implying that $T \subseteq W \setminus \llbracket\varphi \wedge \psi\rrbracket$ and the latter implying that $T \cap F \neq \varnothing$. Moreover, $F \subseteq \llbracket\varphi\rrbracket$, so $T \cap F \subseteq (W \setminus \llbracket\varphi \wedge \psi\rrbracket) \cap \llbracket\varphi\rrbracket = \llbracket\varphi \wedge \neg\psi\rrbracket$; together with $T \cap F \neq \varnothing$, $T \cap F \in \tau$, and Proposition 2, it implies that $M, w \models \widehat{\mathrm{K}}\Box(\varphi \wedge \neg\psi)$.

As for the completeness of the system, the reader can find (for space reasons, an abridged version of) the proof in the Appendix A.

Theorem 1. *The axiom system of Table 1 is sound and strongly complete for the language $\mathcal{L}_{\Box, \mathrm{K}, \mathfrak{T}}$ w.r.t. topological argumentation models.*

4   Further Epistemic Notions

We have seen in Sect. 3 that the language $\mathcal{L}_{\Box, \mathrm{K}, \mathfrak{T}}$ can express several epistemic notions, such as arguments ($\widehat{\mathrm{K}}\Box$), grounded belief ($\widehat{\mathrm{K}}\mathfrak{T}$) and justified belief ($\mathrm{K}\Diamond\Box\varphi$). This section applies the logic of belief, evidence and argumentation to explore further epistemic notions and the way they relate to each other.

Recall the result on the definability of justified belief and grounded belief: $\mathrm{B}^j\varphi := \mathrm{K}\Diamond\Box\varphi$ and $\mathfrak{B}^g\varphi := \widehat{\mathrm{K}}\mathfrak{T}\varphi$. Thus, while justified belief is defined by infallible knowledge and factive combined evidence, grounded belief is defined by infallible knowledge and correctly grounded belief.

From these definitions, one may wonder about the relationship of the given concepts with the ones given by

$$\mathsf{K}\,\widehat{\mathfrak{I}}\,\mathfrak{I}\,\varphi \quad \text{and} \quad \widehat{\mathsf{K}}\,\mathfrak{I}\,\Diamond\,\Box\,\varphi$$

The first substitutes $\widehat{\mathfrak{I}}\,\mathfrak{I}$ for $\Diamond\,\Box$ in $\mathsf{K}\,\Diamond\,\Box\,\varphi$; it can be intuitively read as *"the agent knows that it is consistent with her correctly grounded beliefs that she has a correctly grounded belief of φ"*. The second substitutes $\Diamond\,\Box\,\varphi$ for φ in $\widehat{\mathsf{K}}\,\mathfrak{I}\,\varphi$; it can be read as *"the agent has a grounded belief of the possibility of having a correct argument for φ"*.

Now, do these two formulas describe epistemic notions different from justified belief and grounded belief? The axiom system shows us that the answer is no: they are just two alternative ways of characterising grounded belief:

Proposition 8. $\vdash \mathsf{K}\,\widehat{\mathfrak{I}}\,\mathfrak{I}\,\varphi \leftrightarrow \widehat{\mathsf{K}}\,\mathfrak{I}\,\varphi$ *and* $\vdash \widehat{\mathsf{K}}\,\mathfrak{I}\,\Diamond\,\Box\,\varphi \leftrightarrow \widehat{\mathsf{K}}\,\mathfrak{I}\,\varphi$.

Proof. For space reasons, here we only prove the first.

(1) $\vdash \widehat{\mathfrak{I}}\,\mathfrak{I}\,\varphi \to \widehat{\mathsf{K}}\,\mathfrak{I}\,\varphi$ | (1) $\vdash \widehat{\mathsf{K}}\,\mathfrak{I}\,\varphi \to \widehat{\mathsf{K}}\,\mathfrak{I}\,\mathfrak{I}\,\varphi$
(2) $\vdash \mathsf{K}\,\widehat{\mathfrak{I}}\,\mathfrak{I}\,\varphi \to \mathsf{K}\,\widehat{\mathsf{K}}\,\mathfrak{I}\,\varphi$ | (2) $\vdash \widehat{\mathsf{K}}\,\mathfrak{I}\,\mathfrak{I}\,\varphi \to \neg\widehat{\mathsf{K}}\,\mathfrak{I}\,\neg\mathfrak{I}\,\varphi$
(3) $\vdash \mathsf{K}\,\mathsf{K}\,\widehat{\mathsf{K}}\,\mathfrak{I}\,\varphi \to \widehat{\mathsf{K}}\,\mathfrak{I}\,\varphi$ | (3) $\vdash \neg\widehat{\mathsf{K}}\,\mathfrak{I}\,\neg\mathfrak{I}\,\varphi \to \mathsf{K}\,\widehat{\mathfrak{I}}\,\mathfrak{I}\,\varphi$
(4) $\vdash \mathsf{K}\,\widehat{\mathfrak{I}}\,\mathfrak{I}\,\varphi \to \widehat{\mathsf{K}}\,\mathfrak{I}\,\varphi$ | (4) $\vdash \widehat{\mathsf{K}}\,\mathfrak{I}\,\varphi \to \mathsf{K}\,\widehat{\mathfrak{I}}\,\mathfrak{I}\,\varphi$

We close this section with a new epistemic notion, obtained on the basis of a semantic argument:

Definition 10 (Full-support belief). *Let* $M = (W, \mathcal{E}_0, \tau, \hookleftarrow, V)$ *be a topological argumentation model. The agent has* full-support belief *of a proposition* $P \subseteq W$ *(notation:* $\mathcal{B}^f P$*) if and only if every argument in* LFP_τ *can be strengthened to an argument in* LFP_τ *which supports* P.

Compare the definition of full-support belief with the definition of justified belief (Definition 5): the only difference is that all the arguments involved in defining full-support belief need to be members of LFP_τ. On one hand, the similarities between the definitions of these two concepts suggest that they may share the same properties, and indeed this is the case: within TA models, full-support belief $\mathcal{B}^f$ is a $KD45$ operator. Here we only prove that it is closed under conjunction introduction.

Proposition 9. *Given a TA model, for any* $P, Q \subseteq W$ *we have that.*

$$(\mathcal{B}^f P \wedge \mathcal{B}^f Q) \to \mathcal{B}^f (P \wedge Q)$$

Proof. Given a topological argumentation model, assume that, for all $F \in \mathsf{LFP}_\tau$, not only there is $F' \subseteq F$ such that $F' \in \mathsf{LFP}_\tau$ and $F' \subseteq P$, but also there is $F'' \subseteq F$ such that $F'' \in \mathsf{LFP}_\tau$ and $F'' \subseteq Q$.

Take an arbitrary $T \in \mathsf{LFP}_\tau$. By the assumption, there is $T' \subseteq T$ such that $T' \in \mathsf{LFP}_\tau$ and $T' \subseteq P$. By the assumption again, from $T' \in \mathsf{LFP}_\tau$ it follows that there is $T'' \subseteq T'$ such that $T'' \in \mathsf{LFP}_\tau$ and $T'' \subseteq Q$. But $T'' \subseteq T'$ and $T' \subseteq P$ imply $T'' \subseteq P$. Hence, $T'' \subseteq P \wedge Q$, Thus, for all $F \in \mathsf{LFP}_\tau$, we can find an argument $F' \subseteq F$ such that $F' \in \mathsf{LFP}_\tau$ and $F' \subseteq P \wedge Q$.

On the other hand, the similarity between the semantic definition of full-support belief (Definition 10) and the semantic definition of justified belief (Definition 5) seems to suggest that because the latter can be expressed as $K \Diamond \Box \varphi$, also the former can be written as $K \widehat{\mathfrak{T}} \mathfrak{T} \varphi$. However, this is not the case. By Proposition 8, $K \widehat{\mathfrak{T}} \mathfrak{T} \varphi$ is a syntactical definition of grounded belief. Full-support belief $\mathcal{B}^f$ and grounded belief $\mathfrak{B}^g$ are different, as the latter is not closed under conjunction [17] while the former is (Proposition 9). So full-support belief cannot be syntactically defined by $K \widehat{\mathfrak{T}} \mathfrak{T} \varphi$. Why is there such a discrepancy? It is due to the lack of closure under finite intersection in LFP_τ, as the following proposition shows.

Proposition 10. *Given a TA model, $\mathcal{B}^f P \leftrightarrow \mathfrak{B}^g P$ holds for any $P \subseteq W$ if and only if LFP_τ is closed under finite intersections.*

Proof. From left to right: if grounded belief and full-support belief are equivalent in the given model, then grounded beliefs should be closed under conjunction: for any $P, Q \subseteq W$, if $\mathfrak{B}^g P \wedge \mathfrak{B}^g Q$ holds then $\mathfrak{B}^g (P \wedge Q)$ also holds. Now, if LFP_τ is not closed under finite intersection, it is easy to find P and Q such that the above fact fails. Thus, LFP_τ has to be closed under finite intersection.

From right to left: by Proposition 8, we only need to prove that $\mathcal{B}^f P \leftrightarrow K \widehat{\mathfrak{T}} \mathfrak{T} \varphi$ holds when LFP_τ is closed under finite intersection. For the first direction, assume $\mathcal{B}^f P$; then, for all $F \in \mathsf{LFP}_\tau$ there is $F' \in \mathsf{LFP}_\tau$ such that $F' \subseteq F$ and $F' \subseteq P$. Now take an arbitrary $w \in W$ and an arbitrary $F \in \mathsf{LFP}_\tau$ with $w \in F$; then there is $F' \in \mathsf{LFP}_\tau$ such that $F' \subseteq F$ and $F' \subseteq P$. But $\varnothing \notin \mathsf{LFP}_\tau$ so $F' \neq \varnothing$; there is $v \in F'$ with $F' \in \mathsf{LFP}_\tau$ and $F' \subseteq P$. Hence, for all $w \in W$ and all $F \in \mathsf{LFP}_\tau$ such that $w \in F$, we can find a $v \in F'$ such that $\mathfrak{T} P$ holds on v; then, $K \widehat{\mathfrak{T}} \mathfrak{T} P$ holds in the model. Note how we did not use LFP_τ's closure under finite intersections.

For the second direction, assume $K \widehat{\mathfrak{T}} \mathfrak{T} P$; then, for all $w \in W$ and all $F \in \mathsf{LFP}_\tau$ with $w \in F$, there is $v \in F$ such that there is an argument $F' \in LFP_\tau$ such that $v \in F'$ and $F' \subseteq P$. Note that F' is not required to be a subset of F; still, LFP_τ is closed under finite intersections, so $F \cap F'$ is also in LFP_τ, which gives us an argument in LFP_τ that is a subset of F ($F \cap F' \in \mathsf{LFP}_\tau$) and supports P ($F \cap F' \subseteq P$). So, for all $F \in \mathsf{LFP}_\tau$, we can find an $F' \in \mathsf{LFP}_\tau$ such that $F' \subseteq F$ and $F' \subseteq P$, which implies that $\mathcal{B}^f P$ holds in the model.

A more detailed study of the relationship between grounded, justified and full-support belief will be provided in this paper's full-version, where we can show that B^j implies $\mathcal{B}^f$, which in turn implies $\mathfrak{B}^g$, but not the other way around.

5   Conclusion and Future Work

Continuing the series of works [14,16,17] on models providing an explicit representation of the 'reasons' supporting an agent's beliefs, this paper focuses explicitly on the concepts of evidence and argumentation. On the first, it relies on the topological extension [16] of the so-called evidence models [14,15], representing a piece of evidence as a set of possible worlds. With respect to 'argumentation', we use tools from abstract argumentation theory [9] to single out arguments on which a notion of *grounded belief* is defined. This combination of topological semantics with abstract argumentation theory gives raise to a wide spectrum of epistemic notions, including not only known concepts as *evidence*, *argument*, *justified belief* and *infallible knowledge*, but also new ones, such as *(correctly) grounded belief* and *full support belief*.

The main technical contribution of this paper is the logic of belief, evidence and argumentation, via which the semantic analysis on grounded belief and its relationship with justified belief of [17] is fully characterized. The logic is useful not only for characterising the relationship between the mentioned epistemic notions; it also helps to find new epistemic concepts, deepening our understanding of the notion of grounded belief that is central to this work.

The presented setting opens several interesting alternatives for further research. An immediate one follows from the fact that *full-support belief* has been semantically characterized but not syntactically defined in $\mathcal{L}_{\square,\mathrm{K},\mathfrak{T}}$. Further research on its syntactic definability is necessary and may as well require an extension of the language. Moreover, other interesting notions of belief may arise by using further tools from abstract argumentation theory. Indeed, grounded belief relies on the grounded extension of the argumentation framework, but other extensions might be considered, as *preferred* extension, *stable* extension and so on. They would give raise to further types of belief that can be compared with the ones studied here.

Equally interesting is a move to a multi-agent scenario, with different agents considering possibly different attack relations. This would give rise to a more 'real' argumentation setting, with argumentation taking place not only within an agent's mind, but also between different agents. In turn, this emphasises the importance of a further dynamic layer, exploring the different epistemic actions that might affect the agent's epistemic state. In line with other work on evidence-dynamics in [14], the emergence of new evidence is interesting (as is the dismissal of existing ones); our setting also allows for changes in an agent's attack relation (arising, e.g., from her interaction with others). By providing the formal tools to study such scenarios, one will be able to truly understand how interaction in multi-agent argumentation affects the epistemic state of the involved agents.

A   Completeness for $\mathcal{L}_{\square,\mathrm{K},\mathfrak{T}}$

The proof shows that any $\mathsf{L}_{\square,\mathrm{K},\mathfrak{T}}$-consistent set of $\mathcal{L}_{\square,\mathrm{K},\mathfrak{T}}$-formulas is satisfiable. Satisfiability will be proved in an *Alexandroff qTA models* (see below), which is $\mathcal{L}_{\square,\mathrm{K},\mathfrak{T}}$-equivalent to its corresponding *TA* model.[8] Here are the details.

Definition 11 (qTA model). *A* quasi-topological argumentation model *(qTA) is a tuple* $\mathcal{M} = (W, \mathcal{E}_0, \leqslant, \leftarrowtail, V)$ *in which* $(W, \mathcal{E}_0, \tau, \leftarrowtail, V)$ *is a TA model (with*

---

[8] A similar strategy is used in [16]: show that any consistent set of formulas is satisfiable in a quasi-model, then turn it into a modally-equivalent topological evidence model.

τ generated by $\mathcal{E}_0$, as before) and $\leqslant\ \subseteq (W \times W)$ a preorder such that, for every $E \in \mathcal{E}_0$, if $u \in E$ and $u \leqslant v$, then $v \in E$.

Formulas in $\mathcal{L}_{\square,\mathrm{K},\mathfrak{T}}$ are interpreted in qTA models just as in TA models. The only difference is $\square$, which becomes a normal universal modality for $\leqslant$. More precisely, $\mathcal{M}, w \models \square\varphi$ iff for all $v \in W$, if $w \leqslant v$ then $\mathcal{M}, w \models \square\varphi$. Now, two topological definitions, a refined qTA model, and the connection.

Definition 12 (Specification preorder). *Let* (X, τ) *be a topological space. Its specification preorder* $\sqsubseteq_\tau\ \subseteq (X \times X)$ *is defined, for any* $x, y \in X$, *as* $x \sqsubseteq_\tau y$ *iff for all* $T \in \tau$, $x \in T$ *implies* $y \in T$.

Definition 13 (Alexandroff space). *A topological space* (X, τ) *is* Alexandroff *iff* τ *is closed under arbitrary intersections (i.e.,* $\bigcap T \in \tau$ *for any* $T \subseteq \tau$).

Definition 14 (Alexandroff qTA model). *A qTA-model* $\mathcal{M} = (W, \mathcal{E}_0, \leqslant, \hookleftarrow, V)$ *is called* Alexandroff *iff (i)* $(W, \tau_{\mathcal{E}_0})$ *is Alexandroff, and (ii)* $\leqslant\ =\ \sqsubseteq_\tau$.

Proposition 11. *Given an Alexandroff qTA model* $\mathcal{M} = (W, \mathcal{E}_0, \leqslant, \hookleftarrow, V)$, *take* $M = (W, \mathcal{E}_0, \tau, \hookleftarrow, V)$. *Then,* $[\![\varphi]\!]_{\mathcal{M}} = [\![\varphi]\!]_M$ *for every* $\varphi \in \mathcal{L}_{\square,\mathrm{K},\mathfrak{T}}$.

Proof. Exactly as that of [19, Proposition 5.6.14] for topological evidence models and $\mathcal{L}_{\square,\mathrm{K}}$, as $\mathfrak{T}$ has the same truth condition in qTA and TA models.

For notation, define $\Gamma^{\bigcirc} = \{\varphi \in \mathcal{L}_{\square,\mathrm{K},\mathfrak{T}} \mid \bigcirc\varphi \in \Gamma\}$ for $\Gamma \subseteq \mathcal{L}_{\square,\mathrm{K},\mathfrak{T}}$ and $\bigcirc \in \{\square, \mathrm{K}, \mathfrak{T}\}$. For the proof, let Φ_0 be a $\mathsf{L}_{\square,\mathrm{K},\mathfrak{T}}$-consistent set of $\mathcal{L}_{\square,\mathrm{K},\mathfrak{T}}$-formulas. A slightly modified version of Lindenbaum Lemma shows that it can be extended to a maximal consistent one. Let MCS be the family of all maximally $\mathsf{L}_{\square,\mathrm{K},\mathfrak{T}}$-consistent sets of $\mathcal{L}_{\square,\mathrm{K},\mathfrak{T}}$-formulas; let Φ be an element of MCS extending Φ_0.

Definition 15 (Canonical qTA model). *The canonical qTA model for* Φ, $\mathcal{M}^\Phi = (W^\Phi, \mathcal{E}_0^\Phi, \leqslant^\Phi, \hookleftarrow^\Phi, V^\Phi)$, *is defined as follows.*

- $W^\Phi := \{\Gamma \in \mathrm{MCS} \mid \Gamma^{\mathrm{K}} = \Phi^{\mathrm{K}}\}$ *and* $V^\Phi(p) := \{\Gamma \in W^\Phi \mid p \in \Gamma\}$.
- *For* $\Gamma, \Delta \in W^\Phi$, $\Gamma \leqslant^\Phi \Delta$ *iff$_{def}$ for any* $\varphi \in \mathcal{L}_{\square,\mathrm{K},\mathfrak{T}}$, $\square\varphi \in \Gamma$ *implies* $\varphi \in \Delta$.
- *For any* $\Gamma \in W^\Phi$, *define the set* $\leqslant^\Phi[\Gamma] := \{\Omega \in W^\Phi \mid \Gamma \leqslant^\Phi \Omega\}$. *Then, let* $\mathcal{E}_0^\Phi := \{\bigcup_{\Gamma \in U} \leqslant^\Phi[\Gamma] \mid U \subseteq W^\Phi\} \setminus \{\varnothing\}$.

While $\leqslant^\Phi$ *and* V^Φ *are standard (recall:* $\square$ *is a normal universal modality for* $\leqslant$), *each* $E \in \mathcal{E}_0^\Phi$ *is a non-empty union of the* $\leqslant^\Phi$-*upwards closure of the elements of some subset of* W^Φ. *The last component, the attack relation* $\hookleftarrow^\Phi$, *is the novel one in this model, and it requires more care. First, define* $\{\!|\varphi|\!\}_M := \{\Gamma \in W^\Phi \mid \varphi \in \Gamma\}$. *Then, by taking* τ^Φ *to be the topology generated by* $\mathcal{E}_0^\Phi$ *define, for any* $T, T' \in \tau^\Phi$,

- $T \hookleftarrow^\Phi T'$ *iff$_{def}$* $\begin{cases} T = \varnothing & \text{if } T' = \varnothing \\ T \cap T' = \varnothing \text{ and there is no } \varphi \in \mathcal{L}_{\square,\mathrm{K},\mathfrak{T}} \text{ s.t.} & \text{otherwise} \\ \quad \text{both } \{\!|\mathfrak{T}\varphi|\!\} \subseteq T \text{ and } \widehat{\mathrm{K}}\,\mathfrak{T}\varphi \in \Phi \end{cases}$

In the rest, and when no confusion arises, the superscript Φ *will be omitted.*

Note how $\mathcal{M}^\Phi$ is indeed a qTA model (Definition 11). First, it is clear that $\varnothing \notin \mathcal{E}_0$ and $W \in \mathcal{E}_0$. Moreover, $\leqslant$ is indeed a preorder (see its axioms) satisfying the extra condition. Finally, it can be proved that $\hookleftarrow$ satisfies the three conditions.

Lemma 1. *Let* $\mathcal{M}^\Phi = (W, \mathcal{E}_0, \leqslant, \hookleftarrow, V)$ *be the model of Definition 15. Then,*

1. *for every* $T_1, T_2 \in \tau$: $T_1 \cap T_2 = \varnothing$ *if and only if* $T_1 \hookleftarrow T_2$ *or* $T_2 \hookleftarrow T_1$;
2. *for every* $T, T_1, T_1' \in \tau$: *if* $T_1 \hookleftarrow T$ *and* $T_1' \subseteq T_1$, *then* $T_1' \hookleftarrow T$;
3. *for every* $T \in \tau \setminus \{\varnothing\}$: $\varnothing \hookleftarrow T$ *and* $T \not\hookleftarrow \varnothing$.

Thus, $\mathcal{M}^\Phi$ is a qTA model. The next proposition (standard proof) provides existence lemmas for the standard modality $\square$ and the global modality $\widehat{\mathrm{K}}$.

Proposition 12. *For any* $\varphi \in \mathcal{L}_{\square,\mathrm{K},\mathfrak{T}}$ *and any* $\Gamma \in W$:

- $\Diamond \varphi \in \Gamma$ *iff there is* $\Delta \in W$ *s.t.* $\Gamma \leqslant \Delta$ *and* $\varphi \in \Delta$.
- $\widehat{\mathrm{K}} \varphi \in \Gamma$ *iff there is* $\Delta \in W$ *s.t.* $\varphi \in \Delta$.

Now, tools to prove a similar result for the operator $\mathfrak{T}$, whose truth clause relies on LFP, given by $\hookleftarrow$. First, some useful properties of the model.

Fact 1. *(1)* $\tau = \mathcal{E}_0 \cup \{\varnothing\}$. *(2) If* $\widehat{\mathrm{K}} \square \varphi \in \Phi$, *then* $\{\!|\square \varphi|\!\} \in \tau$. *(3) If* $\widehat{\mathrm{K}} \mathfrak{T} \varphi \in \Phi$, *then* $\{\!|\mathfrak{T} \varphi|\!\} \in \tau$. *(4) For any* $T \in \tau$ *and any* $\varphi \in \mathcal{L}_{\square,\mathrm{K},\mathfrak{T}}$: *if* $T \subseteq \{\!|\varphi|\!\}$, *then* $T \subseteq \{\!|\square \varphi|\!\}$.

Here are the first steps towards locating LFP.

Definition 16 (Semi-acceptable and Acceptable). *Define* $\mathcal{C}_1$ *as*

$$\mathcal{C}_1 = \{T \in \tau \mid \text{there exists } \varphi \in \mathcal{L}_{\square,\mathrm{K},\mathfrak{T}} \text{ such that } \{\!|\mathfrak{T} \varphi|\!\} \subseteq T \text{ and } \widehat{\mathrm{K}} \mathfrak{T} \varphi \in \Phi\}$$

- *An open* $T \in \tau$ *is* semi-acceptable *if and only if, for any* $\psi \in \mathcal{L}_{\square,\mathrm{K},\mathfrak{T}}$ *with* $T \subseteq \{\!|\square \psi|\!\}$, *there is* $\xi \in \mathcal{L}_{\square,\mathrm{K},\mathfrak{T}}$ *such that* $\{\!|\mathfrak{T} \xi|\!\} \subseteq \{\!|\square \psi|\!\}$ *and* $\widehat{\mathrm{K}} \mathfrak{T} \xi \in \Phi$.
- *An open* $T \in \tau$ *is* acceptable *if and only if* T *is semi-acceptable and there is no* $T' \in \tau$ *such that* $T \cap T' = \varnothing$ *and* $T' \cap T'' \neq \varnothing$ *for all* $T'' \in \mathcal{C}_1$.

Define $\mathcal{C}_2$ *as* $\mathcal{C}_2 = \{T \in \tau \setminus \mathcal{C}_1 \mid T \text{ is acceptable}\}$.

Note that no element of $\mathcal{C}_1$ is attacked by elements of τ. Moreover,

Fact 2. *(i) For any* $T \in \tau$, *if* $T \in \mathcal{C}_1$, *then* T *is acceptable. (ii) If* $T \in \tau$ *is semi-acceptable, then* $T \cap T' \neq \varnothing$ *for all* $T' \in \mathcal{C}_1$.

Lemma 2. *Let* $\mathcal{C} = \mathcal{C}_1 \cup \mathcal{C}_2$. *Then,* $\mathsf{LFP} = \mathcal{C}$.

Proof. ($\supseteq$) The proof of this direction can be fulfilled by checking two cases *(i)* $T \in \mathcal{C}_1$ and *(ii)* $T \in \mathcal{C}_2$, which is relatively simple, so we turn to the details of the other direction's proof.

($\subseteq$) Take now $T \in \tau$ such that $T \notin \mathcal{C}$; it will be shown that $T \notin \mathsf{LFP}$. The case with $T = \varnothing$ is immediate, as $\varnothing \hookleftarrow \varnothing$. Thus, suppose $T \neq \varnothing$.

From $T \notin \mathcal{C}$ it follows that $T \notin \mathcal{C}_1$, so there is no $\phi \in \mathcal{L}_{\square,\mathrm{K},\mathfrak{T}}$ such that $\{\!|\mathfrak{T} \phi|\!\} \subseteq T$ and $\widehat{\mathrm{K}} \mathfrak{T} \phi \in \Phi$; hence, from $\hookleftarrow$'s definition, every $T' \in \tau$ with

$T \cap T' = \varnothing$ is such that $T \hookleftarrow T'$. It can be proved by using axiom $K \diamond \square \varphi \to \hat{K} \mathfrak{T} \varphi$ that there is at least one $T' \in \tau$ with $T \cap T' = \varnothing$. Thus, the rest of the proof is divided into two cases: either there is $T' \in \tau$ with $T \cap T' = \varnothing$ and $T' \in \mathcal{C}$ (at least one T' contradicting T is in $\mathcal{C}$), or else for any $T' \in \tau$ with $T \cap T' = \varnothing$, $T' \notin \mathcal{C}$ (no T' contradicting T is in $\mathcal{C}$). In the first case, take any $T' \in \tau$ such that $T \cap T' = \varnothing$ and $T' \in \mathcal{C}$. Then, as it has been argued, $T \hookleftarrow T'$; moreover, as it has been proved, $\mathcal{C} \subseteq$ LFP. Thus, $T \notin$ LFP, as LFP has to be conflict-free.

In the second case, it follows that $C \in \mathcal{C}$ implies $T \cap C \neq \varnothing$. Now, consider the following two sub-cases: either T is semi-acceptable, or it is not. Next, we prove that in both two cases, $T \notin d(\mathcal{C})$. The case where T is semi-acceptable is relatively easy, so we focus on the case where T is not semi-acceptable.

If T is not semi-acceptable, there is $\varphi_T \in \mathcal{L}_{\square, K, \mathfrak{T}}$ such that $T \subseteq \{\!| \square \varphi_T |\!\}$ and there is no $\psi \in \mathcal{L}_{\square, K, \mathfrak{T}}$ such that both $\{\!| \mathfrak{T} \psi |\!\} \subseteq \{\!| \square \varphi_T |\!\}$ and $\hat{K} \mathfrak{T} \psi \in \Phi$. In particular, φ_T itself cannot be such ψ, so either $\{\!| \mathfrak{T} \varphi_T |\!\} \not\subseteq \{\!| \square \varphi_T |\!\}$ or else $\hat{K} \mathfrak{T} \varphi_T \notin \Phi$. But axiom $\mathfrak{T} \varphi \to \square \varphi$ implies $\{\!| \mathfrak{T} \varphi_T |\!\} \subseteq \{\!| \square \varphi_T |\!\}$, so $\hat{K} \mathfrak{T} \varphi_T \notin \Phi$. Now, take any $C \in \mathcal{C}_1$; let $\varphi_C \in \mathcal{L}_{\square, K, \mathfrak{T}}$ be one of the formulas satisfying both $\{\!| \mathfrak{T} \varphi_C |\!\} \subseteq C$ and $\hat{K} \mathfrak{T} \varphi_C \in \Phi$ (by $\mathcal{C}$'s definition, there is at least one). From theorem $\mathfrak{T} \varphi \to \square \mathfrak{T} \varphi$, it follows that $(\square \varphi_T \wedge \mathfrak{T} \varphi_C) \to (\square \varphi_T \wedge \square \mathfrak{T} \varphi_C)$ is a theorem too, and thus so are $(\square \varphi_T \wedge \mathfrak{T} \varphi_C) \to (\square \square \varphi_T \wedge \square \mathfrak{T} \varphi_C)$ (by axiom $\square \varphi \to \square \square \varphi$) and $(\square \varphi_T \wedge \mathfrak{T} \varphi_C) \to \square(\square \varphi_T \wedge \mathfrak{T} \varphi_C)$ (axiom K for $\square$). Hence, by Proposition 12, $K\big((\square \varphi_T \wedge \mathfrak{T} \varphi_C) \to \square(\square \varphi_T \wedge \mathfrak{T} \varphi_C)\big) \in \Phi$.

So far we have $\hat{K} \mathfrak{T} \varphi_T \notin \Phi$ and, for every $C \in \mathcal{C}_1$, not only $\hat{K} \mathfrak{T} \varphi_C \in \Phi$ but also $K\big((\square \varphi_T \wedge \mathfrak{T} \varphi_C) \to \square(\square \varphi_T \wedge \mathfrak{T} \varphi_C)\big) \in \Phi$. The first and theorem $\mathfrak{T} \varphi \leftrightarrow \mathfrak{T} \square \varphi$ imply $\hat{K} \mathfrak{T} \square \varphi_T \notin \Phi$; the second and axiom $\mathfrak{T} \varphi \to \mathfrak{T} \mathfrak{T} \varphi$ imply $\hat{K} \mathfrak{T} \mathfrak{T} \varphi_C \in \Phi$. These two, the third, and axiom $\big(\hat{K} \mathfrak{T} \varphi \wedge \neg \hat{K} \mathfrak{T} \psi \wedge K((\varphi \wedge \psi) \to \square(\varphi \wedge \psi))\big) \to \hat{K} \square(\varphi \wedge \neg \psi)$ imply $\hat{K} \square(\mathfrak{T} \varphi_C \wedge \neg \square \varphi_T) \in \Phi$. For the final part, take the union of $\{\!| \square(\mathfrak{T} \varphi_C \wedge \neg \square \varphi_T) |\!\}$ for all $C \in \mathcal{C}_1$, i.e.

$$S = \bigcup_{C \in \mathcal{C}_1} \{\!| \square(\mathfrak{T} \varphi_C \wedge \neg \square \varphi_T) |\!\}$$

The following two facts about S (whose proof we omit here) are key to what we want to prove ($T \notin d(\mathcal{C})$): **(i)** $S \cap T = \varnothing$. **(ii)** For any $C' \in \mathcal{C}$, $C' \cap S \neq \varnothing$. Since $S \cap T = \varnothing$ and T is not semi-acceptable (so there is no $\varphi \in \mathcal{L}_{\square, K, \mathfrak{T}}$ s.t. both $\{\!| \mathfrak{T} \varphi |\!\} \subseteq T$ and $\hat{K} \mathfrak{T} \varphi \in \Phi$), we have found an open S in τ with $T \hookleftarrow S$, according to the definition of $\hookleftarrow$. But $S \cap C \neq \varnothing$ for all $C \in \mathcal{C}$, so $S \not\hookleftarrow C$ for all $C \in \mathcal{C}$: no open in $\mathcal{C}$ attacks S. Hence, $T \notin d(\mathcal{C})$.

Therefore, regardless of whether T is semi-acceptable or not, we have $T \notin d(\mathcal{C})$. Since d is monotonic and $\mathcal{C} \subseteq$ LFP (as it has been shown), it follows that $d(\mathcal{C}) \subseteq d(\text{LFP}) = \text{LFP}$, which implies $T \notin$ LFP.

Thus, in both cases $T \notin \mathcal{C}$ implies $T \notin$ LFP. This completes the proof.

Proposition 13 (Truth lemma). *For any* $\varphi \in \mathcal{L}_{\square, K, \mathfrak{T}}$ *and any* $\Gamma \in W$,

$$\Gamma \in \{\!| \varphi |\!\}_{\mathcal{M}^\Phi} \text{ if and only if } \Gamma \in [\![\varphi]\!]_{\mathcal{M}^\Phi}$$

Proof. The proof proceeds by induction, with the cases for atomic propositions and Boolean connectives being routine, and those for and $\Box$ and K relying on Proposition 12. Here we focus on the case for $\mathfrak{T}$.

From left to right, suppose $\Gamma \in \{\!|\,\mathfrak{T}\varphi\,|\!\}$. Then, $\mathfrak{T}\varphi \in \Gamma$ so, by Proposition 12, $\widehat{\mathrm{K}}\,\mathfrak{T}\varphi \in \Phi$ which, by item 3 of Fact 1, implies $\{\!|\,\mathfrak{T}\varphi\,|\!\} \in \tau$. Now, let $T = \{\!|\,\mathfrak{T}\varphi\,|\!\}$. Then, *(i)* from $\{\!|\,\mathfrak{T}\varphi\,|\!\} \subseteq T$ and $\widehat{\mathrm{K}}\,\mathfrak{T}\varphi \in \Phi$, it follows that $T \in \mathcal{C}_1$ which, by Lemma 2, implies $T \in \mathsf{LFP}$; *(ii)* $\Gamma \in T$, as $\Gamma \in \{\!|\,\mathfrak{T}\varphi\,|\!\}$; *(iii)* from axiom $\mathfrak{T}\varphi \to \varphi$ it follows that $T \subseteq \{\!|\varphi|\!\}$ which, by inductive hypothesis $\{\!|\varphi|\!\} = [\![\varphi]\!]$, implies $T \subseteq [\![\varphi]\!]$. Hence, by $\mathfrak{T}$'s truth condition, $\Gamma \in [\![\mathfrak{T}\varphi]\!]$.

From right to left, suppose $\Gamma \in [\![\mathfrak{T}\varphi]\!]$. Then, by $\mathfrak{T}$'s truth condition, there is $T \in \mathsf{LFP}$ with $\Gamma \in T$ and $T \subseteq [\![\varphi]\!]$. But, from LFP's definition, $\Gamma \in T$ implies $\leqslant[\Gamma] \subseteq T$; hence, by $\Box$'s truth condition, $T \subseteq [\![\Box\varphi]\!]$. The inductive hypothesis implies $[\![\Box\varphi]\!] = \{\!|\Box\varphi|\!\}$, so then we have $\Gamma \in T$ and $T \subseteq \{\!|\Box\varphi|\!\}$.

By Lemma 2, $\mathsf{LFP} = \mathcal{C}_1 \cup \mathcal{C}_2$; thus, $T \in \mathcal{C}_1 \cup \mathcal{C}_2$. Suppose $T \in \mathcal{C}_1$; then there is $\psi \in \mathcal{L}_{\Box,\mathrm{K},\mathfrak{T}}$ with $\{\!|\,\mathfrak{T}\psi\,|\!\} \subseteq T$ and $\widehat{\mathrm{K}}\,\mathfrak{T}\psi \in \Phi$. Thus, $\{\!|\,\mathfrak{T}\psi\,|\!\} \subseteq T \subseteq \{\!|\Box\varphi|\!\}$, so $\mathrm{K}(\mathfrak{T}\psi \to \Box\varphi) \in \Phi$. Now, take any $\Delta \in \{\!|\,\mathfrak{T}\psi\,|\!\}$; then, $\mathrm{K}(\mathfrak{T}\psi \to \Box\varphi) \in \Delta$. This, together with theorem $\mathrm{K}(\varphi \to \psi) \to (\mathfrak{T}\varphi \to \mathfrak{T}\psi)$ (Proposition 4), implies $\mathfrak{T}\mathfrak{T}\psi \to \mathfrak{T}\Box\varphi \in \Delta$. Moreover: $\Delta \in \{\!|\,\mathfrak{T}\psi\,|\!\}$ implies $\Delta \in \{\!|\,\mathfrak{T}\mathfrak{T}\psi\,|\!\}$, so $\Delta \in \{\!|\,\mathfrak{T}\Box\varphi|\!\}$, that is, $\mathfrak{T}\Box\varphi \in \Delta$. The latter, together with theorem $\mathfrak{T}\varphi \leftrightarrow \mathfrak{T}\Box\varphi$ and axiom $\mathfrak{T}\varphi \to \mathrm{K}(\Box\varphi \to \mathfrak{T}\varphi)$, imply $\mathrm{K}(\Box\varphi \to \mathfrak{T}\varphi) \in \Delta$, and thus $\mathrm{K}(\Box\varphi \to \mathfrak{T}\varphi) \in \Phi$. Hence, $\{\!|\Box\varphi|\!\} \subseteq \{\!|\,\mathfrak{T}\varphi\,|\!\}$ and thus, since $\Gamma \in T$ and $T \subseteq \{\!|\Box\varphi|\!\}$, we have $\Gamma \in \{\!|\,\mathfrak{T}\varphi\,|\!\}$. Otherwise, $T \in \mathcal{C}_2$, and hence for any $\psi \in \mathcal{L}_{\Box,\mathrm{K},\mathfrak{T}}$ with $T \subseteq \{\!|\Box\psi|\!\}$ there is $\xi \in \mathcal{L}_{\Box,\mathrm{K},\mathfrak{T}}$ with $\{\!|\,\mathfrak{T}\xi\,|\!\} \subseteq \{\!|\Box\psi|\!\}$ and $\widehat{\mathrm{K}}\,\mathfrak{T}\xi \in \Phi$. Thus, since φ is such that $T \subseteq \{\!|\Box\varphi|\!\}$, there is $\eta \in \mathcal{L}_{\Box,\mathrm{K},\mathfrak{T}}$ such that $\{\!|\,\mathfrak{T}\eta\,|\!\} \subseteq \{\!|\Box\varphi|\!\}$ and $\widehat{\mathrm{K}}\,\mathfrak{T}\eta \in \Phi$. From here we can repeat the argument used in the case of $T \in \mathcal{C}_1$ in order to get $\Gamma \in \{\!|\,\mathfrak{T}\varphi\,|\!\}$ again. Thus, in both cases, $\Gamma \in \{\!|\,\mathfrak{T}\varphi\,|\!\}$, which completes the proof.

Lemma 3. $\mathcal{M}^\Phi$ *is Alexandroff.*

Proof. Whether $\mathcal{M}^\Phi$ is Alexandroff has nothing to do with $\leftthreetimes$; thus, we can apply Proposition 5.6.15 in [19], which states that if $\tau = \{\bigcup_{\Gamma \in U} \leqslant[\Gamma] \mid U \subseteq W\}$ then $\mathcal{M}^\Phi$ is Alexandroff. But item 1 of Fact 1 and the definition of $\mathcal{E}_0$ imply the required condition; then, $\mathcal{M}^\Phi$ is Alexandroff.

Since $\mathcal{M}^\Phi$ is Alexandroff, Proposition 11 tells us it has a modally equivalent topological argumentation model. Hence, the $\mathsf{L}_{\Box,\mathrm{K},\mathfrak{T}}$-consistent set of $\mathcal{L}_{\Box,\mathrm{K},\mathfrak{T}}$-formulas Φ_0 is satisfiable in a topological argumentation model.

References

1. Hintikka, J.: Knowledge and Belief. Cornell University Press, Ithaca (1962)
2. Board, O.: Dynamic interactive epistemology. Games Econ. Behav. **49**(1), 49–80 (2004)
3. Baltag, A., Smets, S.: A qualitative theory of dynamic interactive belief revision. Texts Log. Games **3**, 9–58 (2008)

4. Spohn, W.: Ordinal conditional functions: a dynamic theory of epistemic states. In: Harper, W.L., Skyrms, B. (eds.) Causation in Decision, Belief Change, and Statistics, pp. 105–134. Kluwer, Dordrecht (1988)
5. van Fraassen, B.C.: Fine-grained opinion, probability, and the logic of full belief. J. Philos. Log. **24**(4), 349–377 (1995)
6. Baltag, A., Smets, S.: Probabilistic dynamic belief revision. In: Johan van Benthem, S.J., Veltman, F. (eds.) A Meeting of the Minds: Proceedings of the Workshop on Logic, Rationality and Interaction, vol. 8 (2007)
7. Doyle, J.: A truth maintenance system. Artif. Intell. **12**(3), 231–272 (1979)
8. de Kleer, J.: An assumption-based TMS. Artif. Intell. **28**(2), 127–162 (1986)
9. Dung, P.M.: On the acceptability of arguments and its fundamental role in non-monotonic reasoning, logic programming and n-person games. Artif. Intell. **77**, 321–357 (1995)
10. Grossi, D.: Abstract argument games via modal logic. Synthese **190**, 5–29 (2013)
11. Caminada, M.W.A., Gabbay, D.M.: A logical account of formal argumentation. Stud. Log. **93**(2–3), 109 (2009)
12. Grossi, D., van der Hoek, W.: Justified beliefs by justified arguments. In: Proceedings of the Fourteenth International Conference on Principles of Knowledge Representation and Reasoning, pp. 131–140. AAAI Press (2014)
13. Artemov, S.N.: The logic of justification. Rev. Symb. Log. **1**(4), 477–513 (2008)
14. van Benthem, J., Pacuit, E.: Dynamic logics of evidence-based beliefs. Stud. Log. **99**(1–3), 61–92 (2011)
15. van Benthem, J., Duque, D.F., Pacuit, E.: Evidence and plausibility in neighborhood structures. Ann. Pure Appl. Log. **165**(1), 106–133 (2014)
16. Baltag, A., Bezhanishvili, N., Özgün, A., Smets, S.: Justified belief and the topology of evidence. In: Väänänen, J., Hirvonen, Å., de Queiroz, R. (eds.) WoLLIC 2016. LNCS, vol. 9803, pp. 83–103. Springer, Heidelberg (2016). https://doi.org/10.1007/978-3-662-52921-8_6
17. Shi, C., Smets, S., Velázquez-Quesada, F.R.: Argument-based belief in topological structures. In: Lang, J. (ed.) Proceedings Sixteenth Conference on Theoretical Aspects of Rationality and Knowledge, TARK 2017, University of Liverpool, Liverpool, UK, 24–26 June 2017, Volume 251 of Electronic Proceedings in Theoretical Computer Science, pp. 489–503. Open Publishing Association (2017)
18. van Benthem, J., Fernández-Duque, D., Pacuit, E.: Evidence logic: a new look at neighborhood structures. In: Bolander, T., Braüner, T., Ghilardi, S., Moss, L. (eds.) Proceedings of Advances in Modal Logic, vol. 9, pp. 97–118. King's College Press (2012)
19. Özgün, A.: Evidence in epistemic logic: a topological perspective. Ph.D. thesis. Institute for Logic, Language and Computation, Amsterdam, The Netherlands, ILLC Dissertation series DS-2017-07, October 2017
20. Knaster, B.: Un théorème sur les fonctions d'ensembles. Annales de la Société Polonaise de Mathématiques **6**, 133–134 (1928)
21. Tarski, A.: A lattice-theoretical theorem and its applications. Pac. J. Math. **5**(2), 285–309 (1955)

The Effort of Reasoning: Modelling the Inference Steps of Boundedly Rational Agents

Sonja Smets$^{(\boxtimes)}$ and Anthia Solaki

ILLC, University of Amsterdam, Amsterdam, The Netherlands
{s.j.l.smets,a.solaki2}@uva.nl

Abstract. In this paper we design a new logical system to explicitly model the different deductive reasoning steps of a boundedly rational agent. We present an adequate system in line with experimental findings about an agent's reasoning limitations and the cognitive effort that is involved. Inspired by Dynamic Epistemic Logic, we work with dynamic operators denoting explicit applications of inference rules in our logical language. Our models are supplemented by (a) impossible worlds (not closed under logical consequence), suitably structured according to the effect of inference rules, and (b) quantitative components capturing the agent's cognitive capacity and the cognitive costs of rules with respect to certain resources (e.g. memory, time). These ingredients allow us to avoid problematic logical closure principles, while at the same time deductive reasoning is reflected in our dynamic truth clauses. We finally show that our models can be reduced to awareness-like plausibility structures that validate the same formulas and a sound and complete axiomatization is given with respect to them.

Keywords: Logical omniscience · Bounded rationality · Inference
Dynamic Epistemic Logic · Impossible worlds

1 Introduction

We place the work in this paper against the background of investigations in AI, Game Theory and Logic on bounded rationality and the problem of logical omniscience [14]. Models for agents with unlimited inferential powers, work well for certain types of distributed systems but are not sufficient to model real human reasoning and its limitations. A number of empirical studies on human reasoning reveal that subjects are *systematically* fallible in reasoning tasks [32, 33]. These provide us with evidence for the fact that humans hold very nuanced propositional attitudes and performing deductive reasoning steps can only be done within a limited time-frame and at the cost of some real cognitive effort. In this context a case can be made for logically competent but not infallible agents who adhere to a standard of *Minimal Rationality* [9]. Such an agent can

© Springer-Verlag GmbH Germany, part of Springer Nature 2018
L. S. Moss et al. (Eds.): WoLLIC 2018, LNCS 10944, pp. 307–324, 2018.
https://doi.org/10.1007/978-3-662-57669-4_18

make *some*, but not necessarily all, of the apparently appropriate inferences. In specifying what makes inferences (in)feasible, empirical facts pertaining to the availability of cognitive resources are crucial; for example, it is natural to take into account limitations of time and memory, when setting the standard of what the agent should achieve. As we approach this topic from the context of logic, we aim to design a normative model, rather than a purely descriptive one.

As an illustration we consider the standard *Muddy Children Puzzle* [14] which is based on the unrealistic assumption that children are unbounded and perfect logicians, who can perform demanding deductive steps at once.

Suppose that n children are playing together and k of them get mud on their foreheads. Each child can see the mud on the others but not on her own forehead. First their father announces "at least one of you is muddy" and then asks over and over "does any of you know whether you are muddy?" Assuming that the kids are unbounded reasoners, the first k − 1 times the father asks, everybody responds "no" but the k-th time all the muddy children answer "yes".

We support the argument in [21] stating that the limited capacity of humans, let alone children, can well lead to outcomes of the puzzle that are not in agreement with the standard textbook analysis. The mixture of reasoning steps a child has to take, needs to be "situated" in specific bounds of time, memory etc. As such, it is our aim in this paper to design a cognitively informed model of the dynamics of inference. To achieve this, we use tools from Dynamic Epistemic Logic (DEL) [3,4,7,11]. DEL is equipped with dynamic operators, which can well be used to denote applications of inference rules. We give a semantics of these operators via plausibility models [4]. Our models are supplemented by (a) impossible worlds (not closed under logical consequence), suitably structured according to the effect of inference rules, and (b) quantitative components capturing the agent's cognitive capacity and the costs of rules with respect to certain resources (e.g. memory, time). Note that our work, while building further on the early approaches based on impossible worlds [16] to address logical omniscience, tries to overcome their main criticism of ignoring the agents' logical competence and lacking explanatory power in terms of what really comes into play whenever we reason. In our work, deductive reasoning is reflected in the dynamic truth clauses. These include resource-sensitive 'preconditions' and utilize a model update mechanism that modifies the set of worlds and their plausibility, but also reduces cognitive capacity by the appropriate cost. We therefore show that an epistemic state is not expanded effortlessly, but, instead, via applications of rules, to the extent that they are cognitively affordable. We illustrate this formal setting on the above mentioned muddy children scenario for bounded rational children for the case $k = 2$. We further show that our models can be reduced to awareness-like plausibility structures that validate the same formulas and a sound and complete axiomatization is given with respect to them.

An arbitrary syntactic awareness-filter, used to discern *explicit* attitudes as in [13], cannot work for our purposes for it cannot be associated with logical competence, and even if ad-hoc modifications are imposed on standard awareness models, by e.g. awareness closure under subformulas, some forms of the problem are retained. A notable exception where awareness is affected by reasoning is given in [34]; we will pursue a similar rule-based approach in this paper. In relation to other work on tracking a fallible agent's reasoning and cognitive effort, we refer to [1,25,26]. The first of these papers accounts for reasoning processes through, among others, inference-based state-transitions but their composition is not specified. The second includes operators for the agent's applications of inference rules, accompanied by cognitive costs, but no semantic interpretation is given. The third uses operators standing for a *number* of reasoning steps, and an impossible-worlds semantics, but it is not clear how the number of steps can be determined nor what makes reasoning halt after that. In contrast, our work combines the benefits of plausibility models and impossible worlds in the realistic modelling of competent but bounded reasoners, and also suggests how their technical treatment can be facilitated.

The paper is structured as follows: in Sect. 2 we introduce our framework and discuss its contribution to the highlighted topics. The reduction laws (i.e. rewrite-rules) and the axiomatization are given in Sect. 3, followed by our conclusions and directions for further work in Sect. 4.

2   The Logical Framework to Model the Effort of Inference Steps

Our framework has two technical aims: (a) invalidating the closure properties of logical omniscience, and (b) elucidating the details of agents engaging in a step-wise, orderly, effortful reasoning process.

2.1   Syntax

Let $\mathcal{L}_p$ denote a standard propositional language based on a set of atoms Φ. Using this notation we first define *inference rules*:

Definition 1 (Inference rule). *Given* $\phi_1, \ldots, \phi_n, \psi \in \mathcal{L}_p$, *an inference rule* ρ *is a formula of the form* $\{\phi_1, \ldots, \phi_n\} \rightsquigarrow \psi$, *read as "whenever every formula in* $\{\phi_1, \ldots, \phi_n\}$ *is true,* ψ *is also true".*

We use $pr(\rho)$ and $con(\rho)$ to abbreviate, respectively, the set of premises and the conclusion of a rule ρ and $\mathcal{L}_R$ to denote the set of all inference rules. To identify the truth-preserving rules, we define:

Definition 2 (Translation). *The translation of a rule* ρ *is given by the following implication in* $\mathcal{L}_p$, *i.e.* $tr(\rho) := \bigwedge_{\phi \in pr(\rho)} \phi \rightarrow con(\rho)$.

We introduce the language $\mathcal{L}$, extending $\mathcal{L}_p$ with two epistemic modalities: K for conventional knowledge, and $\Box$ for *defeasible knowledge*. As argued in [4], it is philosophically interesting to include both attitudes in one system. While K represents an agent's full introspective and factive attitude, $\Box$ is factive but not fully introspective. This weaker notion satisfies the S4-properties and is inspired by the defeasibility analysis of knowledge [19,31], while K satisfies the S5-properties and is considered to be infallible and indefeasible. Regarding the changes of the agent's epistemic state, induced by deductive reasoning, we introduce dynamic operators labeled by inference rules, of the form $\langle \rho \rangle$.

Definition 3 (Language $\mathcal{L}$). *The set of terms T is defined as $T := \{c_\rho \mid \rho \in \mathcal{L}_R\} \cup \{cp\}$ with elements for all the cognitive costs c_ρ of inference rules $\rho \in \mathcal{L}_R$, and the cognitive capacity cp. Given a set of propositional atoms Φ, the language $\mathcal{L}$ is defined by:*

$$\phi ::= p \mid z_1 s_1 + \ldots + z_n s_n \geq c \mid \neg \phi \mid \phi \wedge \phi \mid K\phi \mid \Box\phi \mid A\rho \mid \langle \rho \rangle \phi$$

where $p \in \Phi$, $z_1, \ldots, z_n \in \mathbb{Z}$, $c \in \mathbb{Z}^r$, $s_1, \ldots, s_n \in T$, and $\rho \in \mathcal{L}_R$.

The language comprises linear inequalities of the form $z_1 s_1 + \ldots + z_n s_n \geq c$, to deal with cognitive effort via comparisons of costs and capacity.[1] The modalities K and $\Box$ represent infallible and defeasible knowledge, respectively. The operator A indicates the agent's availability of inference rules, i.e. $A\rho$ denotes that the agent has acknowledged rule ρ as truth-preserving (and is capable of applying it). The dynamic operators of the form $\langle \rho \rangle$ are such that $\langle \rho \rangle \phi$ reads "after applying the inference rule ρ, ϕ is true". In $\mathcal{L}$, formulas involving $\leq, =, -, \vee, \rightarrow$ can be defined as usual. Moreover, a formula of the form $s_1 \geq s_2$ abbreviates $s_1 - s_2 \geq \overline{0}$.

2.2 Plausibility Models

Our semantics is based on plausibility models [4]. In line with [30] we use a mapping from the set of worlds to the class of ordinals Ω to derive the plausibility ordering. The model is augmented by impossible worlds, which need not be closed under logical consequence. However, while the agent's fallibility is not precluded – it is in fact witnessed by the inclusion of impossible worlds – it is reasoning, i.e. applications of rules, that gradually eliminates the agent's ignorance. As a starting point, we adopt a *Minimal Consistency* requirement, ruling out 'explicit contradictions' that are obvious cases of inconsistency for any (minimally) rational agent.

In order to capture the increasing cognitive load of deductive reasoning in line with empirical findings, we first introduce two parameters: (i) the agent's cognitive resources, and (ii) the cognitive cost of applying inferential rules. Regarding (i), we will use *Res* to denote the set of resources, which can contain *memory, time, attention* etc. and let $r := |Res|$ be the number of resources considered in the modelling attempt. Regarding (ii), the cognitive effort of the agent with respect to each resource is captured by a function $c : \mathcal{L}_R \rightarrow \mathbb{N}^r$ that assigns a *cognitive cost* to each inference rule. As the results of experiments show, not all inference rules require equal cognitive effort: [17,27,33] claim that the asymmetry

---

[1] Notice that c is an r-tuple. The choice of r is discussed in the next subsection.

in performance observed when a subject uses *Modus Ponens* and *Modus Tollens* is suggestive of an increased difficulty to apply the latter.[2]

Every model that we consider comes equipped with the parameters *Res* and *c*. We also introduce a *cognitive capacity* component to capture the agent's available power with respect to each resource. As resources are depleted while reasoning evolves, capacity is not constant, but it changes after each reasoning step.

Concrete assignments of the different cognitive costs and capacity rely on empirical research. We hereby adopt a simple numerical approach to the values of resources because this seems convenient in terms of capturing the availability and cost of *time* and it is also aligned with research on *memory* [10,20].[3]

Definition 4 (Plausibility model). *A plausibility model is a tuple* $M = \langle W^P, W^I, ord, V, R, cp \rangle$ *consisting of* W^P, W^I, *non-empty sets of possible and impossible worlds respectively. ord is a function from* $W := (W^P \cup W^I)$ *to the class of ordinals* Ω *assigning an ordinal to each world.* $V : W \to \mathcal{P}(\mathcal{L})$ *is a valuation function mapping each world to a set of formulas.* $R : W \to \mathcal{P}(\mathcal{L}_R)$ *is a function indicating the rules the agent has available (i.e. has acknowledged as truth-preserving) at each world. Cognitive capacity is denoted by cp, i.e.* $cp \in \mathbb{Z}^r$, *indicating what the agent is able to afford with regard to each resource.*

Regarding possible worlds, the valuation function assigns the set of atoms that are true at the world. Regarding impossible worlds, the function assigns all formulas, atomic or complex, true at the world.[4] The function *ord* induces a plausibility ordering, i.e. a binary relation on W: for $w, u \in W$: $w \geq u$ iff $ord(w) \geq ord(u)$, its intended reading being "w is no more plausible than u". Hence, the smaller the ordinal, the more plausible the world. The induced relation $\geq$ is reflexive, transitive, connected and conversely well-founded.[5] We define $\sim$, representing epistemic indistinguishability: $w \sim u$ iff either $w \geq u$ or $u \geq w$.

To ensure that the rules available to the agent are truth-preserving, and assuming that propositional formulas are evaluated as usual in possible worlds,

---

[2] We will focus on *sound* inference rules, i.e. rules whose translation is a tautology, because (a) the agent's state is naturally built on truth-preserving inferences, and (b) it would be infeasible to (empirically) assign a cost to arbitrary arrays of premises and conclusions. This task is meaningful due to the experimental results on how humans handle rule-schemas and on how the logical complexity of the formulas involved in their instantiations relates to their difficulty (although determining the *exact* relation between the complexity of formulas and the cognitive difficulty of a rule-application depends on empirical input and is left for future work). The cost assigned to non-sound rules is thus irrelevant and will not affect our constructions.

[3] Numerical assignments might also pertain to the use of pupil assessment and eye-tracking as measures of attention and indicators of cognitive effort [18,23,37].

[4] We will assume that worlds are valuation-wise unique, i.e. we view the valuation as $V := V_p \cup V_i$, where the functions V_p and V_i taking care of possible and impossible worlds are injective. This assumption is not vital but it serves the simplicity of the setting because we avoid a multiplicity of worlds unnecessary for our purposes.

[5] These properties, which follow from the definition of *ord*, will not force unnecessarily strong (introspective) validities for non-ideal agents because of the presence of impossible worlds.

we impose *Soundness of Rules*: for every $w \in W^P$, if $\rho \in R(w)$ then $M, w \models tr(\rho)$. We also need a condition to hardwire the effect of deductive reasoning in the model. To that end, we take:

Definition 5 (Propositional truths). *Let M be a model and $w \in W$ a world of the model. If $w \in W^P$, its set of propositional truths is $V^*(w) = \{\phi \in \mathcal{L}_p \mid M, w \models \phi\}$. If $w \in W^I$, $V^*(w) = \{\phi \in \mathcal{L}_p \mid \phi \in V(w)\}$.*

Based on V^*, which is determined by V, we impose *Succession* on the model: for every $w \in W$, if (i) $pr(\rho) \subseteq V^*(w)$, (ii) $\neg con(\rho) \notin V^*(w)$, (iii) $con(\rho) \neq \neg\phi$ for all $\phi \in V^*(w)$ then there is some $u \in W$ such that $V^*(u) = V^*(w) \cup \{con(\rho)\}$.

Definition 6 (ρ-radius). *The ρ-radius of a world w is given by*[6]*:*

$$w^\rho := \begin{cases} \{w\}, \text{ if } pr(\rho) \not\subseteq V^*(w) \\ \emptyset, \text{ if } pr(\rho) \subseteq V^*(w) \text{ and } (\neg con(\rho) \in V^*(w) \text{ or } con(\rho) = \neg\phi \\ \quad for\ some\ \phi \in V^*(w)) \\ \{u \mid u\ the\ successor\ of\ w\}, if\ pr(\rho) \subseteq V^*(w)\ and\ \neg con(\rho) \notin V^*(w) \\ \quad and\ con(\rho) \neq \neg\phi\ for\ all\ \phi \in V^*(w) \end{cases}$$

The ρ-radius, inspired by [26], represents how the rule ρ triggers an informational change and its element, if it exists, is called *ρ-expansion*. A rule whose premises are not true at a world does not trigger any change, this is why the only expansion is the world itself. A rule that leads to an explicit contradiction forms the empty radius as is arguably the case for minimally rational agents. If the conditions of *Succession* are met, the radius contains the new "enriched" world. Due to the injectiveness of V_p and V_i, a world's ρ-expansion is unique. As ρ-expansions expand the state from which they originate, inferences are not defeated as reasoning steps are taken, hence *Succession* warrants monotonicity, to the extent that *Minimal Consistency* is respected. Note that w's ρ-expansion amounts to itself for $w \in W^P$ (due to the deductive closure of possible worlds) while an impossible world's ρ-expansion is another impossible world.

2.3   Model Transformations and Semantic Clauses

To evaluate $\langle\rho\rangle\phi$, we have to examine the truth value of ϕ in a transformed model, defined in such a way to capture the effect of applying ρ. Roughly, a pointed plausibility model (M', w) (which consists of a plausibility model and a point indicating the real world) is the *ρ-update* of a given pointed plausibility model, whenever the set of worlds is replaced by the worlds reachable by an application of ρ on them, while the ordering is accordingly adapted. That is, if a world u was initially entertained by the agent, but after an application of ρ does not "survive", then it is eliminated. This world must have been an impossible world and a deductive step uncovered its impossibility. Once such worlds are

---

[6] Note that $=$ between formulas stands for syntactic identity. It is used due to *Minimal Consistency* and the fact that V^* is given directly by V in impossible worlds.

ruled out, the initial ordering is preserved to the extent that it is unaffected by the application of the rule. More concretely, let $M = \langle W^P, W^I, ord, V, R, cp \rangle$ be a plausibility model and (M, w) the pointed model based on w:

Step 1. Given a rule ρ, $W^\rho := \bigcup_{v \in W} v^\rho$. In words, the ρ-expansions of the worlds initially entertained by the agent. So the ρ-updated pointed model (M^ρ, w) should be such that its set of worlds is W^ρ. As observed above, any elimination of worlds is in fact an elimination affecting the set W^I.

Step 2. We now develop the new ordering ord^ρ following the application of the inference rule. Let $u \in W^\rho$. This means that there is at least one $v \in W$ such that $\{u\} = v^\rho$. Denote the set of such v's by N. Then $ord^\rho(u) = ord(z)$ for $z \in min(N)$. Therefore, if a world is in W^ρ, then it takes the position of the most plausible of the worlds from which it originated.

Step 3. V and R are simply restricted to the worlds in W^ρ and $cp^\rho := cp - c(\rho)$. Again, for $u, v \in W^\rho$, we say: $u \geq^\rho v$ iff $ord^\rho(u) \geq ord^\rho(v)$. It is easy to check that all the required properties are preserved.

Prior to defining the truth clauses we need to assign interpretations to the terms in T. Their intended reading is that those of the form c_ρ correspond to the cognitive costs of inference rules whereas those of the form cp correspond to the agent's cognitive capacity. This is why cp is used both as a model component and as a term of our language. The use can be understood from the context.

Definition 7 (Interpretation of terms). *Given a model M, the terms of T are interpreted as follows: $cp^M = cp$ and $c_\rho^M = c(\rho)$.*

Our intended reading of $\geq$ is that $s \geq t$ iff *every i-th component of s is greater or equal than the i-th component of t.* The semantic clause for a rule-application should reflect that the rule must be "affordable" to be executable; the agent's cognitive capacity must endure the resource consumption caused by firing the rule. The semantics is finally given by:

Definition 8 (Plausibility semantic clauses). *The following clauses inductively define when a formula ϕ is true at w in M (notation: $M, w \models \phi$) and when ϕ is false at w in M (notation: $M, w \dashv \phi$). For $w \in W^I$: $M, w \models \phi$ iff $\phi \in V(w)$, and $M, w \dashv \phi$ iff $\neg\phi \in V(w)$. For $w \in W^P$, given that the boolean cases are standard:*

$M, w \models p$ *iff* $p \in V(w)$, *where* $p \in \Phi$	$M, w \dashv \phi$ *iff* $M, w \not\models \phi$
$M, w \models K\phi$ *iff* $M, u \models \phi$ *for all* $u \in W$	$M, w \models A\rho$ *iff* $\rho \in R(w)$

$$M, w \models \Box\phi \text{ iff } M, u \models \phi \text{ for all } u \in W \text{ such that } w \geq u$$
$$M, w \models \langle \rho \rangle \phi \text{ iff } M, w \models (cp \geq c_\rho), M, w \models A\rho \text{ and } M^\rho, w \models \phi$$
$$M, w \models z_1 s_1 + \ldots + z_n s_n \geq c \text{ iff } z_1 s_1^M + \ldots + z_n s_n^M \geq c$$

Validity is defined with respect to possible worlds only. The truth clause for knowledge is standard, except that it also quantifies over impossible worlds. The truth of rule-availability is determined by the corresponding model function. It is then evident that the truth conditions for epistemic assertions prefixed by a rule ρ are sensitive to the idea of resource-boundedness, unlike plain assertions.

The latter require that ϕ is the case throughout the quantification set, even at worlds representing inconsistent/incomplete scenarios. The former ask that the rule is affordable, available to the agent, and that ϕ follows from the accessible worlds via ρ. Since the agent also entertains impossible worlds, she has to take a step in order to gradually minimize her ignorance.

2.4   Discussion

These constructions overcome logical omniscience, while still accounting for how we perform inferences lying within suitable applications of rules. In particular, the argument of impossible worlds suffices to invalidate the closure principles. Moreover, the truth conditions for $\langle\ddagger\rangle\spadesuit\phi$, where $\langle\ddagger\rangle$ abbreviates a sequence of inference rules and $\spadesuit$ stands for a propositional attitude such as K or $\square$, demonstrate that an agent can come to know ϕ via following an *affordable* and *available* reasoning track. In fact, the rule-sensitivity, the measure on cognitive capacity and the way it is updated allow us to practically witness to what extent reasoning evolves. Besides, running out of resources depends not only on the *number* but also on the *kind* and *chronology* of rules. Our approach takes these factors into account and explains how the agent exhausts her resources while reasoning.

Unlike [12, 26] we abstain from a generic notion of reasoning process and we do not presuppose the existence of an arbitrary cutoff on reasoning. Instead, we account explicitly for (a) specific rules available to the agent, (b) their individual applications, (c) their chronology, and (d) their cognitive consumption. This elaborate analysis is crucial in bridging epistemic frameworks with empirical facts for it exploits studies in psychology of reasoning that usually study *individual* inference rules in terms of cognitive difficulty.[7] Furthermore, the enterprise of providing a semantics contributes to [25]'s attempt, who tracks reasoning processes, but lacks a principled way to defend his selection of axioms. Constructing a semantic model that captures the change triggered by rule-applications allows for a definition of validity important in assessing the adequacy of the solution.

We will illustrate our framework on the *Muddy Children Puzzle*, highlighted in the introduction. We analyze the failure of applying a sequence of rules in the $k = 2$ scenario, attributed to the fact that the first rule applied is so cognitively costly for a child that her available time expires before she can apply the next. It thus becomes clear why in even more complex cases (e.g. for $k > 2$) human agents are likely to fail, contrary to predictions of standard logics, whereby demanding reasoning steps are performed at once and without effort. Our attempt models the dynamics of inference and the resource consumption each step induces.

Example 1 (Bounded muddy children). Take m_a, m_b as the atoms for "child a (resp. b) is muddy" and n_a, n_b for "child a (resp. b) answers no to the first question". Let $M = \langle W^P, W^I, ord, V, R, cp \rangle$ be as depicted in Fig. 1. For simplicity, take two rules, transposition of the implication and modus ponens, so

---

[7] See, for example, [9, 17, 27–29]. In fact, different schools in psychology of reasoning attribute inferential asymmetries to different causes. However, the very observation that not all inferences require equal cognitive effort is common ground.

that $R = \{TR, MP\}$ where $TR = \{\neg m_a \to \neg n_b\} \rightsquigarrow n_b \to m_a$, $MP = \{n_b, n_b \to m_a\} \rightsquigarrow m_a$, $Res = \{time, memory\}$, $c(TR) = (5,2), c(MP) = (2,2)$, $cp = (5,7)$.

Analyzing the reasoning of child a (see Fig. 1) after the father's announcement and after child b answered "no" to the first question, we verify that $\Box(\neg m_a \to \neg n_b)$ and $\Box n_b$ are valid, i.e. child a initially knows that if she is not muddy, then child b should answer "yes" (as in that case only b is muddy), and that b said "no". Following a TR-application, the world w_0 is eliminated and its position is taken by its TR-expansion, i.e. w_2 and $cp^{TR} = (5,7) - (5,2) = (0,5)$. In addition, $A(TR)$, and $cp \geq c_{TR}$. Therefore $\langle TR \rangle \Box(n_b \to m_a)$ is also valid. But now the cost of the next step is too high, i.e. $M^{TR}, w_1 \not\models cp \geq c_{MP}$ (compare cp^{TR} and $c(MP)$), so overall the formula $\langle TR \rangle \neg \langle MP \rangle \Box m_a$ is indeed valid.

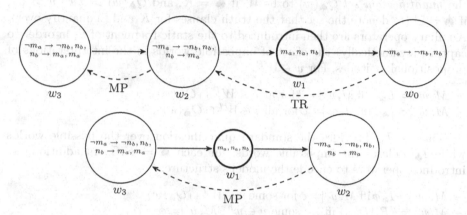

Fig. 1. The reasoning of boundedly rational child a. Thicker borders are used for deductively closed possible worlds. In impossible worlds, we write all propositional formulas satisfied and indicate (non-trivial) expansions via dashed arrows.

3   Reduction and Axiomatization

Work in [35] shows how various models for knowledge and belief, including structures for awareness [13], can be viewed as impossible-worlds models (more specifically, *Rantala models*, [24]), that validate precisely the same formulas (given a fixed background language). In the remainder, we explore the other direction and show that *our* impossible-worlds framework can be reduced to an *awareness-like* one, that only involves possible worlds. In the absence of impossible worlds, standard techniques used in axiomatizing DEL settings (via reduction axioms) can be used. This reduction is a technical contribution; the components of the reduced model lack the intuitive readings of the original framework, but allow us to prove completeness. Further, this method has the advantage of combining the benefits of impossible worlds in modelling non-ideal agents and the technical treatment facilitated by awareness-like DEL structures.

First, we show how the *static* part of the impossible-worlds setting can be transformed into one that merely involves possible worlds and captures the effect of impossible worlds via the introduction of auxiliary modalities and syntactic, *awareness-like* functions. Second, we construct a *canonical model* to obtain a sound and complete axiomatization for the static part. Third, we give DEL-style *reduction axioms* that reduce formulas involving the dynamic rule-operators to formulas that contain no such operator. In this way, we use the completeness of the static part to get a complete axiomatization for the dynamic setting.

3.1   The (Static) Language for the Reduction

We first fix an appropriate language $\mathcal{L}_r$ as the "common ground" needed to show that the reduction is successful, i.e. that the same formulas are valid under the original and the reduced models. As before, let $\spadesuit$ stand for K or $\Box$ and take the *quantification set* $Q_\spadesuit(w)$ to be W if $\spadesuit = K$, and $Q_\spadesuit(w) := \{u \mid w \geq u\}$, if $\spadesuit = \Box$ (to denote the set that the truth clauses for K and $\Box$ quantify over). Auxiliary operators are then introduced to the static fragment of $\mathcal{L}$, in order to capture (syntactically) the effect of impossible worlds in the interpretations of propositional attitudes. For $w \in W^P$:

- $M, w \models L_\spadesuit \phi$ iff $M, u \models \phi$ for all $u \in W^P \cap Q_\spadesuit(w)$
- $M, w \models I_\spadesuit \phi$ iff $M, u \models \phi$ for all $u \in W^I \cap Q_\spadesuit(w)$

That is, $L_\spadesuit$ provides the standard quantification over the possible worlds while $I_\spadesuit$ isolates the impossible words, for each $\spadesuit = K, \Box$. In addition, we introduce operators to encode the model's structure:

- $M, w \models \hat{I}_\spadesuit \phi$ iff $M, u \models \phi$ for some $u \in W^I \cap Q_\spadesuit(w)$
- $M, w \models \langle RAD \rangle_\rho \phi$ iff for some $u \in w^\rho : M, u \models \phi$

The operators $\langle RAD \rangle_\rho$, labeled by inference rules are such to ensure that there is some ϕ-satisfying ρ-expansion. To express that *all* ρ-expansions are ϕ-satisfying, we use $[RAD]_\rho \phi := \langle RAD \rangle_\rho \top \rightarrow \langle RAD \rangle_\rho \phi$, because once an expansion exists, it is unique. Indexed operators of this form provide information on the model's structure; they are introduced syntactically only as temporal-style projections of connections induced by inference rules on the model. This is why their interpretation should be independent of the distinction between possible and impossible worlds. For example, for $w \in W^I : M, w \models \langle RAD \rangle_\rho \phi$ iff for some $u \in w^\rho : M, u \models \phi$. We also use the following abbreviation: if ϕ is of the form $\neg \psi$, for some formula ψ, then $\bar{I}_\spadesuit \phi := \hat{I}_\spadesuit \psi$, else $\bar{I}_\spadesuit \phi := \bot$.

3.2   Building the Reduced Model

Towards interpreting the auxiliary operators in the reduced model, we construct *awareness-like* functions. Take $V^+(w) := \{\phi \in \mathcal{L}_r \mid M, w \models \phi\}$ for $w \in W^I$ and:

- $I_\spadesuit : W^P \rightarrow \mathcal{P}(\mathcal{L}_r)$ such that $I_\spadesuit(w) = \bigcap\limits_{v \in W^I \cap Q_\spadesuit(w)} V^+(v)$. Intuitively, $I_\spadesuit$ takes a possible world w and yields the set of those formulas that are true at all impossible worlds in its quantification set.

$- \hat{I}_\spadesuit \colon W^P \to \mathcal{P}(\mathcal{L}_r)$ such that $\hat{I}_\spadesuit(w) = \bigcup\limits_{v \in W^I \cap Q_\spadesuit(w)} V^+(v)$. Intuitively, $\hat{I}_\spadesuit$ takes a possible world w and yields the set of those formulas that are true at some impossible world in its quantification set.

The function ord captures plausibility and the "world-swapping" effect of rule-applications. Since the latter will be treated via reduction axioms, we provide a reduced model equipped with a standard binary plausibility relation (to serve as an *awareness-like plausibility structure* (ALPS), with respect to which the static logic will be developed). Given the original model $M = \langle W^P, W^I, ord, V, R, cp \rangle$, our reduced model is the tuple $\mathbf{M} = \langle \mathrm{W}, \geq, \sim, \mathrm{V}, \mathrm{R}, cp, \mathrm{I}_\spadesuit, \hat{\mathrm{I}}_\spadesuit \rangle$ where:

$\mathrm{W} = W^P$ $\qquad\qquad\qquad\qquad\qquad\quad$ $\mathrm{V}(w) = V(w)$ for $w \in \mathrm{W}$

$u \geq w$ iff $ord(u) \geq ord(w)$, for $w, u \in \mathrm{W}$ $\qquad$ $\mathrm{R}(w) = R(w)$ for $w \in \mathrm{W}$

$u \sim w$ iff $u \geq w$ or $w \geq u$, for $w, u \in \mathrm{W}$ $\qquad$ $\mathrm{I}_\spadesuit, \hat{\mathrm{I}}_\spadesuit$ as explained before

The clauses based on the reduced model are such that the auxiliary operators are interpreted via the awareness-like functions. They are presented in detail in the Appendix, along with the proof that the reduction is indeed successful:

Theorem 1 (Reduction). *Given a model M, let $\mathbf{M}$ be its (candidate) reduced model. Then $\mathbf{M}$ is indeed a reduction of M, i.e. for any $w \in W^P$ and formula $\phi \in \mathcal{L}_r\colon M, w \models \phi$ iff $\mathbf{M}, w \models \phi$.*

3.3 Axiomatization

We have reduced plausibility models to ALPS. We now develop the (static) logic Λ, showing that it is sound and complete w.r.t. them.

Definition 9 (Axiomatization of Λ). *Λ is axiomatized by Table 1 and the rules Modus Ponens, Necessitation$_K$ (from ϕ, infer $L_K\phi$) and Necessitation$_\square$ (from ϕ, infer $L_\square\phi$).*

Table 1. The axioms of Λ

AXIOMS		
PC	All instances of classical propositional tautologies	
$Ineq$	All instances of valid formulas about linear inequalities	
L_K	The S5 axioms for L_K	*Soundness of Rules* $\;A\rho \to tr(\rho)$
$L_\square$	The S4 axioms for $L_\square$	*Minimal Consistency* $\;\neg(I_\square\phi \wedge I_\square\neg\phi)$
$Succession_1$	$(\bigwedge_{\psi \in pr(\rho)} I_\spadesuit\psi \wedge \neg\hat{I}_\spadesuit\neg con(\rho) \wedge \neg\overline{I}_\spadesuit con(\rho)) \to$ $I_\spadesuit\langle RAD\rangle_\rho con(\rho) \wedge (I_\spadesuit\phi \to I_\spadesuit\langle RAD\rangle_\rho\phi)$, for $\phi \in \mathcal{L}_p$	
$Succession_2$	$I_\spadesuit\langle RAD\rangle_\rho\phi \to I_\spadesuit\phi$, for $\phi \in \mathcal{L}_p$ and $\phi \neq con(\rho)$	
$Succession_3$	$\neg\bigwedge_{\psi \in pr(\rho)} I_\spadesuit\psi \to (I_\spadesuit\phi \leftrightarrow I_\spadesuit\langle RAD\rangle_\rho\phi)$, for $\phi \in \mathcal{L}_p$	
$Succession_4$	$\bigwedge_{\psi \in pr(\rho)} I_\spadesuit\psi \wedge (\hat{I}_\spadesuit\neg con(\rho) \vee \overline{I}_\spadesuit con(\rho)) \to I_\spadesuit[RAD]_\rho\bot$	
$Local\ Connectedness$	$L_K(\phi \vee L_\square\psi) \wedge L_K(\psi \vee L_\square\phi) \to (L_K\phi \vee L_K\psi)$	
$Red_\spadesuit$	$\spadesuit\phi \leftrightarrow (L_\spadesuit\phi \wedge I_\spadesuit\phi)$	*Indefeasibility* $\;L_K\phi \to L_\square\phi$
$Radius_1$	$\langle RAD\rangle_\rho\phi \leftrightarrow \phi$	$\;I_K\phi \to I_\square\phi$
$Radius_2$	$I_\spadesuit[RAD]_\rho\phi \leftrightarrow (I_\spadesuit\langle RAD\rangle_\rho\top \to I_\spadesuit\langle RAD\rangle_\rho\phi)$	

Ineq, described in [15], is introduced to accommodate the linear inequalities.[8] The S5 axioms for L_K and S4 axioms for $L_\square$ mimic the behaviour of K and $\square$ in the usual plausibility models: these operators quantify over possible worlds only. The (clusters of) axioms about *Soundness of Rules*, *Minimal Consistency* and *Succession* take care of the respective model conditions (to the extent that these affect our language, given its expressiveness). The same holds for *Indefeasibility* and *Local Connectedness*, which also mimic their usual plausibility counterparts. To capture the behaviour of radius, we also introduce the *Radius* axioms. Finally, $Red_\spadesuit$ expresses K and $\square$ in terms of the corresponding auxiliary operators. We now move to the following theorems; the proofs can be found in the Appendix.

Theorem 2 (Soundness). Λ *is sound w.r.t. ALPS.*

Theorem 3 (Completeness). Λ *is complete w.r.t. non-standard[9] ALPS.*

Given the static logic, it suffices to reduce formulas involving $\langle \rho \rangle$ in order to get a dynamic axiomatization. It is useful to abbreviate updated terms in our language as follows: $cp^\rho := cp - c_\rho$ and $c_\rho^\rho := c_\rho$.

Theorem 4 (Reducing $\langle \rho \rangle$). *The following are valid in the class of our models*

$$\langle \rho \rangle p \leftrightarrow (cp \geq c_\rho) \wedge A\rho \wedge p$$
$$\langle \rho \rangle (\phi \wedge \psi) \leftrightarrow (cp \geq c_\rho) \wedge A\rho \wedge \langle \rho \rangle \phi \wedge \langle \rho \rangle \psi$$
$$\langle \rho \rangle I_\spadesuit \phi \leftrightarrow (cp \geq c_\rho) \wedge A\rho \wedge I_\spadesuit [RAD]_\rho \phi$$
$$\langle \rho \rangle A\sigma \leftrightarrow (cp \geq c_\rho) \wedge A\rho \wedge A\sigma$$
$$\langle \rho \rangle \langle RAD \rangle_\rho \phi \leftrightarrow (cp \geq c_\rho) \wedge A\rho \wedge \langle RAD \rangle_\rho \phi$$
$$\langle \rho \rangle \neg \phi \leftrightarrow (cp \geq c_\rho) \wedge A\rho \wedge \neg \langle \rho \rangle \phi$$
$$\langle \rho \rangle L_\spadesuit \phi \leftrightarrow (cp \geq c_\rho) \wedge A\rho \wedge L_\spadesuit \phi$$
$$\langle \rho \rangle \spadesuit \phi \leftrightarrow (cp \geq c_\rho) \wedge A\rho \wedge \spadesuit [RAD]_\rho \phi$$
$$\langle \rho \rangle \hat{I}_\spadesuit \phi \leftrightarrow (cp \geq c_\rho) \wedge A\rho \wedge \hat{I}_\spadesuit \langle RAD \rangle_\rho \phi$$
$$\langle \rho \rangle (z_1 s_1 + \ldots + z_n s_n \geq c) \leftrightarrow (cp \geq c_\rho) \wedge A\rho \wedge (z_1 s_1^\rho + \ldots + z_n s_n^\rho \geq c)$$

Theorem 5 (Dynamic axiomatization). *The axiomatic system given by Table 1 and the reduction axioms of Theorem 4 is sound and complete w.r.t. non-standard ALPS.*

4   Conclusions and Further Research

By combining DEL and an impossible-wolds semantics, we modeled fallible but boundedly rational agents who can in principle eliminate their ignorance as long as the task lies within cognitively allowed applications of inference rules. We discussed how this framework accommodates epistemic scenarios realistically and how it fits in the landscape of similar attempts put against logical omniscience. It was finally shown that this combination can be reduced to a syntactic, possible-worlds structure that allows for useful formal results.

We have focused on how deductive reasoning affects the agent's epistemic state. As observed in [6], apart from "internal elucidation", external actions, e.g.

---

[8] *Ineq* is of course slightly adapted as terms are interpreted as *r*-tuples. This makes no difference for the axioms in [15], with the exception of *dichotomy* which is not needed given our reading of inequality.

[9] More on this terminology can be found in the Appendix.

public announcements [2, 22], also enhance the agent's state. The mixed tasks involved in bounded reasoning and in revising epistemic states (also discussed in [36]) require an account of both sorts of actions and of the ways they are intertwined. The various policies of dynamic change triggered by interaction (public announcement, radical upgrades etc., [4, 5]) fit in our framework, provided that suitable dynamic operators and model transformations are defined. For instance, public announcements eliminate the worlds that do not satisfy the announced sentence and radical upgrades prioritize those that satisfy it.

Note that while factivity of knowledge is indeed warranted by the reflexivity of our models, the correspondence between other properties (such as transitivity) and forms of introspection is disrupted by the impossible worlds. Avoiding unlimited introspection falls within our wider project to model non-ideal agents. Just as with factual reasoning though, we propose a principle of moderation, achieved via the introduction of effortful introspective rules whose semantic effect is similarly projected on the structure of the model. Furthermore, it is precisely along these lines that a multi-agent extension of this setting can be pursued.

Apart from extending the *logical* machinery in order to capture richer reasoning processes, another natural development is towards fine-tuning elements of the model hitherto discussed, in order to better align it with the experimental findings in the literature on rule-based human reasoning. We have already indicated that the function c, which is responsible for the assignment of cognitive costs, should be sensitive to both the rule-schemas in question and the complexity of their particular instances. The *well-ordering of inferences* that [9] suggests, is supported by the literature we referred to so far, but, at this stage, the evidence fits a qualitative ordering of schemas while a precise quantitative assignment calls for more empirical input. Specifying the intuitive assumption that the more complex an instance, the more cognitively costly it is, breaks into two tasks (i) choosing some appropriate measure of logical complexity: number of literals, (different) atoms, connectives etc., (ii) using experimental data to fix coefficients that associate the measure with the performance of agents (w.r.t. our selected resources). Such a procedure will be pursued in a future paper and it can illuminate whether there are classes of inferences, sharing properties in terms of our measure, that should be assigned equal cognitive costs, as one might intuitively expect.

Appendix

Due to the construction of awareness-like functions, properties of the original model, concerning *Soundness of Rules*, *Minimal Consistency*, and *Succession* are inherited by the reduced model. Clearly, the new quantification sets are $Q_K(w) = W$ and $Q_\Box(w) = \{u \in W \mid w \geq u\}$. The semantic clauses, based on $\mathbf{M}$, are standard for the boolean connectives; the remaining are given below:

- $\mathbf{M}, w \models p$ iff $p \in V(w)$
- $\mathbf{M}, w \models z_1 s_1 + \cdots + z_n s_n \geq c$ iff $z_1 s_1^{\mathbf{M}} + \ldots + z_n s_n^{\mathbf{M}} \geq c$
- $\mathbf{M}, w \models L_\spadesuit \phi$ iff $\mathbf{M}, u \models \phi$ for all $u \in Q_\spadesuit(w)$

- $\mathbf{M}, w \models I_{\spadesuit}\phi$ iff $\phi \in \mathrm{I}_{\spadesuit}(w)$
- $\mathbf{M}, w \models \spadesuit\phi$ iff $\mathbf{M}, w \models L_{\spadesuit}\phi$ and $\mathbf{M}, w \models I_{\spadesuit}\phi$
- $\mathbf{M}, w \models A\rho$ iff $\rho \in \mathrm{R}(w)$
- $\mathbf{M}, w \models \hat{I}_{\spadesuit}\phi$ iff $\phi \in \hat{\mathrm{I}}_{\spadesuit}(w)$
- $\mathbf{M}, w \models \langle RAD \rangle_{\rho}\phi$ iff $\mathbf{M}, w \models \phi$

Proof for Theorem 1:

Proof. The proof goes by induction on the complexity of ϕ. Recall that validity is defined with respect to the possible worlds in the original model.

- For $\phi := p$: $M, w \models p$ iff $p \in V(w)$ iff $p \in V(w)$ iff $\mathbf{M}, w \models p$.
- For inequalities, $\neg$, $\wedge$ and A, the claim is straightforward.
- For $L_{\spadesuit}\psi$: $M, w \models L_{\spadesuit}\psi$ iff for all $u \in W^P \cap Q_{\spadesuit}(w)$: $M, u \models \psi$ iff (by I.H.) for all $u \in W \cap Q_{\spadesuit}(w)$: $\mathbf{M}, u \models \psi$ iff $\mathbf{M}, w \models L_{\spadesuit}\psi$.
- For $\phi := I_{\spadesuit}\psi$: $M, w \models I_{\spadesuit}\psi$ iff for all $u \in W^I \cap Q_{\spadesuit}(w)$: $M, u \models \psi$ iff for all $u \in W^I \cap Q_{\spadesuit}(w)$: $\psi \in V^+(u)$ iff $\psi \in \mathrm{I}_{\spadesuit}(w)$ iff $\mathbf{M}, w \models I_{\spadesuit}\psi$.
- For $\phi := \spadesuit\psi$: $M, w \models \spadesuit\psi$ iff for all $u \in Q_{\spadesuit}(w)$: $M, u \models \psi$. Since $u \in W^P \cup W^I$, this is the case iff $M, w \models L_{\spadesuit}\psi$ and $M, w \models I_{\spadesuit}\psi$. Given the previous steps of the proof, this is the case iff $\mathbf{M}, w \models L_{\spadesuit}\psi$ and $\mathbf{M}, w \models I_{\spadesuit}\psi$, iff $\mathbf{M}, w \models \spadesuit\psi$.
- For $\phi := \hat{I}_{\spadesuit}\psi$: $M, w \models \hat{I}_{\spadesuit}\psi$ iff for some $u \in W^I \cap Q_{\spadesuit}(w)$: $M, u \models \psi$ iff for some $u \in W^I \cap Q_{\spadesuit}(w)$: $\psi \in V^+(u)$ iff $\psi \in \hat{\mathrm{I}}_{\spadesuit}(w)$ iff $\mathbf{M}, w \models \hat{I}_{\spadesuit}\psi$.
- For $\phi := \langle RAD \rangle_{\rho}\psi$: $M, w \models \langle RAD \rangle_{\rho}\psi$ iff for some $u \in w^{\rho}$: $M, u \models \psi$ iff (by I.H. and $w^{\rho} = \{w\}$ since $w \in W^P$) $\mathbf{M}, w \models \psi$ iff $\mathbf{M}, w \models \langle RAD \rangle_{\rho}\psi$.

Proof for Theorem 2:

Proof. Standard arguments suffice regarding PC, *Ineq*, L_K, $L_{\square}$ and *Modus Ponens*, *Necessitation$_K$*, *Necessitation$_{\square}$* preserve validity as usual. The axioms for *Soundness of Rules*, *Minimal Consistency*, *Succession* are valid due to the respective conditions placed on the model. The validity of *Local Connectedness* is due to the connectedness of the model. The validity of *Indefeasibility*, *Red$_{\spadesuit}$*, *Radius$_2$* is a direct consequence of the semantic clauses for $\spadesuit, L_{\spadesuit}, I_{\spadesuit}$. *Radius$_1$* is valid due to the deductive closure of possible worlds.[10]

Proof for Theorem 3:

Towards showing completeness, we use a suitable canonical model. Taking (maximal) Λ-consistent sets and showing Lindenbaum's lemma follow the standard paradigm.

Definition 10 (Canonical model). *The canonical model for the logic Λ is $\mathcal{M} := \langle \mathcal{W}, \geq, \sim, \mathcal{V}, \mathcal{R}, cp, \mathcal{I}_{\spadesuit}, \hat{\mathcal{I}}_{\spadesuit} \rangle$ where:*

- *$\mathcal{W}$ is the set of all maximal Λ-consistent sets.*

---

[10] Notice that the fact that the interpretations of $\langle RAD \rangle_{\rho}$ and $[RAD]_{\rho}$ are not arbitrary in impossible worlds is important in this proof.

- $\geq$ *is such that for* $w, u \in \mathcal{W}$: $w \geq u$ *iff* $\{\phi \mid L_\Box \phi \in w\} \subseteq u$.
- $\sim$ *is such that for* $w, u \in \mathcal{W}$: $w \sim u$ *iff* $\{\phi \mid L_K \phi \in w\} \subseteq u$.
- $\mathcal{V}(w) = \{p \mid p \in w\}$, *with* $w \in \mathcal{W}$.
- $\mathcal{R}(w) = \{\rho \mid A\rho \in w\}$, *with* $w \in \mathcal{W}$.
- $\mathcal{I}_\spadesuit(w) = \{\phi \mid I_\spadesuit \phi \in w\}$, *with* $w \in \mathcal{W}$.
- $\hat{\mathcal{I}}_\spadesuit(w) = \{\phi \mid \hat{I}_\spadesuit \phi \in w\}$, *with* $w \in \mathcal{W}$.

There are alternative but equivalent definitions of $\geq$ and $\sim$ in terms of the duals $\hat{L}_\Box$ and $\hat{L}_K$, i.e. $\hat{L}_\Box \phi := \neg L_\Box \neg \phi$ and $\hat{L}_K \phi := \neg L_K \neg \phi$. Then $w \geq u$ iff $\{\hat{L}_\Box \phi \mid \phi \in u\} \subseteq w$. The existence lemma is obtained by the traditional routine. That is, for any $w \in \mathcal{W}$, if $\hat{L}_\Box \phi \in w$ then there is some $v \in \mathcal{W}$ such that $w \geq v$ and $\phi \in v$. Analogous claims can be made for $\sim$ and $\hat{L}_K$. Furthermore, due to L_K, $L_\Box$, *Indefeasibility*, *LocalConnectedness* and modal logic results on correspondence [8] the canonical model is reflexive, transitive and (locally) connected (with respect to $\geq$) and $\sim$ is the symmetric extension of $\geq$ (these properties yield the so-called *non-standard* plausibility models). The axioms on *Soundness of Rules*, *Minimal Consistency*, *Succession* and *Radius* are such to ensure that the model has the desired properties.

We then perform induction on the complexity of ϕ to show our truth lemma: $\mathcal{M}, w \models \phi$ iff $\phi \in w$. The claim for propositional atoms, the boolean cases, linear inequalities, and A holds, due to the construction of the canonical model (namely, $\mathcal{V}$ and $\mathcal{R}$), *Ineq*, the properties of maximal consistent sets and I.H. The claim for $\langle RAD \rangle_\rho$ follows by the I.H. and *Radius*$_1$. The claims for $\hat{L}_\Box$ and $\hat{L}_K$ follow with the help of the existence lemmas and I.H., while for $I_\Box$, I_K, $\hat{I}_\Box$ and $\hat{I}_K$ we rely on the construction of the awareness-like functions and then the result is immediate. For $K\phi$ and $\Box\phi$, we make use of Red_K and $Red_\Box$, the I.H. and the results of the previous steps on L_K, I_K and $L_\Box$, $I_\Box$.[11]

Proof for Theorem 4:

Proof

- The claim is easy for the atoms, the boolean cases, the inequalities, A, $\langle RAD \rangle_\rho$, L_K and $L_\Box$. We will only show why the claim holds for I_K and $I_\Box$, $\hat{I}_K$ and $\hat{I}_\Box$ because the claims involving K, $\Box$ will then follow from the clause for $\spadesuit$ and the distribution of $\langle \rho \rangle$ over conjunction.
- Let M be an arbitrary model and w an arbitrary possible world of the model. Suppose $M, w \models \langle \rho \rangle I_K \phi$. Therefore $M, w \models (cp \geq c_\rho)$, $M, w \models A\rho$ and $M^\rho, w \models I_K \phi$. Recall that $W^\rho = \bigcup_{u \in W} u^\rho$. Therefore for all $v \in W^\rho \cap W^I$, $M^\rho, v \models \phi$ (1). Take arbitrary $u \in W^I$ and arbitrary $v \in u^\rho$. Then, $v \in W^\rho \cap W^I$, and by (1) and the definitions of V and radius: $M, v \models \phi$. Overall, $M, w \models I_K [RAD]_\rho \phi$ and by $M, w \models (cp \geq c_\rho)$, $M, w \models A\rho$, we finally get $M, w \models (cp \geq c_\rho) \wedge A\rho \wedge I_K [RAD]_\rho \phi$. For the other direction, suppose

---

[11] In fact, we can claim that this logic is weakly complete with respect to ALPS where $\geq$ is conversely well-founded. This is because our structures have the finite model property (via filtration theorem, [8]) so there are no infinite $>$ chains of more and more plausible worlds.

that $M, w \models (cp \geq c_\rho) \wedge A\rho \wedge I_K[RAD]_\rho\phi$. Take arbitrary $v \in W^\rho \cap W^I$, i.e. there is some $u \in W^I$ such that $v \in u^\rho$. By the truth conditions of $M, w \models I_K[RAD]_\rho\phi$, for all $u \in W^I$, $M, u \models [RAD]_\rho\phi$, i.e. for all $v \in u^\rho$: $M, v \models \phi$. Therefore, for our arbitrary v, it is the case that $M, v \models \phi$, and by definitions of V and radius, $M^\rho, v \models \phi$. Overall, $M^\rho, w \models I_K\phi$ and finally $M, w \models \langle\rho\rangle I_K\phi$.

- Let M be an arbitrary model and w an arbitrary possible world of the model. Suppose $M, w \models \langle\rho\rangle I_\square\phi$. Therefore $M, w \models (cp \geq c_\rho)$, $M, w \models A\rho$ and $M^\rho, w \models I_\square\phi$. Since $W^\rho = \bigcup_{u \in W} u^\rho$, for all $v \in W^\rho \cap W^I$ such that $w \geq^\rho v$: $M^\rho, v \models \phi$ (1). Then, take arbitrary $u \in W^I \cap Q_\square(w)$ and arbitrary $v \in u^\rho$. Since $ord^\rho(v) \leq ord(u)$ (by Step 2 of transformation) and $w \geq u$, we get that $w \geq^\rho v$. Therefore $v \in W^\rho \cap W^I$, and by (1) and the definitions of V and radius: $M, v \models \phi$. Hence $M, w \models I_\square[RAD]_\rho\phi$ and by $M, w \models (cp \geq c_\rho)$, $M, w \models A\rho$, we finally get $M, w \models (cp \geq c_\rho) \wedge A\rho \wedge I_\square[RAD]_\rho\phi$. For the other direction, suppose that $M, w \models (cp \geq c_\rho) \wedge A\rho \wedge I_\square[RAD]_\rho\phi$. Take arbitrary $v \in W^\rho \cap W^I$ such that $w \geq^\rho v$, i.e. there is some $u \in W^I$ such that $v \in u^\rho$. Take the most plausible of these worlds (from which v originated). For this u, since $ord^\rho(v) = ord(u)$ and $w \geq^\rho v$ then $w \geq u$. By the truth conditions of $M, w \models I_\square[RAD]_\rho\phi$, for all $u \in W^I$ such that $w \geq u$: $M, u \models [RAD]_\rho\phi$, i.e. for all $v \in u^\rho$: $M, v \models \phi$. Therefore, for our arbitrary v, it is the case that $M, v \models \phi$, and by definitions of V and radius, $M^\rho, v \models \phi$ too. Overall, $M^\rho, w \models I_\square\phi$ and finally $M, w \models \langle\rho\rangle I_\square\phi$.

- Let M be an arbitrary model and w an arbitrary possible world of the model. Suppose $M, w \models \langle\rho\rangle \hat{I}_K\phi$. Therefore $M, w \models (cp \geq c_\rho)$, $M, w \models A\rho$ and $M^\rho, w \models \hat{I}_K\phi$. Since $W^\rho = \bigcup_{u \in W} u^\rho$, for some $v \in W^\rho \cap W^I$: $M^\rho, v \models \phi$. That is, there is some $u \in W^I$ such that $v \in u^\rho$ and $M^\rho, v \models \phi$. Therefore, for some $u \in W^I$ there is some $v \in u^\rho$ such that (by definitions of V and radius) $M, v \models \phi$. This amounts to $M, w \models \hat{I}_K\langle RAD\rangle_\rho\phi$, and overall $M, w \models (cp \geq c_\rho) \wedge A\rho \wedge \hat{I}_K\langle RAD\rangle_\rho\phi$. For the other direction, suppose $M, w \models (cp \geq c_\rho) \wedge A\rho \wedge \hat{I}_K\langle RAD\rangle_\rho\phi$. From $M, w \models \hat{I}_K\langle RAD\rangle_\rho\phi$, we get that there is some $u \in W^I$ such that for some $v \in u^\rho$: $M, v \models \phi$. But then $v \in W^\rho \cap W^I$ and by definitions of V and radius, $M^\rho, w \models \hat{I}_K\phi$. So overall $M, w \models \langle\rho\rangle \hat{I}_K\phi$.

- Let M be an arbitrary model and w an arbitrary possible world of the model. Suppose $M, w \models \langle\rho\rangle \hat{I}_\square\phi$. Therefore $M, w \models (cp \geq c_\rho)$, $M, w \models A\rho$ and $M^\rho, w \models \hat{I}_\square\phi$. Since $W^\rho = \bigcup_{u \in W} u^\rho$, for some $v \in W^\rho \cap W^I$ with $w \geq^\rho v$: $M^\rho, v \models \phi$. That is, there is $u \in W^I$ such that $v \in u^\rho$, and $w \geq^\rho v$ and $M^\rho, v \models \phi$. Take the most plausible such u. Since $ord(u) = ord^\rho(v)$, $w \geq u$. By these, and definitions of V and radius: there is $u \in W^I \cap Q_\square(w)$ and $v \in u^\rho$ with $M, v \models \phi$, which is precisely $M, w \models \hat{I}_\square\langle RAD\rangle_\rho\phi$. Overall: $M, w \models (cp \geq c_\rho) \wedge A\rho \wedge \hat{I}_\square\langle RAD\rangle_\rho\phi$. For the other direction, suppose $M, w \models (cp \geq c_\rho)$, $M, w \models A\rho \wedge \hat{I}_\square\langle RAD\rangle_\rho\phi$. This means that there is some $u \in W^I \cap Q_\square(w)$ and some $v \in u^\rho$ with $M, v \models \phi$. It suffices to show that $M^\rho, w \models \hat{I}_\square\phi$. But from $w \geq u$ and $v \in u^\rho$, we obtain that $w \geq^\rho v$, and $v \in W^\rho$. Due to this and definitions of V and radius: $M^\rho, v \models \phi$ and then $M^\rho, w \models \hat{I}_\square\phi$. Overall indeed $M, w \models \langle\rho\rangle \hat{I}_\square\phi$.

References

1. Alechina, N., Logan, B.: A logic of situated resource-bounded agents. J. Log. Lang. Inf. **18**, 79–95 (2009)
2. Baltag, A., Moss, L.S., Solecki, S.: The logic of public announcements, common knowledge, and private suspicions. In: Proceedings of the 7th Conference on Theoretical Aspects of Rationality and Knowledge, TARK 1998, pp. 43–56. Morgan Kaufmann Publishers Inc., San Francisco (1998)
3. Baltag, A., Renne, B.: Dynamic epistemic logic. In: Zalta, E.N. (ed.) The Stanford Encyclopedia of Philosophy. Metaphysics Research Lab, Stanford University (2016). Winter 2016 edn
4. Baltag, A., Smets, S.: A qualitative theory of dynamic interactive belief revision. In: Logic and the Foundations of Game and Decision Theory, Texts in Logic and Games, vol. 3, pp. 9–58 (2008)
5. van Benthem, J.: Dynamic logic for belief revision. J. Appl. Non-Class. Log. **17**(2), 129–155 (2007)
6. van Benthem, J.: Tell it like it is: information flow in logic. J. Peking Univ. (Humanit. Soc. Sci. Ed.) **1**, 80–90 (2008)
7. van Benthem, J.: Logical Dynamics of Information and Interaction. Cambridge University Press, Cambridge (2011)
8. Blackburn, P., de Rijke, M., Venema, Y.: Modal Logic. Cambridge University Press, New York (2001)
9. Cherniak, C.: Minimal Rationality. MIT Press, Bradford (1986)
10. Cowan, N.: The magical number 4 in short-term memory: a reconsideration of mental storage capacity. Behav. Brain Sci. **24**, 87–114 (2001)
11. van Ditmarsch, H., van der Hoek, W., Kooi, B.: Dynamic Epistemic Logic, 1st edn. Springer, Heidelberg (2007). https://doi.org/10.1007/978-1-4020-5839-4
12. Duc, H.N.: Reasoning about rational, but not logically omniscient, agents. J. Log. Comput. **7**(5), 633 (1997)
13. Fagin, R., Halpern, J.Y.: Belief, awareness, and limited reasoning. Artif. Intell. **34**(1), 39–76 (1987)
14. Fagin, R., Halpern, J.Y., Moses, Y., Vardi, M.Y.: Reasoning About Knowledge. MIT Press, Cambridge (1995)
15. Fagin, R., Halpern, J.Y.: Reasoning about knowledge and probability. J. ACM **41**(2), 340–367 (1994)
16. Hintikka, J.: Impossible possible worlds vindicated. J. Philos. Log. **4**(4), 475–484 (1975)
17. Johnson-Laird, P.N., Byrne, R.M., Schaeken, W.: Propositional reasoning by model. Psychol. Rev. **99**(3), 418–439 (1992)
18. Kahneman, D., Beatty, J.: Pupillary responses in a pitch-discrimination task. Percept. Psychophys. **2**(3), 101–105 (1967)
19. Lehrer, K.: Theory of Knowledge. Westview Press, Boulder City (2000)
20. Miller, G.: The magical number seven, plus or minus 2: some limits on our capacity for processing information. Psychol. Rev. **63**, 81–97 (1956)
21. Parikh, R.: Knowledge and the problem of logical omniscience. In: Proceedings of the Second International Symposium on Methodologies for Intelligent Systems, pp. 432–439. North-Holland Publishing Co., Amsterdam (1987)
22. Plaza, J.: Logics of public communications. Synthese **158**(2), 165–179 (2007)
23. Sears, C.R., Pylyshyn, Z.: Multiple object tracking and attentional processing. Can. J. Exp. Psychol. **54**, 1–14 (2000)

24. Rantala, V.: Impossible worlds semantics and logical omniscience. Acta Philos. Fenn. **35**, 106–115 (1982)
25. Rasmussen, M.S.: Dynamic epistemic logic and logical omniscience. Log. Log. Philos. **24**, 377–399 (2015)
26. Rasmussen, M.S., Bjerring, J.C.: A dynamic solution to the problem of logical omniscience. J. Philos. Log. (forthcoming)
27. Rips, L.J.: The Psychology of Proof: Deductive Reasoning in Human Thinking. MIT Press, Cambridge (1994)
28. Schroyens, W., Schaeken, W.: A critique of Oaksford, Chater, and Larkin's (2000) conditional probability model of conditional reasoning. J. Exp. Psychol. Learn. Mem. Cogn. **29**, 140–149 (2003)
29. Schroyens, W.J., Schaeken, W., D'Ydewalle, G.: The processing of negations in conditional reasoning: a meta-analytic case study in mental model and/or mental logic theory. Think. Reason. **7**(2), 121–172 (2001)
30. Spohn, W.: Ordinal conditional functions. A dynamic theory of epistemic states. In: Harper, W.L., Skyrms, B. (eds.) Causation in Decision, Belief Change, and Statistics, vol. 2. Kluwer Academic Publishers (1988)
31. Stalnaker, R.: On logics of knowledge and belief. Philos. Stud. **128**(1), 169–199 (2006)
32. Stanovich, K.E., West, R.F.: Individual differences in reasoning: implications for the rationality debate? Behav. Brain Sci. **23**(5), 645–665 (2000)
33. Stenning, K., van Lambalgen, M.: Human Reasoning and Cognitive Science. MIT Press, Boston (2008)
34. Velázquez-Quesada, F.R.: Small steps in dynamics of information. Ph.D. thesis, Institute for Logic, Language and Computation (ILLC), Universiteit van Amsterdam (UvA), Amsterdam, The Netherlands (2011)
35. Wansing, H.: A general possible worlds framework for reasoning about knowledge and belief. Stud. Logica **49**(4), 523–539 (1990)
36. Wassermann, R.: Resource bounded belief revision. Erkenntnis **50**(2), 429–446 (1999)
37. Xu, Y., Chun, M.M.: Selecting and perceiving multiple visual objects. Trends Cogn. Sci. **13**(4), 167–174 (2009)

Author Index

Printed in the United States
By Bookmasters